Aeronautical Engineering

Aeronautical Engineering

Edited by **Margaret Ziegler**

Willford Press
New York

Published by Willford Press,
118-35 Queens Blvd., Suite 400,
Forest Hills, NY 11375, USA
www.willfordpress.com

Aeronautical Engineering
Edited by Margaret Ziegler

International Standard Book Number: 978-1-68285-222-4 (Hardback)

Printed in China.

Contents

Preface

This book is an extensive source of knowledge on aeronautical engineering. It provides a detailed account of the new technologies in the areas of aviation and aerospace exploration. The book presents some innovative insights into aeronautical science and its different elements such as control engineering, aircraft designs, etc. Some of the researches covered within the book highlight the aerodynamic aspects and their applications. Detailed information is compiled in this book to carry out new innovations in manufacturing aircrafts and space-crafts. It aims to serve as a valuable reference for graduates, researchers and professionals.

After months of intensive research and writing, this book is the end result of all who devoted their time and efforts in the initiation and progress of this book. It will surely be a source of reference in enhancing the required knowledge of the new developments in the area. During the course of developing this book, certain measures such as accuracy, authenticity and research focused analytical studies were given preference in order to produce a comprehensive book in the area of study.

This book would not have been possible without the efforts of the authors and the publisher. I extend my sincere thanks to them. Secondly, I express my gratitude to my family and well-wishers. And most importantly, I thank my students for constantly expressing their willingness and curiosity in enhancing their knowledge in the field, which encourages me to take up further research projects for the advancement of the area.

Editor

Compiler Optimizations Impact the Reliability of the Control-Flow of Radiation-Hardened Software

Ronaldo Rodrigues Ferreira[1], Rafael Baldiati Parizi[1], Luigi Carro[1], Álvaro Freitas Moreira[1]

ABSTRACT: This paper discusses how compiler optimizations influence software reliability when the optimized application is compiled with a technique to enable the software itself to detect and correct radiation-induced control-flow errors. Supported by a comprehensive fault-injection campaign using an established benchmark suite in the embedded systems domain, we show that the compiler is a non-negligible source of noise when hardening the software against radiation-induced soft errors.

KEYWORDS: Compilers, Radiation effects, Single event upsets, Software reliability, Software engineering.

INTRODUCTION

Compiler optimizations are taken for granted in modern software development, enabling applications to execute more efficiently in the target hardware architecture. Modern architectures have complex inner structures designed to boost performance, and if the software developer were to be aware of all those inner details, performance optimization would jeopardize the development processes. Compiler optimizations are transparent to the developer, who picks the appropriate ones to the results s/he wants to achieve, or, as it is more common, allowing this task to be performed by the compiler itself by flagging if it should be more or less aggressive in terms of performance.

Industry already offers microprocessors built with 22 nm transistors, with a prediction that by 2026, the size of the transistor will reach 5.9 nm (ITRS, 2012). This aggressive technology scaling creates a big challenge concerning the reliability of microprocessors using newest technologies. Smaller transistors are more likely to be disrupted by transient sources of errors caused by radiation, known as *soft-errors* (Borkar, 2005). Radiation particles originated from cosmic rays when strikes a circuit induces bit flips during software execution, and because transistors are becoming smaller in size, there is a higher probability that these transistors will be disrupted by a single radiation particle with smaller transistors requiring a smaller amount of charge to disrupt their stored logical value. The newest technologies are so

1. Universidade Federal do Rio Grande do Sul – Porto Alegre/RS – Brazil

Author for correspondence: Ronaldo Rodrigues Ferreira | Instituto de Informática, Universidade Federal do Rio Grande do Sul | Avenida Bento Gonçalves, 9500, Campus do Vale – Bloco IV – Agronomia | CEP 91.509-900 Porto Alegre/RS – Brazil | Email: rrferreira@inf.ufrgs.br

sensitive to radiation that their usage will be compromised at the sea level, as predicted in the literature (Normand, 1996). Rech *et al.*, (2012) have shown that modern graphics processing unit (GPU) cards are susceptible to such an error rate that makes their usage unfeasible in critical embedded systems. However, industry is already investing in GPU architectures as the platform of choice for high performance and low power embedded computing, such as the ARM Mali® embedded GPU (ARM, n.d.).

The classical solution to harden systems against radiation is the use of *spatial redundancy*, i.e., the replication of hardware modules. However, spatial redundancy is prohibitive for embedded systems, which usually cannot afford extra costs of hardware area and power. The increase on power is a severe problem, because it is expected that 21% of the entire chip area must be turned off during its operation to meet the available power budget, and an impressive chip area of 50% at 8 nm (Esmaeizadeh, 2011). This creates the *dark silicon* problem (Esmaeizadeh, 2011), i.e., a huge area of the circuit cannot be used during its lifecycle. This problem gets worse when the microprocessor has redundant units, because system's reliability could be compromised if redundant units were turned off. The current solution to this problem is to use radiation-hardened microprocessors, which are designed to endure radiation. The problem with this approach is the low availability, high unit pricing, and International Traffic in Arms Regulations (ITAR) restrictions of those radiation-hardened components. For instance, a 25 MHz microprocessor has a unitary price of $ 200,000.00 (Mehlitz and Penix, 2005). This high unit pricing makes the use of radiation-hardened microprocessors unfeasible for embedded systems used in aircrafts, not to say about cars and low-end medical devices, such as pacemakers. For these critical embedded systems, where cost and ITAR restrictions are hard constraints, a cheaper, but yet effective approach for reliability against radiation, is necessary.

Software-Implemented Hardware Fault-Tolerance (SIHFT) (Goloubeva, 2006) is an approach for radiation reliability that adds redundancy in terms of extra instructions or data to the application, keeping the hardware unchanged. SIHFT techniques work by modifying the original program by adding *checking mechanisms* to it. SIHFT techniques are classified either as *control-flow* or as *data-flow*. The former is designed to detect when an illegal jump has occurred during application execution to possibly proceed with the resolution of the correct jump address or at least signaling that such an error has occurred. The latter checks if a data variable being read is correct or not. While the effects of data-flow SIHFT methods are clear (usually, the duplication of program variables or the addition of variable checksums solves the problem), the impacts of the control-flow ones, are yet not well understood. Because the control-flow methods modify the program's control-flow graph (CFG), which happens to be the same artifact used by compiler optimizations, the efficiency of control-flow reliability techniques might be influenced by the optimizations in an unpredictable way.

In this paper, we evaluate how the cumulative usage of compiler optimizations influence reliability of applications hardened with the state-of-the-art Automatic Correction of Control-flow Errors (ACCE) (Vemu, Gurumurthy and Abraham, 2007) control-flow SIHFT technique, which is selected, because it is the current most efficient method in terms of reliability, attaining an error correction rate of ~70%. The application set we use in this paper is drawn from the MiBench suite (Guthaus *et al.*, 2001).

RADIATION EFFECTS ON SOFTWARE RELIABILITY

Highly energized radiation particles are known hazard sources in electronics since the 1970s (Binder *et al.*, 1975), as well as the mitigation schemes for such sources. Single-Event Transient (SET) is the observed physical effect of radiation on electronics, corresponding to voltage glitches in circuitry, which by itself does not incur on system hazards. System hazards originate when SETs are caught up by memories (e.g., SRAM's) and sequential logic (e.g., registers), thus, becoming a Single-Event Upset (SEU). An SEU is a non-permanent damage to the systems (i.e., transient) that results in a bit with logical value 1 that flips to 0 and vice-versa. Mitigation approaches might be as follows:

- *Substrate and gate-level*: The reduction of charge generation and collection.
- *Hardware design*: The modification of circuit response through the addition of logical elements or even the storage of data on spatially separated nodes.

- *System level*: The addition of redundancy at system level, e.g., software hardening.

Definitely, the system level mitigation approach is the most feasible for components off-the-shelf and to overcome ITAR restrictions.

A bit flip caused by a radiation-induced SEU can compromise the software in two different ways. Firstly, an SEU can corrupt *data*, i.e., the values of program variables. Secondly, an SEU can corrupt *control*, i.e., the program flow of execution. To illustrate these two situations, consider the program presented in Fig. 1, which corresponds to the Bubble Sort algorithm. Bubble Sort is a naïve $O(n^2)$ solution for sorting an array of arbitrary numbers.

The Bubble Sort algorithm is divided in labeled regions named basic blocks (identified by the gray and white regions in the Bubble Sort source code). A *basic block* is a region of a program where the contained program instructions does not contain any branch, i.e., iteration loops (e.g., for and while commands), if-conditionals, function call, and return. Therefore, a basic block only contains *variable assignments* and *logical evaluations*. The CFG of a program P is a graph G_p=(V, E), where the set V of vertices contains the program's basic blocks and the set E of edges contains the transitions in the execution flow. To illustrate this, the CFG of the Bubble Sort algorithm is presented in Fig. 2.

An executed branch of the program P (represented by an arrow in Fig. 2) is said to be *legal*, if and only if, it is an element of the set E of G_p and the condition to execute it is satisfied (e.g., the blue arrow in Fig. 2); it is *wrong* if the executed branch is an element of the set E, but its condition to execute cannot be satisfied (e.g., the purple arrow in Fig. 2); and an executed branch is *wrong* if it is not an element of the set E (e.g., the red arrow in Fig. 2).

```
Algorithm Bubblesort(input: n, V)
def n : number of values to sort;
def V[n] : array of size n;
def temp, i, j: integer variables;
i := n - 1;
while ( i >= 1 ) do
   j := 0;
   while ( j < i ) do
         if ( V[j] < V[j+1] )
               temp   := V[j];
               V[j]   := V[j+1];
               V[j+1]:= temp;
         end if
         j := j + 1;
   end while
   i := i - 1;
end while
return V;
```

Figure 1. Bubble Sort algorithm with explicit basic-blocks, which are represented by the grouped numbered lines.

A control-flow error (CFE) occurs when either a wrong or illegal branch is executed. Notice that in these two cases, a CFE cannot exist if the program execution is not corrupted, i.e., an illegal branch cannot exist in a correct program execution, because it only executes branches from the set E; a wrong branch cannot exist, because it is always possible to satisfy the logical conditions of all branches if program execution is correct. A CFE can be created by a radiation-induced SEU in any of the following three scenarios:

- A non-branch instruction being executed changes into a non-valid branch, i.e., the operation code data is corrupted.
- The target address of a valid branch is corrupted.
- One of the variables composing a logical expression that activates a branch is corrupted.

Scenarios (1) and (2) leads to an illegal branch, and scenario (3) leads to a wrong branch.

A data-flow error (DFE) is caused by a radiation-induced SEU that corrupts variables within a basic block. A DFE might lead to erroneous results or even to a CFE, in case the corrupted variable is used in a logical expression controlling a branch. The focus of this paper is CFE's. For an extensive review of mitigation techniques of CFE and DFE, interested readers may refer to the work of Goloubeva *et al.*, (2006).

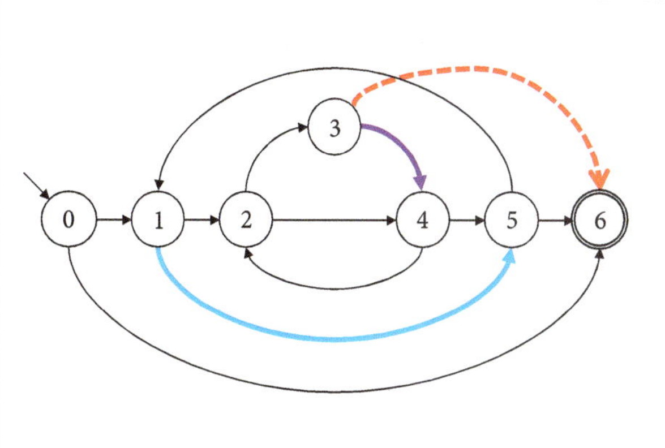

Figure 2. Control-flow graph of the Bubble Sort algorithm. The blue arrow is a legal branch (together with the black arrows), the purple arrow is a wrong branch, and the red arrow is an illegal branch.

The detection of transient CFE was established in the literature with techniques that check assertions during runtime. The general idea is to compute signatures identifying each basic block, and checking the signatures generated during compilation and runtime. If they do not match, an error is signaled. CFE's were first identified by the usage of watchdog processors, which are intrusive in the hardware design (Saxena and McCluskey, 1990). Lately, techniques based on the signature checking scheme in software, such as the Control-flow Checking Approach (CCA) (Kanawati *et al.*, 1996), were identified, but with a coverage rate of only 38% and a performance overhead of 50%.

Advances in the signature checking method offered some improvements on coverage and performance, such as the Control-Flow Checking by Software Signatures, which incurs in 50% of overhead in execution time and program size (Oh *et al.*, 2002). The most efficient technique of signature checking capable of correcting errors is the ACCE (Vemu *et al.*, 2007), which incurs in approximately 20% of overhead in execution time to produce an average 70% of correct answers in fault-injection campaigns. However, ACCE is not capable of correcting errors that occur within a basic block, i.e., in the data flow; hence, the use of complementary techniques is required. When ACCE is enhanced with data-flow correction, its coverage rate achieves the average of 91.6% (Vemu *et al.*, 2007).

Because the CFG is one of the most important software artifacts used by compilers when analyzing and modifying programs, it is important to measure how the compiler impacts the reliability of the software mitigation techniques for radiation-induced CFE. The understanding of these impacts is imperative to employ software mitigation techniques in real systems. This paper presents a study using the ACCE mitigation technique, which is briefly reviewed in the next section.

AUTOMATIC CORRECTION OF CONTROL-FLOW ERRORS

ACCE (Vemu *et al.*, 2007) is a software technique for reliability that detects and corrects CFE's due to random and arbitrary bit-flips that might occur during software execution. The hardening of an application with ACCE is done at compilation, because it is implemented as a transformation pass in the compiler. ACCE modifies the applications' basic blocks with the insertion of extra instructions that perform the error detection and correction during software execution. In this section, we briefly explain how the ACCE works in two separate subsections, one dedicated to error detection and the other to error correction are discussed in the subsequent subsections. The reader should refer to the ACCE article for a detailed presentation and experimental evaluation (Vemu *et al.*, 2007). The fault model that ACCE assumes is further described in the "Fault Model and Methodology" section.

Control-Flow Error Detection

ACCE performs online detection of CFE-s by checking the signatures in the beginning and in the end of each basic block of the CFG, thus, ACCE is classified as a *signature checking* SIHFT technique as termed in the published literature. The basic block signatures are computed and generated during compilation; the signature generation is critical, because it requires computing non-aliased signatures between the basic block, i.e., each block must be unambiguously identified. In addition, for each basic block found in the CFG, two additional code regions are added, the *header* and the *footer*. The signature checking during execution takes place inside these code regions. Figure 3 shows two basic blocks (labeled as **N2** and **N6**) with the additional code regions. The top region corresponds to the header and the bottom to the footer. Still, at compilation, the ACCE creates two additional blocks for each function, namely the function entry block and the Function Error Handler (FEH). For instance, Fig. 3 depicts a portion of two functions, *f1* and *f2*, both owning entry blocks labeled as **F1** and **F2**, and FEHs, labeled as **FEH_1** and **FEH_2**, respectively. Finally, ACCE creates a last extra block, the Global Error Handler (GEH), which can only be reached from a FEH block. The role of these blocks will be presented soon.

At runtime, the ACCE maintains a global signature register (represented as S), which is constantly updated to contain the signature of the basic block that the execution has reached. Therefore, during the execution of the *header* and *footer* code regions of each basic block, the value of the signature register is compared with the signatures generated during compilation for those code regions, and if those values do not match, a CFE is detected and then the control should be transferred to the corresponding FEH block of the function where the execution takes place at that time. The ACCE also maintains the current function register (represented as **F**), which stores the unique identifier of the function currently being executed. The current function register is only assigned at the extra entry function block. This process encompasses the detection of an illegal and erroneous branch due to a soft error.

Figure 3 depicts an example of the checking and update of signatures performed in execution time that occurs in a basic block. In this example, the CFE occurs in the block **N2** of function **F1**, where an illegal jump incorrectly transfers the control flow to the basic block **N6** of function **F2**. When the execution reaches the footer of the block **N6**, the signature register **S** is checked against the signature generated at compilation. In this case, **S** = 0111 (i.e., the previous value assigned in the header of the block **N2**). Thus, the branch test in the **N6** footer will detect that the expected signature does not match with the value of S, and thus, the CFE must be signaled (step 1 in Fig. 3). In this example, the application branches to the address *f2_err*, making the application enter the **FEH_2** block (because the error was detected by a block owned by the function **F2**, the FEH invoked is the **FEH_2**). At this point, the CFE is detected and ACCE can proceed with the correction of the detected CFE.

Control-Flow Error Correction

The correction process starts as soon as an illegal jump is detected by the procedure described in the last subsection,

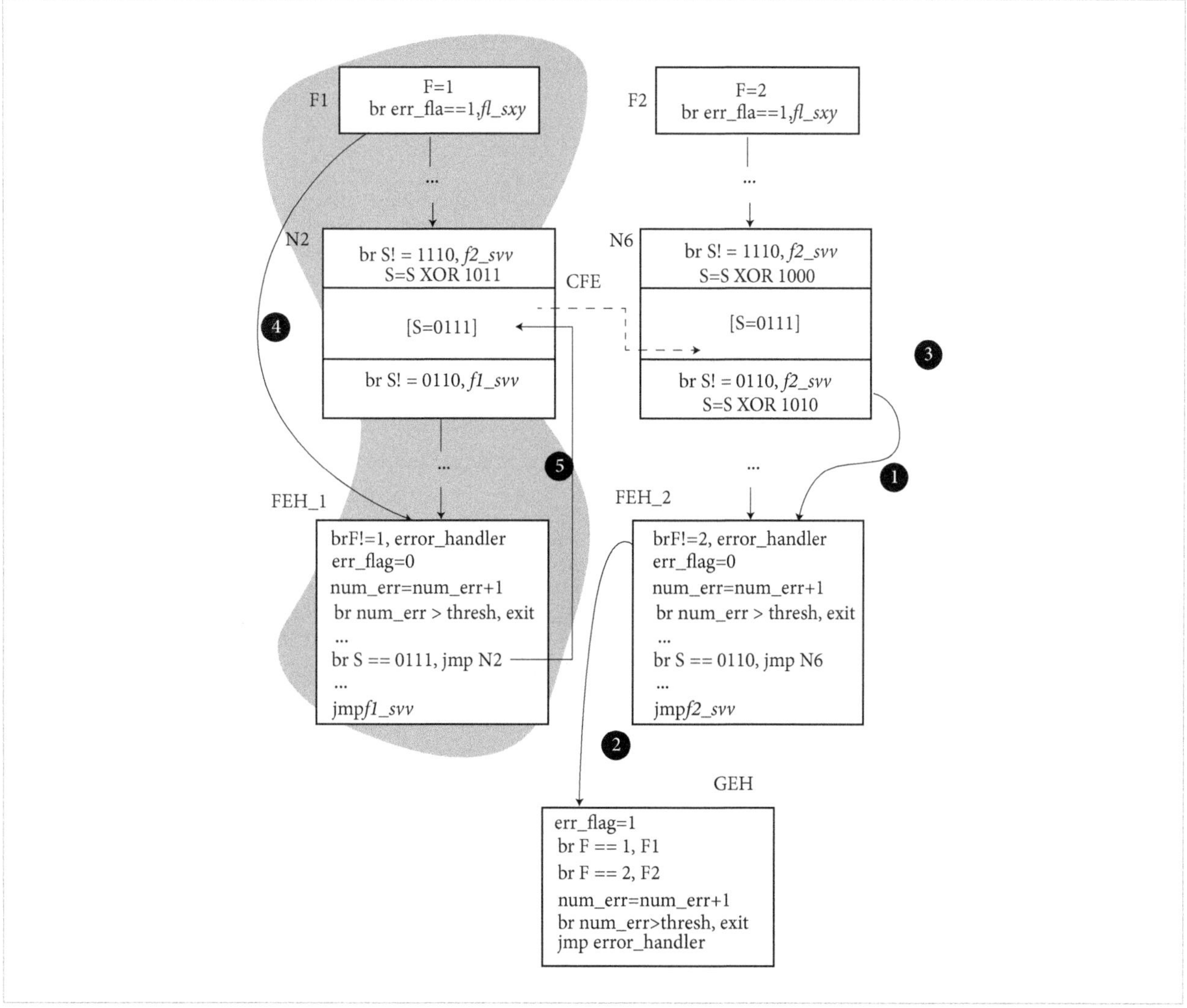

Figure 3. Depiction of how the control is transferred from a function to the basic blocks that ACCE has created when a control-flow error occurs during software execution. In this figure, there is a control-flow error (dashed arrow) causing the execution to jump from the block **N2** of function **F1** to the block **N6** of function **F2**.

with the control flow transferred to the FEH corresponding to the function where the CFE is found. The FEH checks if the illegal jump was originated in the function it is responsible to handle its detected errors by comparing the value of the function's identifier (**F1** or **F2**, in the example of Fig. 3) with the current function register **F**. If the error occurred in the function stored in the **F** register, FEH evaluates the current value of the signature register and then transfers the control to the basic block that is the origin of the illegal jump (this origin is stored in the **S** register). On the other hand, if the illegal jump is not originated in the function where the detection has occurred, the FEH then transfers the control flow to the GEH. In this case, the GEH is responsible for identifying the function where the CFE has occurred and to transfer the control flow back to this function, so that the error is correctly treated by the function's FEH. The GEH searches the function where the error has occurred and transfers the control to its entry block, which then sends the control flow to the proper FEH so that the error can be corrected, i.e., branching the control to the basic block where the CFE has occurred.

Recalling the example depicted in Fig. 3, after the CFE is detected and the control is transferred to **FEH_2** (step 1), the **F** register is matched against the function identifier of the function from where the control originated. However, because the CFE originated in the basic block **N2** of the function **F1**, **F** = 1. Therefore, **FEH_2** is not capable of finding the basic block where the CFE originated, and then it transfers the control to the GEH so that the correct FEH can be found (step 2). The GEH searches for the function identifier stored in **F**, until it finds that it should branch to **F1** (step 3). Upon reaching the entry block **F1**, the variable *err_flag*=1, because it is assigned to 1 in the GEH, meaning that there is an error that should be fixed, thus, the control branches to **FEH_1** (step 4). Now, because **F**=1, **FEH_1** knows that it is the FEH capable of handling the CFE and, as such, it sets the variable *err_flag* to 0. Finally, it searches for the basic block that has the signature equal to the register **S**. Upon finding it, the control branches to this basic block, i.e., **N2** in Fig. 3 (step 5). This last branch restores the control flow to the point of the program right before the occurrence of the CFE. Notice that inside all the FEH and the GEH, there is the variable *num_error*, counting how many times the control has passed through a FEH or a GEH. This acts as a threshold for the number of how many times the correction must be attempted, which is necessary to avoid an infinite loop in case the registers **F** or **S** get corrupted for any reason. This process concludes the correction of a CFE with the ACCE.

FAULT MODEL AND METHODOLOGY

The fault model we assume in the experiments is the *single bit flip*, i.e., only one bit of a word is changed when a fault is injected. The ACCE is capable of handling multiple bit flip as long as the bits flipped is within a same word. Because the fault injection, as it will be discussed later, guarantees that the injected fault ultimately turned into a manifested error, it does not matter how many bits are flipped, i.e., there is no silent data corruption, meaning the faults that cause a word to change its value neither change the behavior of the program nor its output. This could happen in the case that the fault flipped the bits of a dead variable.

The ACCE technique was implemented as a transformation pass in the Low Level Virtual Machine (LLVM; Lattner and Adve, 2004) production compiler, which performs all the modifications in the CFG using the LLVM Intermediate Representation (LLVM-IR). The LLVM was selected as our compilation platform, because of its increased use in the industry, accompanied with a very detailed documentation and quality of its source code. The ACCE transformation pass was applied *after* the set of compiler optimizations, because executing in the opposite order, a compiler optimization could invalidate the ACCE generated code and semantics. Table 1 presents the LLVM optimization passes used in the experiments.

Because the ACCE is a SIHFT technique to detect and correct CFE-s, the adopted fault model simulates three distinct control-flow disruptions that might occur due to a CFE. Remember that a CFE is caused by the execution of an illegal branch to a possibly wrong address. The branch errors considered in this paper are as follows:

- *Branch creation*: The program counter is changed, transforming an arbitrary instruction (e.g., an addition) into an unconditional branch.
- *Branch deletion*: The program counter is set to the next program instruction to execute independently if the current instruction is a branch.
- *Branch disruption*: The program counter is disrupted to point to a distinct and possibly wrong destination instruction address.

We implemented a software fault injector, using the GDB (GNU Debugger), in a similar fashion as implemented by Krishnamurthy *et al.*, (1998), which is an accepted fault-injection methodology in the embedded systems domain, to perform the fault-injection campaigns. The steps of the fault-injection process are the following:

- The LLVM-IR program resulting from the compilation with a set of optimization and with ACCE is translated to the assembly language of the target machine.
- The execution trace in assembly language is extracted from the program execution with GDB.
- A branch error (branch creation, deletion, or disruption) is randomly selected. On an average, each branch error accounts for 1/3 of the amount of injected errors.
- One of the instructions from the trace obtained in step 2 is chosen at random for fault injection. In this step, a histogram of each instruction is computed because instructions that execute more often have a higher probability to be disrupted.
- If the chosen instruction in step 4 executes n times, choose at random an integer number k with $1 \leq k \leq n$.
- Using GDB, a breakpoint is inserted right before the k-th execution of the instruction selected in step 4.
- During program execution, upon reaching the breakpoint inserted in step 6, the program counter is intentionally corrupted by flipping one of its bits to reproduce the branch error chosen in step 3.
- The program continues its execution until it finishes.

A fault is only considered valid, if it has generated a CFE, i.e., silent data corruption and segmentation faults were not considered to measure the impacts of the compiler optimizations on reliability. All the experiments in this paper were performed in a 64-bit Intel Core i5 2.4 GHz desktop with 4 GB of RAM and the LLVM compiler version 2.9. For all program versions, where each version corresponds to the program compiled with a set of optimizations plus the ACCE pass, 1,000 faults were injected using the aforementioned fault-injection scheme. In the experiments we considered ten benchmark applications from the MiBench (Guthaus *et al.*, 2001) embedded benchmark suite as follows: *basicmath*, *bitcount*, *crc32*, *dijkstra*, *fft*, *patricia*, *quicksort*, *rijndael*, *string search*, and *susan* (comprising *susan* corners, edge, and smooth).

IMPACT OF COMPILER OPTIMIZATIONS ON SOFTWARE RELIABILITY

This section studies the impacts on software reliability when an application is compiled with a set of compiler optimizations and further hardened with the ACCE method. Throughout this section, the baseline for all comparisons is an application compiled with the ACCE method without any other compiler optimization. The ACCE performs detection and correction of CFE-s,

Table 1. Set of Low Level Virtual Machine optimization passes used for experimental evaluation in this paper.

-adce	-loop-reduce
-always-inline	-loop-rotate
-argpromotion	-loop-simplify
-block-placement	-loop-unroll
-break-crit-edges	-loop-unswitch
-codegenprepare	-loweratomic
-constmerge	-lowerinvoke
-constprop	-lowerswitch
-dce	-mem2reg
-deadargelim	-memcpyopt
-deadtypeelim	-mergefunc
-die	-mergereturn
-dse	-partial-inliner
-functionattrs	-prune-eh
-globaldce	-reassociate
-globalopt	-reg2mem
-gvn	-scalarrepl
-indvars	-sccp
-inline	-simplifycfg
-instcombine	-simplify-libcalls
-internalize	-sink
-ipconstprop	-sretpromotion
-ipsccp	-strip
-jump-threading	-strip-dead-debug-info
-lcssa	-strip-dead-prototypes
-licm	-strip-debug-declare
-loop-deletion	-strip-non-debug
-loop-extract	-tailcallelim
-loop-extract-single	-tailduplicate

thus all data discussed in this section considers the *correction rate* as the data to compute the efficiency metric. In this analysis, we use 58 optimizations provided by the LLVM production compiler. Finally, the results were obtained using the fault model and fault injection methodology described in the section "Fault Model and Methodology".

The impact of the compiler optimizations when compiling for reliability is measured in this paper using the metric Relative Improvement Percentage (RIP; Pan and Eigenmann, 2006). The RIP is presented in Eq. 1, where F_i is a compiler optimization, $E(F_i)$ is the error correction rate obtained for a hardened application compiled with F_i, and E_B is the error correction rate obtained for the baseline, i.e., the application compiled only with ACCE and without any optimization.

$$RIP_B(F_i) = \frac{E(F_i) - E_B}{E_B} \cdot 100\% \quad (1)$$

Figure 4 shows a scatter plot of the obtained RIP for each application, with each of the 58 LLVM optimizations being a point in the *y*-axis. Each point represents the hardened application compiled with a single LLVM optimization at a time, with each application compiled with 58 distinct optimizations. Figure 4 shows that several optimizations increase the RIP considerably, sometimes reaching a RIP of ~10%. This is a great result, which shows that reliability can be increased for free by just picking appropriate optimizations that facilitates for ACCE the process of error detection and correction. However, we also find that some optimizations totally jeopardize reliability, reaching a RIP of -73.27% (bottom filled red circle for *bitcount*).

It is also possible to gather evidence that the structure of the application also influences how an optimization has an impact on the RIP of reliability. Let us consider the *block-placement* optimization, which is represented by the white diamond in Fig. 4. In the case of the *qsort* application, *block-placement* has a RIP of -42.75% and a RIP of +11.68%. The reader can notice that other optimizations also show this behavior (increasing RIP for some applications and decreasing it for others). It also happens that some hardened applications are less sensitive to compiler optimizations, as it is the case of the *crc_32* one, where the RIP is within the ±5% interval around the baseline.

Figure 5 depicts the RIP of a selected subset of the 58 LLVM optimizations, making it clear that even within a small subset, the variation in the RIP for reliability is far from negligible. For instance, the *always-inline* LLVM optimization has an error correction RIP interval of [-4.55%, +9.24%].

Usually compiler optimizations are applied in bulk, using several of them during compilation. Therefore, it is also important to examine if successive optimization passes could compromise or increase software reliability of a hardened application. Figure 6 presents the error correction rate RIP, where the hardened application was compiled with a subset of the 58 LLVM optimizations. In this experiment, we used six sizes of subsets: 10, 20, 30, 40, 50, and 58. The RIP shown in Fig. 6 is the average of five random subsets, i.e., it is an average of distinct subsets of the same size. Taking the average and picking the optimizations at random, reproduces the effects of indiscriminately picking the compiler optimizations, or at least, selecting optimizations with the object of optimizing performance without previous knowledge of how the selected optimizations together influences the software reliability.

It is possible to see that the cumulative effect of compiler optimizations in the error correction RIP is in most of the cases deleterious, but for a few exceptions. Figure 6 confirms that some applications are less sensitive to the effects of compiler optimizations, e.g., the *crc32* has its RIP within the interval [-1.11%–0.73%]. On the other hand, *basicmath*, *bitcount*, and *patricia*, are jeopardized. It is interesting to notice that the RIP in case of picking a subset of optimizations is not subject to the much severe reduction that was measured when only a single optimization was used (Fig. 4), providing an evidence that the composition of distinct optimization may be beneficial for reliability.

Based on the data and experiments discussed in this section, it is clear that selection of compiler optimizations requires the software designer to take into the consideration that some optimizations may not be adequate in terms of reliability for a given application. Moreover, data also shows that a given optimization is not only by itself a source of reliability reduction; reliability is also dependent of the application being hardened, and how a given optimization facilitates or not the work of the ACCE technique.

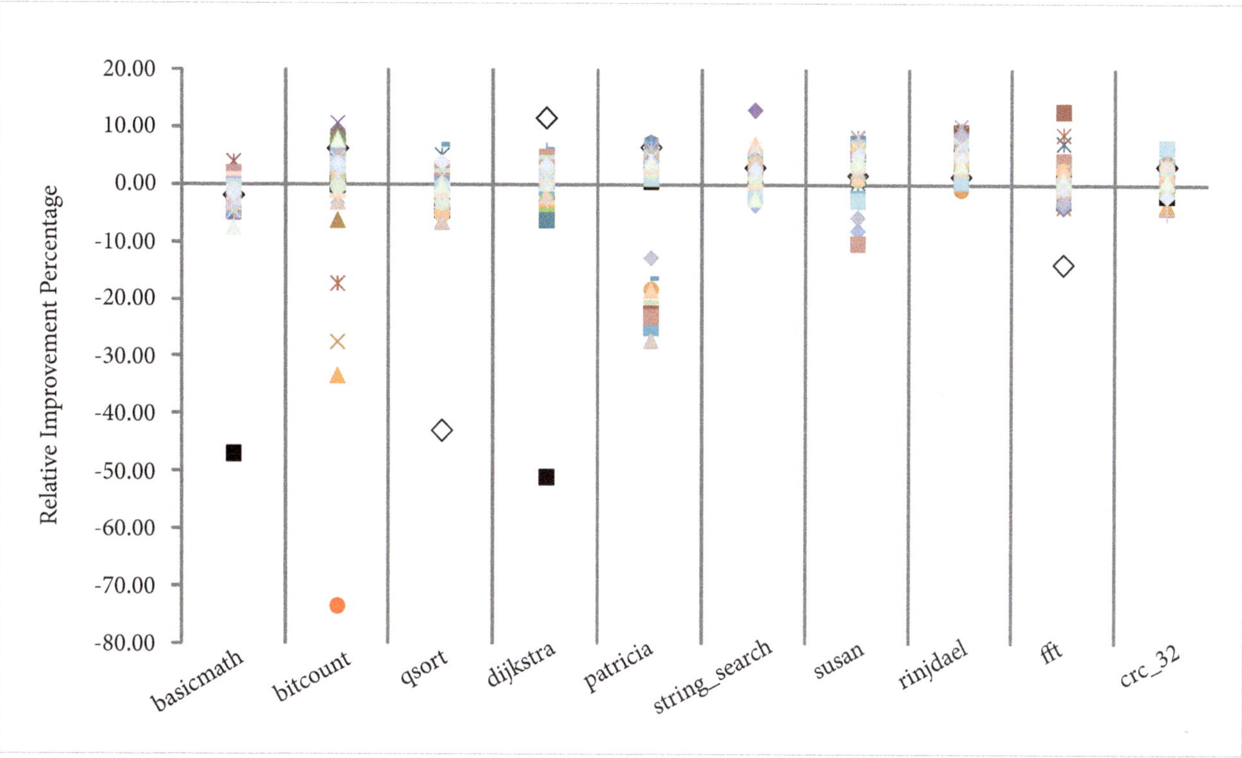

Figure 4. Relative Improvement Percentage for the error correction rate of applications hardened with ACCE under further compiler optimization. Each hardened application was compiled with a single optimization at a time, but all applications were compiled with the 58 Low Level Virtual Machine optimizations, thus, each hardened application has 58 versions. The baseline (Relative Improvement Percentage = 0%) is the error correction rate of the hardened application compiled without any Low Level Virtual Machine optimization.

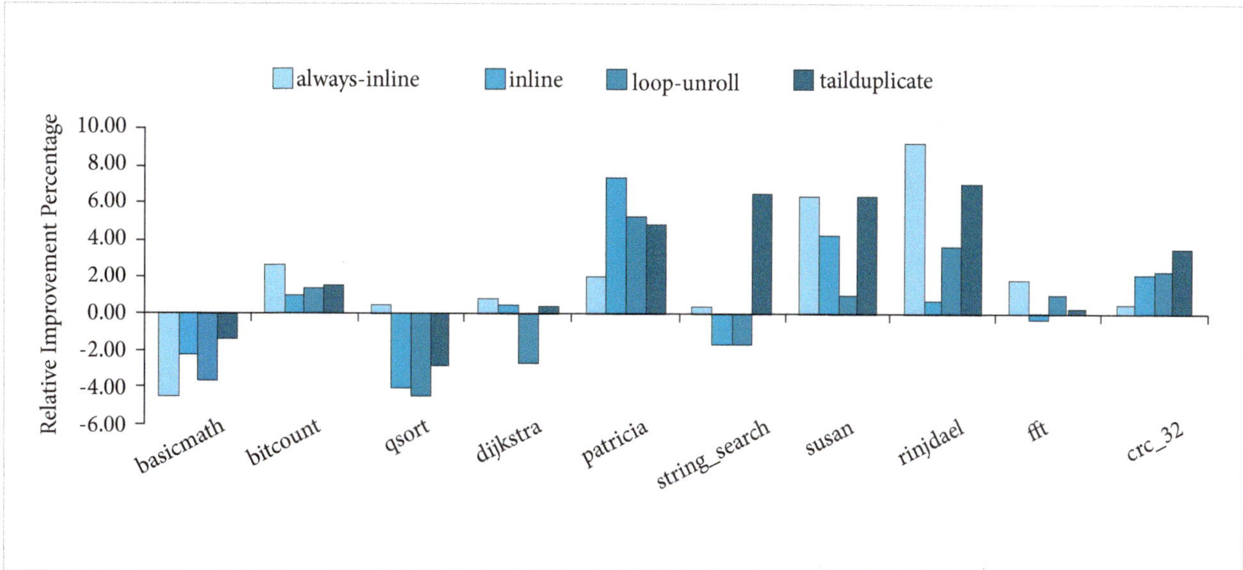

Figure 5. Relative Improvement Percentage of a selected subset of the 58 Low Level Virtual Machine optimizations. The baseline (Relative Improvement Percentage=0%) is the error correction rate of the hardened application compiled without any Low Level Virtual Machine optimization.

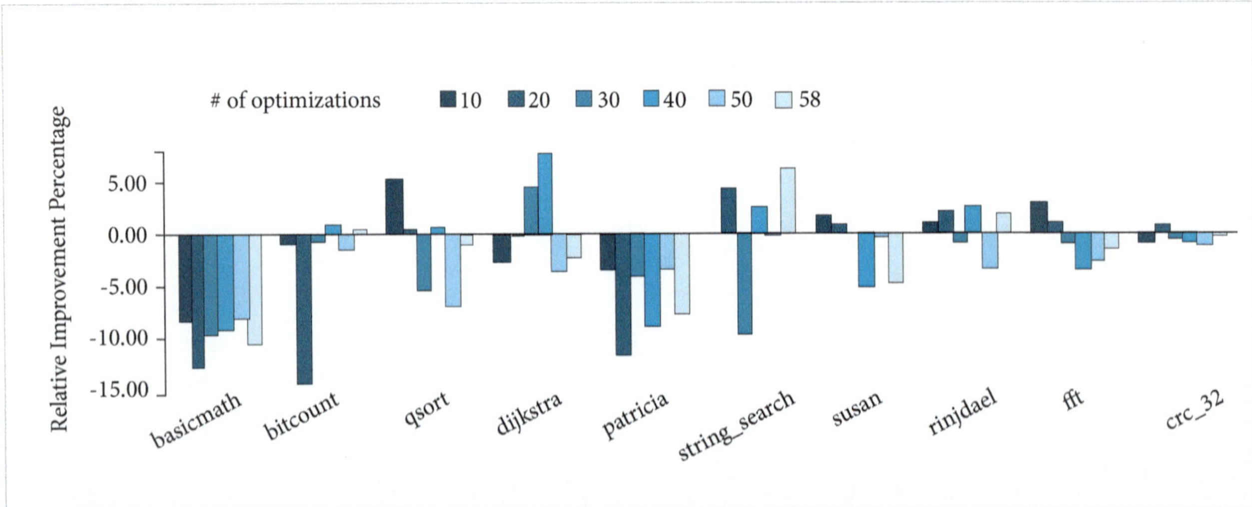

Figure 6. RIP of random subsets of the 58 Low Level Virtual Machine optimizations with a varying number of optimizations for each different subset: 10, 20, 30, 40, 50, and 58 optimizations. The Relative Improvement Percentage for each subset was measured taking the average of six random subsets for each subset size. Hence, distinct possible optimizations for subsets were considered. The baseline (Relative Improvement Percentage=0%) is the error correction rate of the hardened application compiled without any Low Level Virtual Machine optimization.

RELATED WORK

Much attention has been devoted to the impact of compiler optimizations on program performance in the published literature. However, the understanding of how those optimizations work together and how they influence each other is a rather recent research topic. The Combined Elimination (CE) (Pan and Eigenmann, 2006) is an analysis approach to identify the best sequence of optimizations for a given application set using the GNU Compiler Collection (GCC). The authors discuss that simple *orchestration* schemes between the optimizations can achieve near-optimal results as if it has performed an exhaustive search in all the design space created by the optimizations. CE is a greedy approach that first compiles the programs with a single optimization, using this version as the baseline. From those baseline versions, the set of RIP is calculated, which is the percentage that the program's performance is either reduced or increased. With the RIP at hand for all baselines, the CE starts removing the optimizations with negative RIP, until the total RIP of all optimizations applied into a program do not reduce. CE was evaluated in different architectures, achieving an average RIP of 3% for the SPEC2000, and up to 10% in case of the Pentium IV for the floating point applications.

The Compiler Optimization Level Exploration (COLE) (Hoste and Eeckhout, 2008) is another approach to achieve performance increase by selecting a proper optimization sequence. COLE uses a population-based multi-objective optimization algorithm to construct a Paretto optimal set of optimizations for a given application using the GCC compiler. The data found with COLE give some insightful results about how the compiler optimizations behave when they are applied with several of them at the same time. For instance, 25% of the GCC optimizations appear in at most one Paretto set, and some of them appear in all sets. Therefore, 75% of all the optimizations do not contribute to improve the performance, meaning that they can be safely ignored! COLE also shows that the quality of an optimization is highly tied with the application set.

The Architectural Vulnerability Factor (AVF) (Mukherjee *et al.*, 2003) is a metric to estimate the probability that the bits in a given hardware structure will be corrupted by a soft error when executing a certain application. The AVF is calculated as the total time the vulnerable bits remains in the hardware architecture. For example, the register file has a 100% AVF, because all of its bits are vulnerable in case of a soft error. The AVF metric is highly influenced by the application due to liveness of program's variables. For instance, a dead variable has a 0%

AVF, because it is not used in a computation. The impact of the GCC optimizations in the AVF metric is evaluated by trying to reduce the AVF-delay-square-product (ADS), introduced by the authors (Jones *et al.*, 2008). The ADS considers a linear relation of the AVF between the square of the performance in cycles, clearly prioritizing performance over reliability. It is reported that the $-O_3$ optimization level is detrimental both to the AVF and performance, because the benchmarks that are considered (MiBench) have increased the number of loads executed. Again, it was found that the *patricia* application was the one with the highest reduction in the AVF at 13%.

Bergaoui and Leveugle (2011) analyzed the impact of compiler optimizations on data reliability in terms of variable liveness. *Liveness* of a variable is the time period between the variable that is written and it is last read before a new write operation. The authors concluded that the liveness is not related only with the compiler optimization, but it also depends on the application being compiled, which is in accordance with the discussion of this paper. This paper shows that some optimizations tend to extend the time a variable is stored in a register instead of memory. The goal behind this is obvious, i.e., it is much faster to fetch the value of a variable when it is in the register than in the memory. However, the memory is usually more protected than registers because of cheap and efficient error correction code (ECC) schemes, and thus, thinking about reliability, it is not a good idea to expose a variable in a register for a longer time. The solution to that could be the application of ECC, such as Huffman to the program variables itself. Decimal Hamming (DH) (Argyrides *et al.*, 2011) is a software technique that performs this for a class of programs where the program's output is a linear function of the input. The generalization of the efficient data-flow SIHFT techniques, such as DH (i.e., ECC of program variables) is still an open research problem.

CONCLUSION

In this paper, we characterized the problem of compiling embedded software for reliability, given that compiler optimizations impact the coverage rate. The study presented in this paper makes clear that selecting optimizations indiscriminately, can decrease software reliability to unacceptable levels, probably avoiding the software to be deployed as originally planned. Embedded software and systems deployed in space applications must always be certified with evidence that they support harsh radiation environments, and given the increasing technology scaling, other safety critical embedded systems might have to tolerate radiation-induced errors in a near future. Therefore, the embedded software engineers must be very careful while compiling the safety critical embedded software.

Future research work is focused on the formalization of the ACCE transformation pass to generate automatic proofs about the correctness of programs compiled with the ACCE. This step is important to allow the certification of software hardened with ACCE.

ACKNOWLEDGEMENTS

This work is supported by the CAPES foundation of the Ministry of Education, CNPq research council of the Ministry of Science and Technology, and FAPERGS research agency of the State of Rio Grande do Sul, Brazil. R. Ferreira was supported with a doctoral research grant from the Deutscher Akademischer Austauschdienst (DAAD) and from the Fraunhofer-Gesellschaft, Germany.

REFERENCES

Argyrides, C., Ferreira, R., Lisboa, C. and Carro, L., 2011, "Decimal Hamming: A Novel Software-Implemented Technique to Cope with Soft Errors", Proceedings of the 26th IEEE International Symposium on Defect and Fault Tolerance in VLSI and Nanotechnology Systems, pp. 11-17, doi: 10.1109/DFT.2011.35

ARM, n.d. 2012, "ARM Mali Graphics Hardware", Retrieved in December 21, 2012, from http://www.arm.com/products/multimedia/mali-graphics-hardware/index.php.

Bergaoui, S. and Leveugle, R., 2011, "Impact of Software Optimization on Variable Lifetimes in a Microprocessor-Based System", Proceedings of the 6th IEEE International Symposium on Electronic Design, Test and Application, pp. 56-61, doi: 10.1109/DELTA.2011.20

Binder, D., Smith, E.C. and Holman, A.B., 1975, "Satellite Anomalies from Galactic Cosmic Rays", IEEE Transactions on Nuclear Science, Vol. 22, No. 6, pp. 2675-2680, doi: 10.1109/TNS.1975.4328188

Borkar, S., 2005, "Designing Reliable Systems from Unreliable Components: The Challenges of Transistor Variability and Degradation", IEEE Micro, Vol. 25, No. 6, pp. 10-16, doi: 10.1109/MM.2005.110

Esmaeizadeh, H., Emily, B., Renee, A. and Sankaralingam, K., 2011, "Dark Silicon and the End of Multicore Scaling", IEEE Micro, Vol. 32, No. 3, pp. 122-134, doi: 10.1109/MM.2012.17

Goloubeva, O., Rebaudengo, M., Sonza Reorda, M. and Violante, M., 2006, "Software-Implemented Hardware Fault Tolerance", Ed. Springer, New York, NY, USA, p 228.

Guthaus, M.R., Ringenberg, J.S., Ernst, D., Austin, T.M., Mudge, T. and Brown, R.B., 2001, "MiBench: A Free, Commercially Representative Embedded Benchmark Suite", Proceedings of the IEEE International Workshop of Workload Characterization, pp. 3-14, doi: 10.1109/WWC.2001.990739

Hoste, K. and Eeckhout, L., 2008, "Cole: Compiler Optimization Level Exploration", Proceedings of the 6th Annual IEEE/ACM International Symposium on Code Generation and Optimization, pp. 165-174, doi: 10.1145/1356058.1356080

ITRS, 2012, "ITRS 2009 Roadmap", International Technology Roadmap for Semiconductors.

Jones,T.M., O'Boyle, M.F.P. and Ergin, O., 2008, "Evaluating the Effects of Compiler Optimisations on AVF", Proceedings of the Workshop on Interaction Between Compilers and Computer Architecture, 6p.

Kanawati, K., Krishnamurthy, N., Nair, S. and Abraham, J.A., 1996, "Evaluation of Integrated System-level Checks for On-Line Error Detection", Proceedings of the 2nd International Computer Performance and Dependability Symposium, pp. 292-301, doi: 10.1109/IPDS.1996.540230

Krishnamurthy, N., Jhaveri, V. and Abraham, J.A., 1998, "A Design Methodology for Software Fault Injection in Embedded Systems", Proceedings of the Workshop on Dependable Computing and its applications, pp. 12

Lattner, C. and Adve, V., 2004, "LLVM: A Compilation Framework for Lifelong Program Analysis & Transformation", Proceedings of the International Symposium on Code Generation and Optimization, pp. 75-86, doi: 10.1109/CGO.2004.1281665

Mehlitz, P.C. and Penix, J., 2005, "Expecting the unexpected – radiation hardened software", NASA Ames Research Center, pp. 10.

Mukherjee, S.S., Shrewsbury, M.A., Weaver, C., Emer, J. and Reinhardt, S.K., 2003, "A Systematic Methodology to Compute the Architectural Vulnerability Factors for a High-Performance Microprocessor", Proceedings of the 36th Annual IEEE/ACM International Symposium on Microarchitecture, pp. 29-40, doi: 10.1109/MICRO.2003.1253181

Normand, E., 1996, "Single Event Upset at Ground Level", IEEE Transactions on Nuclear Science, Vol. 43, No. 6, pp. 2742-2750, doi: 10.1109/23.556861

Oh, N., Shirvani, P.P. and McCluskey, E.J., 2002, "Control-flow Checking by Software Signatures", IEEE Transactions on Reliability, Vol. 51, No. 1, pp. 111-122, doi: 10.1109/24.994926

Pan, Z. and Eigenmann, R., 2006, "Fast and Effective Orchestration of Compiler Optimizations for Automatic Performance Tuning", Proceedings of the International Symposium on Code Generation and Optimization, pp. 319-332, doi: 10.1109/CGO.2006.38

Rech, P., Aguiar, C., Ferreira, R., Silvestri, M., Griffoni, A., Frost, C. and Carro, L., 2012, "Neutron-Induced Soft Errors in Graphic Processing Units", IEEE Radiation Effects Data Workshop, pp. 1-6, doi: 10.1109/REDW.2012.6353714

Saxena, N. and McCluskey, E., 1990, "Control Flow Checking using Watchdog Assists and Extended-Precision Checksums", IEEE Transactions on Computers, Vol. 39, No. 4, pp. 554-559, doi: 10.1109/12.54849

Vemu, R., Gurumurthy, S. and Abraham, J.A., 2007, "ACCE: Automatic Correction of Control-Flow Errors", IEEE International Test Conference, pp. 1-10, doi: 10.1109/TEST.2007.4437639

Analyzing the Unscented Kalman Filter Robustness for Orbit Determination Through Global Positioning System Signals

Paula Cristiane Pinto Mesquita Pardal[1], Hélio Koiti Kuga[2], Rodolpho Vilhena de Moraes[1]

ABSTRACT: The nonlinear unscented Kalman filter (UKF) is evaluated for the satellite orbit determination problem, using Global Positioning System (GPS) measurements. The assessment is based on the robustness of the filter. The main subjects for the evaluation are convergence speed and dynamical model complexity. Such assessment is based on comparing the UKF results with the extended Kalman filter (EKF) results for the solution of the same problem. Based on the analysis of such criteria, the advantages and drawbacks of the implementations are presented. In this orbit determination problem, the focus is to analyze UKF convergence behavior using different sampling rates for the GPS signals, where scattering of measurements will be taken into account. A second aim is to evaluate how the dynamical model complexity affects the performance of the estimators in such adverse situation. After solving the real-time satellite orbit determination problem using actual GPS measurements, through EKF and UKF algorithms, the results obtained are compared in computational terms such as complexity, convergence, and accuracy.

KEYWORDS: Orbit determination, Nonlinear Kalman filters, GPS measurements, Real time.

INTRODUCTION

This work points out the nonlinear unscented Kalman filter (UKF) robustness assessment for a real-time satellite orbit determination problem, using Global Positioning System (GPS) measurements. This evaluation is based on comparing the UKF performance with the extended Kalman filter (EKF) for different sampling rates of the measurement from GPS signals. One-second analysis takes into account the dynamical model complexity effects on the performance of the two estimation techniques. The main subjects for the comparisons between the estimators are: convergence speed, divergence occurrence, faults, and statistical shortcomings. Based on the analysis of such criteria, the advantages and drawbacks of each estimator are exhibited.

The orbit determination of an artificial satellite is done using real data from the GPS receivers. In the orbit determination process of artificial satellites, the nature of both the dynamic system and the measurements equations are nonlinear. As a result, here it is necessary to manage a fully nonlinear problem in which the disturbing forces as well as the measurements are not easily modeled. This orbit determination problem lies in estimating the variables that completely specify a satellite trajectory in the space, processing a set of information (in this case, pseudo-range measurement) related to such body. As far as this work is concerned, the more accurate GPS phase measurements are not used here, because the main goal is not the search for accuracy, but a comparison of performance under

1.Instituto de Ciência e Tecnologia – São José dos Campos/SP – Brazil 2.Instituto Nacional de Pesquisa Espacial – São José dos Campos/SP – Brazil
Author for correspondence: Paula Cristiane Pinto Mesquita Pardal | Rua Eng. João Fonseca dos Santos, 158, Apto. 163B – Vila Adyana | CEP 12.243-620 São José dos Campos/SP – Brazil | Email: paulacristiane@gmail.com

different sampling rates of the measurements from GPS. Furthermore, if carrier phase measurements were used, the ambiguity resolution algorithm or any other artifacts to overcome such hindrance could eventually mask the results, misguiding the conclusions.

A spaceborne GPS receiver is a powerful resource to determine orbits of artificial Earth satellites by providing many redundant measurements which ultimately yields high degree of the observability to the problem. The Topex/Poseidon (T/P) satellite is a nice example of using GPS for space positioning. Through an onboard GPS receiver, the pseudo-ranges (error corrupted distance from satellite to each of the tracked GPS satellites) can be measured and used to estimate the full orbital state.

The EKF is very likely the most widely used real-time estimation algorithm for nonlinear systems (Maybeck, 1982). However, the experience from the estimation community has shown that the EKF is difficult to implement, requiring some skill to get tuned since depends very much on the nearness of the initial conditions to the true values; and the linearity on the time scale of the filter working updates. Many of these difficulties arise from the linearization required by the EKF method. Specifically for the orbit estimation problem, under inaccurate initial conditions (Pardal *et al.*, 2011) and scattered measurements, the EKF implementation can lead to unstable or diverging solutions. Therefore, there is a strong need for a method that is probably more accurate than linearization, but that does not be liable to neither the implementation nor additional computational costs of other higher order filtering schemes. To overcome this limitation, the unscented transform (UT) was developed as a technique to propagate mean and covariance information through nonlinear transformations. The UKF is one of the sigma-point Kalman filters (SPKF), a new family of estimators that claims to yield equivalent or better performance than the EKF and elegantly to extend to nonlinear systems, without the linearization steps (van der Merwe, 2004; Julier and Uhlmann, 1997, 2004). This family of algorithms presents a new approach to generalize the KF for nonlinear dynamics and observation models.

Assessment between EKF and UKF was studied before by these and other authors, with different focus. Soken and Hajiyev (2011) compared two different robust Kalman filtering algorithms: Robust EKF and Robust UKF for the case of measurement malfunctions. In both filters, by the use of defined variables named as measurement noise scale factor, the faulty measurements were taken into the consideration with a small weight and the estimations were corrected without affecting the characteristic of the accurate ones. Proposed robust KFs were applied for the attitude estimation process of a pico satellite and the results are compared. El-Sheimy *et al.* (2006) studied which Kalman filtering design works best for GPS and micro-electro-mechanical (MEMS) inertial systems, since both have complementary qualities that make integrated navigation systems more robust. Jose (2009) implemented an UKF for integrating inertial navigation system (INS) with GPS and compared the results with the EKF approach, in performance and robustness. In a loosely coupled integrated INS/GPS system, inertial measurements from an inertial measurement unit IMU (angular velocities and accelerations in body frame) were integrated by the INS to obtain a complete navigation solution and the GPS measurements were used to correct for the errors and avoid the inherent drift of the pure INS system. Pardal *et al.* (2011) compared between the EKF and the nonlinear SPKF for a real-time satellite orbit determination problem, using GPS measurements for degraded initial conditions. The main subjects for the comparison between the estimators are convergence speed and computational implementation complexity. The aim was: to analyze the filters robustness; and to know the way such inaccuracies affect the performance of the estimators.

In this orbit determination problem the core is to analyze the convergence behavior for each filter in situations where there are different sampling rates of the measurement. Indeed there are small up to larger intervals between the processing of two GPS signals, and such intervals affect the performance of the estimators. A second goal evaluates the dynamical model complexity effects in the performance of the estimators for this case, especially during the adverse situations of larger reception intervals between two measurements. Therefore, the performance evaluation of the EKF (the most widely used estimation algorithm) and the UKF (supposedly the most appropriate estimation algorithm for nonlinear problems) to orbit determination problems in real time is due and justified.

SIGMA-POINT KALMAN FILTERS

When the system dynamics and the observation model are of linear nature, the conventional KF is the optimal solution and must be used fearlessly. However, not rarely, the system dynamics and/or the measurement models are nonlinear, and convenient extensions of the KF, like the EKF, have been used.

The SPKF is a new family of estimators that allows similar performance to the KF for linear systems and elegantly extends to nonlinear systems, without need of the linearization procedures. This family of algorithm is a new approach to generalize the KF for nonlinear process and observation models (Julier and Uhlmann, 1997, 2004; van der Merwe *et al.*, 2004). A set of weighted samples, the sigma-points, is used for computing mean and covariance of a probability distribution. Such algorithms include the UKF that is based on the UT, which is a nonlinear transformation of mean and covariance.

The SPKF represents a technique claimed as to lead to a more accurate and easier way to implement filter than the EKF or a second order Gaussian filter. Its approach is described, as follows (van der Merwe, 2004):

- A set of weighted samples is calculated deterministically based on the decomposition of the covariance and mean of a random variable.
- The sigma-points are propagated through the real nonlinear function, using only functional estimation, that is, analytical derivatives are not used to generate a posteriori set of sigma points.
- The later statistics are calculated using propagated sigma-points functions and weights. In general, they assume the form of a simple weighted average of the mean and the covariance.

Following, it will be separately explained the UT and the UKF, the filter stemming from this transformation.

UNSCENTED TRANSFORM

Essentially this is a manner of calculating the statistics of a random variable that passes through a nonlinear transformation. The UT approach is illustrated in Fig. 1 (Julier and Uhlmann, 1997; van der Merwe, 2004): select a suitable set of points (sigma-points) so that their mean and covariance are $\bar{\mathbf{x}}$ and $\mathbf{P}_{xx}$, respectively (Julier and Uhlmann, 1997, 2004). In turn, the

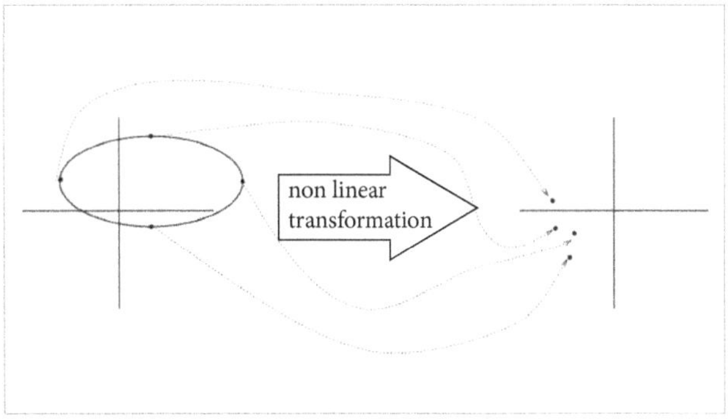

Figure 1. Unscented transform.

nonlinear function is applied to each point of the set to yield a cloud of transformed points. The statistics of the transformed points (mean $\bar{\mathbf{y}}$ and covariance $\mathbf{P}_{yy}$) can then be calculated to form an estimate of the nonlinearly transformed mean and covariance.

The sigma-points are carefully and deterministically chosen so that they exhibit certain specific properties, that is, they are not drawn at random like common Monte Carlo methods. Besides, they can be weighted in ways that are inconsistent with the distribution interpretation of sample points like in a particle filter (Julier and Uhlmann, 1997; van der Merwe, 2004).

The n-dimensional random variable x, with $\bar{\mathbf{x}}$ mean and $\mathbf{P}_{xx}$ covariance, is approximated by $2n+1$ weighted points, the so known sigma-points, given by:

$$\begin{aligned}
\chi_0 &= \bar{\mathbf{x}} \\
\chi_i &= \bar{\mathbf{x}} + \left(\sqrt{(n+\lambda)\,\mathbf{P}_{xx}}\right)_i \\
\chi_{i+n} &= \bar{\mathbf{x}} - \left(\sqrt{(n+\lambda)\,\mathbf{P}_{xx}}\right)_i
\end{aligned} \tag{1}$$

in which $\lambda=\alpha^2(n+k)-n$ includes scaling parameters. The constant parameter α controls the size of the sigma-points distribution ($0\leq\alpha\leq1$), and k provides an extra degree of freedom used to fine-tune the higher order moments; $k = 3 - n$ for a Gaussian distribution (Wan and van der Merwe, 2001). In Eq. 1, each element of the n-dimensional random variable x is replaced by a set of sigma-points generated from the mean and the covariance of x. Thus, the vector variable x has become a matrix of n x i dimension.

The transformation occurs as follows:

- Transform each point through the nonlinear function to yield the set of transformed sigma-points:

$$y_i = \mathbf{f}\,[\chi_i] \tag{2}$$

- The observations mean is given by the weighted average of the transformed points:

$$\bar{y} = \sum_{i=0}^{2n} W_i\, y_i \tag{3}$$

- The covariance is the weighted outer product of the transformed points:

$$P_{yy} = \sum_{i=0}^{2n} W_i\,[y_i - \bar{y}][y_i - \bar{y}]^T \tag{4}$$

W_i is the weight associated to the i-th point given by:

$$W_0 = \frac{k}{(n+k)}$$
$$W_i = \frac{1}{2(n+k)},\ i = 1,\ldots,n \tag{5}$$
$$W_{i+n} = \frac{1}{2(n+k)},\ i = 1,\ldots,n$$

UNSCENTED KALMAN FILTER

Using UT, the following steps are processed in the KF:

- Predict the new state system and its associated covariance, taking into account the effects of the Gaussian white noise process.
- Predict the expected observation and its residual innovation matrix considering the effects of the observation noise.
- Predict the cross correlation matrix.

In order to lead to the new filter, the UKF, these steps are arranged in the EKF, re-structuring: the dynamics; the state vector; and the observations model.

Table 1 presents an algorithm for the UKF. In the filter initialization, the mean, $\hat{\mathbf{x}}_{k-1}$, and the covariance matrix, $\hat{\mathbf{P}}_{k-1}$, of the state vector x are calculated, in reference to the prior instant, $\mathbf{t}_{k-1}$. Following, the set of sigma-points is generated, from the mean and the covariance matrix, previously calculated. In the propagation step, the generated state sigma-point set is propagated to the instant $\mathbf{t}_k$, using the nonlinear dynamics equation (a), and the predict mean and covariance matrix are calculated (b). During the update cycle: the observations sigma-points are generated (a), propagated through the nonlinear observations equation (b), and its mean is obtained (c); the predict matrices of innovation, $\mathbf{P}_k^{vv}$, and correlation, $\mathbf{P}_k^{xy}$, are computed (d); and finally the Kalman gain is calculated, in order to update the state $\hat{\mathbf{x}}_k$, and the covariance matrix, $\hat{\mathbf{P}}_k$. They are used as a priori information in the next instant, $\mathbf{t}_{k+1}$, to generate the new set of sigma-points.

Table 1. Unscented Kalman filter algorithm.

1. Initialization (t = k–1):	
	$\hat{\mathbf{x}}_{k-1} = E[\mathbf{x}_{k-1}] = E[\mathbf{x}_{k-1}^{\mathrm{T}}\mathbf{w}_{k-1}^{\mathrm{T}}]^{\mathrm{T}} \in \Re^n$ $\hat{\mathbf{P}}_{k-1} = E[(\mathbf{x}_{k-1} - \hat{\mathbf{x}}_{k-1})(\mathbf{x}_{k-1} - \hat{\mathbf{x}}_{k-1})^{\mathrm{T}}]$
2. Sigma-points:	
	$\chi_{0,k-1} = \hat{\mathbf{x}}_{k-1}$ $\chi_{i,k-1} = \hat{\mathbf{x}}_{k-1} + \left(\sqrt{(n+\lambda)\hat{\mathbf{P}}_{k-1}}\right)_i$; $\chi_{i+n,k-1} = \hat{\mathbf{x}}_{k-1} - \left(\sqrt{(n+\lambda)\hat{\mathbf{P}}_{k-1}}\right)_i$
3. Propagation:	
a	$\chi_{0,k} = \mathbf{f}[\chi_{0,k-1}]$ $\chi_{i,k} = \mathbf{f}[\chi_{i,k-1}], i = 1,\ldots,2n$
b	$\bar{\mathbf{x}}_k = \sum_{i=0}^{2n} W_i\,\chi_{i,k}$ $\bar{\mathbf{P}}_k = \sum_{i=0}^{2n} W_i[\chi_{i,k} - \bar{\mathbf{x}}_k][\chi_{i,k} - \bar{\mathbf{x}}_k]^{\mathrm{T}} + \mathbf{Q}_k$
4. Update:	
a	$\chi_{0,k} = \bar{\mathbf{x}}_k$ $\chi_{i,k} = \bar{\mathbf{x}}_k + \left(\sqrt{(n+\lambda)\bar{\mathbf{P}}_k}\right)_i$; $\chi_{i+n,k} = \bar{\mathbf{x}}_k - \left(\sqrt{(n+\lambda)\bar{\mathbf{P}}_k}\right)_i$
b	$y_{0,k} = \mathbf{h}[\chi_{0,k}]$ $y_{i,k} = \mathbf{h}[\chi_{i,k}], i = 1,\ldots,2n$
c	$\bar{y}_k = \sum_{i=0}^{2n} W_i\, y_{i,k}$
d	$\mathbf{P}_k^{vv} = \sum_{i=0}^{2n} W_i\,[y_{i,k} - \bar{y}_k][y_{i,k} - \bar{y}_k]^{\mathrm{T}} + \mathbf{R}_k$ $\mathbf{P}_k^{xy} = \sum_{i=0}^{2n} W_i\,[\chi_{i,k} - \bar{\mathbf{x}}_k][y_{i,k} - \bar{y}_k]^{\mathrm{T}}$
e	$\mathcal{K}_k = \mathbf{P}_k^{xy}(\mathbf{P}_k^{vv})^{-1}$ $\hat{\mathbf{x}}_k = \bar{\mathbf{x}}_k + \mathcal{K}_k(\mathbf{y}_k - \bar{\mathbf{y}}_k)$ $\hat{\mathbf{P}}_k = \bar{\mathbf{P}}_k - \mathcal{K}_k\,\mathbf{P}_k^{vv}\,\mathcal{K}_k^{\mathrm{T}}$

EXTENDED KALMAN FILTER

If the dynamical system and the observations model are linear, the KF is the recursive estimator most used at the present day since it is easy to implement and to use on digital computers. Its recursiveness leads to lesser memory storage, which makes it ideal for real-time applications. The EKF is a nonlinear version of the KF that generates reference trajectories which are updated at each measurement processing, at the corresponding instant (Maybeck, 1982; Brown and Hwang, 1985).

Because it is very difficult to accurately model the artificial satellites orbit dynamics, the EKF is generally used in works of such nature. Its algorithm always brings updated reference trajectory around the most current available estimate.

Exploiting the assumption that all transformations are quasi-linear, the EKF simply linearizes all nonlinear transformations and substitutes the Jacobian matrices for the linear transformations in the KF equations. The EKF consists of phases of time and measurement updates. In the first one, state and covariance are propagated from one precedent instant to a posterior one, which means that they are propagated between discrete instants of the system dynamics model. In the second one, state and covariance are corrected from the measurement obtained in the posterior instant of time, through the observations model. Therefore, the method nature is recursive, so it does not need to store previously the measurements in large matrices.

Following, the step for the EKF time update (or propagation) cycle is presented:

$$\dot{\overline{\mathbf{x}}}_k = \mathbf{f}(\hat{\mathbf{x}}_{k-1})$$
$$\overline{\mathbf{P}}_k = \varphi_{k,k-1}\hat{\mathbf{P}}_{k-1}\varphi^T_{k,k-1} + \mathbf{Q}_k \tag{6}$$

where **f** is a nonlinear vector function modeling the orbit motion, $\overline{\mathbf{x}}_k$ and $\overline{\mathbf{P}}_k$ are respectively the propagated state and the covariance for t_k; $\varphi_{k,k-1}$ is the state transition matrix between t_{k-1} e t_k; $\mathbf{Q}_k$ is the dynamics noise matrix given in Eq. 7. It is required the Jacobian matrix ($\partial \mathbf{f}/\partial \mathbf{x}$) for the transition matrix computation, which can be either simplified or very difficult to obtain.

$$\mathbf{Q}_k = \int_{t_{k-1}}^{t_k} \varphi(t, t_{k-1})\mathbf{G}(t)\mathbf{Q}(t)\mathbf{G}^T(t)\varphi^T(t, t_{k-1})dt \tag{7}$$

The equations for the EKF measurement update cycle are:

$$\mathbf{K}_k = \overline{\mathbf{P}}_k\mathbf{H}_k^T(\mathbf{H}_k\overline{\mathbf{P}}_k\mathbf{H}_k^T + \mathbf{R}_k)^{-1}$$
$$\hat{\mathbf{P}}_k = (\mathbf{I}-\mathbf{K}_k\mathbf{H}_k)\,\overline{\mathbf{P}}_k \tag{8}$$
$$\hat{\mathbf{x}}_k = \overline{\mathbf{x}}_k + \mathbf{K}_k\,[\mathbf{y}_k-\mathbf{h}_k(\overline{\mathbf{x}}_k)]$$

In Eq. 7, **G**(t) is the white noise addition matrix. In Eq. 8, $\mathbf{h}_k$ is a nonlinear vector function modeling the measurements; $\mathbf{H}_k$ is the corresponding partial derivative matrix $(\partial \mathbf{h}_k/\partial \mathbf{x})$; $\mathbf{K}_k$ is the Kalman gain; $\mathbf{R}_k$ is the observations noise matrix; $\hat{\mathbf{x}}_k$ and $\hat{\mathbf{P}}_k$ are respectively the state vector and the covariance updated for the instant *k*; $\mathbf{y}_k$ is the observations vector corresponding to the instant *k*.

Notwithstanding, the EKF has limitations. First: linearization can produce highly unstable filters if the assumptions of local linearity are violated; second: the derivation of the Jacobian matrices is nontrivial in most applications, and often leads to significant implementations difficulties (Julier and Uhlmann, 1997); third: analytical Jacobian matrices can be a very difficult and error-prone process.

Summarizing, linearization, as applied in the EKF, is widely recognized to be inadequate, but the alternatives incur substantial costs in terms of derivation and computational complexity. Hence, there is a strong need for a method that is probably more accurate than linearization but does not incur costs of implementation and computational of higher order than the other filters. The sigma-point algorithms were developed to meet these needs (Julier and Uhlmann, 1997).

ORBIT DETERMINATION

The orbit determination process consists of obtaining values of the parameters that completely specify the motion of an orbiting body (here, an artificial satellite), based on a set of observations of the body. It involves nonlinear dynamics and nonlinear measurement systems, which depends on the tracking system, and estimation technique (for instance, KF or least squares (Maybeck, 1982; Brown and Hwang, 1984)). The dynamical system model consists of the description for the dynamics of the satellite orbital motion, measurements models, Earth's rotation effects, and perturbation models. Furthermore the state variables defining the initial conditions, these models depend on a variety of parameters which affect both the dynamic motion as the measurement process (Montenbruck and Gill, 2001). Due to the complexity of the applied models it is hardly possible to solve such models equations directly for any of these parameters from a given set of observations.

The observation may be obtained from the ground station networks using laser, radar, Doppler, or by space navigation systems, as the GPS. The choice of the tracking system depends on a compromise between the goals of the mission and the available tools. In the case of the GPS, the advantages are global coverage, high precision, low cost, and autonomous navigation resources. The GPS may provide orbit determination with accuracy at least as good as methods using ground tracking networks. The later provides standard precision around tens of meters and the former can provide precision as tight as some centimeters. The GPS provides, at a given instant, a set of many redundant measurements, which makes the orbit position observable geometrically.

After some advances of technology, the single frequency GPS receivers provide a good basis to achieve fair precision at relatively low cost, still attaining the accuracy requirements of the mission operation. The GPS allows the receiver to determine its position and time geometrically anywhere at any instant with data from at least four satellites. The principle of navigation by satellites is based in sending signals and data from the GPS satellites to a receiver located onboard the satellite that needs to have its orbit determined. This receiver measures the travel time of the signal and then calculates the distance between the receiver and the GPS satellite. If the clocks are not synchronized, four measurements are required to obtain its position. Those measurements of distances are called pseudo-ranges.

The instantaneous orbit determination using GPS satellites is based on the geometric method. In such method, the observer knows the set of GPS satellites position in a reference frame, obtaining its own position in the same reference frame.

However, sequential orbit determination makes use of the orbital motion modeling to predict between measurement times and measurement model to update the orbit by processing of measurements from GPS. This gives rise to recursive and real-time KF estimator for the orbit determination (Brown and Hwang, 1985).

FILTER DYNAMIC MODEL

In the case of orbit determination via GPS, the ordinary differential equations which represent the dynamic model are in its simplest form, given traditionally as follows:

$$\begin{aligned} \dot{\mathrm{r}} &= \mathrm{v} \\ \dot{\mathrm{v}} &= -\mu \frac{\mathrm{r}}{\mathrm{r}^3} + \mathrm{a} + \mathrm{w}_v \\ \dot{\mathrm{b}} &= \mathrm{d} \\ \dot{\mathrm{d}} &= 0 + \mathrm{w}_\mathrm{d} \end{aligned} \tag{9}$$

wherein the variables are placed in the inertial reference frame. In Eq. 9, **r** is the vector of the position components (x, y, z); **v** is the velocity vector; **a** represents the modeled perturbing accelerations; $\mathbf{w}_v$ is the white noise vector with covariance **Q**; b is the user satellite GPS clock bias; d, the user satellite GPS clock drift; and w_d the noise associated with the GPS clock. The GPS receiver clock offset was not taken into account, so as not to obscure the conclusions drawn in this paper due to introduction of clock offset models in the filters. Indeed, the receiver clock offset was beforehand obtained and used to correct the GPS measurements, so that the measurements are free from the error derived from receiver clock offset.

FORCE MODEL

The main disturbing forces of gravitational nature that affect the orbit of an Earth's artificial satellite are: the non-uniform distribution of Earth's mass; ocean and terrestrial tides; and the gravitational attraction of the Sun and the Moon. There are also the non-gravitational effects, such as: Earth atmospheric drag; direct and reflected solar radiation pressure; electric drag; emissivity effects; relativistic effects; and meteorites impacts.

The disturbing effects are in general included according to the physical situation presented and to the accuracy that is intended for the orbit determination. Here, we include only a minimum set of perturbations which enable us to assess the performance of both filters, namely geopotential and third body point mass effect of Sun and Moon.

The Earth is not a perfect sphere with homogeneous mass distribution, and cannot be considered as a material point. Such irregularities disturb the orbit of an artificial satellite and the keplerian elements that describe the orbit do not behave ideally. The geopotential function can be given by (Kaula, 1966):

$$U(\mathrm{r}, \phi, \lambda) = \frac{\mu}{r} \sum_{n=0}^{\infty} \sum_{m=0}^{n} \left(\frac{R_\mathrm{T}}{\mathrm{r}}\right)^n P_{nm} (\sin \phi)(C_{nm} \cos m\lambda + S_{nm} \sin m\lambda) \tag{10}$$

where μ is Earth gravitational constant; R_T is mean Earth radius; r is the spacecraft radial distance; ϕ is the geocentric latitude; λ is the longitude on Earth fixed coordinates system; C_{nm} and S_{nm} are the harmonic spherical coefficients of degree n and order m; P_{nm} are the associated Legendre functions. The constants μ, R_T, and the coefficients C_{nm}, and S_{nm} determine a particular gravitational potential model.

Another gravitational perturbation source is due to the Sun and Moon attraction. They are more meaningful at larger distance from Earth. As the orbital variations are of the same type, be the Sun or the Moon the attractive body, they are normally studied without distinguishing the third body. The Sun–Moon gravitational attraction mainly acts on node and perigee, causing precession of the orbit and on the orbital plane. The general three-body problem model is here simplified to the circular restricted three-body problem, where the orbital motion of a third body (satellite), which mass can be neglected, around two other massive bodies is studied. The force acting on the third body (the satellite) in the inertial reference frame can be expressed as (Prado and Kuga, 2001; Guan, 2013):

$$\ddot{\mathbf{r}}_3 = -Gm_1\frac{\mathbf{r}_{13}}{\mathbf{r}_{13}^3} - Gm_2\frac{\mathbf{r}_{23}}{\mathbf{r}_{23}^3} \tag{11}$$

where $\mathbf{r}_{13} = \mathbf{r}_3 - \mathbf{r}_1$, $\mathbf{r}_{23} = \mathbf{r}_3 - \mathbf{r}_2$, and $\mathbf{r}_i$, i=1,2,3 corresponds to the i-th body distance vector to the center of mass of the system; and m_1 and m_2 are the masses of the Sun and the Moon, respectively.

OBSERVATIONS MODEL

The nonlinear equation of the observation model is:

$$\mathbf{y}_k = \mathbf{h}_k(\mathbf{x}_k, t) + \nu_k \tag{12}$$

where, at time t_k, $\mathbf{y}_k$ is the vector of m observations; $\mathbf{h}_k(\mathbf{x}_k)$ is the nonlinear function of state $\mathbf{x}_k$, with dimension m; and ν_k is the observations errors vector, with dimension m and covariance $\mathbf{R}_k$. For the present application, one only uses the ion-free pseudo-range measurements from the GPS receiver of T/P satellite. Also, the receiver clock offset was computed before and used to correct the pseudo-range measurements. In addition, the nonlinear pseudo-range measurement was modeled according to Chiaradia *et al.* (2003).

MEASUREMENTS SAMPLING RATES IMPACT IN THE ORBIT DETERMINATION

Previous presented studies (Pardal et al., 2009b, 2010 and Pardal, 2011) showed that the accuracy improvement for the dynamics models did not better the errors resulting between the references from precise orbit ephemeris from JPL/NASA (POE/JPL) and the values stemming from filters estimation process. This means that the magnitude of the errors obtained through the UFK or EKF is not reduced when increasing the complexity of the dynamic model: from a model for the geopotential with high degree and order to a complex model containing the three major disruptive effects of the orbit. Further, such results showed an equivalent competitiveness between the estimators, since the errors are of the same order of magnitude. It might have occurred because, in those results, the orbit determination process was done for small sampling intervals of the measurements.

Taking this into account, before concluding that a simpler dynamics modeling can be indiscriminately adopted, there is a need for another test (Pardal *et al.*, 2010). Such test has two well determined purposes: to examine carefully the benefits of increasing the adopted dynamics model accuracy; and to investigate the apparent competitiveness between the estimators. This test rested in executing the orbit determination process considering different intervals between the GPS observations (pseudo-range) sampling. The intervals of sampling were 10, 30, 60, 300, 600, 1200, and 1800 seconds, in other words, conditions were swept from one very small range (10 seconds) to a range extremely high (1800 seconds) between the prediction and the subsequent correction of the predicted values to complete the cycle of the estimation process. With the gradual spacing of the interval between two measurements, the intention was to verify the dynamics models complexity, and the application of each filter in the time update cycle. In such situation, the propagation has it effects raised and the modeling accuracy becomes more significant in the accuracy of both filters estimative.

RESULTS

The tests and the analysis for the EKF and the UKF algorithms are presented. To validate and to analyze the methods, real GPS data from the T/P satellite are used. The filters estimated position and velocity are compared with T/P POE from JPL/NASA. The test conditions consider real ion-free pseudo-range data, collected by the GPS receiver onboard T/P, on November 19, 1993, at different sampling rates, presenting on average between 5 and 6 GPS satellites tracked. The GPS data were previously preprocessed to remove the outliers so they cannot mislead the filters or mask different data rejection policies of each filter. The tests have covered a long one day period of orbit determination.

The force model goes from a simple geopotential up to order and degree (2×2), with harmonic coefficients from JGM-2 model, to a model including perturbations due to geopotential up to order and degree (28×28) and due to the Sun–Moon gravitational attraction (Pardal *et al.*, 2009a, 2010, 2011). The pseudo-range measurements were corrected to the first order with respect to ionosphere.

As already pointed, this work is not a search for results accuracy. It aims at UKF robustness assessment, which is done through the comparison of performance between UKF and EKF estimators under different sampling rates. There are peculiar interest for speed convergence, and divergence occurrence.

Table 2 shows the analysis for the predicted pseudo-range residuals convergence, which is measured in terms of time span, Dt_{sam}, of data processed. The convergence is assumed when the residuals achieve similar statistics of the reference solution residuals. When small samplings intervals, such as 10, 30, and 60 seconds are used, convergence occurs instantaneously after the estimation process starts, for both UKF and EKF algorithms,

Table 2. Pseudo-range residuals convergence speed.

Model	Δtsam (s)	UKF	EKF
		convergence time (h)	convergence time (h)
Geo2	10-30-60	0	0
	300	no convergence observed	no convergence observed
	600	–	–
	1200	–	–
	1800	–	–
Geo5	10-30-60	0	0
	300	0	no convergence observed
	600	no convergence observed	–
	1200	–	–
	1800	–	–
Geo10	10-30-60	0	0
	300	0	no convergence observed
	600	0	–
	1200	0	–
	1800	2	–
Geo28	10-30-60	0	0
	300	0	0
	600	0	1.5
	1200	0	5
	1800	2	no convergence observed
S-M	10-30-60	0	0
	300	0	0
	600	0	1.5
	1200	0	5
	1800	2	no convergence observed

and for any dynamics model adopted. For this reason, results obtained for 10, 30, and 60 seconds of sampling interval will be placed in the same line of Table 1. The model of geopotential up to low order and degree 2×2 (Geo2) starts diverging for 300 seconds of sampling interval, regardless of the estimator applied. The improved geopotential up to order and degree 5×5 (Geo5) dynamics model stops converging when UKF is the filter for 600 seconds, while EKF is not able to converge at all since 300 seconds. From an improved geopotential model and on, there is always occurrence of divergence when EKF is the algorithm, and convergence keeps occurring if UKF is the chosen algorithm. That is to say: for geopotential up to order and degree 10×10 (Geo10), divergence is detect at 300 seconds; and for geopotential up to high order and degree 28×28 (Geo28), and for a model that compounds geopotential up to order and degree 28 with Sun-Moon gravitational attraction (S-M), there is occurrence of divergence at 1800 seconds. The convergence time (consequently the convergence speed) is the same for the Geo28 and the S-M models up to 300 seconds of sampling interval, for both estimators, as shown in Table 1. The filters convergence time starts to be different at 600 seconds of interval between two measurements, and this difference keeps the same for each test case (600, 1200, or 1800 seconds), whether the model is Geo28 or S-M. At this point it is possible to pinpoint a model limitation for convergence analysis.

Table 3 shows the convergence analysis for the position RNT (radial, normal, and along-track) components error, which is again measured in terms of data time of processing. When small intervals of sampling are used, such as 10, 30, and 60 seconds, convergence is detected instantaneously after the starts of the estimation process, for the two estimators, and for any choice of dynamics model. From 300 seconds on, the behavior is the same described in Table 2. The Geo2 model errors start diverging for 300 seconds of sampling interval, regardless of the estimator applied. The improved Geo5 dynamics model errors stop converging when the UKF is the filter for 600 seconds, while

Table 3. Errors in position convergence speed.

Model	Δt sam (s)	UKF convergence time (h)			EKF convergence time (h)		
		R	N	T	R	N	T
Geo2	10-30-60	0	0	0	0	0	0
	300		no convergence observed			no convergence observed	
	600		–			–	
	1200		–			–	
	1800		–			–	
Geo5	10-30-60	0	0	0	0	0	0
	300	0	0	0		no convergence observed	
	600		no convergence observed			–	
	1200		–			–	
	1800		–			–	
Geo10	10-30-60	0	0	0	0	0	0
	300	0	0	0		no convergence observed	
	600	0	0	0		–	
	1200	0	0	0		–	
	1800	1	1	0		–	
Geo28	10-30-60	0	0	0	0	0	0
	300	0	0	0	0	0	0
	600	0	0	0	0.5	0.2	0.3
	1200	0	0	0	1	3.5	3.2
	1800	1	1	0	9	9.3	no convergence observed
S-M	10-30-60	0	0	0	0	0	0
	300	0	0	0	0	0	0
	600	0	0	0	0.5	0.2	0.3
	1200	0	0	0	1	3.5	3.2
	1800	1	1	0	9	9.3	no convergence observed

the EKF is not able to converge at all since 300 seconds. For the Geo10 dynamic model, divergence is detected at 300 seconds; and for the Geo28, and S-M models, there is occurrence of divergence at 1800 seconds. Again, for models Geo28 and S-M, it can be noticed that filters convergence time difference starts at 600 seconds of sampling interval, and this difference keeps the same for each test case (600, 1200, or 1800 seconds), no matter the model. So far, a model limitation for convergence analysis might be clear.

Another statistical check is done, in order to confirm that the algorithms effectively reached convergence. The reference pseudo-range residuals statistics (mean and standard deviation) for each model and filter are available in the yellow lines of Table 4. As the three most improved dynamics models (Geo10, Geo28, and S-M) statistics for 10, 30, and 60 seconds considerably resemble (Pardal *et al.*, 2009b, 2010), "reference" in Table 4 refers to a 60-s sampling interval, and is representative of the three intervals. Now, in the analysis of the two poorest dynamics models (Geo2 and Geo5), all the sampling intervals are explicit in Table 4 in order to show poor dynamics behavior as the sampling intervals increase. It is clear that poor dynamics models are more sensitive to the intervals enlargement, since Geo2 model stops converging at 300 seconds, and Geo5, at 300 or 600 seconds, depending on the estimator applied. Through Table 4 it is also noticeable that if the model is too poor (e.g., Geo2), neither UKF nor EKF are able to keep convergence for larger sampling intervals, and divergence behavior is detected at 300 seconds for both. However, if the model is slightly improved (for instance, Geo5), UKF implementation shows more robustness than EKF one, because still converges at 300 seconds, while EKF starts diverging. From Table 3, it becomes evident that the estimators really reached convergence, since their statistical values remain nearly the same as the reference ones.

In order to portray such findings, Fig. 2 illustrates the reference residuals (small sampling intervals of 60 seconds) behavior, and the 1800 seconds sampling interval case behavior for both the EKF and the UKF estimators, using S-M as the dynamics model. It clearly indicates clues of EKF's divergence for such larger sampling intervals of measurements.

Proceeding the investigation, Table 5 shows total Root Mean Square (RMS) position error, where the reference values are again listed in the first row (yellowed). Again, in the analysis of the two poorest dynamics models (Geo2 and Geo5), all the sampling intervals are registered in Table 5 in order to show poor dynamical behavior as the sampling intervals increase. It is clear that poor dynamics models are more responsive to the intervals increasing: the Geo2 model results stop converging at 300 seconds while the Geo5, at 300 or 600 seconds, according to the estimator applied. Through Table 5 it is also perceptible that for an excessively poor model (Geo2, for example), UKF and EKF are unable to converge for larger sampling intervals, and divergence occurs at 300 seconds for both. However, if the model is slightly improved (for instance, Geo5), UKF implementation still

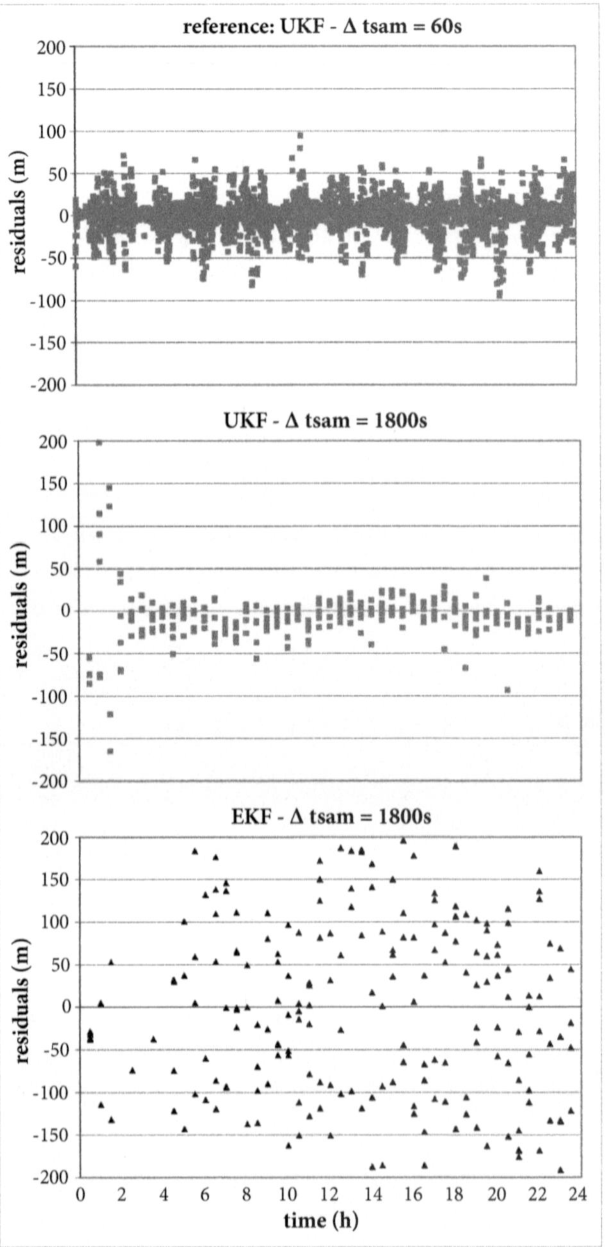

Figure 2. Pseudo-range residuals convergence and divergence occurrences.

converges at 300 seconds, while EKF starts diverging, which indicates more robustness of UKF when compared to EKF. UKF and EKF resulting RMS errors are only computed after assumed convergence time. For Table 5, it is also verifiable that the estimators really reached convergence, since their RMS values remain nearly close to the reference ones.

Next, Fig. 3 depicts a particular case where a relatively poor Geo5 dynamics model is adopted. While the UKF filter

Table 4. Pseudo-range residual statistics, after convergence.

Model	Δtsam (s)	UKF residuals mean±SD (m)	EKF residuals mean±SD (m)
Geo2	10	0.218±18.504	0.155±18.531
	30	0.081±26.992	-0.003±27.103
	60	-0.091±37.687	-0.276±37.916
	300 to 1800	no convergence observed	no convergence observed
Geo5	10	0.138±13.064	0.076±13.133
	30	-0.007±14.602	-0.087±15.058
	60	-0.251±16.575	-0.426±17.914
	300	-1.604±25.216	no convergence observed
	600 to 1800	no convergence observed	–
Geo10	reference	-0.187±14.690	-0.277±16.221
	300	-1.272±18.766	no convergence observed
	600	-2.401±19.786	–
	1200	-4.193±23.873	–
	1800	-7.714±33.229	–
Geo28	reference	-0.091±14.363	-0.180±15.942
	300	-1.234±17.489	1.861±45.886
	600	-2.379±18.242	-3.557±27.425
	1200	-4.216±22.295	-19.831±90.423
	1800	-7.929±32.163	no convergence observed
S-M	reference	-0.109±14.021	-0.201±15.639
	300	-1.294±15.872	-1.021±43.197
	600	-2.397±16.542	-3.6126±25.945
	1200	-4.163±20.518	-19.814±91.237
	1800	-7.641±31.156	no convergence observed

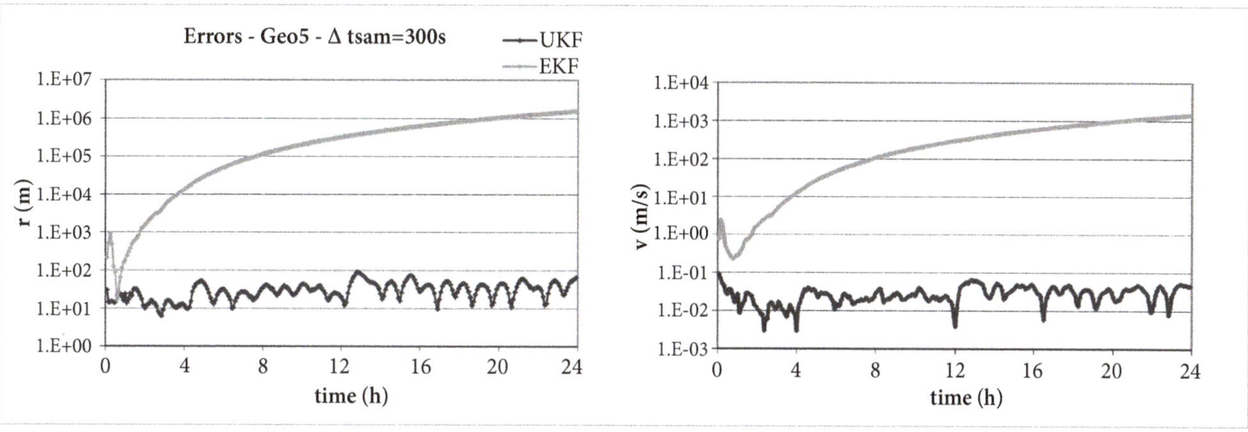

Figure 3. Convergence and divergence behavior for Geo5 dynamics model.

Table 5. Error in position total Root Mean Square, after convergence.

Model	Δtsam (s)	UKF	EKF
		total RMS(m)	total RMS(m)
Geo2	10	34.336	34.391
	30	49.834	49.703
	60	68.774	69.837
	300 to 1800	no convergence observed	no convergence observed
Geo5	10	20.521	20.642
	30	20.050	20.301
	60	22.481	27.303
	300	35.562	no convergence observed
	600 to 1800	no convergence observed	–
Geo10	reference	18.150	23.953
	300	22.273	no convergence observed
	600	22.658	–
	1200	24.000	–
	1800	28.962	–
Geo28	reference	17.135	23.159
	300	19.217	21.798
	600	19.386	29.487
	1200	20.943	no convergence observed
	1800	26.420	–
S-M	reference	16.253	22.520
	300	15.437	18.416
	600	15.398	26.489
	1200	16.403	no convergence observed
	1800	23.990	–

remains converging, a divergence behavior is shown in the EKF implementation, for 300 seconds of sampling interval between two measurements. This result indicates that even if the model is not adequately chosen, the EKF believes that the model and linearization are correct. The UKF does not have linearizations and, in adverse situations, such as the ones of larger sampling intervals between two data samples or inaccurate initial conditions (Pardal *et al.*, 2011), behaves more adequately. However, if the dynamics model is extremely truncated, such as the Geo2 analyzed in this work, neither the UKF nor the EKF will reach convergence for large sampling intervals, as shown in Tables 1–4. In Fig. 3, Δr and Δv represents, respectively, the absolute value of the errors in position and in velocity, in the inertial reference frame coordinates.

Figure 4 shows the errors in the RNT components for the UKF and EKF reference cases (small 60-s sampling interval, left side) and the larger 1800 seconds error case results for the EKF and the UKF estimators (right side). The outstanding behavior in the right side happens again in any "no convergence observed" case pointed out in Table 4. It indicates signs of the EKF divergence for such a very large sampling interval, while UKF reaches the convergence zone, not much later than the left side results. So far, the results showed that the performance of the filters decreases as increasing the sampling intervals (according to assays of the error in position, the pseudorange residuals, and the convergence presented previously). These results point to the advantage of using the nonlinear theory for orbital dynamics, in small intervals of the UKF and the EKF algorithms.

In order to finish the results analysis, it is to be said that even considering the convergence for small sampling intervals, where UKF and EKF present similar performance, the algorithm is very sensitive to the initialization of the covariance matrix. This means that the algorithm convergence depends on the proper choice of such matrix. Therefore, the

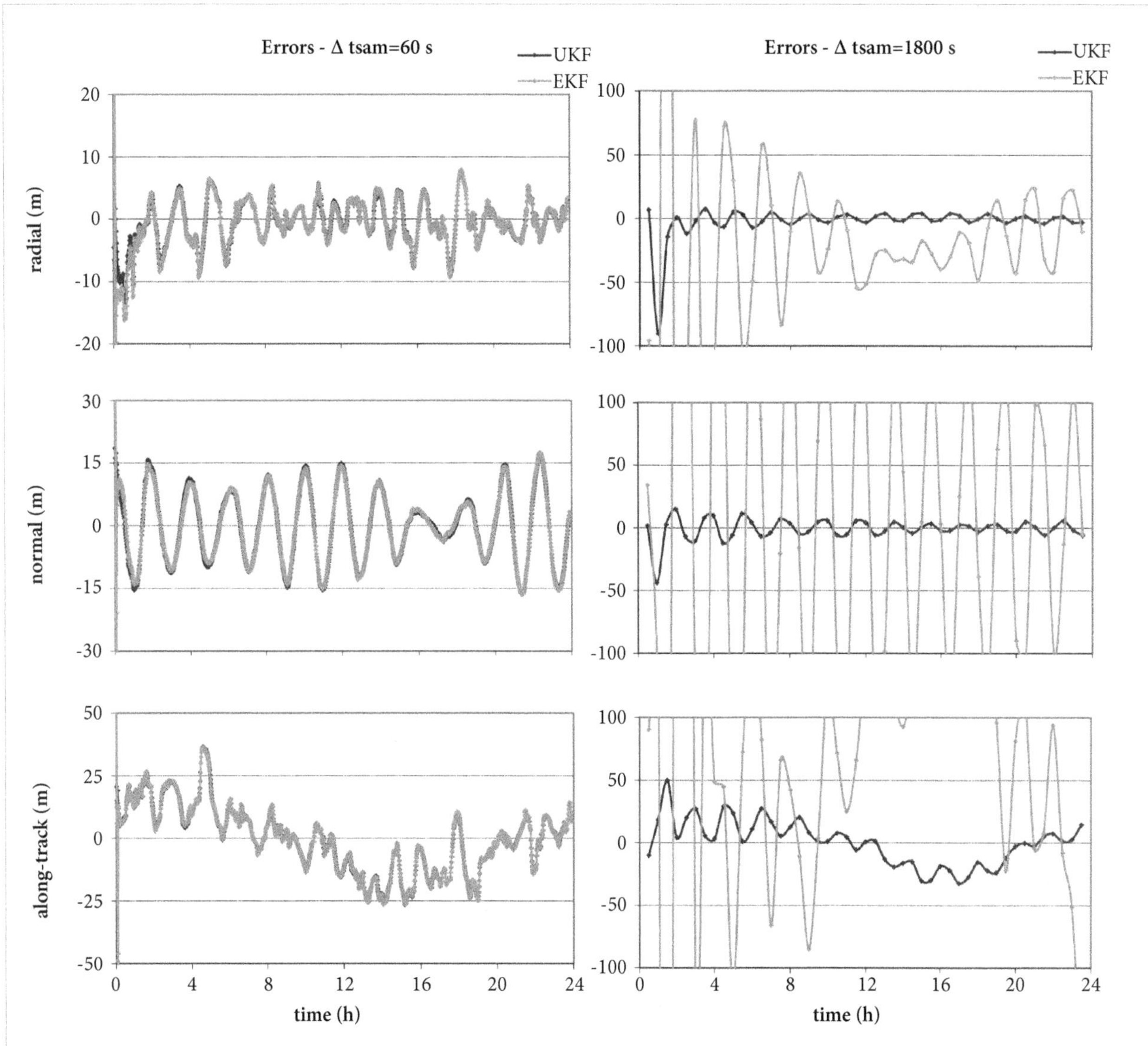

Figure 4. Convergence and divergence behavior of the errors in RNT (radial, normal, and along-track) components.

results are not general, and the algorithm was adjusted for this type of orbit determination application.

CONCLUSIONS

The robustness to increasing sampling intervals of two nonlinear estimators, namely the EKF and UKF was assessed for a real-time satellite orbit determination problem using real GPS measurements. One day (24h) of GPS receiver measurements of T/P satellite at different sampling rates were processed. The emphasis was to characterize each filter convergence behavior in situations where the sampling rates vary from small to larger intervals between two measurements. Different dynamical models were analyzed, in order to establish the modeling effects in the orbit determination process.

Results showed that when small sampling intervals are used, UKF and EKF yield similar performance, with high performance, for almost all dynamics models. The exception is made if the model is extremely truncated (Geo2) or slightly improved (Geo5), where in the later Geo5 case, UKF maintains convergence and in the former Geo2 case, none of the estimators reach convergence even for 60-s sampling interval. As expected, increasing the sampling intervals decreases the filters performance. As larger is the interval more difficult is for EKF and UKF to reach convergence. When UKF is compared with EKF, in all cases of larger intervals, the UKF always attains convergence first. The rupture threshold for this application in particular occurs to all modeling complexities if EKF was the used algorithm. Only in two situations (the two poorest models adoption) for the UKF implementation, divergence occurred. Therefore it is to be said that the UKF is more robust than the EKF for larger sampling interval between measurements, for this type of orbit determination application.

ACKNOWLEDGMENTS

The authors wish to express their consideration to the Instituto Nacional de Pesquisas Espaciais (INPE) that kindly provided everything necessary for this paper to be developed. The authors are also grateful to Fundação de Amparo à Pesquisa do Estado de São Paulo (FAPESP) for the support, under contracts No. 2013/00960-4, No. 2012/21023-6; to CAPES; and to CNPq.

REFERENCES

Brown, R. G. and Hwang, P. Y. C., 1984, A Kalman filter approach to precision GPS geodesy, Navigation: Global Positioning System, Vol. II, pp. 155-166.

Brown, R. G. and Hwang, P. Y. C., 1985, Introduction to random signals and applied Kalman filtering, 3rd ed. New York: John Wiley & Sons. 502 p.

Chiaradia, A. P. M., Kuga, H. K. Kuga and Prado, A. F. B. A., 2003, Single frequency GPS measurements in real-time artificial satellite orbit determination, Acta Astronautica, Vol. 53, n. 2, pp. 123-133.

El-Sheimy, N., Shin, E-H. and Niu, X., 2006, Extended vs. Unscented Kalman Filters for Integrated GPS and MEMS Inertial. Available from: http://www.insidegnss.com/auto/0306%20Kalman.pdf

Guan, T., 2013, Special cases of the three body problem, [online Colorado University database]. Available from: http://inside.mines.edu/fs_home/tohno/teaching/PH505_2011/Paper_TianyuanGuan.pdf

Jose, J. M., 2009, Performance comparison of Extended and Unscented Kalman Filter implementation in INS-GPS integration, Ph.D thesis. Czech Technical University in Prague, Prague (in English). Available from: http://epubl.ltu.se/1653-0187/2009/095/LTU-PB-EX-09095-SE.pdf

Julier, S. J. and Uhlmann, J. K., 1997, A new extension of the Kalman filter for nonlinear systems, International Symposium on Aerospace/Defense Sensing, Simulation and Controls. SPIE.

Julier, S. J.and Uhlmann, J. K., 2004, Unscented filtering and nonlinear estimation. IEEE Transactions on Automatic Control, Vol. 92, n. 3, Mar 2004.

Kaula, W.M., 1966, Theory of Satellite Geodesy. Blasdell Pub. Co. Walthmam, Mass.

Maybeck, P.S., 1982, Stochastic models, estimation, and control. Vol. 2, Academic Press, NY.

Montenbruck, O. and Gill, E., 2001, Satellite orbits: models, methods, and applications, Springer-Verlag, Berlin Heidelberg NewYork. 369 p.

Pardal, P. C. P. M., Kuga, H. K. and Vilhena de Moraes, R., 2009a, A discussion related to orbit determination using nonlinear sigma point Kalman filter, Mathematical Problems in Engineering, v. 2009, 12 p. Hindawi Publishing Corporation, doi:10.1155/2009/140963.

Pardal, P. C. P. M., Kuga, H. K. and Vilhena de Moraes, R., 2009b, Non linear sigma point Kalman filter applied to orbit determination using GPS measurements, 22nd International Meeting of the Satellite Division of the Institute of Navigation (ION GNSS 2009), 22–25 Sept., Savannah, USA.

Pardal, P. C. P. M., Kuga, H. K. and Vilhena de Moraes, R., 2010, Comparing the extended and the sigma point Kalman filters for orbit determination modeling using GPS measurements,23rd International Meeting of the Satellite Division of the Institute of Navigation (ION GNSS 2010), 21-24 Sept., Portland, USA.

Pardal, P. C. P. M., Real time orbit determination through non linear sigma-point Kalman filter, 2011, Ph.D thesis, National Institute for Space Research (INPE), São José dos Campos. (In Portuguese).

Pardal, P. C. P. M., Kuga, H. K.and Vilhena de Moraes, R., 2011, Robustness assessment between sigma point and extended Kalman filter for orbit determination,. Journal of Aerospace Engineering, Sciences and Applications, v. III, pp. 35-44. doi: 10.7446/jaesa.0303.04

Prado, A. F. B. A. and Kuga, H. K. (Eds.), 2001, Space Technology Fundaments, São José dos Campos: INPE. 220 p. (In Portuguese).

Soken, H. E. and Hajiyev, C., 2011, REKF and RUKF Development for Pico Satellite Attitude Estimation in the Presence of Measurement Faults, 5th International Conference on Recent Advances in Space Technologies (RAST), 9–11 June 2011. pp. 891-896. doi: 10.1109/RAST.2011.5966972

van der Merwe, R., 2004, Sigma-Point Kalman filters for probabilistic inference in dynamic state-space models, Ph.D thesis. Oregon Health & Science University, Portland.

van der Merwe, R., Wan, E.A. and Julier, S.J., 2004, Sigma-point Kalman filters for nonlinear estimation and sensor-fusion applications to integrated navigation, AIAA Guidance, Navigation, and Control Conference and Exhibit, 16-19 Aug. 2004, Rhode Island.

Wan, E. A. and van der Merwe, R., 2001, The unscented Kalman filter, Kalman Filtering and Neural Networks. Haykins, S. (Ed.), John Wiley & Sons, New York, Chap. 7.

Effect of Varying Design Options on the Transient Behavior of a Hybrid Rocket Motor

Raed Kafafy[1], Muhammad Hanafi Azami[1], Moumen Idres[1]

ABSTRACT: Hybrid rockets provide compelling features for use in atmospheric and space rocket propulsion. One of the prominent applications of hybrid rockets which foster on its characteristics is the propulsion of micro air launch vehicles. In this paper, a set of design options of a hybrid rocket motor is evaluated for propulsion of micro air launch vehicles. In order to evaluate the various design options of a hybrid rocket, we developed design and performance simulation codes. A simulation code is based on a legacy interior ballistic model. MATLAB® environment was used to develop the design and performance analysis codes and to visualize the temporal variation of performance characteristics and grain geometry during burning. We employ the developed codes to assess the replacement of solid rocket motors which are typically used in Air Launch Vehicles by hybrid rocket motors. A typical Micro Air Launch Vehicle mission to launch a 20-kg payload into a 400-km circular polar orbit is assumed. The results show that a hybrid rocket is a suitable candidate for micro air launch vehicles. The performance is improved in terms of specific impulse and thrust with smaller size in the same mission. Several design parameters of hybrid rocket motors were also evaluated and analyzed, including different fuel port geometry, type of fuels and oxidizers, number of ports, nozzle design and initial mass flux. These design parameters bring a significant effect on hybrid rocket performance and size.

KEYWORDS: Hybrid rocket, Micro air launch vehicle, Internal ballistics, Regression rate.

INTRODUCTION

Hybrid rockets are featured with restarting capability, increased safety, high performance and relatively moderate cost. This unique combination of compelling features poses them as prominent candidates to replace solid rocket motors in air-launched tactical missiles and launch vehicles, besides liquid rocket engines in ground-launched strategic missiles and launch vehicles in the near future. A hybrid rocket is comprised of fuel and oxidizer in different physical states, typically a solid fuel and a liquid or gaseous oxidizer. It represents a compromise between a solid rocket motor and a liquid rocket engine (Mingireanu, 2009). In general, hybrid rockets have specific impulse higher than solid rockets and specific impulse density greater than liquid bi-propellant rockets (Kannalath *et al.*, 2003). Wax based hybrid rocket motors are non-toxic, non-hazardous, shippable as freight cargo, potentially carbon neutral and they can be throttled for thrust control or shut down in case of in-flight anomaly then restart on demand (DeSain *et al.*, 2009). In 2004, the Ansari X Prize has put hybrid technology to a brighter future with the successful launch of SpaceShipOne (David, 2004). SpaceShipOne was succeeded by SpaceShipTwo, which is as nearly twice in size and passed the first powered test flight in April 2013.

Despite the advantages of hybrid rockets, there are several shortcomings which are specifically related to overall performance, reliability and cost effectiveness. The overall performance of a hybrid rocket is affected slightly by combustion instabilities and low solid fuel regression rates which require a relatively large fuel surface area to attain the required thrust level (Connell-Jr *et al.*, 2009).

1.International Islamic University Malaysia – Selangor – Malásia
Author for correspondence: Raed Kafafy | IIUM-Mechanical Engineering Department | Jalan Gombak, 50728 | Selayang/Selangor – Malásia | Email: rkafafy@iium.edu.my

The source of combustion instabilities is due to a complex coupling of thermal transients in the solid fuel, the wall heat transfer blocking due to fuel regression rate, and the transients in the boundary layer that forms on the fuel surface (Karabeyoglu *et al.*, 2005). The oxidizer and fuel are unable to mix quickly in a typical hybrid rocket motor, which results in low propellant burning rates causing a reduction in performance (Jacob, 2007).

Problems related to regression enhancement have become a main focus of many researchers. Moreover, there are issues of fuel web burn out, combustion efficiency, combustion stability, throttling characteristics, and nozzle throat material response (Venuopal *et al.*, 2011). Mainly, the performance of a hybrid rocket depends on the regression rate of the motor, which is defined as the rate at which the solid fuel regresses normal to the surface (Kumar and Kumar, 2012). The regression rate can hardly be accurately measured due to high scattering effects. According to interior ballistic model, during combustion, the heat is transferred to the solid surface by convection and radiation, where the solid phase fuel is decomposed. Zilliac and Karabeyoglu (2006) developed a model to predict the regression rate behavior. However, their model is applicable to vaporizing fuels in a cylindrical grain configuration that do not form significant char or melt layers.

In addition to regression rate, there is a fundamental need to evaluate the pressure and the temporal changes of the combusting gas in the engine chamber because they are related to various phenomena such as ignition and extinction, shape change of a fuel grain due to the combustion, instability due to vortices, acoustics, and injectors, and so on.

Several methods were proposed to improve hybrid rockets regression rate such as adopting swirl flow of oxidizers, doping metal additives in fuel composition, using fuel which has melted layer on the fuel surface and designing optimal grain geometry. The swirl flow can increase the residence time (or contact time) of oxidizer stream with fuel surface along the port. Results showed that the average regression rate increases up to 200% as swirl number increases. However, the enhancement of regression rate is severely localized near the inlet of fuel port. Later, the enhancement of regression technique is by employing vortex tube, which have 150% increase in regression rate where the swirl flow dominates over the entire fuel port (Lee *et al.*, 2005). Employing fuel additives, such as ammonium perchlorate (AP), ammonium nitrate (AN) and other nitro-organic compounds, will lower the heat vaporization effectively. This will allow more solid fuel to combust into the flow of oxidizer. The regression rates were also affected by the addition of activated aluminum powder with 20% by weight to increase the fuel mass flux by 70% over that of pure hydroxyl-terminated polybutadiene (HTPB) (Chiaverini *et al.*, 2000). Studies by Karabeyoglu *et al.* (2004) showed that the regression rate of paraffin-based fuels can be increased up to 400% compared to classic fuels such as HTPB (Nakagawa and Hikone, 2011). It is observed that melted-fuel droplets entraining to the gas phase and melted fuel are flowing along the solid-fuel surface. When the oxidizer flows at high speed over the upper side of the melting fuel surface, the surface of the liquid layer becomes unstable, minute wave is formed and tiny droplets are produced at the tips of the wave and supplied to flame zone. This process occurs almost without vaporization (Ishiguro *et al.*, 2011). It is proven experimentally that the melted-fuel flow layer is created on the solid-phase fuel surface and entraining droplets actually exist. The amount of this melted fuel contributes to the high regression rate because it needs less heat flux from the combustion gas flow to promote regression (Nakagawa and Hikone, 2011). The regression rates of the paraffin-based fuels increase as the melted-fuel viscosity decreases (Nakagawa and Hikone, 2011).

Numerous analytical works have been conducted to describe the gaseous flow during the combustion process. Culick (1996) described the mean gaseous flow through the combustion chamber. Saad and Majdalani (2009) developed an analytical model for basic flowfield in hybrid rockets by employing an arbitrary headwall injection which can be used for benchmarking to test large-scale numerical simulations. However, there are limitations because of burning rate sensitivity, complex fluid dynamics and interactions with heat transfer from the flame zone and the fuel surface, viscosity effects on pressure and the mixing of the two streams (Chiaverini and Kuo, 2007).

At Stanford, a group of postgraduate students had launched their 3-inch diameter nitrous oxide/aluminized paraffin hybrid rocket. They have optimized their design by using Gauss-Newton Nonlinear Least Squares (NLSQ) algorithm and used a 4th-order Runge-Kutta method to integrate the equation of motion (McCormick *et al.*, 2004). A genetic algorithm called HYROCS code, developed by researchers at Purdue University, used a different approach to design and optimize hybrid rockets (Schoonover *et al.*, 2000). Therefore, designing a hybrid rocket engine is not a straight forward process. It is remains challenging because of complexity which involve numerous continuous parameters and variables. There are few published propulsion system design codes available for the industry but many of them employ exhaustive searches in order to optimize continuous variables (Schoonover *et al.*, 2000). During engine operations, both the exposed surface area and perimeter of fuel ports change with time, the mixture ratio will tend to shift even if the oxidizer mass flow is held fixed. High mass flux is desirable in performance to achieve higher volumetric efficiency within the combustion chamber. However, this desire is restricted by the fact that efficient combustion may not be possible at these high mass fluxes and that mixture-ratio shifts become more severe under these conditions (Vonderwell *et al.*, 1995).

We used another approach to examine the performance of a wheel-typed hybrid rocket motor through simulation of different design parameters. This can be integrated with the micro launch air vehicle used in Pegasus. We used the same mission requirement such as same payload, initial mass, required velocity change (ΔV) and altitude to test the hybrid rocket code. We also have evaluated the performance of wheel-typed grain design with the recent used of multiple circular ports. The result showed an improvement on the regression rate. The following section provides a brief description of the design code, followed by a description of the generic simulation algorithm to estimate the performance of the designed hybrid rocket. The next section describes the analysis of the designed hybrid rocket performance through simulation of varying designed parameters. Evaluation section discloses the effects on the regression rate with different geometries.

PRELIMINARY DESIGN OF HYBRID ROCKET

The preliminary design of a hybrid rocket is based on a legacy internal ballistic model (Humble *et al.*, 1995). The algorithm can handle fuel grains with multiple ports of both circular and wheel types. At first, the basic design requirement and reasonable design margins are selected. The design module starts with thermochemical analysis of the hybrid rocket propellant to estimate the main properties of the propellant including specific heat ratio, the molecular mass of combustion products, flame temperature and characteristic velocity. The thermochemical data for LOx/HTPB is obtained from software developed by the National Aeronautics and Space Administration (NASA) which is known as the Chemical Equilibrium with Application (CEA). Frozen-flow approximation is adopted in the model. This should give an optimum oxidizer to fuel ratio (O/F). We further assume that the regression rate is constant at every location during burning while the oxidizer mass flow rate will remain fixed throughout the combustion process. The flowchart of design procedure is shown in Fig. 1.

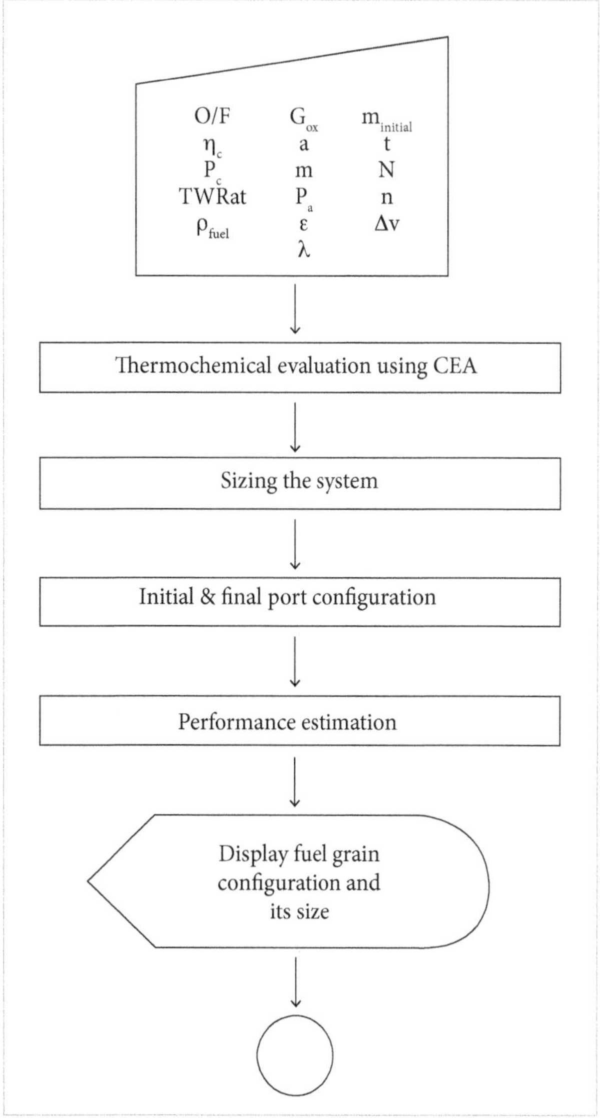

Figure 1. Hybrid rocket preliminary design flowchart.

CEA software requires input of the major and minor species of the chemical reaction, changes of enthalpy and entropy formation of both fuel and oxidizer. The output is the variation of specific heat ratio, molecular weight of the species and the flame temperature with the propellant O/F. Polynomial functions are fitted to the data calculated by CEA for later use in the design and simulation algorithms.

The values of specific heat ratio, molecular weight and flame temperature obtained from CEA are used to calculate the characteristic velocity assuming 98% combustion efficiency.

$$c^* = \frac{\eta_{c^*}\sqrt{\gamma R T_f}}{\gamma\left(\frac{2}{\gamma+1}\right)^{\frac{\gamma+1}{2\gamma-2}}} \tag{1}$$

where c^* is the characteristic velocity, η_{c^*} is the combustion efficiency, γ is the specific heat ratio, R is the gas constant (J/kg.K) and T_f is the flame temperature (K).

The nozzle exit Mach number (M_e) is calculated from the relation of nozzle expansion ratio (ε):

$$\varepsilon = \frac{1}{M_e}\sqrt{\left\{\frac{2}{\gamma+1}\left(1+\frac{\gamma-1}{2}M_e^2\right)\right\}^{\frac{\gamma+1}{\gamma-1}}} \tag{2}$$

The exit pressure (P_e) is calculated using isentropic of chamber pressure (P_c) relation:

$$P_e = \frac{P_c}{\left(1+\frac{\gamma-1}{2}M_e^2\right)^{\frac{\gamma}{\gamma-1}}} \tag{3}$$

Given g_0 is the standard acceleration of gravity, the rocket specific impulse (I_{sp}) is calculated from:

$$I_{sp} = \lambda\left\{\frac{c^*\gamma}{g_0}\sqrt{\left(\frac{2}{\gamma-1}\right)\left(\frac{2}{\gamma+1}\right)^{\frac{\gamma+1}{\gamma-1}}\left[1-\left(\frac{P_e}{P_c}\right)^{\frac{\gamma-1}{\gamma}}\right]}+\frac{c^*\varepsilon}{g_0 P_c}(P_e-P_a)\right\} \tag{4}$$

where λ is the nozzle efficiency and P_a is the ambient pressure.

The combustion chamber pressure is assumed here as a design parameter of the hybrid rocket. The sizing of the overall propulsion system needs the initial mass and total ΔV. For the mission selected here, the mission analysis of the micro air launch vehicle gives a total initial mass of 1,287.34 kg and ΔV of 2,624 m/s (Aldheeb *et al.*, 2012). These values are used to calculate the final mass, propellant mass, fuel mass and fuel mass flux of the hybrid rocket stage using the following relations:

$$m_{final} = m_{initial}e^{\frac{-\Delta v}{I_{sp}g_o}} \tag{5}$$

where m_{final} is the final mass (kg) and $m_{initial}$ is the initial mass (kg).

$$m_{prop} = m_{initial} - m_{final} \tag{6}$$

where m_{prop} is the propellant mass (kg).

$$m_{fuel} = \frac{m_{prop}}{O/F+1} \tag{7}$$

where m_{fuel} is the fuel mass (kg) and O/F is the oxidizer to fuel ratio.

$$G_f = \frac{G_{ox}}{O/F} \tag{8}$$

where G_f is the fuel mass flux (kg/m².s) and G_{ox} is the oxidizer mass flux (kg/m².s).The grain design of wheel type port configurations can be calculated from the initial oxidizer mass flux and the number of ports. The geometric design of a wheel type fuel grain can be calculated from the following geometric relations. Figure 2 illustrates the geometric parameters used in the geometric relations.

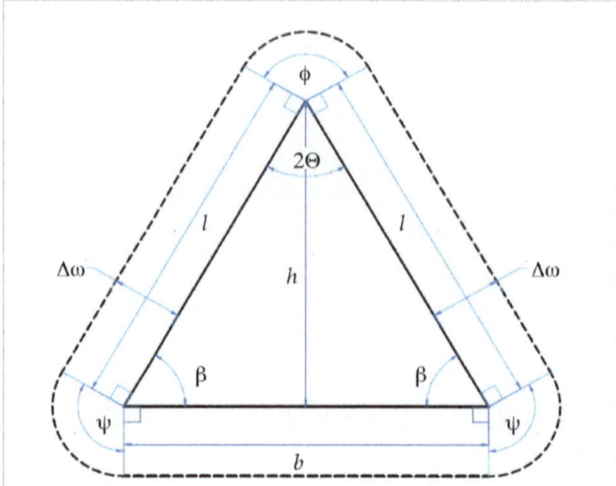

Figure 2. Geometry of a wheel type grain with triangular ports.

$$A_{pi} = \frac{\dot{m}_{oxpp}}{G_{ox}} \tag{9}$$

where A_{pi} is the initial port area (m²) and $\dot{m}_{oxpp}$ is the oxidizer mass flow rate in each port (kg/s).

$$\theta = \frac{\pi}{N} \tag{10}$$

where θ is the port half angle (degree) and N is the number of ports.

$$h = \sqrt{\frac{A_{pi}}{\tan\theta}} \tag{11}$$

where h is the triangle height (m).

$$b = 2h\tan\theta \tag{12}$$

where b is the length of triangle base (m).

$$l = h/\cos\theta \tag{13}$$

where l is the length of the side triangle (m).

$$P_{pi} = 2l + b \tag{14}$$

where P_{pi} is the initial port perimeter.

$$w^2 + \frac{2l+b}{\pi}w - \frac{m_{fuel}}{N\rho_{fuel}L_p\pi} = 0 \tag{15}$$

where w is the web thickness (m), ρ_{fuel} is the fuel density (kg/m^3) and L_p is the port length (m).

$$r_h = w/\sin\theta - w \tag{16}$$

where r_h is the center hole radius (m).

$$r_g = h + 2w + r_h \tag{17}$$

where r_g is the grain radius (m).

$$A_{throat} = \dot{m}_{prop}\, c^*/P_c \tag{18}$$

where A_{throat} is the throat area (m^2) and $\dot{m}_{prop}$ is the propellant mass flow rate (kg/s).

The length of the fuel grain can be calculated from the following empirical relation:

$$L_p = \left(\frac{\dot{m}_{fuel}}{N\rho_{fuel}a(G_{ox}+G_f)^n P_{pi}}\right)^{\frac{1}{m+1}} \tag{19}$$

where $\dot{m}_{fuel}$ is the fuel mass flow rate (kg/s), m is the fuel length exponent, a is the regression rate coefficient, and n is the mass flux exponent.

The layout of a wheel typed fuel grain with 8 ports, which is obtained by the design code, is shown in Fig. 3.

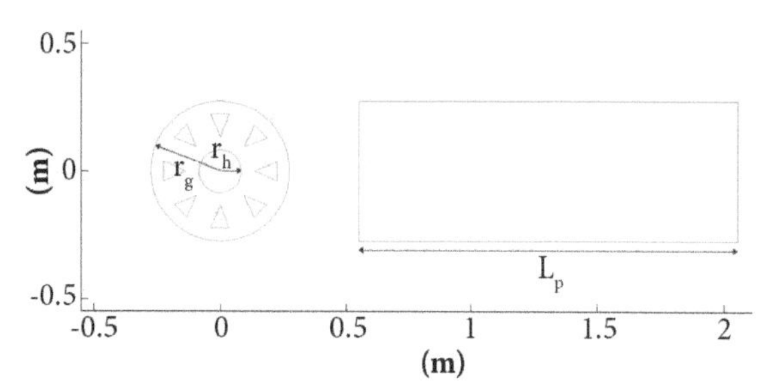

Figure 3. Layout of a wheel type hybrid rocket fuel grain with 8 ports obtained by the design code.

Performance parameters such as the regression rate, exit velocity and thrust are calculated using the following relations. Empirical constants such regression rate coefficient (a), mass flux exponent (n) and length exponent (m) are taken from previous experiment. These values, a = 0.0002, n = 0.75, m = -0.15 and λ=0.9 are obtained from (Humble *et al.*, 1995).

$$Rdot = a(G_{ox} + G_{fuel})^n L_p^m \tag{20}$$

where Rdot is the fuel regression rate (m/s).

$$v_e = \sqrt{\frac{2\gamma RT_o}{(\gamma-1)}\left\{1 - \left(\frac{P_e}{P_o}\right)^{\frac{\gamma-1}{\gamma}}\right\}} \tag{21}$$

where v_e is the exit velocity, T_o is the chamber temperature and P_o is the stagnation pressure (Pa).

$$F = \lambda\left[\dot{m}_{prop}v_e + (P_e - P_a)A_e\right] \tag{22}$$

where F is the thrust force (N) and A_e is the nozzle exit area (m^2).

MICRO AIR LAUNCH VEHICLE USING HYBRID ROCKET MOTOR

We have evaluated hybrid rocket motor performance and size as shown in Table 1. We set the same initial condition and mission requirement as in existing solid rocket motor in MALV for Pegasus XL. For an optimal design of HTPB/LOx, the oxidizer-to-fuel ratio is 2.1, as shown in the thermochemical analysis using CEA. The volume of the oxidizer tank was calculated from mass and density equation and by assuming that the tank is cylindrical. The details of the initial condition and the performance of the designed hybrid rocket motor are given in Table 1.

Table 1. Parameters of the designed hybrid rocket motor for MALV for the first stage.

Parameter	Hybrid Rocket Motor
Optimal O/F	2.1
Specific heat ratio	1.231
Flame temperature (K)	3593.0
Characteristic velocity (m/s)	1747.3
Chamber pressure (MPa)	7.515
Oxidizer mass flux (kg/s/m^2)	300
Thrust (kN)	39.27
Motor length (m)	3.01
Throat area (m^2)	0.0029
Exit area (m^2)	0.1708
Nozzle expansion ratio	58.40
Burning time (s)	45
Fuel diameter (m)	0.66
Initial mass (kg)	1287.34
Propellant mass (kg)	806.97
Specific impulse (s)	316.42
Average regression rate (m/s)	0.00194

SIMULATION OF HYBRID ROCKET PERFORMANCE

The generic simulation algorithm shows the analytical performance of the designed hybrid rocket. Over the period of time, the surface fuel port area will increase due to burning. We assumed that the burning surface will burn perpendicularly and the corner of each port started to have a fillet shape. According to the interior ballistic model, O/F and the geometry of the ports will vary with time. The flowchart of the code is illustrated in Fig. 4.

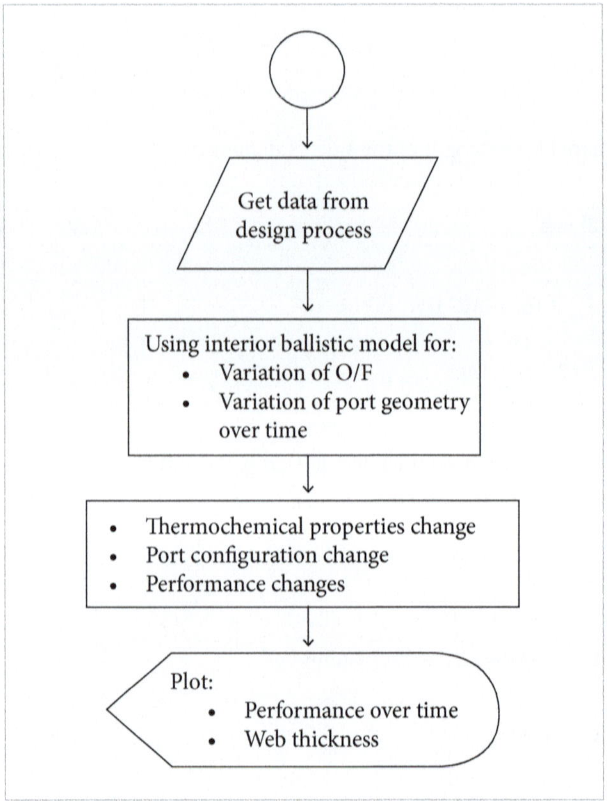

Figure 4. Flowchart of hybrid rocket simulation algorithm.

The simulation algorithm starts by getting values from the output of the design algorithm. During burning, the web thickness decreases continuously and the change in the web section can be estimated from:

$$\Delta w = w_i - \dot{r}\Delta t \qquad (23)$$

where Δw is the change of web thickness (m), w_i is the initial web thickness (m) and Δt is the period of time (s).

Angles are formed by assuming that the two side lines and the base is burnt perpendicular leaving each corner a small fillet. The following equations give the angle at each corner of the triangle.

$$\phi = 180 - 2\theta, \tag{24}$$

where ϕ is the top fillet angle formation (degree).

$$\psi = 90 + \theta, \tag{25}$$

where ψ is the side fillet angle formation (degree).

As time progresses, the web decreases, forming a new burning surface with varying perimeter and cross sectional area in each port that can be calculated from:

$$Peri_{new} = \left(\frac{\phi}{360} \cdot 2\pi\Delta w\right) + \left(\frac{\psi}{360} \cdot 2\pi\Delta w \cdot 2\right) + (2l) + b \tag{26}$$

where $Peri_{new}$ is the new perimeter formed (m).

$$Ap_{new} = N\left\{\left(\frac{\phi}{360} \cdot \pi\Delta w^2\right) + \left(\frac{\psi}{360} \cdot \pi\Delta w^2 \cdot 2\right) + (b\Delta w) + (2l\Delta w) + A_{pi}\right\} \tag{27}$$

where Ap_{new} is the new total port formed (m^2).

When the port geometry changes, the O/F will vary and, accordingly, the thermochemical properties will change. By using an internal ballistic model for a non-circular type of fuel, the new O/F can be obtained from Eq (28).

$$O/F_{new} = \frac{\dot{m}_{ox}^{1-n} Ap_{new}^{\ n}}{\rho a L_p Peri_{new} N} \tag{28}$$

where O/F_{new} is the new oxidizer to fuel ratio, $\dot{m}_{ox}$ is the oxidizer mass flow rate and ρ is the density of the fuel.

The variation of O/F will affect the performance of the hybrid rocket as well. For simplicity, we assume the oxidizer mass flow rate remains constant throughout burning whereas the fuel mass flow rate is allowed to vary according to this relation:

$$\dot{m}_{fuel} = Rdot \,.\, \rho Peri_{new} L_p N \tag{29}$$

The hybrid rocket O/F, instantaneous web, exit velocity, thrust-to-weight ratio, regression rate and specific impulse are plotted with respect to the time as shown in Fig. 5.

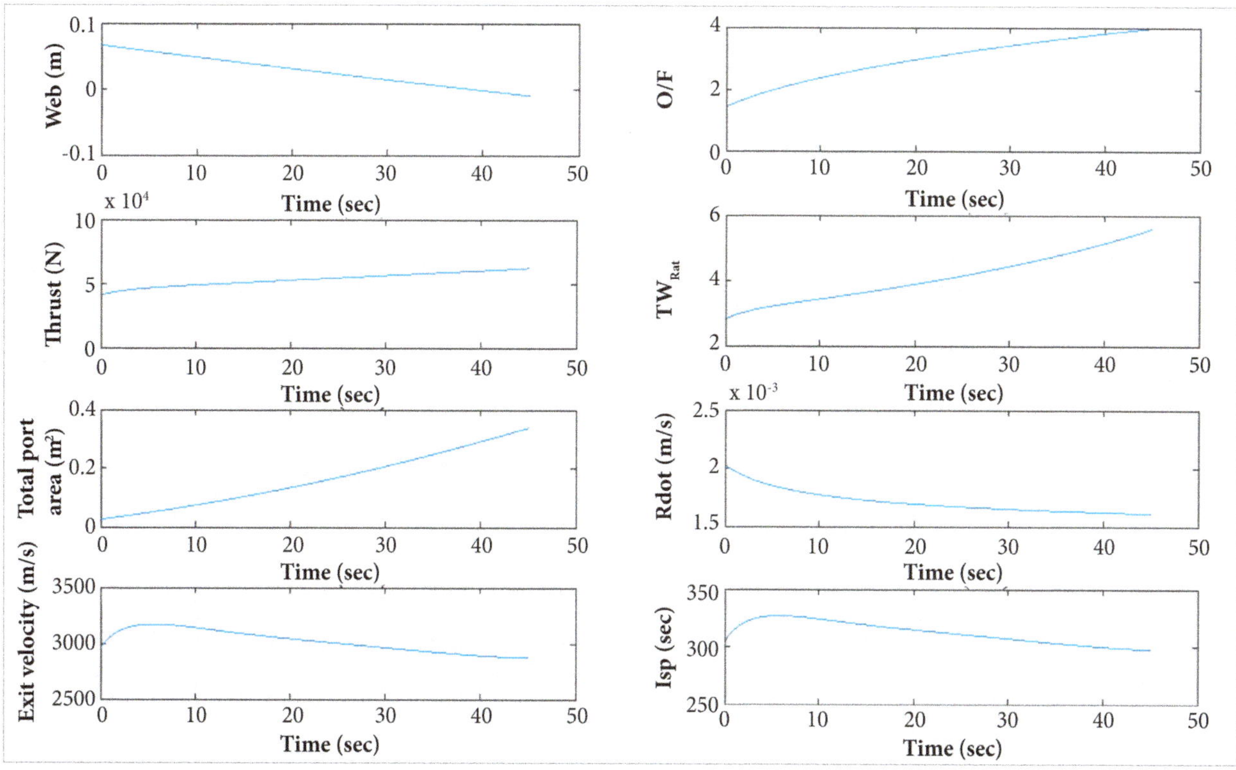

Figure 5. Transient performance of a wheel type hybrid rocket.

A decrease in web thickness is plotted in the top left hand side of the figure. The decreasing in web thickness results in the increase of port area. The changes in the geometry will result in the variation of oxidizer-to-fuel ratio. Notice that the oxidizer-to-fuel ratio increases although the oxidizer mass flow rate is held constant because the port diameter increases with time. The burning rate depends on the regression rate and it is the function of the total mass flux. The regression rate decreases exponentially with time due to changes in mass flux. It is convenient to maintain the regression rate constant along the fuel length at each time step to obtain a small variation of port diameter. The thrust-to-weight ratio increases sharply during the early stage because of the rapid increase in O/F and corresponds to the characteristic velocity. This progressive burning is mainly due to the increase in burning area. However, both specific impulse and exit velocity have a maximum point then started to decrease. The increase in specific impulse at the early stage is dominated by the propellant mass flow rate which is low and, after the increase in the port area, this mass flow rate will start to rise, resulting in a decrease in the total specific impulse. The same goes to the variation of exit velocity which is highly dependent on the chamber pressure and exit conditions.

EFFECT OF VARYING INITIAL DESIGN PARAMETERS ON HYBRID ROCKET PERFORMANCE

We have investigated some of other design features such as nozzle designs with different expansion ratio, different initial oxidizer mass flux, number of ports used and different oxidizers and types of fuels used. These design parameters are based on the user's input. For the purpose of comparison, we maintained the same mission requirement as for micro air launch vehicle. From the results, it proved that these designs will vary its basic performance. Figures 6 and 7 show that, the higher the nozzle expansion ratio, the higher the specific impulse as well as exit velocity. An increase in the nozzle expansion ratio would reduce the exit pressure towards ambient pressure, increasing the thrust coefficient, providing a higher level of thrust, and forcing the system to converge with the predicted results (Terrence *et al.*, 2009).

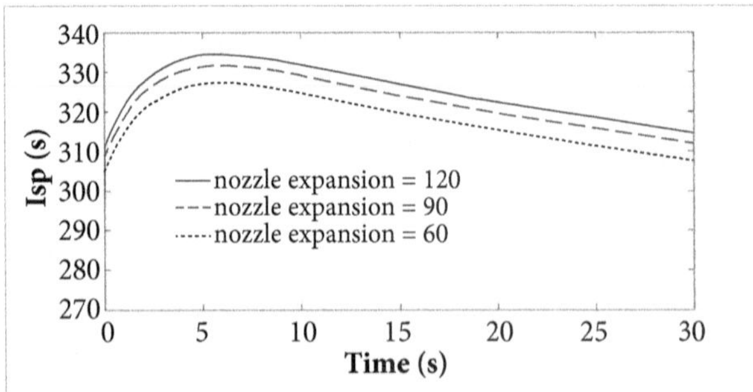

Figure 6. Transient behavior of specific impulse for various nozzle expansion ratios.

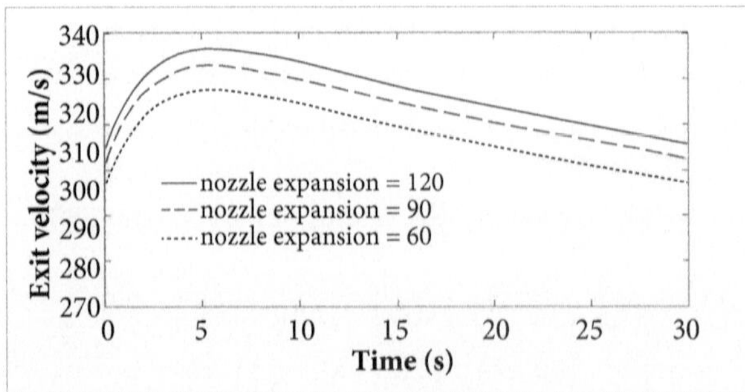

Figure 7. Transient behavior of exit velocity for different expansion ratios.

The variation of initial oxidizer mass flux, G_{oi}, will affect the fuel dimension the most. Based on Fig. 8, specific impulse will have less sustainability at higher initial mass flux. The fuel mass flux will have a rapid increase at the early stage and will decrease at some period of time. According to Eq. 20, the regression rate is the function of total mass flux. The total mass flux is the summation of both fuel and oxidizer mass flux. Figure 9 also shows that at higher initial mass flux gives a better thrust-to-weight ratio.

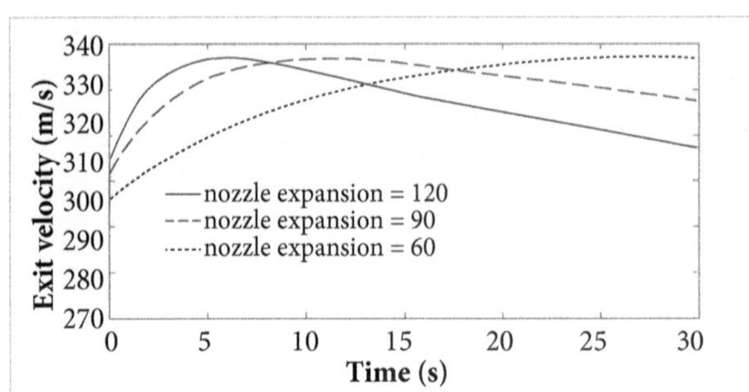

Figure 8. Transient behavior of specific impulse variation for different initial mass fluxes.

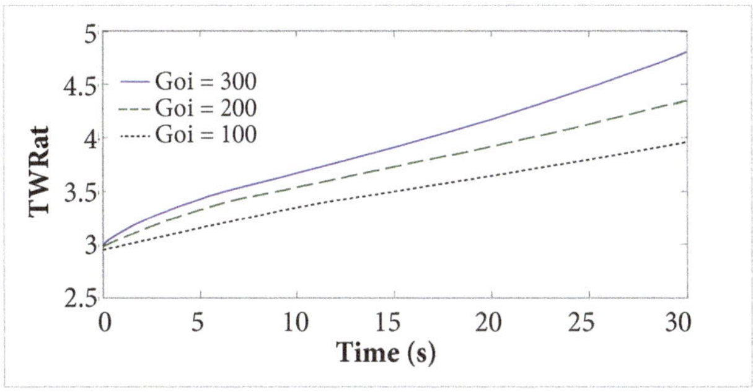

Figure 9. Transient behavior of thrust-to-weight ratio for different initial mass fluxes.

The effect of number of ports on a wheel fuel grain design is discussed and analyzed. The results are shown in Table 2. Multiple ports in fuel grain are desirable to produce larger fuel surface area. The overall performance of hybrid rocket motor is critically depends on the flow mixing degree in the combustion chamber. These multiple combustion ports promote better combustion efficiency due to turbulent mixing environment in the mixing chamber downstream (Sutton and Biblarz, 2001). For many ports design, it showed that the regression rate is improved. However, this requires an increase in the design of grain diameter. Significant increase in chamber diameter indicates a strong increase in regression rate due to the possibility of boundary layer growth with less temperature gradient and more intense radiation heat transfer (Chiaverini and Kuo, 2007).

Figure 10 showed specific impulse variation of different port number. At early burning stage, specific impulse tends to increase rapidly at higher number of ports but it will not sustain longer. The same trend is also shown in Fig. 11 for the exit velocity. Fewer number of ports have its own capability to prolong its impulse but had a slightly decrease. Less number of ports will have larger web thickness, thus have more fuel to keep the fuel burning. The pressure slightly decreased throughout the major portion of the hybrid engine test firing due to volumetric expansion within the combustion chamber and nozzle erosion. The change in cross-section grain area also reduced oxidizer mass flux, which in turn reduces burning rate. Thus, gas production is reduced (Terrence *et al.*, 2009).

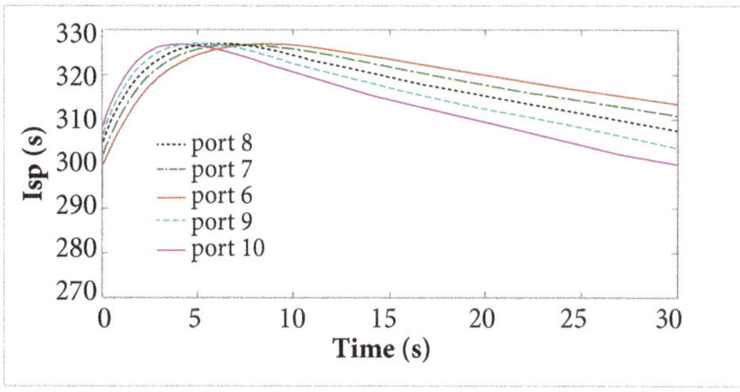

Figure 10. Transient behavior of specific impulse for different number of ports.

Table 2. Variation of performance parameters with number of ports.

Performance parameter	6 ports	7 ports	8 ports	9 ports	10 parts
Fuel length (m)	1.826	1.656	1.506	1.376	1.264
Web (m)	0.069	0.068	0.067	0.067	0.067
Fuel volume per port (m^3)	0.064	0.055	0.048	0.043	0.039
Final area of port (m^2)	0.042	0.037	0.035	0.034	0.033
Final perimeter of port (m)	0.705	0.682	0.666	0.654	0.645
Final surface area of port (m^2)	1.288	1.129	1.003	0.881	0.816
Radius of center hole (m)	0.068	0.088	0.109	0.129	0.153
Radius of grain (m)	0.292	0.311	0.332	0.352	0.373
Regression rate (cm/s)	0.208	0.211	0.214	0.217	0.219

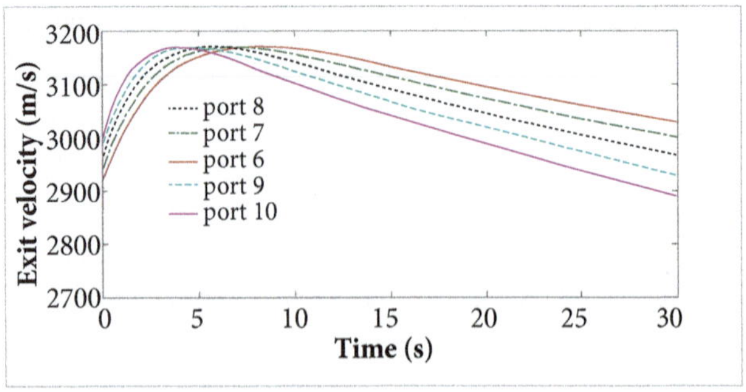

Figure 11. Transient behavior of exit velocity for different number of ports.

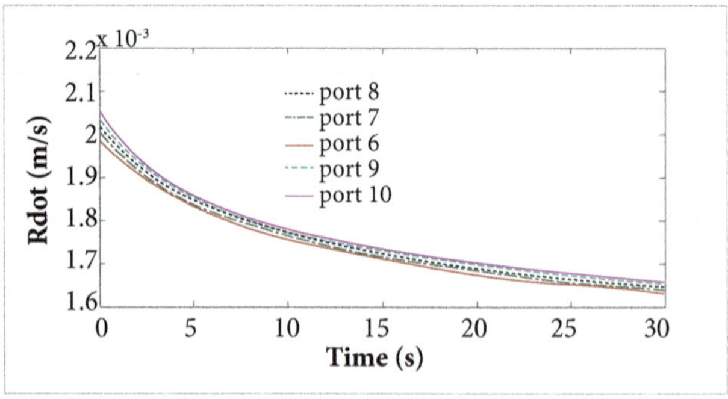

Figure 12. Transient behavior of regression rate for different number of ports.

Many ports have better regression rate and more fuel can be decomposed as shown in Fig. 12, but it requires more weight for the structural design. Fewer ports have lighter weight on its structural design, but the regression rate is lower. These conditions resulted in little variation between different port designs.

The effect of varying the oxidizer on performance is analyzed. Different oxidizers with HTPB fuel were considered, namely oxygen (LOx), hydrogen peroxide (H_2O_2) and nitrogen tetroxide (N_2O_4). The results show that the performance was mainly influenced by the thermo chemical properties of the fuel/oxidizer mixture. Liquid oxygen has lower boiling point and density compared to hydrogen peroxide. The HTPB and liquid oxygen provides a relatively high regression rate with a relatively stable burn (Mingireanu, 2009). Thermochemistry properties and curve fit relation for hydrogen peroxide and nitrogen tetroxide adopted from Humble *et al.* (1995) and CEA. The properties of candidate oxidizers and fuels are given in Tables 3 and 4.

Figure 13 shows that liquid oxygen oxidizer gives the highest specific impulse. Nevertheless, it has an optimum value until it drops at a certain time. However, hydrogen peroxide has better thrust-to-weight ratio compared to other oxidizers as shown in Fig. 14. Liquid oxygen gives a better thrust during the early stage while nitrogen tetroxide gives the lowest value during the burn. Liquid oxygen and hydrogen peroxide give a better improvement on the regression rate. This proved that the choice of oxidizers bring impact on regression enhancement.

Table 3. Properties of oxidizers.

Oxidizer	Chemical formula	T_{FP} (K)	T_{BP} (K)	P_{vap} (MPa)	Density (kg/m³)	Heat of formation (cal/g)
Oxygen	O_2	54	90	5.07 @ 154K	1142	0
Hydrogen peroxide	H_2O_2	267.4	419	0.345 @ 298K	1414	-1440 (90%HP) -1340 (98%HP)
Nitrogen tetroxide	N_2O_4	261	294	0.765 @ 344K	1440	-50.9
Nitrous oxide	N_2O	182.29	184.67	5.15@ 293K	102	443

T_{FP}: freezing point temperature; T_{BP}: boiling point temperature; P_{vap}: vapour pressure.

Table 4. Properties of fuels.

Fuel	Chemical formula	T_{Melt} (K)	T_{BP} (K)	Density (kg/m³)	Enthalpy of formation (kJ/mol)
HTPB	$C_{10}H_{15}.4\ O_{0}.07$	514.15	N/A	920	-51.9
Paraffin wax	$C_{28}H_{58}$	327.55	438.89 @ 1atm	809	-1438.2

T_{Melt}: melting temperature; T_{BP}: boiling point temperature.

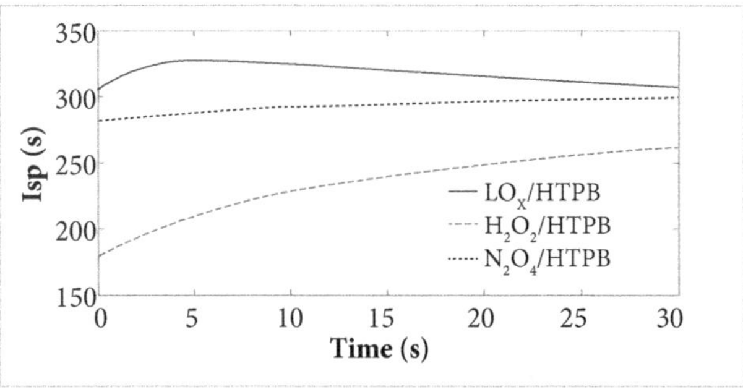

Figure 13. Transient behavior of specific impulse for different types of oxidizer.

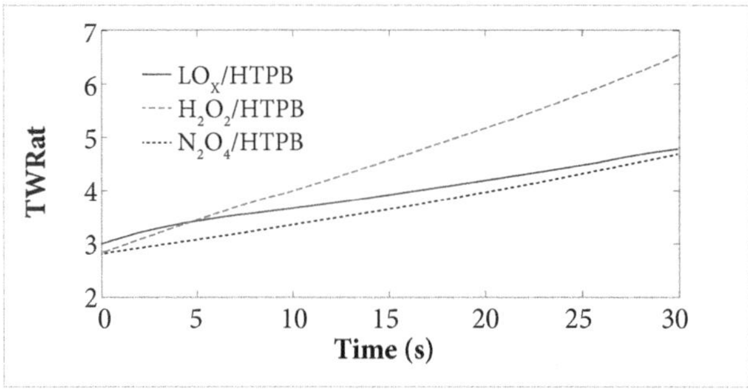

Figure 14. Transient behavior of thrust-to-weight ratio for different types of oxidizer.

We have also compared the performance with different types of fuels such as Paraffin and conventionally used HTPB with nitrous oxide as an oxidizer. Paraffin based fuel shows regression rate enhancement as in Fig. 15. This is numerically proven on experimental works carried out by previous researchers at Stanford University. This fast burning fuels form a hydro dynamically unstable liquid layer over the surface which encourages high regression rate (Karabeyoglu *et al.*, 2004).

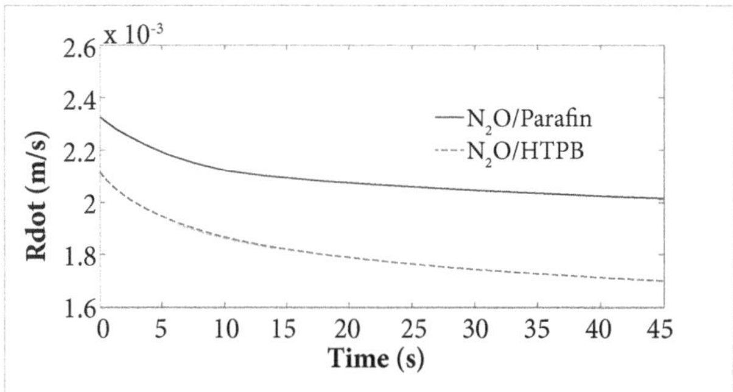

Figure 15. Transient behavior of regression rate for different types of fuels.

Figure 16 shows an inverse effect of using different types of fuels. Paraffin fuel gives a decrease in specific impulse over time. This trend occurs because of thermochemical properties of both types of fuels. Values of a and n parameters, appearing in Eq. 20 for N_2O/Paraffin and N_2O/HTPB are a = 0.488, n = 0.62 (Karabeyoglu *et al.*, 2004) and a = 0.198, n = 0.325 (Lee and Tsai, 2009).

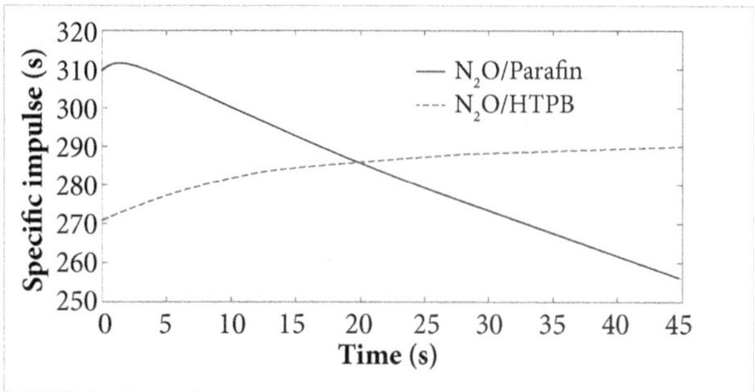

Figure 16. Transient behavior of specific impulse for different types of fuels.

EVALUATION OF WHEEL-TYPED AND MULTIPLE CIRCULAR PORTS PERFORMANCE

Table 5 shows that wheel-typed port requires a larger grain diameter but less fuel length for the desired thrust. Wheel-typed port with thicker web has an advantage for longer burning time. Moreover, wheel-typed ports have better burning port surface with at least 30% increase. This results show 5% increase of the average regression rate. However, studies showed that for short fuel grains, the boundary layer will not fully develop along the port longitudinal axis and some of the downstream parts of the fuel grain will gasify without complete burning while long grains, the entire oxidizer will be consumed before reaching the end of the grain (Kannalath *et al.*, 2003).

To estimate the final grain diameter at each port we use:

$$D_{pf} = \sqrt{\frac{4m_{fuel}}{\pi L_p \rho N} + D_{pi}^2} \tag{30}$$

where D_{pf} is the final diameter of port (m) and D_{pi} is the initial diameter of port (m).

Table 5. Comparison of multiple-circular ports and wheel-typed ports design for 8 ports.

	Multiple-circular	Wheel-typed
Grain diameter (m)	0.6	0.66
Fuel length (m)	2.073	1.506
Web thickness (m)	0.06	0.0673
Final area of each ports (m^2)	0.027	0.035
Final surface area of each ports (m^2)	1.197	1.003
Average regression rate (m/s)	0.00204	0.00214

$$w = \frac{D_{pf} - D_{pi}}{2} \quad (31)$$

The analytical solution for circular ports with constant oxidizer mass flow is:

$$\int_{D_{pi}}^{D_p} D_p^{2n} dD_p = 2a\left(\frac{4\dot{m}_{ox}}{\pi}\right)^n L_p^m \int_0^t dt \quad (32)$$

where D_p is the diameter of port.

For plotting at any time, the variation of grain diameter uses this relation:

$$D_p = \left(a(4n+2)\left(\frac{4\dot{m}_{ox}}{\pi}\right)^n L_p^m t + D_{pi}^{2n+1}\right)^{\frac{1}{2n+1}} \quad (33)$$

Using the ballistic model, the variation of O/F for circular ports is:

$$O/F_{new} = \frac{\dot{m}_{ox}^{1-n} D_p^{2n-1}}{4^n \rho \pi^{1-n} a L_p N} \quad (34)$$

Figures 17, 18 and 19 show a comparison of both grain designs in terms of its specific impulse, the regression rate and thrust-to-weight ratio. In Fig. 17, multiple circular ports show better improvement of specific impulse. This is probably caused by the fast change in the geometrical area.

Figure 18 shows a clear improvement on the regression rate of the wheel-typed grain. As the time goes, the surface burning area becomes much larger compared to the circular ports, but both designs show a decline trend due to a decrease of mass flux. Both designs show a regressive burning where the thrust decreases over the time. However, wheel-typed grain has an increase on thrust-to-weight ratio compared to the circular ports. More fuel is burnt in the wheel-typed grain due to a larger burning surface area.

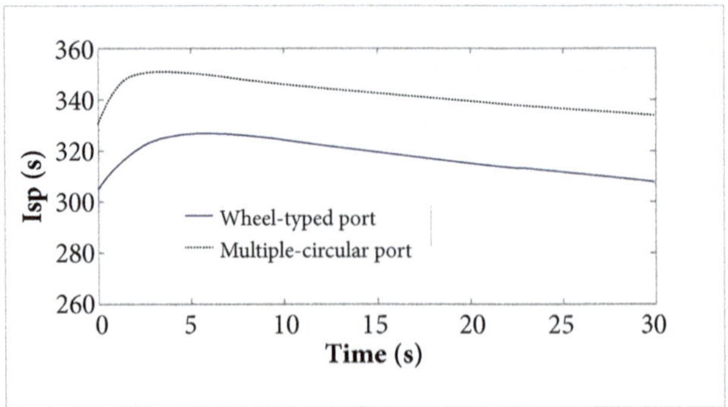

Figure 17. Transient behavior of specific impulse for different fuel grain types.

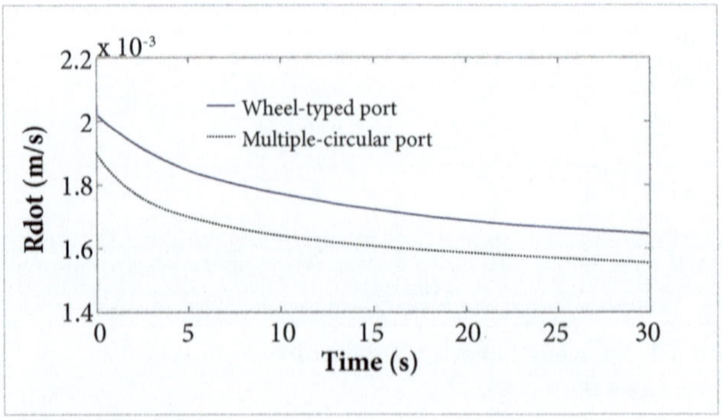

Figure 18. Transient behavior of regression rate for different fuel grain types.

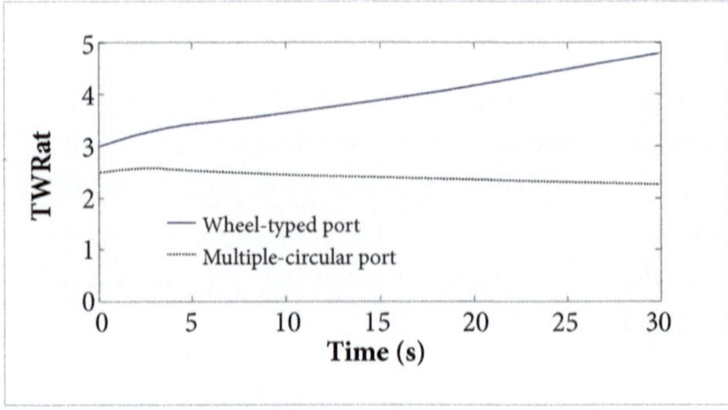

Figure 19. Transient behavior of thrust-to-weight ratio for different fuel grain design.

CONCLUSION

The designed code and generic algorithm analytical performance simulation of hybrid rockets was developed and successfully demonstrated. Prior to the code, we have shown the effect of varying design parameters which can continuously change the performance. Hybrid rocket designers have to properly select and decide to meet the mission requirements. Since this code is developed for the design of propulsion system, it will require further modifications to include all factors needed by space mission analysts and designers.

Based on the obtained results, the choice of initial mass flux determines its performance sustainability. According to the temporal variation, higher initial mass flux will give a rapid increase on its performance, but will not sustain longer. The choice of number of ports also plays an important role in determining the hybrid rocket performance. Many ports have better regression rate and thrust but encounter structural integrity problem.

In general, the wheel-typed grain design has higher regression rate and higher thrust-to-weight ratio compared to the multiple circular ports, but it requires a larger grain diameter. As the results showed, grain geometry has a profound effect on regression rate. Larger surface area is preferable but it has structural limitations. A conventionally used multiple-circular port has better structural integrity but lack of performance. Hybrid rocket designers have to compromise these effects to achieve the mission requirement for a micro air-launch vehicle.

REFERENCES

Aldheeb, M.A., Kafafy, R.I., Idres, M., Omar, H.M. and Abido, A.M., 2012, "Design Optimization of Micro Air Launch Vehicle Using Different Evolution", Journal of Aerospace Technology and Management, Vol. 4, No. 2, pp. 185-196. doi: 10.5028/jatm.2012.04020112.

Chiaverini, M.J. and Kuo, K.K., 2007, "Fundamentals of Hybrid Rocket Combustion and Propulsion", American Institute of Aeronautics and Astronautics, Virgina, USA.

Chiaverini, M. J., Serin, N., Johnson, D. K., Lu, Y.-C., Kuo, K. K. and Risha, G.A., 2000, "Regression Rate Behaviour of Hybrid Rocket Solid Fuels", Journal of Propulsion and Power, Vol. 16, No. 1, pp. 125-132.

Connell-Jr, T.L., Santi, S.A., Risha, G.A., Muller, B.A. and Batzel, T.D., 2009, "Experiment and Semi-Experiment Modeling of Lab-Scale Hybrid Rocket Performance", 45th AIAA/ASME/SAE/ASEE Joint Propulsion Conference & Exhibit. Denver, Colorado.

Culick, F. E. C., 1966, " Rotational axisymmetric mean flow and damping of acoustic waves in a solid propellant rocket", AIAA Journal, Vol. 4, No 8, pp. 1462-1464.

David, L., 2004, "SpaceShipOne Wins $10 Million Ansari X Prize in Historic 2nd Trip to Space", Retrieved in January 10, 2012, from http://www.space.com/403-spaceshipone-wins-10-million-ansari-prize-historic-2nd-trip-space.html.

DeSain, J.D., Brady, B.B., Metzler, K.M., Curtiss, T.J. and Albright, T.V., 2009, "Tensile Tests of Paraffin Wax for Hybrid Rocket Fuel Grains", 45th AIAA/ASME/SAE/ASEE Join Propulsion Conference & Exhibit, Denver, Colorado.

Humble, R.W., Henry, G.N. and Larson, W.J., 1995, "Space Propulsion Analysis and Design", Ed. McGraw-Hill, USA.

Ishiguro,T., Keizi Sinohara, K.S. and Nakagawa, I., 2011, "A Study on Combustion Efficiency of Paraffin-based Hybrid Rockets", 47th AIAA/ASME/SAE/ASEE Joint Propulsion Conference & Exhibit, San Diego, California.

Jacob, E., 2007, "The Effect of Oxidizer Laced Hybrid Rocket Regression Rates and Performance", American Institute of Aeronautics and Astronautics (AIAA) Southeastern Regional Student Conference, Savannah, Georgia.

Kannalath, R., Kuznetsov, A. and Natan, B., 2003, "Design of a Lab-scale Hydrogen Peroxide/Hydroxyl Terminated Polybutadiene Hybrid Rocket Motor", 39th AIAA/ASME/SAE/ASEE Joint Propulsion Conference & Exhibit. Alabama.

Karabeyoglu, A., Zilliac, G., Cantwell, B. J., DeZilwa, S. and Castellucci, P., 2004, "Scale-Up Tests of High Regression Rate Paraffin-Based Hybrid Rocket Fuels", Journal of Propulsion and Power, Vol. 20, No. 6, pp. 1037-1045. doi: 10.2514/1.3340.

Karabeyoglu, M., Zilwa, S., Cantwell, B. and Zilliac, G., 2005, "Modelling of Hybrid Rocket Low Frequency Instabilities", Journal of Propulsion and Power, Vol. 21, No. 6, pp. 1107-1116. doi: 10.2514/1.7792.

Kumar, C. P. and Kumar, A., 2012, "A Numerical Study on the Regression Rate of Hybrid Rocket Motors Using a Combination of Enhancement Techniques", 48th AIAA/ASME/SAE/ASEE Joint Propulsion Conference & Exhibit, Atlanta, Georgia.

Lee, C., Na, Y. and Lee, G., 2005, "The Enhancement of Regression Rate of Hybrid Rocket Fuel by Helical Grain Configuration and Swirl Flow", 41st AIAA/ASME/SAE/ASEE Joint Propulsion Conference & Exhibit, Tucson, Arizona.

Lee, J., Moon, H., Sung, H., Kim, J. and Park, S., 2011, "Combustion Characteristics of Initial Port Diameter Variation of Solid Fuel in Hybrid Rocket Motor", 47th AIAA/ASME/SAE/ASEE Joint Propulsion Conference & Exhibit. San Diego, California.

Lee, T.-S. and Tsai, H.-L., 2009, "Fuel Regression Rate in a Paraffin-HTPB Nitrous Oxide Hybrid Rocket", 7th Asia-Pacific Conference on Combustion, Taipei.

McCormick, A., Hultgren, E., Lichtman, M., Smith, J., Sneed, R. and Azimi, S., 2004, "Design, Optimization, and Launch of a 3" Diameter N2O/Aluminized Paraffin Rocket", AIAA.

Mingireanu, F., 2009, "Hybrid rocket motor internal ballistic model and oxidizer dopping. Applications", 4th International Conference on Recent Advances in Space Technologies – RAST '09, Istanbul.

Nakagawa, I. and Hikone, S., 2011, "Study on the Regression Rate of Paraffin-Based Hybrid Rocket Fuels", Journal of Propulsion and Power, Vol. 27, No. 6, pp. 1276-1279. doi: 10.2514/1.B34206.

Saad, T. and Majdalani, J., 2009, "Rotational flowfields in porous channels with arbitrary headwall injection", Journal of Propulsion and Power, Vol.25, No 4, pp. 921-929.

Schoonover, P.L., Crossley, W.A. and Heister, S.D., 2000, "Application of a Genetic Algorithm to the Optimization of Hybrid Rockets", Journal of Spacecraft and Rockets, Vol. 37, No. 5, pp. 622-629. doi: 10.2514/2.3610.

Sutton, G.P. and Biblarz, O., 2001, "Rocket Propulsion Element", Ed. John Wiley & Sons, Canada.

Terrence, L., Connell, J., Santi, S.A., Risha, G.A., Muller, B.A. and Batzel, T.D., 2009, "Experiment and Semi-Experiment Modeling of Lab-Scale Hybrid Rocket Performance", 45th AIAA/ASME/SAE/ASEE Joint Propulsion Conference & Exhibit, Denver, Colorado.

Venuopal, S., Rajesh, K. and Ramanujachari, V., 2011, "Hybrid Rocket Technology", Defense Science Journal, Vol. 61, No. 3, pp. 193-200.

Vonderwell, D.J., Murray, I.R. and Heister, S.D., 1995, "Optimization of Hybrid-Rocket-Booster Fuel-Grain Design", Journal of Spacecraft and Rockets, Vol. 32, No. 6, pp. 964-969. doi: 10.2514/3.26716.

Zilliac, G. and Karabeyouglu, M., 2006, "Hybrid Rocket Fuel Regression Rate Data and Modelling", 42nd AIAA/ASME/SAE/ASEE Joint Propulsion Conference & Exhibit, California.

4

Combustion Modeling of Aluminum Incorporated in Low-Explosive Formulations such as Solid Propellants

Rene Francisco Boschi Gonçalves[1], Koshun Iha[1], José Atílio Fritz Fidel Rocco[1]

ABSTRACT: Aluminum that is incorporated in an energetic material such as a propellant plays a significant role in the combustion process by means of stabilization with regard to the burning and generation of additional energy. The use of simulation softwares to model the combustion mechanism and kinetic parameters of the elementary reactions that compose the oxidation were used as the pressure variation of the combustion chamber of a rocket motor conditions. The behavior of the molar fraction of the chemical species during the combustion and its posterior stabilization were observed. The systems submitted to higher pressures tend to stabilize more rapidly, according to the greater chemical speed of the elementary reactions.

KEYWORDS: Aluminum, Combustion simulation, Chemkin.

INTRODUCTION

FLAME STRUCTURE OF A COMPOSITE SOLID PROPELLANT GRAIN

Most of the current available knowledge on chemical aspects involving combustion of composite solid propellants regards studies of flame structures like, for example, the spatial distribution of temperatures and concentrations of chemical species involved in the combustion. Data analysis of the flame structure can provide information on the composition (generated from reactions in the condensed phase) of the products that have originated from thermal decomposition on the burning surface of a solid propellant grain. Therefore, it is possible to identify the reactions that occur in the condensed phase, in addition to the reaction pathways involved. Furthermore, the chemical structure of the flame provides information on the mechanism and kinetics of the chemical reactions in the gas phase and subsequent processing of the products, with reactions considered responsible for the generation of heat from the gas phase. The development of combustion models for solid propellant formulations requires information on the reactions involved, in both the condensed and gas phases of the process. Without this information, it is impossible to develop a model that is able to predict actual parameters such as burning rate and other ballistic characteristics, which are intrinsic to the material considered Korobeinichev *et al.*, 1974.

In the last four decades, studies have been performed in order to meet the ever more comprehensive process of energetic material combustion in special formulations of solid

1.Instituto Tecnológico de Aeronáutica – São José dos Campos/SP – Brazil

Author for correspondence: Rene Francisco Boschi Gonçalves | Departamento de Química, Instituto Tecnológico de Aeronáutica | Praça Mal. Eduardo Gomes, 50 – Vila das Acácias | CEP 12.228-900 São José dos Campos/SP – Brazil | Email: renefbg@gmail.com

propellants (Barrere, 1968; Williams *et al.*, 1969; Kuo *et al.*, 1984; Chiaverini *et al.*, 1999; Gonçalves *et al.*, 2009). However, the complete combustion of this process is unknown because, in practice, it involves many parameters such as steady-state combustion, erosive burning, combustion in transient state (ignition and combustion instabilities), and combustion of finely divided metal particles such as aluminum or boron (incorporated in the material) (Davenas, 2003).

BURNING INSTABILITIES

Currently, most of the research done in propulsion of solid propellant rocket engines (SPRE) is directed to the understanding of a phenomenon that is widely known as "burning instability." This phenomenon directly affects the burning of such material due to the resonance characteristic of the combustion process. A great mass of information is obtained pertaining to flame structure, heat-release mechanisms, combustion response in acoustical terms (Shusser *et al.*, 2000), and the development of the flow that is established from burning. The term "combustion response" refers to the response given by low-amplitude linear frequency caused by the combustion of solid propellants (harmonic oscillations in the combustion chamber pressure during combustion) (Brewster, 2000). However, despite these contributions, the existing burning models provide limited information about the fluid dynamics behavior in the regions where oscillations are observed in turbulent flow of gases inside the combustion chamber of the SPRE (Flandro *et al.*, 2000).

In the 1960s, when most research in this area was directed to formulations of composites based on ammonium perchlorate (AP), Summerfield *et al.*, (1966) identified the mechanism by which this process occurred and proposed a burning model named granular diffusive flame. In this model, based on the assumption that both the fuel and the oxidant are converted from solid to gas on the burning surface and are diffused together forming the flame front, an understanding of the composite combustion process as a whole was initiated. However, parameters such as large variations in the burning velocity of solid propellants as a function of the pressure in the combustion chamber, distribution of the granulometric profile of the loads (AP, aluminum, and Al-additives), and the concentration of such fillers in the composite formulated required models broader than initially proposed by Summerfield *et al.*, (1966).

In the early 1970s, Beckstead *et al.*, (1970) presented a model called multiple flame, which was based on a complex interaction between the flame generated from the thermal decomposition of the oxidant and two other diffusive flames immediately above the interface between the "binder" (organic part of the propellant) and the oxidant (material responsible for supplying oxygen to the combustion process). This phenomenological interpretation of the combustion process of solid propellant was used for many years to explain the burning behavior of composites based on AP, and also in double-based ones (nitroglycerin/nitrocellulose). Apparently, the diffusive flame generated from the thermal decomposition of AP is dominant in the mechanism present in Beckstead's model and is related to the manner in which the AP decomposes, generating a large amount of chemical species derived from chlorine. These species (extremely reactive) allow a fast increase of the flame temperature and maintain the stability of the whole combustion process of the composite. To date, this phenomenological interpretation of Beckstead *et al.*, (1970) remains valid.

The use of computer packages for burning simulation of energetic materials, with an emphasis on aluminum oxidation, is the focus of this work.

METHODOLOGY

The combustion modeling of solid propellant formulations can be performed through computer simulation using the software Chemkin (Aurora reactor, for the current work), developed by "Sandia National Laboratories, Livermore, CA" (Coltrin *et al.*, 1991). This software is aimed at solving complex problems involving chemical kinetics using complex reaction pathways. The program's architecture offers information on specific problems, as well as models of independent problems and some packages with preselected reaction models that take into account the equations of mass, energy, chemical species and, in some cases, momentum conservation. When solving a specific problem, there is a need to identify the chemical species involved in the process (combustion) and its properties, such as enthalpy (H), entropy (S) and heat capacity at constant pressure and volume (Cp, Cv) beyond the reaction pathways and reaction rates involved.

COMBUSTION OF AN ALUMINUM PARTICLE OXIDIZED BY AIR

Computer simulations provide a very convenient support relative to understand the physics of combustion of many chemical species, particularly aluminium, in processes which take into account the growth of the oxide layer that covers its surface. If the modeling is sufficiently complete, it is possible to obtain results (data) leading to the burning rates involved in aluminum combustion.

Initial simulations of aluminum combustion (drops) were performed by Liang and Beckstead (1998). This two-dimensional model takes into account effects such as forced convection, radiation, growth of the oxide layer, transport properties variation and an extended condensation zone. The model considers a reaction mechanism involving nine steps, including two surface reactions, three gas-phase reactions, a dissociation reaction of the molten Al_2O_3, and three condensation reactions. In order to study the combustion of aluminum (drops) in an environment, for example, in the case of a rocket engine, these researchers (Liang and Beckstead, 1998) modified the kinetic model to consider the reaction of aluminum with H_2O and CO_2, as well as condensation reactions involving each of these oxidizers. Simultaneously, the thermodynamic properties were adjusted to follow changes in pressure.

In this modeling, the process of aluminum combustion in vapor phase occurs in a homogeneous reactor where it is considered that throughout the control volume there is a perfectly homogenous mixture between fuel and oxidant chemical species (PSR – Perfect Stirred Reactor). The reaction mechanism for the oxidation of aluminum in the air consists of nine elementary reactions that are described next:

$$Al + O_2 = AlO + O \quad (1)$$

$$AlO = Al + O \quad (2)$$

$$AlO + O_2 = AlO_2 + O \quad (3)$$

$$AlO_2 = AlO + O \quad (4)$$

$$Al_2O = Al + AlO \quad (5)$$

$$Al_2O_2 = AlO + AlO \quad (6)$$

$$Al_2O_2 = AlO_2 + Al \quad (7)$$

$$Al_2O_2 = Al_2O + O \quad (8)$$

$$O_2 + M = O + O + M, \text{ where M is an inert molecule.} \quad (9)$$

THERMAL DECOMPOSITION MODELING OF ALUMINUM

In order to perform the thermal decomposition modeling of aluminum in a homogeneous reactor (transient state), the software Chemkin/Aurora model was utilized. In this type of reactor, the reacting species (in gaseous form) are homogeneously dispersed throughout the system volume, where the ignition subsequently occurs, resulting in an abrupt increase in temperature (due to heat release by the exothermic combustion reactions of alumina) and volume (due to high gas formation), in addition to changes in mole fractions of the chemical species considered for the modeling.

The temperature used in the simulation (1700°C) resembles the observed temperature in a combustion chamber of a conventional rocket motor. Various values have been used to denote pressure, including 1, 10, 30, and 60 atmospheres, with the last value being the observed pressure in a combustion chamber of a conventional rocket motor. The mechanism used in modeling and Arrhenius data of each elementary reaction are depicted in Table 1.

Table 1. Alumina combustion mechanism.

Reactions	A (mol-cm-s-K)	b	Ea (kJ/mol)
$Al+O_2=AlO+O$	9.72E13	0.0	159.95
$Al+O+M=AlO+M$	3.0E17	-1.0	0.00
$AlO+O_2=OAlO+O$	4.62E14	0.0	19885.90
$Al_2O_3=AlOAlO+O$	3.0E15	0.0	97649.99
$Al_2O_3=OAlO+AlO$	3.0E15	0.0	126999.89
$AlOAlO=AlO+AlO$	1.0E15	0.0	117900.00
$AlOAlO=Al+OAlO$	1.0E15	0.0	148900.00
$AlOAlO=AlOAl+O$	1.0E15	0.0	104249.94
$OAlO=AlO+O$	1.0E15	0.0	88549.86
$AlOAl=AlO+Al$	1.0E15	0.0	133199.94
$Al=Al$	1.0E14	0.0	0.00
$Al_2O_3=Al_2O_3$	1.0E14	0.0	0.00
$Al+H_2O=H+AlOH$	1.14E12	0.0	442.80
$Al+H_2O=AlO+H_2$	9.6E13	0.0	2868.60
$AlOH=Al+OH$	1.0E15	0.0	66431.80
$AlOH=AlO+H$	1.0E15	0.0	57725.20

RESULTS AND DISCUSSION

The combustion simulations of alumina were performed at different pressures (1, 10, 30, and 60 atmospheres). In Fig. 1 it is possible to observe the behavior of the molar fractions of the chemical species involved in the process, according to the elapsed time of combustion from ignition, and also the system volume variation as a function of time.

The predominant chemical species in the simulation at 1 atmosphere are aluminum, gaseous oxygen, aluminum oxide, and elementary oxygen. Due to the low pressure at which the system is submitted, there is little formation of alumina and its intermediates. In this case, therefore, the predominant reaction is the capture of one of the oxygen atoms of O_2 by the aluminum atom, with the formation of aluminum oxide (I) and elemental oxygen.

Figure 2 shows the simulation results depicting the pressure at 10 atmospheres.

In the case where the pressure is 10 atmospheres, a higher reaction rate is initially noticed due to a sudden volume increase as a result of the system ignition. The mole fractions of intermediate species suffered additions, as in this case the higher pressure enables the occurrence of all intermediate reactions present in the mechanism. However, there is still not enough time to stabilize the system considered (2.10^{-4} s), residence time of chemical species in the combustion chamber in question, showing that the reaction does not occur fast enough to ensure burning stability and greater efficiency.

Figure 3 presents the combustion process at 30 atmospheres.

The speed of the process, in terms of the (volume versus time) curve slope, increased even more in this case, as there is a fast increase and stabilization of the volume, shortly after the ignition of the combustion process.

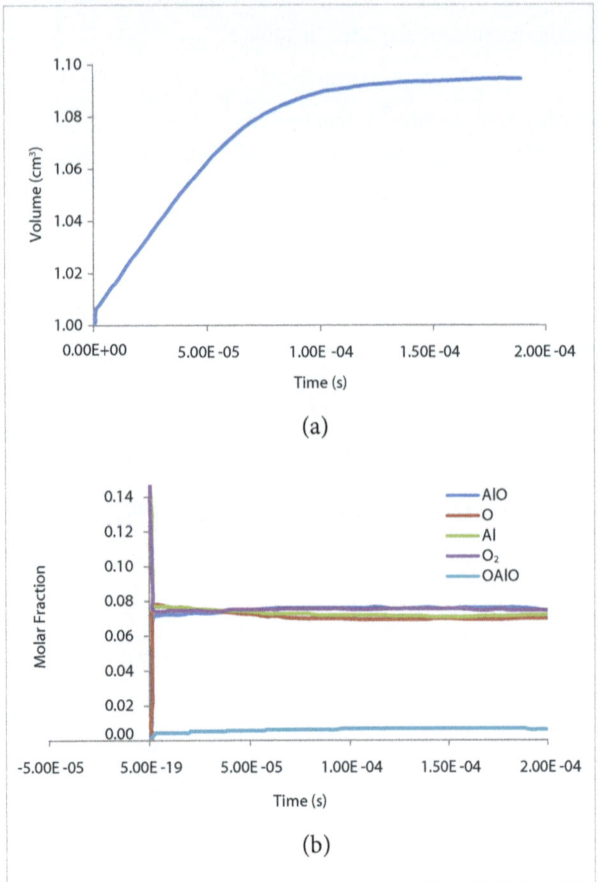

Figure 1. Volume (a) and molar fractions of the chemical species (b) versus time from ignition to the combustion process of alumina with gaseous oxygen at 1 atmosphere.

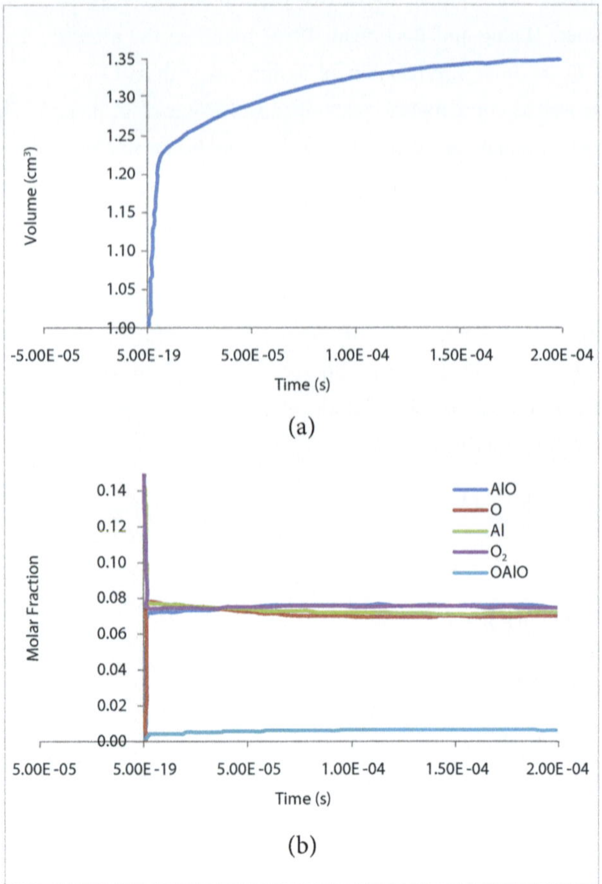

Figure 2. Volume (a) and molar fractions of the chemical species (b) versus time from ignition to the combustion process of alumina with gaseous oxygen at 10 atmospheres.

The molar fractions of the chemical species suffered significant variations and this time the speed of the elementary reactions present in the mechanism is high enough to stabilize the system shortly after ignition. This stabilization ensures a better control of the burning process, providing stability and increasing combustion efficiency.

Another simulation was performed, with the system subjected to a pressure of 60 atmospheres, shown in Fig. 4.

The difference between the aluminum at 30 and 60 atmospheres in the combustion process is based on the speed at which the burning process stabilizes. At 60 atmospheres, the system volume increases instantaneously to its equilibrium (value greater than that achieved at lower pressures), after the ignition stimulation. In the case of chemical species, there is a mole fraction variation that is approximately its equilibrium value, which is reached almost instantly after ignition.

CONCLUSION

The kinetics of alumina formation from aluminum was simulated at different pressures in a condition close to the one found in a combustion chamber of a solid propellant rocket motor, not considering the reacting flow to which the system is subjected in a continuous operation. An increase in the pressure at which the reactants are subjected is a determinant in the early establishment of a steady-state, in which there is a balance between the chemical species that react and their products. It was observed that the systems subjected to greater pressures tend to stabilize faster, due to the greater chemical speed of elementary reactions present in the combustion mechanism of aluminum, enabling a greater burning stability of the material and higher efficiency.

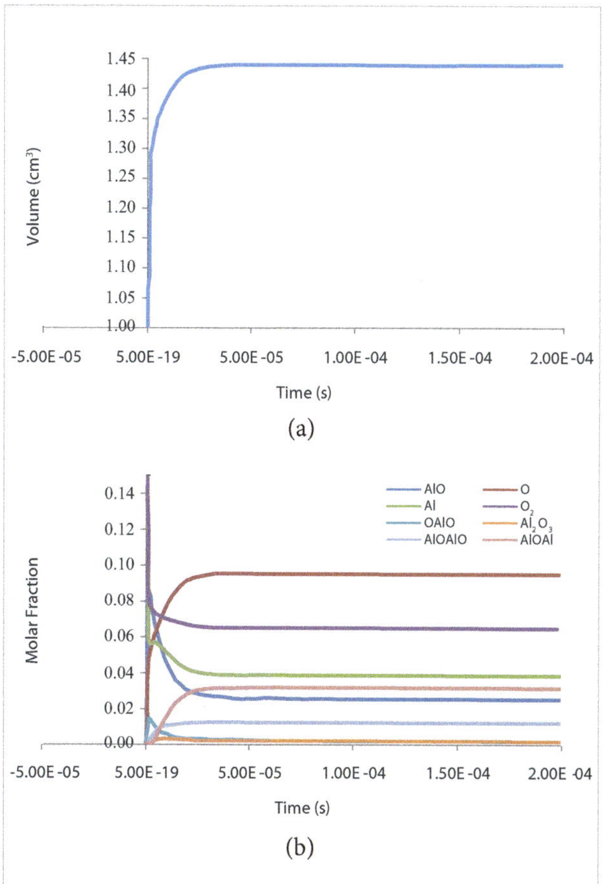

Figure 3. Volume (a) and molar fractions of the chemical species (b) versus time from ignition to the combustion process of alumina with gaseous oxygen at 30 atmospheres.

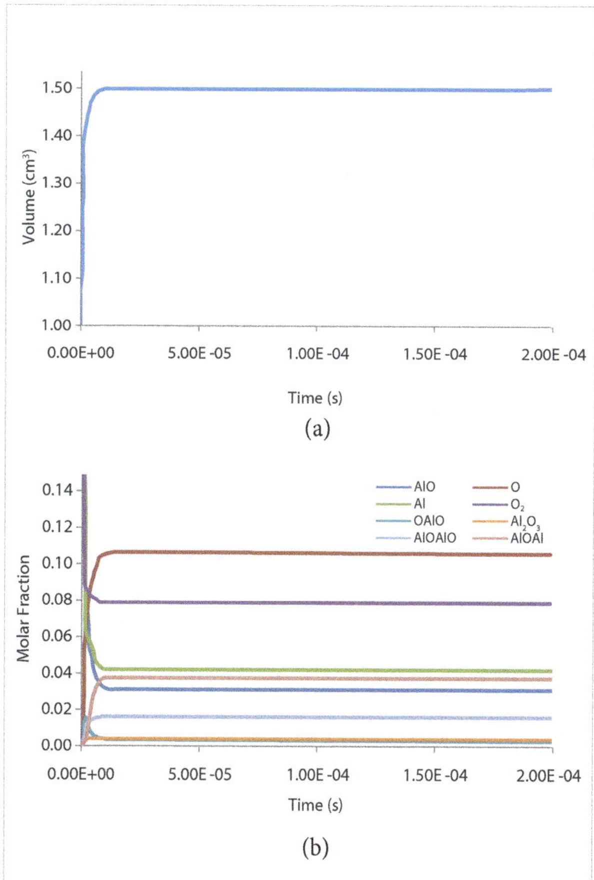

Figure 4. Volume (a) and molar fractions of the chemical species (b) versus time from ignition to the combustion process of alumina with gaseous oxygen at 60 atmospheres.

In chemical propulsion systems, such as the SPRM, increased operating pressure in the combustion chamber leads to displacement of the equilibrium toward the products, thus increasing the yield of the reaction.

ACKNOWLEDGMENTS

The authors thank CAPES, CNPq, and FAPESP for the financial support rendered.

REFERENCES

Barrere, M., 1968, "Solid Propellant Ignition: General Consideration". Research Aerospatiale, pp. 15-28.

Beckstead, M.W., Derr, R.L. and Price, C.F., 1970, "A Model of Composite Solid Propellant Combustion Based on Multiple Flames", AIAA Journal, Vol. 8, pp. 2200-2207.

Brewster, M.Q., 2000, "Solid Propellant Combustion Response: Quasi-Steady (QSHOD) Theory Development and Validation", Progress in Astronautics and Aeronautics, AIAA, Vol. 185.

Chiaverini, M.J., Harting, G.C., Lu, Y.C., Kuo, K.K., Peretz, A., Jones, H.S., Wygle, B.S. and Arves, J.P., 1999, "Pyrolysis behavior of hybrid-rocket solid fuels under rapid heating conditions", Journal of Propulsion and Power, Vol. 6, pp. 888-895.

Coltrin, M.E., Kee, R.J., Evans, G.H., Meeks, E., Rupley, F.M. and Grcar, J.F., 1991, "A Fortran Program for Modeling One-Dimensional Rotating-Disk/Stagnation Flow Chemical Vapor Deposition Reactors", Sandia Report.

Davenas, A., 2003, "Development of Modern Solid Propellants". Journal of Propulsion And Power, Vol. 19, pp. 1108-1128.

Flandro, G.A., Cai, W. and Yang, V., 2000, "Turbulent Transport in Rocket Motor Unsteady Flowfield", Progress in Astronautics and Aeronautics, AIAA, Vol. 185.

Gonçalves, R.F.B, Rocco, J.A.F.F., Iha, K. and Machado, F.B.C., 2009, "Modelagem da combustão da dinitramida de amônio por simulação computacional", Química Nova, Vol. 32.

Korobeinichev, O.P., Anisiforov, G.I. and Tereshchenko A.G., 1974, "Study of high-temperature kinetics and mechanism of thermal decomposition of mixtures of ammonium perchlorate-polymeric binder-catalyst using the time-of-flight mass-spectrometer", 12th Aerospace Science Meeting.

Kuo, K.K., Gore, J.P. and Summerfield, M., 1984, "Transient Burning of Solid Propellant", Progress in Astronautics and Aeronautics, Vol. 90, AIAA, pp. 599-651.

Liang, Y. and Beckstead, M.W., 1998, Numerical Simulation of Quasi-Steady, Single Aluminum Particle Combustion in Air, AIAA Paper 98-254.

Shusser, M., Culick, F.E.C. and Cohen, N.S., 2000, "Combustion Response of Ammonium Perchlorate", 36 th AIAA, SAE, ASME and ASEE Joint Propulsion Conference, Huntsville, A.L.

Summerfield, M., Parker, K.H. and Most, W.J., 1966, "The Ignition Transient in Solid Propellant Rocket Motor", Princeton University, Princeton, N.J., Aerospace and Mechanical Sciences Report 769.

Williams, F.A., Barrere, M. and Hung, N.C., 1969, "Fundamental Aspects of Solid Propellant Rockets", AGARDograph 116, Technivision Services, pp. 510-531, Slough, England.

Cognitive Based Design of a Human Machine Interface for Telenavigation of a Space Rover

Luca De Filippis[1], Enrico Gaia[2], Giorgio Guglieri[1], Marco Re[1], Claudia Ricco[3]

ABSTRACT: Human Machine Interface (HMI) design is a critical field of work because no general guidelines or rules have been assessed. In order to aid practitioners to design effective HMIs, different methodologies have been studied. To understand task objectives and plan goal-oriented actions, human operators exploit specific cognitive processes that have to be supported with advanced interfaces. Including cognitive aspects in HMI design allows generating an information flow that reduces user mental workload, increasing his/her situation awareness. This paper focuses on design and test of a Graphical User Interface (GUI) for the telenavigation of a space rover that makes the cognitive process of the user a priority in relation to the other development guidelines. To achieve this, a Cognitive Task Analysis (CTA) technique, known as Applied Cognitive Work Analysis (ACWA), is combined with a multi-agent empirical test to ensure the GUI effectiveness. The ACWA allows evaluating mission scenarios, i.e. piloting the rover on the Mars surface, in order to obtain a model of the human cognitive demands that arise in these complex work domains. These demands can be used to obtain an effective information flow between the GUI and the operator. The multi-agent empirical test, on the other hand, allows an early feedback on the user mental workload aiming to validate the GUI. The result of the methodology is a GUI that eases the information flow through the interface, enhancing operator's performance.

KEYWORDS: Cognitive engineering, Human machine interface, Space robotics, Space exploration, Graphical User Interface.

INTRODUCTION

In today's world, many systems are remotely operated or supervised by individuals who are decision makers. The planning, monitoring and controling of many of these systems are supported via visual display units. Applications that heavily depend on teleoperation and telenavigation are space exploration missions. Complex and dangerous conditions — deriving from environmental hazards — may occur to humans involved in these missions usually performed with supervised systems. As an example, Biesiadecki *et al.* (2007) state that "successful operation of the Mars Exploration Rover (MER) vehicles has depended on both manually-directed and autonomous driving. The two methods are complementary and careful selection of the right technique leads to better overall performance". The Graphical User Interface (GUI) design becomes a critical task in mission accomplishment, enhancing operator's Situation Awareness (SA) and control capabilities.

Existing design approaches cover only specific aspects of HMI design (Störrle, 2010), bringing to application-oriented solutions which limit the development of complex architectures. Many approaches and tools to tackle individual problems in the interface design have been developed by Hashimoto *et al.* (2011), but any integrated solution addressing the whole design process has been defined. A big effort has been spent to design and implement displays and interfaces to operate rovers in space: as NASA Visual Environment for Remote and Virtual Exploration (VERVE), Predictive and Interactive Graphical Interface (PIGI) (Pedersen *et al.*, 2010, 2012; Burridge and Hambuchen, 2009) or CliffBot Maestro (Norris *et al.*, 2009).

1.Politecnico di Torino – Torino – Italy 2.Thales Alenia Space Italia S.p.A – Torino – Italy 3.Università degli Studi di Torino – Torino – Italy

Author for correspondence: Giorgio Guglieri | Department of Mechanical and Aerospace Engineering, Politecnico di Torino | C.so Duca degli Abruzzi 24, Torino – 10129 – Italy | Email: giorgio.guglieri@polito.it

Assessed modeling approaches are:

- Operator Function Model (OFM) (Chu *et al.*, 1995), used in supervising, monitoring and controling of Unmanned Combat Aerial Vehicles (UCAVs) (Narayanan *et al.*, 2000);
- Direct Manipulation Interfaces (DMI) proposed by Hutchins *et al.* (1985) with the aim of reducing gaps between the user's goals and their knowledge of the system;
- Ecological Interface Design (EID) (Rasmussen and Vicente, 1989), a theoretical framework for designing interfaces in complex human-machine systems based on skills, rules, knowledge (SRK) taxonomy (Rasmussen, 1983) and the abstraction hierarchy (AH);
- Goals, Operators, Methods, and Selection Rules (GOMS) (Card *et al.*, 1983) a formal predictive modeling technique for interface design based on cognitive problem solving behavior (Eberts, 1994);
- Cognitive Task Analysis (CTA) (Gordon and Gill, 1997).

Direct correlation between user's SA and overall performance of the supervised system (Endsley, 2003) make the information flow through the GUI an essential feature in the designing process (Baxter, 2013). This paper proposes an integrated methodology in which SA requirements form the basis of the design phase. This integrated solution is achieved using a CTA approach — Applied Cognitive Work Analysis (ACWA) proposed by Elm (2003)—, coupled with direct testing methodology which involves user's performance assessment (Endsley, 2000) as a feedback on design. This leads to the definition of an integrated approach able to obtain graphical interfaces for space applications that can ensure high standards in terms of users' performance and SA.

CTA methodology can capture expert's knowledge in managing complex, dynamic and changing environments (Mast, 2014). Redding (1992) defines CTA "as an approach in determining the mental processes and skills required in performing a task and the changes that occur as the skill develops". Many of the above-mentioned approaches cannot lead to an explicit representation of the operator goals (Okura *et al.*, 2013) which is one of the major limitations of the OFM, or they are not effective in complex human-machine systems (Corujeira, 2013).

The paper introduces a methodology to design GUI with small, manageable, engineering transformations, each requiring the skilled application of the methodology's principles rather than requiring a design revelation at any point in the process. The design progress, therefore, occurs by generating artifacts that capture the results of each of these intermediate stages. Each of these artifacts also provides an opportunity to evaluate completeness and quality of the analysis and design effort. The design process is then associated with an empirical test (Nielsen, 1994), designed to obtain information about user performance and SA. A comparative analysis is conducted introducing support of experts, in order to extract information about the user's SA and Mental Workload (MW), obtaining a complete methodology that allows the practitioner to be supported from the design phase to tests and analysis.

The remainder of this paper is organized as follows: next section presents an in-depth analysis of the ACWA methodology and covers all the design phases in which it is divided. Further sections describe the testing procedure to obtain all the information needed to correctly assess the GUI. Subsequently, the results of the testing phase is evaluated and further analysis on user perception and performance is made. The paper concludes with a discussion on the overall design approach.

DESIGN METHODOLOGY

In order to model the cognitive process used to accomplish mission tasks and design a Human Machine Interface (HMI) able to support the user to reach mission goals, the ACWA methodology has been exploited. This approach is subdivided in four design processes and it starts with Functional Abstraction Network (FAN) definition to model the functional process required to perform the goal. The FAN captures the essential domain concept and the relationships between the problem-space and the domain practitioners. The next step is overlaying the Cognitive Work Requirements (CWRs) on the functional model as a way of identifying the cognitive demands which arise in the domain and require support. These cognitive demands may be successfully executed after identifying the Information and Relationship Requirements (IRRs). IRRs definition supports Representation Design Requirements (RDRs) which define the "shaping" of the information. The last step of the CTA approach used in this paper is the developing of Presentation Design Concepts (PDCs) to implement these RDRs, producing the correct information transfer to the users in order to fulfill their cognitive demands.

An in depth analysis of each step of the methodology is made in the remainder of this chapter. However, it must be stressed that all the steps of the process are "open" in parallel, so as design thoughts for CWRs, IRRs, or PDCs occur, they can be recorded and aid in the definition of the FAN itself. Figure 1 represents the complete methodology flowchart, where ACWA steps are related to mockup generation and testing in a recursive process.

FUNCTIONAL ABSTRACTION NETWORK

FAN is a function-based goal-means decomposition of the domain. This step has its roots in the formal, analytic goal-means decomposition method pioneered by Rasmussen (1986) for representing cognitive work domains as an AH. The FAN is a structured representation of the functional concepts and their relationship used as the context for the information system to be designed. This produces a multilevel recursive means-ends representation of the work domain structure.

In practice, building a FAN is an iterative process: the FAN starts from an initial base of knowledge regarding the domain that is gradually expanded using complementary techniques as observations, interviews or training (Elm *et al.*, 2003). In order to obtain an artifact that reflects the process depiction without representing only its physical components, a "flow modeling" approach is introduced. This model is based on the definition of different shapes to represent different stages of the process: Sources, Sinks, Storage, Transport and Conversion shapes are used in order to represent the functional operation of both abstract and relatively physical processes (Fig. 2). Figure 3 represents the primary goal formulated in the application here described, to start the development of the FAN.

Then, complementary techniques, as face-to-face interview or verbal protocol techniques are used in order to expand and enrich the domain understanding and to evolve a function-based model. Figure 4 shows the evolution of the FAN from goal 1 to goal 2. It expands the concept of map updating that is necessary to obtain a new map from the old one, merging new environment data obtained during the field-mapping process.

The FAN obtained in this paper starts from a field mapping overall mission objective: this main goal is subsequently expanded through the definition of system deployment functions and the analysis of rover motion functionality. At the very low level of abstraction, the FAN leads to define the functions of gathering

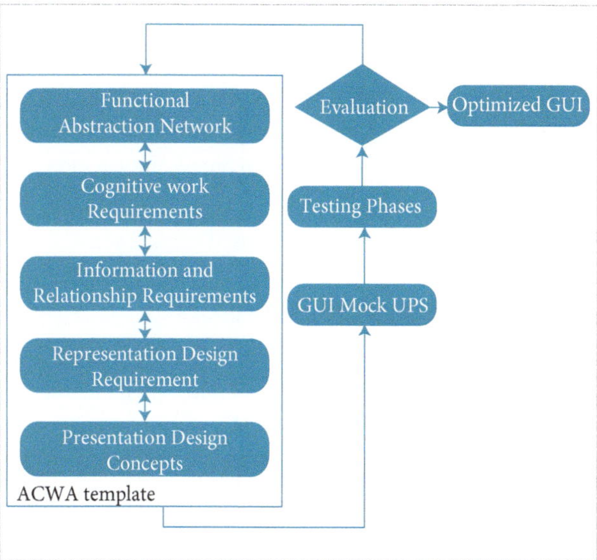

Figure 1. Design methodology.

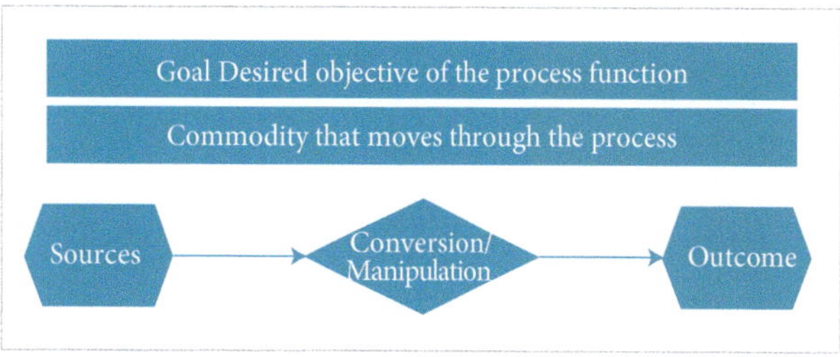

Figure 2. FAN Template.

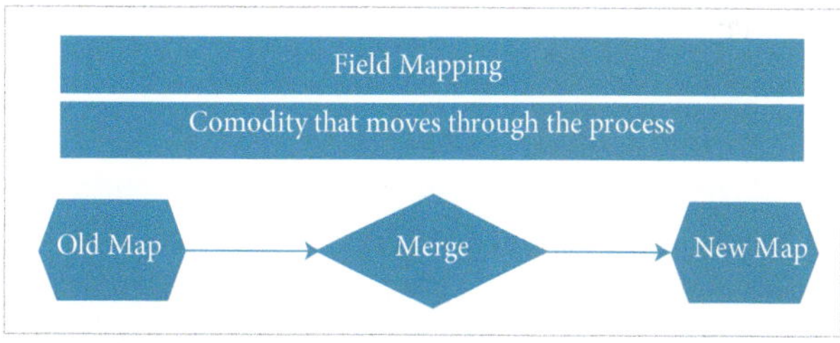

Figure 3. FAN Main Goal.

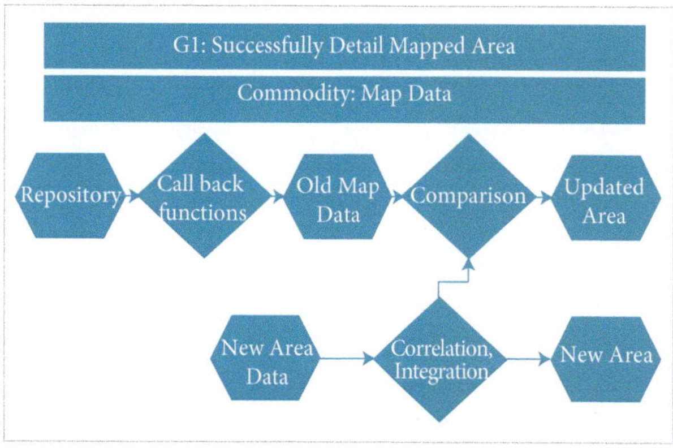

Figure 4. Main Goal Supporting Function.

environment data, successfully recovering command information and evaluating operative mode constrains.

As stated above, the scenario of the project is a field mapping oriented mission, all the other features and mission possibilities will be dropped to simplify the case of study. This assumption is needed to define a work domain not as complex as a whole mission definition, which allows researchers to describe it in minute details. For this reason, only the rover functions directly coupled with its motion capabilities and its deployment abilities are taken into account and investigated during this work. Furthermore, to ease the FAN depiction, the mission is intended to be in nominal mode; failure management will be only expected but it will be not investigated.

The FAN representation of the work domain's concept brings us to the second step of the ACWA process: deriving the CWR.

COGNITIVE WORK REQUIREMENTS

CWR represent the cognitive demands for each part of the domain model, i.e., all type of recognition, decision-making and problem-solving activities.

In the methodology, the addition of CWRs to the design repository is described as "thickening" the analysis (Elm *et al.*, 2003): in fact, each CWR is attached to a node of the FAN as an enrichment of the domain concept understanding. Based on the underlying premises of the CTA, these CWR center around either goal or functional process, monitoring for goal satisfaction and resource availability, planning or selecting among alternative options to achieve goals, and controlling the functional process. By organizing the specification of operator's cognitive requirements around nodes in the FAN rather than organizing requirements around predefined task sequences, the representation helps to ensure a consistent, decision-centered perspective (Table 1).

This perspective implies that the FAN nodes can be associated with different kind of CWRs: some of them can be found across a variety of domains, so a "template" of generic CWR can be tested for each node of the FAN, while others are uniquely coupled with the scenario features and demands. Thus, the FAN forms the basis for the structure of the cognitive demands reflected in the CWRs. For example, every goal node in the FAN has associated "goal monitoring" decisions; likewise, processes have associated "process monitoring" decisions and, similarly, there will always be some "feedback monitoring" decisions related to assessing whether actions are achieving the desired result.

In many ways, this step is where the decision support system requirements are shaped: hence, good CWRs depiction is essential to the final resulting GUI. A portion of the CWR obtained from the FAN are listed below:

- Monitoring field-mapping progress regarding overall map dimension;
- Monitoring successful positioning of the sensor for data acquisition;
- Selecting the optimal motion of the rover to maximize sensor potential;
- Choosing the best command sequence to move the rover to the acquisition site;
- Monitoring current command input to avoid limit crossing;
- Selecting the best navigation option to maximize rover response during mission;
- Monitoring actual rover movement capabilities to maximize its potential;
- Monitoring actual rover movement possibilities to perform defined action;
- Monitoring rover system behavior to obtain information about its actual status;
- Monitoring correct acquisition of sensed commands.

Table 1. Cognitive Work Requirements for Goal 1.

	CWR description
CWR-G1-1	Monitor field mapping progress with respect to the overall map dimension
CWR-P1-1	Monitor the presence of old map of the target area already available
CWR-P1-2	Select source between old and new map that maximize information accuracy
CWR-P1-3	Compare old and new data
CWR-P1-4	Monitor correct correlation of new mapped areas to already mapped ones

INFORMATION AND RELATIONSHIP REQUIREMENTS

While mental demands of every FAN node are gathered through CWRs definition, the information required for each decision to be made is still unidentified. IRRs are defined as the set of informative elements necessary to successfully resolve the associated CWRs. Thus, the focus of this step in the methodology is to identify the ideal and complete set of information for the associated decision-making process. Therefore, IRR forms the basis for the visualization support to be designed in the next phase. However, information is not data: data can require complex transformations to become information. Thus, IRRs have a much deeper impact on the entire system architecture than merely the "look and feel" of the final GUI (Elm *et al.*, 2003).

It is important to note that IRRs are strongly related to mental demands fulfillment (CWRs) and are not limited by data availability in the current system. If the required data is not directly available, ACWA provides a logical basis to obtain that data (e.g. pulling data from a variety of databases, adding additional sensors, creating "synthetic" values); this leads to one of the assumptions of the project: some of the information obtained from IRR requires data not available with current state of the art sensors, but strongly needed for a successful space telenavigated mission.

This approach is different in relation to the one that human-factor engineers have had in the past (designing an interface after the system equipment has been specified). The standard approach, in fact, cannot give any profit in terms of understanding the user's MW involved in the process of acquiring the correct SA.

Consequently, "the ACWA approach is fundamentally broader in scope than other approaches to interface design that do not consider the impact of IRRs on system architecture specifications" (Vicente *et al.*, 1996).

It is easy to notice that a lot of information can be needed by a single CWR to be completely defined. Furthermore, high-level goals — as the ones in Table 2 — are dependent upon supporting information, successfully developed in lower level goals.

A summary table was created to correctly evaluate all the information needed to satisfy the mental demands of our domain. In this table, the information are correlated to the supporting function, in order to define a hierarchy between them. Table 3 collects all the information which need to be provided to the user in order to trigger the cognitive processes related to each functional process. This information is, in many cases, the same for different cognitive processes — as it is shown on the second column of the table, where each number corresponds to a specific cognitive process. The table allows correlating each piece of information to the number of cognitive processes in which it is involved. Information involved in many cognitive processes need to have priority on the GUI, because the user will need them all the time to maintain its SA.

Upon obtaining IRRs, the ensemble containing the FAN, the CWRs and their associated IRRs represent a solid basis for the development of the decision-making graphical support.

Table 2. Information and Relationship Requirement for Goal 1.

	IRR description for Goal 1 listed with respect to CWR
CWR-G1-1	Monitor field mapping progress regarding overall map dimension
IRR-G1-1.1	Deliver mapping progress status with respect to the whole area of interest
IRR-G1-1.2	Deliver typical rate of acquisition (design limits)
IRR-G1-1.3	Actual rate of acquisition with respect to typical rate
CWR-P1-1	Monitor the presence of old map of the target area already available
IRR-P1-1.1	Deliver selected area availability in storage
IRR-P1-1.2	Deliver data accuracy of old map
CWR-P1-2	Select source between old and new map that maximize information accuracy
IRR-P1-2.1	Rate of change between old map data and new map data
IRR-P1-2.2	Data already stored with respect to selected zone
CWR-P1-3	Compare old and new data
IRR-P1-3.1	Actual sensed data accuracy with respect to stored data accuracy
IRR-P1-3.2	Rate of change between stored and sensed data
CWR-P1-4	Monitor correct correlation of new mapped areas to already mapped ones
IRR-P1-4.1	Actual position of the sensing system
IRR-P1-4.2	Deliver nearest position available in the stored map
IRR-P1-4.3	Actual error in position

However, to obtain a Graphical Interface that communicates with operators without effort, the IRRs have to be converted into visual widgets. The last two steps of the ACWA, which are the RDRs and the PDCs, allow the methodology to bridge this gap.

REPRESENTATION DESIGN REQUIREMENTS AND PRESENTATION DESIGN CONCEPTS

RDRs define the goals and scope of the information representation, in terms of cognitive tasks it is intended to support. It also provides a description of the supporting

Table 3. Portion of Information and Relationship Requirement list.

IRR	Goal supported by IRR
Command information: acceleration (actual and maximum)	4-5-6-7.5-8.2
Command information: acquisition sensor check up	7.5-8.2
Command information: command history	7.5-8.2
Command information: command verification	8.2
Command information: commanded torque	4-7.5-8.2
Command information: communication bandwidth	8.2
Command information: communication coverage	8.2
Command information: communication windows	8.2
Command information: electric current needed	4-5-6-7.5-8.2
Command information: maximum torque available	4-5-6-7.5-8.2
Command information: Power available to engines (actual and typical)	4-5-6-7.5-8.2
Command information: speed (actual and maximum)	4-5-6-7.1–7.2–7.5-8.2
Command information: Wheels Differential torque	4-7.5-8.2
Environment data: acquisition data rate (actual and typical)	8.1
Environment data: memory space available	8.1
Environment data: sensor status	7.3-8.1
Environment data: sun exposure (actual and typical)	8.1
Environment data: temperature (actual and typical)	8.1
Environment data: wind speed (actual and typical)	8.1
Guidance and Navigation data: Checkpoint position	3-4-5-6

information required to sustain the cognitive tasks. Furthermore, this step in the process begins shifting the attention from "what" is to be displayed (defined by FAN, CWRs and IRRs) to "how" to display it. It adds a more complete description of the behaviors and features needed to communicate the information effectively, as well as an allocation of the Information/Relationship Resources across the entire set of displays within the workspace.

The visual framework of a Rover Simulation has been taken into account to obtain the RDRs and the PDCs.

A software framework able to create 3D immersive virtual simulations has been used in order to develop a particular Rover Simulation application. The software was developed inside a Piedmont regional funded project, ended in May 2012, called STEPS (*Sistemi e Tecnologie per l'EsPlorazione Spaziale*). STEPS was a project supported by Regione Piemonte and carried out by *Thales Alenia Space Italia*, Small and Medium Enterprises, Universities and public Research Centres belonging to the network "*Comitato Distretto Aerospaziale del Piemonte*". The project objective was to develop hardware and software demonstrators for descent, soft landing and surface mobility of robotic and manned equipment during Moon and Mars exploration. The demonstrator was created in the Collaborative System Engineering (COSE) Centre, a *Thales Alenia Space Italia* facility in Torino which operates within the Engineering & Advanced Studies Directorate of the Domain Exploration and Science Italy. The Centre main mission is to research, develop, integrate and propose new methodologies and tools to enhance the system engineering capabilities and the multidisciplinary collaboration.

Using the Virtual Rover Demonstrator it is possible to simulate a rover motion on a planetary surface.

Three visual displays are used to obtain a wide screen virtual experience of the rover and the terrain (Fig. 5).

Another benefit provided by RDRs is that, as long as the domain remains unchanged, the RDR serves as an explicit documentation of the presentation concept purpose, despite of the technologies available and used to implement it. As newer technologies become available, and as their interaction with human perception becomes better understood, the technologies used to implement the RDR can evolve.

The RDRs are obtained splitting the IRRs into:

- *Essential information*: it represents vital information, which needs to be on the central screen. It gathers information which have to be easily accessed by users;
- *Detailed information*: represents information which may improve user's awareness with an increase in their workload. This information is more accurate and detailed and can be used to improve awareness on a selected feature.

The following data have been evaluated as part of essential information: user commands, speed, rover heading and "out of range" alert of these parameters. Other "alert signals" have to be fit into costless information: proximity hazard, pitch acceleration and roll acceleration. These alerts are needed because the rover stability can be heavily compromised if these parameters exceed their thresholds. This information represents the one related to a greater number of IRRs.

On the other hand, in the detailed information, there are included the pitch and roll, angular velocity and speed of the wheels. It is easy to notice that the user may perform a proper telenavigation task by focusing attention on the central screen. The information on lateral screens plays a detailing role: if the user understands that something is not working as intended (e.g. from no-cost information) he/she will focus on lateral screens in order to obtain more detailed information. The set of information used on the lateral screen are obtained from the one that will affect a smaller number of functional processes.

After the development of the RDRs, the last step in the methodology is obtaining explicit PDCs for the Interface (as GUI mock-ups). This final step requires the knowledge of the human perception and its interaction with the various presentation techniques and attributes. With the RDR as a guide, the sketches, drawings, and brainstorming concepts can all be resolved against the display's intent and requirements. The issues of how it is perceived can best be done with empirical testing of prototypes and often requires considerable tuning and adjustment to achieve the representation capabilities specified in the RDRs.

Figure 5. Visual Display Framework.

During this phase, researchers found, for example, that the information panels couldn't be put on the further edge (in relation to the central screen position) of both the side screens due to their dimensions. In fact, if positioned on the further edge, information becomes accessible only with a larger motion of the head, with a consequent degradation in driving performance. To avoid this issue, while maintaining a good distinction between essential and detailed information, researchers choose to insert the information panel on the nearest edge of the lateral screens.

The availability of two different panels (both the side screens are used to represent detailed information) allows a further classification to aid users in the rationalizing process of the interface. Thus, researchers try to collect on the left screen the detailed information related to kinematic data and system parameters, while on the right screen were implemented navigation information. The final mock-up of the GUI obtained using this methodology is shown below (Figs. 6, 7 and 8):

The availability of an older interface (Fig. 9) with the same purpose of the one created using the ACWA approach, allows researchers to enhance it, decreasing the time spent for this last step.

All the improvements done on the interface are listed in Table 4.

TEST DESCRIPTION

An empirical test was developed in order to obtain information both on the user SA and on their performance in task fulfillment.

Several methods of testing SA have been documented (Endsley *et al.*, 2000; Endsley, 1995a; Endsley, 1995b), usually divided into:

- Knowledge-based measurement techniques; and
- Performance-based measurement techniques.

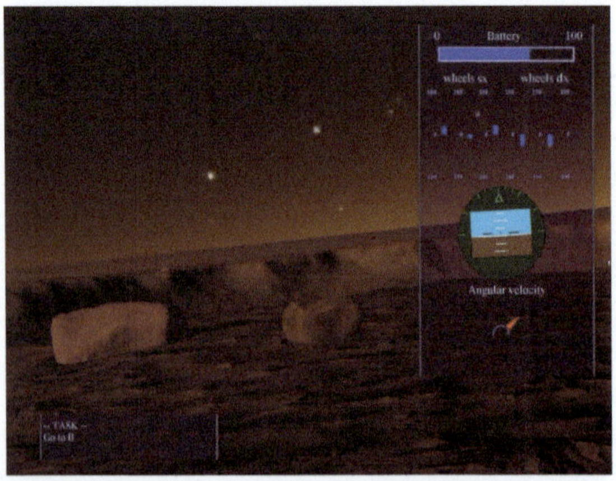

Figure 6. Left screen Graphical User Interface.

Figure 7. Central screen Graphical User Interface.

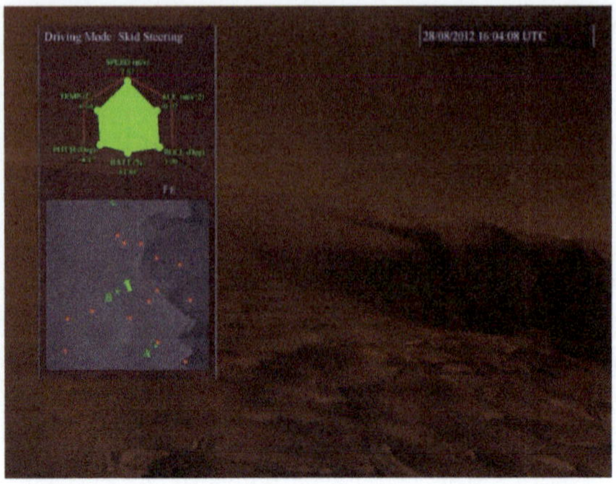

Figure 8. Right screen Graphical User Interface.

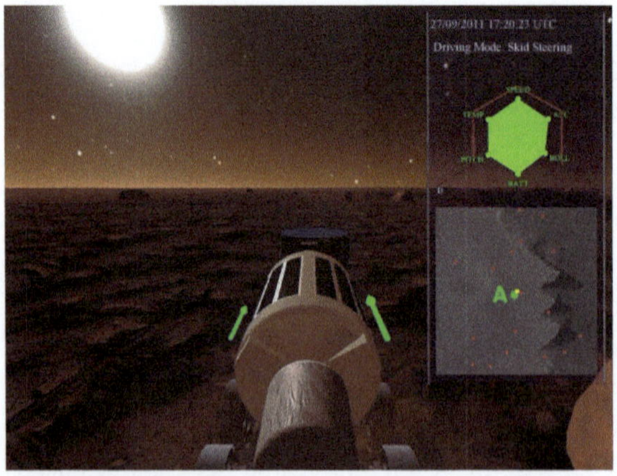

Figure 9. Old Graphical User Interface for remote control.

Table 4. Graphical User Interface improvements.

Screen	Improvement
Lateral-left	Battery status bar added
Lateral-left	Wheel slip indicator added
Lateral-left	Artificial horizon added
Lateral-left	Angular velocity display added
Central	Directional compass added
Central	Ghosting function improved
Central	Torque-related color function added
Central	Rover speed indicator added
Lateral-right	Hexagon position modified
Lateral-right	Map position modified
Lateral-right	Rover heading function added in the map

The knowledge-based measurement techniques are founded on either simulations or real trials. The task is selected according to the level or type of SA being addressed by the experiments. The method identifies independent variables (i.e., the type of display for a GUI, the type of interaction device used to pilot a remote vehicle) and dependent variables, such as objective and subjective measures of testeing knowledge (i.e., understanding) and performances.

Furthermore, there are several complex techniques which attempt to determine or model the subject's knowledge of the situation at different times throughout simulation runs.

For example, the Situation Awareness Global Assessment Technique (SAGAT) freezes the simulator screens at random times during the runs, and queries the subjects about their knowledge of the environment. This knowledge can be at several levels of cognition, from the most basic of facts to complicated predictions of future states.

The performance-based measurement of SA has taken several forms. Some techniques measure the overall final performance of the human-in-the-loop system in any or all of its tasks. Alternatively, Testable Responses can be used in order to evaluate SA: the subjects will face predetermined situations during the simulation that require decisive and identifiable actions, if the subjects have the correct level of SA, they can correctly perform the required actions while, with low awareness, they cannot perform it.

As a general comment, to provide a detailed assessment of the subject's SA, the knowledge-based techniques are more accurate, as they measure these variables directly. Performance-based measurement can only make inferences based upon the particular information the subject acted upon, and how it was interpreted, thus these techniques are very useful when well-determined performance are assessed, while the knowledge-based techniques can be more accurate when a lot of different aspects are observed together, and multiple performance assessments can be made.

Self-rating techniques are used in order to gain a subjective assessment of participant's SA. Typically administered post-trial, self-rating techniques involve participants providing a subjective rating of their perceived SA via an SA related rating scale. As an example, the Situation Awareness Rating Technique (SART) is a subjective rating technique developed for the assessment of the pilot's SA.

The primary advantages of self-rating techniques are their ease of application (easy, quick and of low cost) and their non-intrusive nature (since they are administered post-trial) very useful for the early assessment in the design process. However, subjective self-rating techniques are heavily criticized for several reasons, including the various problems associated with the collection of SA data post-trial (correlation of SA with performance, poor recall) and also issues regarding their sensitivity.

Another method used in order to estimate SA is based on observer's ratings during or at the end of the trial. Observer rating techniques typically involve a Subject Matter Expert (SME) observing participants performing the task under analysis and then providing an assessment or rating of each participant's SA. The SA ratings are based upon observable SA related behavior exhibited by the participants during task performance. The main advantages associated to the use of observer rating scales to measure SA are their non-intrusive nature and their ability to be applied 'in-the-field'. However, the extent to which observers can accurately rate participant SA is questionable, and also multiple SMEs may be required. A five points rating scale (1 = very poor, 5 = very good) and an additional 'not applicable' category can be used for each observable SA related behavior of the tester.

The objectives of the testing phase are different: on one hand a first evaluation of the ACWA have to be made in order to assess its impact on the user perception, on the other hand the GUI itself has to be tested to verify its usability. To obtain a feasible and light testing methodology that can inspect these features, a brand new approach to the testing phase is developed. A performance-based concept is used in order to investigate the GUI usability while self-rating questionnaires are defined to obtain information about operator's SA. These early tests in the overall design phase of the simulator cannot take advantage of the knowledge-based techniques: the overall behavior of the system is still undefined, thus it is impossible to define independent variable to be observed.

While the self-rating questionnaires are administered post-trial in order not to influence the performance, the SMEs will focus their attention on the operator's behavior during the trials and on their accord with the GUI.

The coupling of the self-rating techniques with the observer rating techniques allows practitioners to enhance the testing phase thanks to the ease of application of these methodologies without incurring in the subjective limitations of the user self-assessment. Furthermore, this mixed technique offers the best tradeoff between result's significance and test costs.

The performance-based measurement obtained from the mixed (self-rating and observer) testing technique can be used as an indicator of the effectiveness of the ACWA definition of the interface. The GUI, defined through the ACWA, should enhance the operator's awareness, easing his understanding of the process: if an increase in the user's performances can be found during the test, the ACWA has a positive impact on the interface design.

TEST PROCEDURE

The testing procedure is hereafter shortly described. A number of selected users were expected to drive the rover in remote manual control (i.e., using the joystick) along an assigned path (defined by 6 waypoints, in alphabetical order from A to F, distributed on the map). Each test is completed when the rover reaches the final waypoint F, or when the battery completely discharges. Together with this main task, a second — and less important — goal has been defined to increase the user MW. This second objective referred to a completely different mental process regarding the one involved in the fulfillment of the main goal. To accomplish this, the users were asked to count the number of "skid events" — i.e., to count the number of bars which turn red, Fig. 10 — while performing the main task (to reach the final waypoint F).

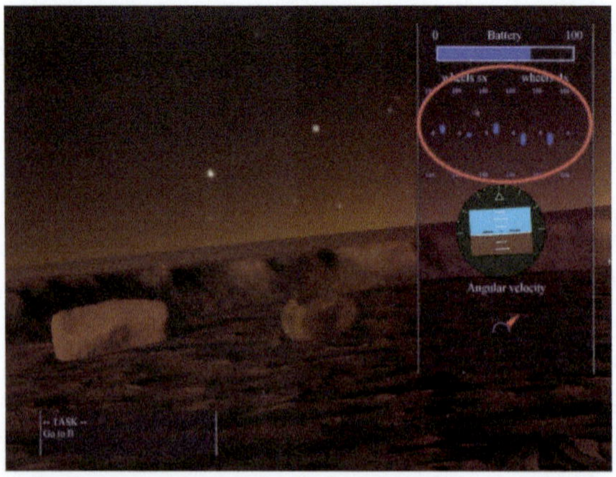

Figure 10. Highlighted Skid events indicator on left screen.

The SA assessment is based on expert observer's ratings (during the test) and user's self-ratings (at the end of the test). Prior to testing, a further questionnaire was given to the user: the aim of this "Relationship Questionnaire" (RQ) (Bartholomew & Horowitz, 1991) was to evaluate the attachment style of every participant.

The RQ is a self-report utility which allows investigating the general orientation of our adult intimate relationships, regarding psychological and emotional intimacy. The test is arranged in four short self-descriptions, each of which summarizes the basic aspects of one of the four main patterns of attachment. This classification system allows the subject to be identified, as a prototype, within one of the four attachment styles: secure attachment style, dismissive-avoidant, preoccupied and fearful-avoidant.

During the test, subjects have to choose the most representative prototype of themselves, furthermore they also have to assess the extent to which each of the four prototypes represents them, using a 1 (strongly disagreement) to 7 (total agreement) scale. This allows us to assess both the image that the subjects have of themselves and the image that they have of the others: obtaining, by the intersection of the data, a "model of self" and a "model of others", which may be positive or negative. In this way, the RQ investigates the hypothesis of Bowlby (1973), that attachment styles reflect the internal working models of self and others, which can be both positive and negative. It follows that the secure attachment, in which subjects are characterized by coherence, autonomy and self-confidence is derived from the combination of a positive model of self and a positive model of others. The dismissive-avoidant attachment results from a positive self-evaluation associated to a negative representation of others: subjects tend not to be coherent, they sacrifice the intimacy and deny the importance of relationships, in "a kind of self-sufficiency affective and existential" (Bruni, 2004). The combination of a negative model of self and a positive model of the other develops instead a preoccupied attachment, in which the subjects, which are distinguished by their relentless pursuit and apprehension about the relations, safeguard their low self-esteem exposing the incoherence and idealization of their own relationships. Finally, fearful-avoidant attachment is derived from the combination of a negative model of self and a negative model of the other; fearful individuals avoid involvement with others for fear of being rejected, even though they wish it fervently, since, because of their low self-esteem, they do not feel worthy and expect others to be ill-disposed, unreliable and rejecting. Attachment styles highlighted by the RQ test also adapt, to some extent, to the relationship of "trust" that is established between the pilot and the interface of the vehicle driven.

In addition to that, researches have extensively demonstrated that the different attachment styles are related to the ability to handle, emotionally and cognitively, different situations and stress that they can generate (Ricco, 2009). The attachment style will directly affect perception and, also, cognitive processes that can be used: consequently it has a key role in the efforts to improve the GUIs.

Thus the RQ could give information about the attachment style required to improve the teleoperation: in fact, shifting from the concept of pilots to the idea of supervisors could change the desired psychological profile of the user. For example, pilots, most of the time, are very self-confident and self-centered people: this could lead to a difficult interaction with automated systems which require the confidence of the pilot to operate, exactly as it happens between people that need mutual trust in order to collaborate on a project.

The tester experience started in a room next to the simulator where he has to fill the RQ isolated from the other users to avoid suggestions; the anonymous questionnaire, sealed inside a blank envelope, were collected by a dedicated practitioner. After the RQ collection, the tester could perform the driving session: in this phase three observers must fill out their questionnaire, looking at the user's behaviors and at their performance (i.e., eyes motion, head motion or collecting defined data during the attempt) while maintaining absolute silence to avoid tester's distraction. The last phase consists of the self-rating questionnaire: the tester answers the questions alone in the adjacent room then the anonymous questionnaires were collected by a dedicated practitioner. These questionnaires will explore the perception of the GUI from the operator point of view: some questions are related to parameter controlled by the observers (e.g., rover speed) in order to have a feedback on performance, while other are related to the cognitive processes that the operator have to exploit to correctly maneuver the rover. These questions can be used as a direct investigation on the ACWA effectiveness: if the operator cannot understand the GUI elements, it means that the process that he exploits is different from the one obtained from the ACWA that defines the GUI shape.

Within three months, two different sessions of tests were scheduled. Both the sessions involved same users and same observers in order to avoid erroneous comparison. In the last test, the GUI was not modified, but a time restriction was given to the testers in order to increase their MW (Paas *et al.*, 2004). The main task (to reach the final point "F") and the secondary objective (to count the skid-events) were maintained, while the execution time was diminished by 20% in relation to the execution time measured during the first test. The aim of this second test was to increase the performance-based knowledge of the user's understanding process in a more dynamic and strained environment, and to verify the evaluations obtained in the previous test.

TEST

The user should demonstrate their ability to detect, understand and respond to events while maintaining a good performance in both assigned tasks. To do this, they have to maneuver the rover through six waypoints positioned on Mars' surface in order to reach the final waypoint F. While the user is performing the test, three observers will judge their behavior without interference.

The observer's ratings are divided into knowledge-based (e.g., distance and speed, randomly collected by the observer) and performance-based (e.g. time required to complete the path, number of collisions, average cross-track error in forward drive or the ability to drive straight, response to the presence of a random obstacle).

When the driving test is completed, the user has to fill a self-rating questionnaire. The self-ratings give the assessment of the overall performance according to user's opinion.

General open comments are collected from both observers and testers, mainly related to the user's SA and the HMI performance. In order to allow comparison within tests, the same questions made in the first test were used in this self-rating questionnaire. Furthermore, the same users were engaged for testing the new GUI.

The number of users involved in the tests is limited (N = 7): this is acceptable because the number of operators is very limited in space applications. Space agencies invest many resources in training operators in order to employ highly specialized professionals. This leads to a deep knowledge of the system on which they are operating: therefore different applications employ users who possess a slightly different training. Thus, statistical survey cannot show direct correlation or general behavior because each system requires different skills and training. For this reason, the results of the tests performed to evaluate the developed GUI will not be statistically useful.

EVALUATION

Testing evaluation was made using two different modules: the first one, a self-rating questionnaire, was given to the user after the end of their performance, while the second one, the Observer-rating questionnaire, was filled during the test by three observers. However, to avoid erroneous or misleading behavior, the observer could not interfere with the user during the tests.

The self-rating questionnaire consisted in different questions about user's perception of the overall test: they are asked to remember detailed parameters (e.g., the average speed maintained during the test or battery level at the end of the test) and to give feedback about their perception (judging several features of the GUI as the directional compass or the proximity hazard feature).

The majority of the questions consisted in a multiple-choice answer to ease the user's fulfillment possibility: an example is shown below (Table 5).

The observer questionnaire tracked down the same parameters, e.g. rover speed, together with indicators that can help understanding user's SA, such as coherence between user's eyes motion and their task or their execution time (Table 6).

The self-rating questionnaire and the observer questionnaire were then compared to obtain information about the overall SA of the user. The answer of observers can be used as neutral criteria to evaluate the personal, thus subjective, ratings of the users.

In this way, pratictioners gathered information about performace, obtained from the observers annotations during the testers trial (e.g. average speed and execution time), and about interface effectiveness (thus the effectiveness of the cognitive approach was used to define the interface itself), obtained comparing the observer's annotation and the subjective rating of the user.

For example, the tester has to remember the level of charge of the battery at the end of their trial: the comparison between their answer and the actual data gathered by the observers allows obtaining information about the information transfer through the interface.

On the other hand, the comparison between the first test and the second one, which has an increased tester's MW, may improve the knowledge about the cognitive process of testers. The increase in MW means a lowering in SA: this will affect tester capability of acquiring information from the system. This could highlight the cognitive processes of the tester: if the operator cognitive process and the cognitive approach used in the interface design exploits the same process this leads to a natural acquisition of the information by the operator (i.e. the tester has not to process the information to understand it). In this way, the test could also verify the results of the previous test about interface effectiveness.

On the other hand, the RQ has been used to deepen the attachment style of the tester in order to obtain information about the tester's individual difference that can significantly affect their performance. The RQ was submitted to the user before the start of the test and consists in four brief descriptions which summarize each of the main patterns of a different attachment style, from secure to fearful.

TEST RESULTS

In this section, the results of the two tests performed on the different GUIs have been reported. The following tables present the attachment style, the personal evaluation and the evaluation of the observer, for each of the seven users. In the second test (that was carried out with an increase MW), only two of the testers reach the final point F within the time restriction. Furthermore, both of them possess a secure attachment style (Table 7).

This means that the secure attachment style, in which subjects are characterized by consistency, autonomy and self-confidence, may have a positive impact on the relationship of "trust" that is

Table 5. Self-rating questionnaire.

How do you judge your performance in terms of cruise speed?					
☐Not applicable	☐ Very poor	☐ Poor	☐ Average	☐ Good	☐ Very good
Looking at the map, how you judge red markers to detect the obstacles?					
☐ Not applicable	☐ Very poor	☐ Poor	☐ Average	☐ Good	☐ Very good

Table 6. Observer questionnaire.

User eyes motion and coherence between current task and attention appointment.					
☐ Not applicable	☐ Very poor	☐ Poor	☐ Average	☐ Good	☐ Very good
Covered distance (number of waypoints)					

established between the pilot and the interface of the vehicle driven. However, there are also other two secure testers that have not completed the task in the given time: this means that a further investigation has to be conducted in order to assess the real correlation between pilots' attachment style and their performance.

On the other hand, the results can give information about the overall effectiveness of the design methodology in terms of users' performance and SA.

Comparing testers' answers given in the two tests with the data collected by observers, it is possible to understand if the tester's perception of the GUI was coherent.

Table 8 shows that users can evaluate and remember correctly the remaining battery level of charge at the end of the test. This means that also in a high MW environment (i.e. the second test), this feature enhance pilot's SA and satisfy the pilot's mental demands.

Also, the future-position indicator (that was improved because of the IRR definition) was perceived correctly by the testers (Table 9).

From this comparison, it appears that the tester perception of the future position indication was better in the second test: this could be derived from the increase in their mental demands, thus the elements of the GUI that reduce their workload are more evident.

Another element that has been evaluated is the map: in Table 10, the testers' subjective assessment of the red marker used for obstacle detection is given. The results highlight a slightly increase in testers usage and perception of these elements in the second test. This leads toward two different assumptions: the first one is that, with an increase in the MW, testers have to improve their perception using the map elements more frequently, while the second one is that, during the first test, the performance pressure on testers was too low to force them to actively use the GUI at its maximum.

In Table 11, the wheel skid counter has been reported: this was the secondary objective of the test to increase the MW of testers without conditioning their performance. If we compared this table to the observed eye motion coherence (Table 12), it is

Table 7. Task performance test number 2.

Tester	Attachment	Collision (Test I)	Collision (Test II)	Time restriction
1	Secure	Yes	No	Yes
2	Reserved	No	No	No
3	Secure	Yes	No	Yes
4	Secure	No	No	No
5	Troubled	No	No	No
6	Reserved	No	Yes	No
7	Secure	No	No	No

Table 8. Battery level, test number 1 and 2.

Tester	Tester (Test I)	Observer (Test I)	Tester (Test II)	Observer (Test II)
1	70	50	70	65
2	50	50	55	30
3	50	58	60	65
4	40	50	50	50
5	45	45	40	30
6	30	30	10	15
7	30	50	27	30

noticeable how testers with a poor coherence (i.e., tester 1 in test I, tester 2 in test II) cannot tell the right number of skid events.

On the other hand, most of the participants have performed the secondary objective with a good level of accuracy: this means that the overall complexity of this task was correctly tuned in both of the tests.

In addition to that, the MW perceived by testers were paired with their eyes movement coherence as detected by observers (Table 12): it is easy to notice that testers subjected to high MW (average/high in the tester charts) most of the time have a low eyes motion coherence, which implies an overall difficulty in gaining the correct SA with regard to the current task. Furthermore, it seems that the time restriction has not influenced the perceived MW of the testers: most of the testers found the first test demanding, thus a further increase in their workload could have not been correctly perceived.

Table 9. Future position perception.

Tester	Test I	Test II
1	Good	Good
2	Poor	Average
3	Average	Average
4	Good	Good
5	Good	Good
6	Poor	Good
7	Average	Good

Table 10. Red marker obstacle detection on map.

Tester	Test I	Test II
1	Poor	Good
2	Good	Good
3	Good	Good
4	Good	Average
5	Average	Average
6	Good	Good
7	Good	Good

Table 11. Wheel Skid counter.

Tester	Tester (Test I)	Observer (Test I)	Tester (Test II)	Observer (Test II)
1	18	40	15	23
2	30	19	35	16
3	25	28	17	14
4	5	14	10	13
5	5	6	15	17
6	21	18	5	2
7	17	18	15	13

Table 12. Mental Workload and tester coherence.

Tester	Tester (Test I)	Observer (Test I)	Tester (Test II)	Observer (Test II)
1	Average	Poor	Average	Average
2	High	Very Poor	Low	Very Poor
3	Average	Very Good	Average	Average
4	Low	Good	Low	Average
5	High	Very Good	High	Poor
6	High	Good	High	Very Poor
7	Average	Average	Average	Average

CONCLUSION AND FUTURE IMPLEMENTATION

This paper presents the development of a GUI for remote control of a rover in a planetary environment and the following testing phase to evaluate its performance regarding users' SA and MW. The methodology used in order to define the GUI stands out against the other approaches, because it makes the cognitive process of the user a priority in relation to the other development guidelines. Furthermore, it allows, through an iterative process, to update the HMI based on the test results. This can be done because there is a strong interconnection between the physical architecture of the system and the HMI: if tests point out that something has to be modified, researchers can easily understand which information has to be changed and where this information is used. On the other hand, the multi-agent tests allow researchers to understand the lacks in the interface using few testers and link these results to a well-defined psychological profile. Furthermore the testing technique has proved as a feasible methodology to obtain information on the GUI effectiveness and on the operator performance in early phase design, while more complex and expensive techniques, as knowledge-based ones, cannot be used.

As research perspective:

- The information which cannot be correctly understood by testers will be revised to improve the GUI during a second and more detailed design overview;
- Involving observers is a useful improvement of test evaluation though it is important that parameters collected during tests are as objective as possible, precisely to prevent them from being distorted by factors purely individual. This will lead to the use of automatic functions that support observer rating (e.g. autonomous eyes motion capture or tracking of commands given by the pilot) in future testing phases;
- A further and more complex (i.e. using intrusive testing methodologies) research about the connection between user's attachment style and their ability to successfully teleoperate a system has to be made.

ACKNOWLEDGEMENT

The authors would like to thank the management, the researchers and the engineering staff *of Thales Alenia Space Italia* (COSE-Center) for the essential technical support during implementation and testing of the HMI concept.

REFERENCES

Bartholomew, K. and Horowitz, L.M., 1991, "Attachment styles among young adults: A test of a four-category model". Journal of Personality and Social Psychology, Vol. 61, No. 2, pp. 226-244. doi:10.1037/0022-3514.61.2.226.

Biesiadecki, J.J., Leger, P.C. and Maimone, M.W., 2007, "Tradeoffs Between Directed and Autonomous Driving on the Mars Exploration Rovers". International Journal Of Robotic Research, Vol. 26, No. 1, pp. 91-104. doi: 10.1177/0278364907073777.

Bowlby, J., 1973, Attachment and loss. Vol. 2, Separation. New York.

Bruni, R., 2004, "Nuovi scenari della genitorialità: la famiglia tra affetti e legami". Istituto Italiano di Medicina Sociale-Azioni di sostegno alla genitorialità, pp. 145-184.

Burridge, R.R. and Hambuchen, K.A., 2009, "Using prediction to enhance remote robot supervision across time delay". IEEE/RSJ International Conference on Intelligent Robots and Systems (IROS). doi: 10.1109/IROS.2009.5354233.

Card, S.K., Moran, T.P. and Newell, A., 1983. The psychology of human-computer interaction. Lawrence Erlbaum Associates Inc. NJ.

Chu, R.W., Mitchell, C.M. and Jones, P.M., 1995 "Using the operator function model and OFMspert as the basis for an intelligent tutoring system: towards a tutor/aid paradigm for operators of supervisory control systems". IEEE Transactions on Systems, Man and Cybernetics, Vol. 25, No. 7, pp. 1054-1074. doi: 10.1109/21.391287.

Corujeira, J., 2013, "A situation awareness interface for a bi-wheeled industrial hovercraft: design, development and evaluation". Master Thesis, Universidade de Madeira, Funchal, Portugal.

Eberts, R.E., 1994, "User Interface Design". Engelwood Cliffs. Prentice Hall, NJ.

Elm, W.C., Potter, S.S., Gualtieri, J.W., Easter, J.R. and Roth, E.M., 2003. "Applied Cognitive Work Analysis: A Pragmatic Methodology for Designing Revolutionary Cognitive Affordances", In: Hollnagel, E. (Ed.), Handbook for Cognitive Task Design, Lawrence Erlbaum Associates, London, pp. 357-382.

Endsley, M.R., 1995a, "Measurement of SA in dynamic systems". Human Factors: The Journal of the Human Factors and Ergonomics Society, Vol. 37, No. 1, pp. 65–84. doi: 10.1518/001872095779049499.

Endsley, M.R., 1995b, "Toward a theory of SA in dynamic systems". Human Factors: The Journal of the Human Factors and Ergonomics Society, Vol. 37, No. 1, pp. 32–64. doi: 10.1518/001872095779049543.

Endsley, M.R., Sollenberger, R. and Stein, E., 2000, "Situation Awareness: a Comparison of Measures". Proceedings of the Human Performance, Situation Awareness and Automation: User-Centered Design for the New Millennium.

Gordon, S.E. and Gill, R.T., 1997, "Cognitive Task Analysis", In: Zsambok, C.E., Klein, G. (Eds.), Naturalistic Decision Making, pp. 131-140.

Hashimoto, S., Ishida, A., Inami, M. and Igarashi, T., 2011, "TouchMe: An Augmented Reality Based Remote Robot Manipulation". The 21st International Conference on Artificial Reality and Telexistence, Proceedings of ICAT2011, Osaka.

Hutchins, E.L., Hollan, J.D. and Norman, D.A., 1985, "Direct manipulation interfaces". Human Computer Interaction, Vol. 1, No. 4, pp. 311-338. doi: 10.1207/s15327051hci0104_2.

Narayanan, S., Ruff, H.A., Edala, N.R., Geist, J.A, Patchigolla, K.K, Draper, M. and Haass, M., 2000, "Human-integrated supervisory control of uninhabited combat aerial vehicles". Journal of Robotics and Mechatronics, Vol. 12, No. 6, pp. 628-639.

Nielsen, J., 1994, "Usability Inspection Methods". Conference Companion on Human Factors in Computing Systems, pp. 413-414. doi: 10.1145/259963.260531.

Norris, J.S., Powell, M.W., Fox, J.M., Crockett, T.M. and Joswig, J.C., 2009, Cliffbot Maestro, NASA Tech Briefs.

Okura, F., Ueda, Y., Sato, T. And Yokoya, N., 2013, "Teleoperation of mobile robots by generating augmented free-viewpoint images". IEEE/RSJ International Conference on Intelligent Robots and Systems (IROS), pp. 665-671. doi: 10.1109/IROS.2013.6696422.

Paas, F., Renkl, A. and Sweller, J., 2004, "Cognitive Load Theory: Instructional Implications of the Interaction between Information Structures and Cognitive Architecture". Instructional Science, Vol. 32, No. 1-2, pp. 1-8. doi: 10.1023/B:TRUC.0000021806.17516.d0.

Rasmussen, J., 1983, "Skills, rules, and knowledge; Signals, signs, and symbols, and other distinctions".IEEE Transactions on Systems, Man and Cybernetics, Vol. 13, No. 3, pp 257-266. doi: 10.1109/TSMC.1983.6313160.

Rasmussen, J., 1986, "Information Processing and Human-machine Interaction: An Approach to Cognitive Engineering". System Science and Engineering, Elsevier Science, New York.

Rasmussen, J. and Vicente, J.K., 1989, "Coping with human error through system design: Implications for ecological interface design". International Journal of Man-Machine Studies, Vol. 31, No. 5, pp. 517-534. doi:10.1016/0020-7373(89)90014-X.

Ricco, C., 2009, "La matrice relazionale del sé: fallimento genitoriale e trauma evolutivo nei figli". In: Borgogno, F., Vigna Taglianti, M. (a cura di), Identificazione e trasmissione psichica fra gli adulti e il bambino. Quaderni di Psicoterapia Infantile, Borla, Roma.

Störrle, H., 2010, "Model Driven Development of User Interface Prototypes: An Integrated Approach". Technical University of Denmark, Copenhagen, Denmark.

Vicente, K.J., Christoffersen, K. and Hunter, C.N., 1996 "Response to Maddox critique". Human Factors, Vol. 38, No. 3, pp. 546-549. doi: 10.1518/001872096778702051.

Extending the Student Qualitative Undertaking Involvement Risk Model

Jeremy Straub[1]

ABSTRACT: The Student Qualitative Undertaking Involvement Risk Model (SQUIRM) was designed to facilitate the determination of the impact of using student (or inexperienced) workers, on a project. The model identifies several prospective categories of risk. It, then, discusses the risk potential and source and provides a limited consideration of how to mitigate this risk. The risk sources considered included those specific to student (inexperienced worker) involvement, standard risks and standard risk sources which are enhanced by the use of student (inexperienced) workers. This paper presents a qualitative assessment framework and begins the process of quantifying the model. The difference between the use of students (in an academic or industrial setting) versus inexperienced workers is also considered. The base model is presented and extended by further tracing the risk sources back, using root cause analysis techniques. The application of the base and extended models to various projects is discussed. Considerations in choosing which model to use for a given application are also presented. The paper concludes by presenting a value model for considering student (inexperienced worker) involvement benefits versus associated risks, and the differences in the risk reward ratio between academic, internship and junior worker scenarios.

KEYWORDS: Risk model, Student workers, Small spacecraft, Aerospace education.

INTRODUCTION

Student involvement in research and other projects is common at universities around the world. Through internships, part-time work and other mechanisms, students also perform limited work for commercial, governmental and other employers. Despite the prevalence of student involvement in the development of key technologies and their performance of numerous duties, the management literature contains little consideration of the specific risk elements introduced by student workers. Inexperienced workers (including students, interns and junior employees) have particular characteristics that may create new risk sources and alter the likelihood and magnitude of typical risks.

An understanding of the impact of using student (and other inexperienced) student workers is particularly important in the case of aerospace projects due to the low defect tolerance, inaccessibility and criticality of many projects. Small spacecraft, for example, are commonly integrated as secondary payloads on rockets carrying other orders-of-magnitude more expensive hardware. They must meet the same (or perhaps even more stringent) integration standards as the primary payload. Some small spacecraft have also been launched via the International Space Station, necessitating their compliance with human safety standards. Once they are in orbit, they are also on their own, with no practical servicing capability. Design and implementation failures can, thus, cause a spacecraft to fail integration testing and not get launched, to fail subsequent to integration and damage expensive equipment or pose a threat to astronauts or fail on orbit, impairing mission performance. The training and the research provided by these efforts is integral to

1.University of North Dakota – Grand Forks/ND – USA

Author for correspondence: Jeremy Straub | University of North Dakota | 3950 Campus Road, Stop 9015, Grand Forks/ ND | 58202-9015 | USA | Email: jeremy.straub@my.und.edu

developing new technologies as well as training the next generation of aerospace professionals. Given this, a better understanding of the risks posed by student and inexperienced staff involvement is necessary.

In prior work (Straub *et al.*, 2013b), a risk model specifically targeted at students (and to some extent, at all inexperienced workers) was presented, called the Student Qualitative Undertaking Involvement Risk Model (SQUIRM). This paper, while discussing and classifying numerous types of risk, did not (due to space limitations) present an evaluation of the model nor demonstrate its application to any particular scenarios.

This paper picks up where the previous work left off. It expands on the prior work in several ways. First, it presents an enhanced model that augments the base SQUIRM framework with root cause analysis, resulting in a more detailed consideration of student status on typical (non-student) risk factors. The use of this model can provide a more robust evaluation of the impact of student participation, as compared to the base model. However, it is not a panacea, and prospective tradeoffs between the use of the two approaches are discussed. Second, it begins the process of quantifying the SQUIRM and extended SQUIRM frameworks, discussing how the models can be used in order to assess risks (considering likelihood, impact and the mitigation techniques employed) on a single project basis or across multiple projects. Third, it presents a value model for evaluating the participation of student (and other inexperienced) workers. This model facilitates the determination of the value proposition of using this type of staff, which can be compared to increased risks and other associated costs. Finally, the differences between types of inexperienced workers are briefly discussed, before concluding.

BACKGROUND

This section provides an overview of areas that the current work benefits from a wealth of prior work in. Despite a growing contemporary interest, the tasking of trainee or inexperienced workers to real-world projects is certainly not a new phenomenon. Apprenticeship-style training has been used throughout history (Elbaum, 1989; Snell, 1996). Modern approaches, however, combine formal and experiential techniques. One relevant technique is project-based learning. In the remainder of this section, the benefits of project-based learning are, first, discussed. Next, prior work, regarding assessment of the value of students to faculty efforts, is briefly considered. Finally, a brief discussion about risk perception is presented.

PROJECT-BASED LEARNING

With project-based learning (PBL), students are involved in hands-on projects that could be developed specifically for a course or which might feature student involvement in faculty research or other real-world projects. PBL has been shown to be an effective instructional tool at all levels of education: from collegiate graduate-level to primary school level (Brodeur *et al.*, 2002; Hall *et al.*, 2002; Mathers *et al.*, 2012; Mountrakis and Triantakonstantis, 2012; Nordlie and Fevig, 2011; Straub *et al.*, 2013a). It has also been demonstrated across a wide variety of subject disciplines, including project management (Pollard, 2012), psychology (Dahlgren and Dahlgren, 2002), physics (Duch, 1996), computer science (Broman *et al.*, 2012; Correll *et al.*, 2013), mathematics (Roh, 2003), engineering entrepreneurship (Okudan and Rzasa, 2006) and aerospace (Jayaram *et al.*, 2010; Saunders-Smits *et al.*, 2012), computer (Qidwai, 2011), electrical (Bütün, 2005; Ribeiro, 2008) and mechanical (Coller and Scott, 2009; Robson *et al.*, 2012) engineering.

In addition to teaching subject-specific skills, PBL projects can teach students how to work with those outside their specific discipline, as is required in the vast majority (Hayne *et al.*, 2012) of workplaces. Gaining a shared prior knowledge base (such as through PBL techniques) can improve team efficiency (Hayne *et al.*, 2012). Workers with interdisciplinary skills are in demand (Sulaiman *et al.*, 2010); PBL also provides students with an opportunity to learn "soft" skills which are required for workplace success (Jackson and Hancock, 2010).

PBL has also been shown to have a beneficial impact on student motivation (Doppelt, 2003), self-image and creativity (Ayob *et al.*, 2012) and material retention (Bauerle and Park, 2012). Field-based/realistic-environment PBL has been shown to increase students' understanding of course materials (Simons *et al.*, 2012). Nagda *et al.* (1998) show that one type of PBL, research participation, can also improve student retention, particularly for at-risk students. The benefits of PBL to student placement, after graduation, have been demonstrated by Hotaling *et al.*, (2012) and Fasse *et al.*, (2012). Gilmore (2013) even argues that techniques such as PBL, for teaching STEM disciplines, are critical to national prosperity.

In aerospace engineering and related disciplines, many students are gaining practical experience working on small spacecraft and high altitude ballooning projects. The SQUIRM framework (Straub *et al.*, 2013b) was created, initially, to assess the risks applicable to student involvement in a small spacecraft project; however, it is useful for many applications beyond this. The utility of PBL for teaching aerospace engineering (Straub *et al.*, 2013a; Straub and Whalen, 2013), software development for aerospace applications (Straub *et al.*, in press) and providing other benefits (Swartwout, 2004; Swartwout, 2011) has been demonstrated. CubeSat projects have been demonstrated to be an effective pedagogical approach (Larsen and Nielsen, 2011; Larsen *et al.*, 2013; Straub, 2013).

The level of the aforementioned benefits, Zydney *et al.* (2002a) proffer, increases with the duration of participation. However, not all students reach these higher levels of benefit; while numerous reasons for premature termination of student participation in a research project exist, manifestation of the risk factors discussed in a subsequent section may explain some of the incomplete experiences.

VALUE OF STUDENT INVOLVEMENT TO FACULTY RESEARCH

If student involvement's benefit was solely student education, the need to characterize and mitigate risks would be dramatically reduced. The impact of a student/inexperience-specific or general risk factor's occurrence can have impact to the student participant's success; it can also have a pronounced effect on the project as well. While students may gain (possibly even enhanced) benefit from risk actualization, the project stands to suffer. To characterize the magnitude of impact, it is important to consider faculty perceptions of student involvement on research projects. Zydney *et al.* (2002b) proffer that faculty see students' participation as valuable, with over half of them indicating that students' contribution to their work was "important" or "very important". Thus, the failure of a student to make progress is a risk that may be comparable to causing damage or other types of impact on prior work. While student participation is valuable to faculty, it appears that project completion may be less important to students, as Prince *et al.* (2007) demonstrated a lack of correlation between the research productivity level of faculty and students' educational benefits.

RISK PERCEPTION

One reason that student workers may be more risk-occurrence prone is a failure to properly assess risk likelihood and impact. However, despite a significant correlation between youth and inexperience, it is important to note the potentially confounding impact of risk perception. Because of this, there may be a performance difference between younger and older individuals with similar experience levels in a field. A full exploration of the topic of risk perception is far beyond the scope of this paper; however, reviews of areas of this topic are readily available. Botterill and Mazur (2004) provide a general overview of the topic, while Slovic *et al.* (1982) consider the value of studying it. Boholm (1998) reviews and compares risk perception research over a twenty-year period and Mitchell (1995) considers risk perception and risk reduction in the context of an organization.

The crux of the risk perception problem is that younger individuals may fail to appreciate the applicability of risk to them and its impact (Weinstein, 1984). This has been documented across multiple areas, including driving (Deery, 1999), sexual (Levinson *et al.*, 1995) and other "health-threatening" (Cohn *et al.*, 1995) behaviors. Steinberg (2004) attributes the greater risk-taking tolerance of youth to "age differences in psychological factors that influence self-regulation". Thus, age may confound the experience/risk correlation, and intensify certain risk factors when both young age and inexperience are the case. Given this, traditional-age undergraduates may have a higher propensity to fail to see how their actions, behaviors or inaction may create risks, or the impact that these risks may have on them or others.

Risk perception, however, is not only affected by age. Correlation has been shown with gender (DeJoy, 1992), culture (Rippl, 2002) and other factors (Sjöberg, 2000; Wildavsky and Dake, 1990). The impact of education in correcting risk perceptions has been demonstrated by Ronan and Johnston (2001). Weber and Milliman's (1997) work suggests that "risk preference" may be a stable aspect of an individual's personality, highlighting the importance of risk perception on the acceptance or rejection of the risk in a given circumstance. Renn (1998) discusses the importance of risk perception in relation to the management of risks.

THE STUDENT QUALITATIVE UNDERTAKING INVOLVEMENT RISK MODEL

The following subsections, reprinted with minimal modification from Straub *et al.* (2013b), provide an overview of the risk categories of the SQUIRM framework (which is depicted in Fig.1). First, technical, schedule and other standard risks will be discussed. Then, the risks posed by student worker involvement will be considered.

TECHNICAL, SCHEDULE AND OTHER STANDARD RISKS

Every project, including those involving students, must deal with numerous possible risk factors. Project managers attempt to control many of these risk factors, assume others, and they are, ultimately, forced to ignore a large set of risks that they have no insight into or control over. Numerous standard risks are well documented in the literature and will not be reviewed in detail here. The impact of student participation on these standard risks is considered. For each risk factor, a brief description of its nature is provided. This is followed by a discussion of how the risk factor is influenced by or may influence student project involvement.

TECHNICAL RISK

The technical risk category is comprised of the set of risks that could result from a failure of hardware and software or its integration and operations to perform as required to meet project's objectives. Three aspects are considered: construction/fabrication of assemblies, failure of purchased components and their integration.

Construction/Fabrication

Construction and fabrication risks are inherent to any manufacturing process. Quality control processes, including those designed to prevent defects as well as those to detect and remediate defects, are generally included to mitigate these risks. In a student project, which generally doesn't involve mass-production, one is confronted with two primary risks. First, standards-based quality control may be cost-prohibitive to implement. Second, students who lack knowledge and understanding of the characteristics of the product may be poorly equipped to detect and evaluate the significance of errors.

Component

Components obtained from suppliers will occasionally be defective, either due to manufacturing or shipping issues. Production processes generally incorporate an acceptance testing procedure or supplier process validation procedure. A student-involved project, generally, suffers from two risk factors regarding components. First, the limited production (in many cases, producing only a single or small quantity of units) precludes the implementation of a standard quality process. Second, student inexperience may result in a failure to properly design acceptance tests or to detect latent issues.

Integration

The process of combining components together introduces risks due to design and implementation failures. Design failures may result in a system, which, regardless of how well it is assembled, cannot perform the desired task. Implementation issues may result in degraded performance, non-operation, or failure after a period of time operating. Student designers and workers generally have traits that significantly increase the probability of these risks happening. Having an incomplete or largely untested understanding of the design process or specific design elements may result in wholly unworkable designs or designs with latent and hard-to-detect flaws. Limited time and resources will generally result in a comparatively lower level of testing being conducted. The fact that this testing will likely be performed by inexperienced (student) testers further exacerbates the problem. Even if a perfect design is produced, inexperience in the techniques required for construction may result in sub-par construction, component attachment and solder connection issues, and so forth. These may cause the assembly to not work initially or to be prone to failure.

SCHEDULE RISK

Every project faces the possibility that its schedule will not be met. External factors, such as the unavailability of key components, and internal factors, such as staff absences or equipment failure, may result in delays. When these delays impact the critical path, the project schedule is impaired. Key areas of consideration for projects involving students include schedule estimation error, critical path risks and schedule creep.

Schedule Estimation Error

Estimation error occurs when the time projected for task completion is different than actual task completion. A certain amount of error is to be expected; however, when tasks are consistently taking longer than projected, the project's schedule is at risk. Estimation error is common, even for experienced estimators. Students, who do not have significant experience, may fail to consider anything other than the best-case scenario. Alternately, they may not completely understand the process that they are estimating and, thus, omit the time required for overlooked process components. Either of these may result in (possibly dramatic) under estimation. On the other hand, students may be overwhelmed and wildly overestimate, (so as to avoid the pitfalls of underestimation). This is, however, problematic, as it may result in the project's momentum being lost, if materials, tools or staff for subsequent phases are not available when a previous phase is completed early.

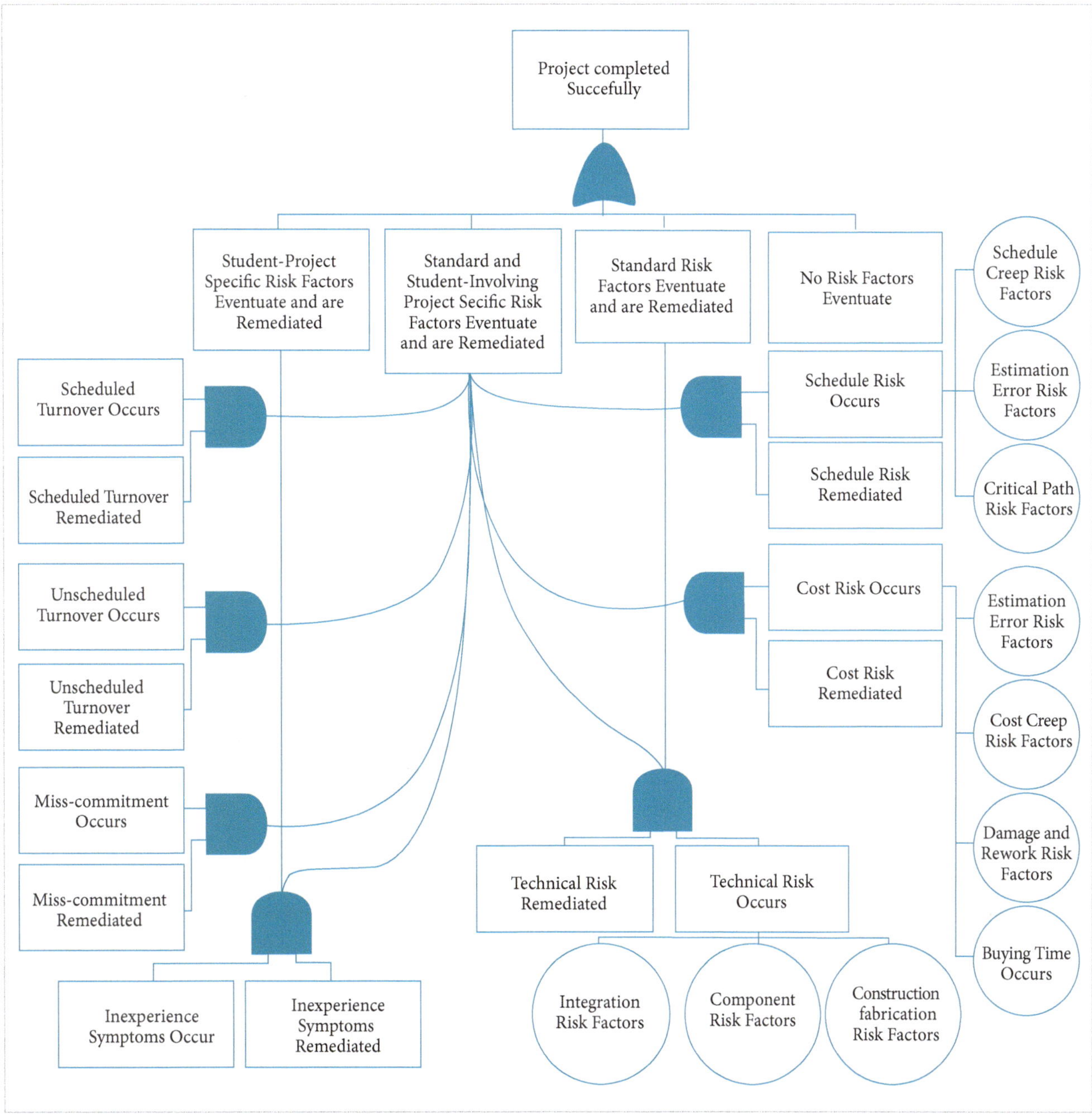

Figure 1. SQUIRM Model Diagram (Straub *et al.*, 2013b).

Critical Path Risk

Critical path risk is a set of risk factors that impact the chain of tasks, which, in succession, take the longest amount of time. As the project is not complete until all of these tasks are done, anything that elongates the schedule of a task on the critical path (or another task, which becomes a critical path task due to schedule overrun) affects the project's overall schedule. Critical path risk can be created by factors that are both external and internal to the project. External factors may include impairment to the availability of supplies, unavailability of key equipment at the needed time, changes in laws or regulations and many other factors. Internal factors, however, are the primary area where projects with student involvement differ from conventional projects. Internal issues that may be exacerbated by student involvement include staff availability issues, delays caused by quality failures — and, thus, the need to repair or recreate the improperly produced items —, and delays caused by poor scheduling. Staff availability and quality issues are discussed in other areas of the model. Poor scheduling may be the result of a failure to identify precursor and successor tasks due to failing to identify required task inputs and outputs or, more simply, error in the actual creation of the schedule. Either of these can easily occur when a schedule is produced by an inexperienced scheduler.

Schedule Creep

Schedule creep is the schedule component of scope creep. Scope creep occurs when changes or documentation issues result in a more robust product being produced than the one called for by planning. The involvement of students, who are generally eager to please and may not understand the impact of accepting changes (or not understand that they are implicitly accepting a change), increases the risk of schedule creep. The fact that most academic projects are run by professors who are trained as researchers - not project managers - and may have limited documentation further exacerbates this risk.

COST RISK

With tight budgets and long-duration funding cycles, cost overrun is a significant risk to student-involved projects. Cost overruns can lead to reduced deliverable utility and/or quality. If severe enough (and supplemental funding cannot be sourced), they can even lead to project termination and failure. Risks that must be considered relative to student involvement include estimation error, cost creep, damage and rework costs, and costs associated with meeting schedule requirements.

Cost Estimation Error

Cost estimation error closely mirrors schedule estimation error. It occurs when the level of cost required to be incurred for a given activity is different from the level forecast. While variation is expected, proper estimation should result in some tasks concluding with small overruns and others being completed under budget. Generally, an allowance for unexpected costs is included in the budget as a separate line item to allow the absorption of additional costs, should the project average out to a slight overrun. As with schedule estimation error, students who may be estimating costs for the first time (or may have limited domain experience, even if they have performed cost estimating before) may be prone to underestimate, due to ignoring complexity or inadvertently omitting various types of costs or specific costs.

Cost Creep

Cost creep is the cost component of scope creep. Scope creep occurs when changes are accepted without commensurate changes in budget and schedule. Due to student inexperience and other factors, scope creep is likely on student projects. If scope creep occurs, it is likely that cost creep will occur.

Damage and Rework

Damage and rework costs are incurred when hardware, facilities, supplies or the item being created are damaged due to carelessness, accident, misuse or otherwise. Damage and rework costs are likely on a student-involved project. First, the lack of a production environment designed for the repetitive production of an item means that construction and integration jigs will be setup on the fly. This may result in inadvertent loss of control, dropping, or the application of unwanted torques or pressure to parts or assemblies. Second, the lack of a repetitive production environment means that there is not a set of well-tested task instructions that can be followed. Third, supply and equipment limitations may result in jury-rigging of various jig-elements, making damage more likely. Forth, horseplay or carelessness may result in damage. All of the aforementioned are exacerbated by having young and/or inexperienced individuals working on the project.

Buying Time

Costs can be incurred to resolve schedule issues. For example, a component could be purchased, at additional expense, to return the project to schedule or an external consultant could be hired to expedite a process. Due to this, schedule issues can become cost issues. Given how student involvement can exacerbate schedule risk, it would seem that student involvement would heighten the possibility of transferring schedule overruns to cost in order to hit a key deadline.

RISKS POSED BY STUDENT WORKER INVOLVEMENT

Several risk factors are impacted so dramatically by student involvement as to deserve separate consideration from their standard counterparts. Each is now discussed in detail.

Scheduled Turnover

Scheduled turnover has a dramatic impact, but can be planned for. It is attributable to the fact that students only participate in a given effort for a period of time. When this participation ends the student may be unavailable to provide documentation or assistance related to their work on the project. As students become task-experts, if documentation is not stressed, understanding can be lost — or a key component of an integrated system can become unserviceable. Compounding this issue is the fact that many students are not adept in documenting their work and lack an understanding of the need for documentation and what needs to be documented. Mitigation strategies for this risk include knowledge distribution, stressing documentation throughout a project's lifecycle, and validating the usefulness of documentation, by requiring its use prior to a student-worker's departure.

Unscheduled Turnover

Unscheduled turnover is a risk factor present in all types of organizations. As in corporate work environments, medical, personal and other factors may necessitate a worker's immediate departure from the workplace. Mitigation techniques for this class of risk include duplication (or responsibility distribution) of key roles, wide knowledge distribution, and stressing documentation and documentation validation.

Miss-commitment

Students' miss-commitment can be more problematic than the occurrence of turnover. With turnover, the project leader has knowledge of the current status of the team member. With miss-commitment, the individual is still present and ostensibly working on their assigned tasks; however, due to conflicting demands for limited time resources (and the academic-trumping of most project duties) the student worker may not have time to make the requisite level of project progress. This is compounded by the cramming-centric work styles learned by many students, which lead to the belief that everything can be 'made up' at the last moment. With student miss-commitment, project leaders may not become aware of the issue, until investigating the cause of a key deadline being missed. Mitigation techniques for this class of risks include defining tasks to have demonstrable milestones, creating an environment where challenges are reported instead of obfuscated, and involving multiple individuals in key tasks.

Inexperience

Inexperience is, of course, a problem that is faced by numerous projects in every sphere. A team member may be new to the workforce, or may lack experience in the specific areas required by a project. However, inexperience is a particular issue in student-centric projects as many students lack practical experience. This translates into misestimating and a lack of experience in problem resolution techniques. This class of risks can be mitigated by training students in the desired behaviors (e.g., how to estimate in a given sphere, how to deal with problems, etc.). This mitigation not only benefits the project, but also prepares the students for workplace entry.

EXTENDING THE MODEL WITH ROOT CAUSE ANALYSIS TECHNIQUES

The original SQUIRM model, presented in prior work (Straub *et al.*, 2013b), expanded upon the causal factors for standard risks, which could be exacerbated by student/inexperienced workers' involvement. While some discussion of the causality of the student-worker-specific risks was included, these were not incorporated into the formal model. The SQUIRM-Extended Model (SQUIRM-E) adds these causal factors to the model, as shown in Fig. 2. This addition is necessary to begin quantitative assessment using the model (which is discussed in a subsequent section). This section begins with a discussion of the value of the use of root cause analysis and then the rest of this section discusses the new elements of the SQUIRM-E model and expands upon the types of risks posed by them and their causes.

ROOT CAUSE ANALYSIS

The premise of root cause analysis (RCA) is that a better understanding of the underlying factors of an exceptional occurrence (either positive or negative) facilitate a better understanding of how negative occurrences can be avoided in the future and positive occurrences brought about. Significant prior work exists in this area; a high-level overview is provided by Rooney and Heuvel (2004). RCA has been used for process analysis (Weidl *et al.*, 2005), investigating medical error (Iedema *et al.*, 2006), improving patient safety (Neily *et al.*, 2003), as well as in analyzing and improving industrial safety and performance (Carroll *et al.*, 2002). A discussion of several tools for RCA was presented by Doggett (2004).

In the context of this work, RCA was used to assess why student-involved projects and student workers could have higher levels of risk actualization than a similar project not incorporating inexperienced workers. In prior work (Straub *et al.*, 2013b), this was applied to seek out causes that were specific to student (and other inexperienced) workers. In this paper, RCA is used to decompose standard risk factors to assess the prospective contribution of inexperience and related factors on these risk areas.

RCA is not the only technique that could be used to assess these types of risks. However, it has several benefits. Unlike some other approaches, for example, it uses a bottom-up approach which makes it suitable for projecting risks instead of analyzing actualized risks. This is particularly valuable in the context of non-operations risk analysis, where prior occurrences in a recurring process cannot be analyzed to project future risk factors and their likelihood. With RCA, the individual factors contributing to each type of prospective risk have been identified. These can, then, inform planning (in order to facilitate avoidance and mitigation) as well as be used to arrive at an understanding of the risk level of a project and its areas of particular risk. To perform RCA, prospective sources of the higher-level risk factors previously presented were identified. These are described in greater detail throughout the remainder of this section.

INEXPERIENCE SYMPTOMS OCCUR

The risk categories related to inexperience are a lack of attention to detail, lack of self-motivation, uncertainty as to how to perform a task, overconfidence that causes failure and problems with the work environment. These are now discussed.

Lack of Attention to Detail

Student workers may lack an understanding of the importance of particular details of a task, lack an understanding of the actual details (i.e., what is a correct implementation at a detailed level versus an incorrect one), or may simply fail to pay the level of attention required. This may be exacerbated due to other time commitments (reducing the amount of time that can be devoted to these details and task performance), the level of strain that the student is under (particularly if the student lacks coping mechanisms), and other factors (such as the amount of time available during the semester, etc.).

Lack of Self-Motivation

Students (particularly lower-level undergraduates) may not yet have developed the skills, habits and work ethic required to self-motivate work when tasks seem unexciting or are in support of a longer-term goal. This may translate into unsatisfactory performance in terms of meeting deadlines, unsatisfactory work product or other deficiencies. It may also trigger or contribute to other risk factors (such as miss-commitment if work piles up due to not starting things until there is an imminent due date, etc.).

Unsure of How to Perform Task

Students may be unsure of how to perform particular tasks or elements of a task. This may translate into delays waiting for clarification, attempts that result in wasted materials and time, obviously defective products or products with latent defects that may impair progress during later phases (e.g., integration, testing). This lack of understanding may decrease motivation, increase frustration and delays may trigger other issues such as miss-commitment.

Overconfidence Causes Failure

Students may underestimate the difficulty of a task or overestimate their own capabilities. This can have several different symptoms, depending on when it occurs. First, it can cause issues with scheduling and costing. Students may underestimate the amount of time that will be required

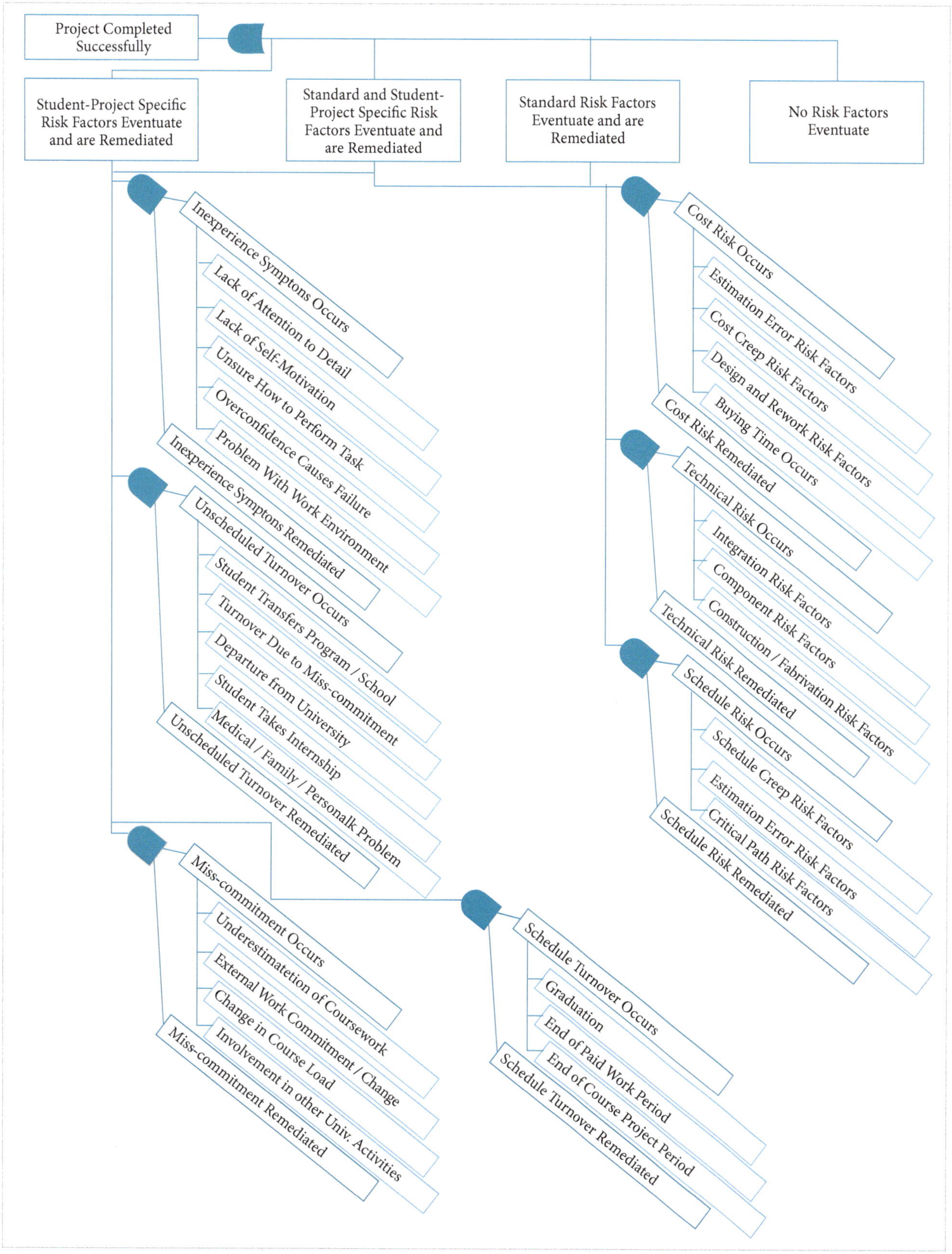

Figure 2. SQUIRM-E Model Diagram.

for learning how to perform a task, experimenting to gain understanding and/or correcting less-than-acceptable products. They may also underestimate the amount of waste material that may be consumed by reattempts to fix defects.

Second, it can result in unsatisfactory performance in terms of meeting deadlines, unsatisfactory work product or other deficiencies due to the aforementioned scheduling and the reality of performance conflicting, or a lack of understanding of what an acceptable product is, triggering a need for significant rework. This may translate into delays waiting for clarification, attempts that result in wasted materials and time, obviously defective products or products with latent defects which may impair progress during later phases (e.g., integration, testing). These issues may trigger other risk factors such as miss-commitment, decreased motivation and increased frustration.

Third, this may result in students responding negatively to feedback, as they think that it is unnecessarily critical (based on their inaccurate assumptions about their own capabilities and what constitutes an acceptable level of performance). This may also increase frustration, decrease motivation and potentially trigger other issues, such as turnover.

Problem with Work Environment

Student workers may lack an understanding of how to cope with difficulties in the workplace environment. For example, they may not understand how to deal with a poor manager (and the, particularly if a student, manager may lack the skills and understanding required to resolve this conflict). They may also lack the skills required to resolve workplace conflict or to collaborate with others in the work environment. This can potentially trigger miss-commitment, if work is left to pile up while issues are being resolved, or if unscheduled turnover occurs.

UNSCHEDULED TURNOVER OCCURS

Unscheduled turnover can be caused by a student transferring between degree programs or colleges/universities, as a result of miss-commitment, because of a student's departure from the university, or even by a student taking an internship or a medical, family or other personal problem. Each is now discussed.

Student Transfers Program/School

In the context of their educational pursuits, students make decisions in light of what they perceive as their own best interests (which may consider short and/or long term goals). The inflexibility of the semester system may limit students' ability to provide notice (even for a paid position), should they decide to transfer between schools or programs. They may also lose interest at the point that they realize that program participation is no longer supporting their goals (framed now in terms of their new school/department). This may result in low or no-notice turnover.

Turnover due to Miss-commitment

Students may miss-commit (reasons for this are discussed subsequently). If this miss-commitment becomes an acute problem, students may terminate their involvement in paid and/or unpaid extracurricular activities in deference to their immediate academic time needs. This may occur with low or no notice or it may simply result in the student failing to show up (without any sort of explanation).

Departure from University

Students may leave (or be dismissed from) the university for a wide variety of reasons. This may also result in low or no-notice turnover.

Student Takes Internship

Students may decide to pursue an internship to increase their skills and/or post-graduation employment opportunities. Internships may pay more than on-campus employment and generally offer work experience benefits and prospective employer contact that on-campus employment cannot. Students may begin an internship with little or no notice (as employers may offer internships at the last minute to meet their needs and funding capabilities); in many cases, however, internships can be a planned absence and a student may be able to/decide to return to the project after its completion.

Medical/Family/Personal Problem

Like any worker, students may suffer from medical family or other personal problems. These may be intensified by students' lack of coping skills and/or the lack of a need to maintain an income, even in the face of a major medical condition. Notice levels, the potential for students to return to the project upon the resolution of the issue and the duration of the issue will, obviously, vary significantly based on the nature of the issue.

SCHEDULED TURNOVER OCCURS

Scheduled turnover is an expected occurrence at a college or university. It can be caused by student graduation, the end of a paid (e.g., extramurally funded) work period or the end of a course project period. Each is now considered.

Graduation

Students enroll in a university with their departure planned (unlike a typical work environment where employees may not plan to make a career out of a job, but also look at it as something to pursue for an indeterminate period of time). Graduation, fortunately, will be an occurrence that is known well in advance and can be planned for to ensure proper handover. Students, however may fail to notify project leaders (either due to a presumption that they should be notified by some other means or to avoid less-interesting handover activities) and/or have a declining level of interest (particularly after they have secured a job or admission into another program for graduate studies, etc.), that may reduce the ability to conduct and/or the quality of handover activities.

End of Paid Work Period

Research grant (or other funding source) work may have a definite cut-off point after which no additional funding is available to continue a position. This creates a known date-of-departure for a student from a project (or a transition from a paid role to continuation on a volunteer basis). This should be known to the investigator (and thus not suffer from the aforementioned failure-to-notify problem) and be able to be planned for. Students may lose interest and/or change their final day if they find an alternate position, as they approach their known final days.

End of Course Project Period

Course projects, like paid work periods, have definite (and known-to-the investigator) end dates. A desire to receive a good final grade, however, may keep students motivated until the end of the period.

MISS-COMMITMENT

Miss-commitment is to be expected with students who may be unable to gauge the level of work required both from their academic, paid work and extramural pursuits. Miss-commitment, thus, can occur due to students' underestimation of coursework time commitments, an external work commitment commencing or changing, a change in a student's course load and/or involvement in other university activities. These are now considered.

Underestimation of Coursework

Students may overcommit to extramural projects or paid on-campus project work, based on an underestimation of the level of time required for their coursework. This may result in delays, turnover or impaired quality.

External Work Commitment/Change

Students who are working on a project in either a paid or volunteer basis may have jobs outside the project or may seek/take a job based on the benefits it may provide (e.g., work experience, employer contact) or due to their personal financial situation. This may result in low or no-notice changes in project involvement levels, turnover or a decline in product quality.

Change in Course Load

Students may change the number or selection of courses they are taking during the semester and this may change somewhat from semester to semester. This may result in turnover, delays, or quality impairment.

Involvement in Other University Activities

Students may decide to pursue other university extracurricular activities in addition to or instead of the project, or the level of involvement required for (or desired in) these activities may change, reducing the students' level of involvement in the project and/or causing delays, quality problems or turnover.

DIFFERENCES BETWEEN AND CHOOSING BETWEEN USING SQUIRM AND SQUIRM-E

With both the SQUIRM framework and its extension presented, the two can now be compared. This section reviews the differences between SQUIRM and its extension, SQUIRM-E. It discusses the benefits of using one versus the other across multiple scenarios.

DISCUSSION OF THE DIFFERENCES BETWEEN SQUIRM AND SQUIRM-E

The fundamental difference between SQUIRM and SQUIRM-E is the addition, in SQUIRM-E, of the decomposition of standard risk classes in order to also consider risk sources attributable to student and inexperienced workers. This has resulted in two models, each of which is better suited for certain applications (as compared to the other). The remainder of this section considers specific benefits of using one model over the other. It begins by discussing the comparative simplicity presented by SQUIRM, versus SQUIRM-E, and where this simplicity may be valuable. Next, it discusses how SQUIRM-E leans further towards student workers, making SQUIRM more suitable for use or adaptation to non-student, inexperienced workers (or students in contexts where the student status is less relevant). Finally, logistical considerations such as project size and assessor environment familiarity are discussed before a concluding discussion regarding model selection.

COMPARATIVE SIMPLICITY

The SQUIRM framework, by abstracting the root causes of the student-specific risk types into larger categories, is comparatively easier to work with. This is particularly useful in cases where real numbers for these risk types are unknown and cannot be accurately estimated, or where data has been collected without sufficient granularity for use with the more granular model. Alternately, those estimating without data may prefer the more detailed model, as it allows them to consider the risk, likelihood and impact for specific prospective problems, without having to consider whole categories at one. The use of the SQUIRM-E framework, thus, would correspond to a bottom-up risk identification strategy, while the SQUIRM framework (for student-specific risk types) would correspond to a top-down risk identification and assessment approach.

TYPES OF INEXPERIENCED WORKERS

While the SQUIRM model contains elements that may be useful for all areas of inexperience, the elaborations in SQUIRM-E have been targeted specifically at student workers (with a particular focus towards student workers working in the context of a university environment). The further that the actual situation diverges from this, the less valuable the SQUIRM-E elaborations may be. Alternately, one might use these as a starting point, removing (and/or replacing) irrelevant topics and making changes as needed to relevant ones that have an incorrect focus for the scenario under consideration.

PROJECT SIZE

For smaller projects or projects that are less critical, there may be less need for and resources with which to perform risk management. In these cases, the use of the simpler model (and, in fact, even simplifying the SQUIRM framework to remove the third-level error sources) may be prudent.

FAMILIARITY WITH PARTICULARS OF STUDENT WORK ENVIRONMENT

Those with greater familiarity with the risks and nature of the student-involved work environment may find less need for the additional granularity of the SQUIRM-E model. However, as some risk types occur infrequently, heuristic models based on past experiences may oversimplify actual risk levels. Alternately, non-university employers that are less familiar with the particulars of student worker risks may desire to use a modified version of the full SQUIRM-E model. This adaptation is discussed in a subsequent section.

CHOOSING A MODEL

While the two models are not that dissimilar, the selection of a model should be based on the complexity of the project as well as particular needs related to assessing student-status-attributable risk factors. Choosing the incorrect version of the model to use may result in oversimplification, under or overstatement of risks and/or unnecessary work.

EXAMPLES, APPLICATION AND COMPARISON

This section presents three examples which are used to aid reader understanding and application of the models, and to compare the SQUIRM and SQUIRM-E frameworks. These examples include a small spacecraft project, a surface rover project and a near-space recovery system development project. Each is now briefly presented, followed by discussion and the presentation of steps for model use.

ENGINEERING A SPACECRAFT PROTOTYPE

A project to build a prototype for a small spacecraft included numerous risk factors, as it was an initial effort for the participants. The project team included students working on it for class credit and several more that were participating as an extracurricular activity. Risk factors for this project included technical factors and personnel factors. There were numerous technical issues that could have presented a problem as the prototype was student designed and fabricated. A printed circuit board ended up being the factor that created a significant schedule impact (and a minor cost impact, which was absorbed by contingency funds). Due to the board not working during testing (after the students had mounted components on the supplier-fabricated-from-student-design board), the prototype was not able to be launched on a high altitude balloon before the end of the spring semester and incurred a significant delay, having to wait until the project team returned in the fall.

BUILDING A SURFACE ROVER

A project to build a rover model, that is in some ways analogous to one that could be used on the moon or on Mars, suffered from significant personnel issues. Students involved in the project, while eager for it to succeed, lacked the knowledge and experience required to bring the project to fruition. When schedule issues occurred (initially with the mechanical design), no strategy was found to rectify them, and the project's schedule continued to slip, impairing numerous successor tasks. While near-heroic efforts were made to attempt to complete the rover during the final days of the schedule, insufficient time was available for testing. The project suffered a final component failure which it was unable to recover from.

TWO ATTEMPTS AT A NEAR SPACE RECOVERY SYSTEM

The Near Space Recovery Technology (NSRT) was proposed as a senior design project for two consecutive years. With senior design projects, students self-select into groups based on selecting a topic that they are interested in working on. The goal of the NSRT project was to create a method to control the descent of a high altitude balloon (HAB) payload. Various approaches for doing this were considered by the NSRT team including timing the balloon burst to maximize the amount of time spent at altitudes with favorable wind patterns and, possibly, to incorporate a mechanism for controlling the rate of descent through various altitudes (again to maximize the time spent at favorable ones). The teams, for both years of the NSRT project, were static, with no students entering late or leaving the team mid-year. During the first year of the NSRT project, the students worked well together and had natural leaders. The first group was competitively more motivated. However, the goals of the project were overly ambitious and were not able to be completed. During the second year, the team was less enthusiastic, not self-driven and lacked commitment to the project. No natural leaders emerged from this group.

The objectives for the second year project were only a subset of those from the first year project; however, in both cases, the team advanced many areas to an approximately 90%-of-completion threshold before the project ended.

DISCUSSION

The examples discussed are all university projects. They, however, span three categories of participant commitment. The first (spacecraft prototype) was largely a volunteer project, though there were a limited number of students working for class credit. The surface rover project was a mixed-mode project, with a significant number of students participating as volunteers, and a significant number participating for class credit. The third (recovery system) was a project for class credit.

All three projects would have been good candidates for the use of the SQUIRM or SQUIRM-E model. The first, due to its scope, duration and complexity would have been able to support the additional overhead of the SQUIRM-E model and benefit from the greater understanding of student/inexperienced staff-specific on standard risk classes that it provides. The second, due to its shorter duration, smaller size and more informal nature, would have been better suited to the qualitative use of the SQUIRM framework. The third falls between the two prior examples and would have required a more detailed analysis of application-specific tradeoffs in determining what model to use. Particularly in the second year, however, an analysis of the risk factors, which eventuated during the first year (while not conclusive), could have aided the decision making process.

The risk models for these projects are, thus, very different. This is suggested by the different types of risks that eventuated. In the first case, high turnover and what eventually became an expected high loss rate between intake and ongoing participant numbers as well as turnover due to students feeling pressures from other areas of their academic pursuits, came to define the risk model. The second suffered from conflicts between the commitment levels of the two groups of students (falling into both the miss-commitment group, from an expectation of participation levels from students in the class participant category with regards to the volunteers, and from unscheduled turnover - for different group-specific reasons - within both groups). Finally, the third suffered largely from technical issues, largely attributable to inexperience and some miss-commitment.

APPLICATION

If similar projects were planned in the future, they could use SQUIRM or SQUIRM-E as appropriate (see above), following a five-step approach.

First, the nature of the project must be defined. A discussion of this is beyond the scope of this article; however, several common frameworks exist, including those by Wertz *et al.* (2011) and Fortescue *et al.* (2011). A simplified version for small high altitude ballooning projects (which could be adapted to other aerospace projects) has also been proposed (Straub and Fevig, 2012). These frameworks incorporate risk analysis in different ways; however, this process — using SQUIRM/SQUIRM-E — should involve the following four steps.

Second, areas of student (inexperienced staff) involvement, areas impacted by student involvement, and areas not impacted by or involving students, should be identified. The use of the SQUIRM/SQUIRM-E model is appropriate for the first two areas; the last one should use conventional risk assessment and management techniques.

Third, a granularity level of risk assessment must be determined, based on the scale and nature of the project. Risk could be assessed at the whole-project-level or at any logical division level thereunder. The granularity level need not be consistent; thus, areas of higher risk or risk impact could be assessed at higher levels of granularity than less risky or impactful areas.

Forth, for each unit of assessment, risk factors should be identified. This will involve application/task-specific brainstorming as well as reviewing the student/inexperienced worker-attributable factors presented by the SQUIRM model. For each factor, a likelihood and impact level should be estimated (based on historic data, experience or other technique).

Finally, any summative assessment should be performed. This may include combining risk data from sub-tasks into task-level assessments (or from tasks into project-level assessment), evaluating student/inexperienced worker participation value and comparing project-level assessments.

The foregoing can be performed qualitatively or quantitatively. Quantitative analysis is discussed in greater detail in the subsequent section.

QUANTIFYING THE MODEL

While the discussion up to this point has been qualitative, both the SQUIRM and SQUIRM-E models lend themselves to being used with quantitative data, if it is available. Figure 3 demonstrates how the identified risk areas, along with mitigation/response strategies identified using the SQUIRM/SQUIRM-E model, can be used to assess the weighted (by likelihood of occurrence) risk impact levels for particular risk sources and for the project overall. The overall project risk levels may serve to facilitate comparison between projects (in conjunction with other metrics such as project importance and cost, etc.).

RISK ASSESSMENT

Risks are assessed both in terms of their likelihood of occurrence and the magnitude of impact that they may have if they eventuate. Risks may be assessed based on probabilities, if sufficient historical data exists or a probabilistic model is known or can be inferred, or they can be categorized (with approximate average probabilities, used to facilitate quantitative comparisons).

The impact can, similarly, be quantified in terms of time, resource and cost (which may be combined into a single cost metric), if data is available. Alternately, they can be categorized and an average value used.

MITIGATION/RESPONSE ASSESSMENT

The risk effect may be altered by the existence (or development) of mitigation and response strategies. Mitigation strategies may reduce likelihood, impact, or both, while response strategies focus solely on reducing

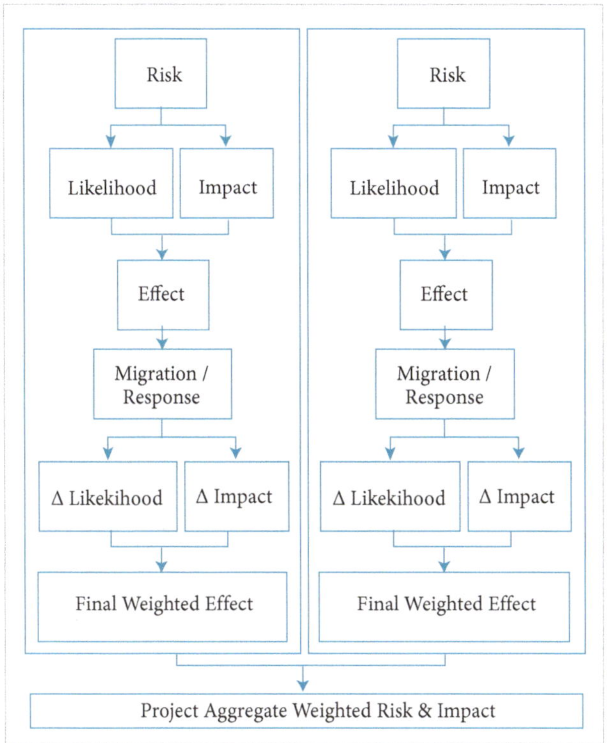

Figure 3. Quantifying SQUIRM & SQUIRM-E.

impact. The change created by the existence of one or more of these strategies should be considered. Again, actual numbers or classifications and average values can be used for this assessment.

COMBINING FOR RESULT

The risk effect and mitigation/response change are combined for each risk factor. Then (if multiple risk factors are present), the final weighted effects are combined, to produce an aggregated risk impact value for the project. It is important, when using this approach, that all values use a comparable scale (e.g., combining average and historical cost values should be done carefully to avoid over or understatement of risks). If risk values are being used to compare projects, then the need for a common scale extends to all items being compared. Thus, it is ideal (but often not practical) to use historical data and (inflation and other factor-adjusted) real costs, as this facilitates direct comparison.

DATA FOR MODEL PARAMETERS

One particular challenge in the use of SQUIRM or SQUIRM-E quantitatively is the collection of the parameters which are required in order to perform the quantitative analysis. Problematically, this data likely varies on an application-specific basis (or general data would need to be validated for application-specific use). While, for small satellites, some relevant data has been collected by Brumbaugh and Lightsey (2013), and they are collecting data (Brumbaugh and Lightsey, 2014) to facilitate a more robust analysis, this doesn't cover all areas required by this model, nor does it help those attempting to assess risk in other application areas. For areas and applications where this data is not available, it will need to be estimated based on past experience and other available information. The collection of data specific to particular applications is an area for future work.

VALUE MODEL FOR INEXPERIENCED WORKERS

The foregoing may lead one to question the value of using inexperienced workers (particularly students) on any project of particular importance. Would the students/junior employees not be better served (and better serve others) by gaining experience through non-impactful learning exercises instead of work on real projects (which could be negatively impacted)? This section considers the value of student (and other inexperienced) workers. Figure 4 presents a diagram of the considerations.

COST OF INEXPERIENCED/STUDENT WORKERS

The cost of inexperienced and student workers is aptly identified by the SQUIRM and (to a greater extent) SQUIRM-E models. Clearly, each prospective risk may impair a project (if it eventuates) incurring time, productivity impairment (including productivity impairment of other more senior workers that may need to help rectify student/inexperienced worker mistakes), material and goodwill costs. Somewhat (in many cases) offsetting, this is the lower wage levels paid to student/inexperienced staff. Thus, for tasks that these individuals can learn to perform effectively and with minimal (or comparable to more experienced staff) oversight, a cost savings may be enjoyed. The assignment of junior staff to these types of tasks, however, may impair their learning process and prevent them from

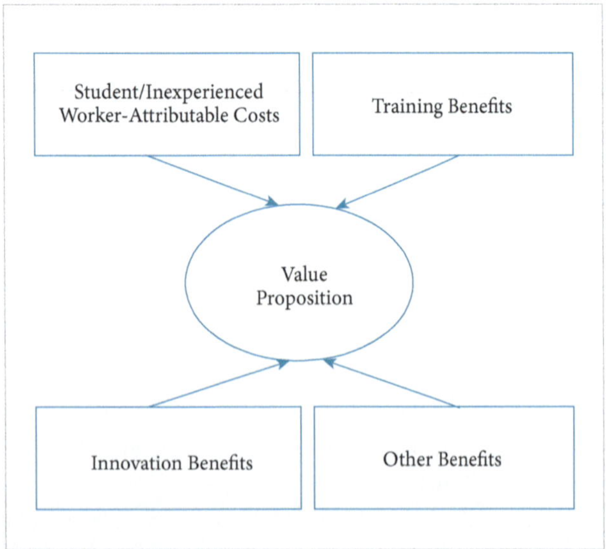

Figure 4. Value Proposition for the Use of Inexperienced and Student Workers.

gaining (or decrease the speed of them gaining) skills that could make them more valuable to their current and future prospective employers.

TRAINING BENEFITS

The proverbial adage of "killing two birds with one stone" can be used in an attempt to justify the use of student/unskilled workers on real projects. If students/unskilled workers can be productively contributing to a project while also gaining experience, it would seem that two types of benefit are being gained for a single cost. While this may certainly be true in some (perhaps many) cases, the oversimplification of the cost model (i.e., the consideration of a "single" cost) may be inaccurate. Costs may be higher to facilitate the student/inexperienced worker participation, which should be taken into account in the comparison.

DISCONTINUOUS INNOVATION BENEFITS

One area where student/inexperienced workers may offer particular benefit is in identifying sources of discontinuous innovation. These workers, who may not fully understand where the proverbial "box" is, may be well-suited to think outside of it. Swartwout (2004; 2011) identifies this, for example, as a key benefit of "university class" small spacecraft programs: the higher level of risk tolerance and the presence of the junior staff make these types of missions well suited to trying innovative ideas and identifying areas for innovation in operations.

DISCUSSION OF THE DIFFERENCES BETWEEN STUDENT VOLUNTEERS, PAID STUDENT WORKERS, INTERNS AND JUNIOR EMPLOYEES

It has been stated, previously, that the SQUIRM and SQUIRM-E frameworks can be used to address risks across several different types of junior employees; however, the risk factor impact posed by these different groups are dissimilar. This section begins the process of considering the differences between the multiple types of workers that the SQUIRM/SQUIRM-E models could be applied to (in some cases with limited modifications). The particulars of each worker type are now discussed, this includes: student volunteers, paid student workers, interns and junior employees.

STUDENT VOLUNTEERS

Student volunteers will (correctly) view their participation as at-will. If they are interested, see benefits being provided and have time, they will continue working on the project. If they lose interest, feel that they are not receiving (or have already received all applicable) benefits or are confronted with other draws on their time, they will stop. Retention of students from semester-to-semester may be difficult, as they may perceive participation as an opt-in activity (like joining a club or taking a class), where a participation decision is made anew each semester. They may fail to realize or understand the impact of their change in participation status on others that have also donated their time to provide benefit to them or the cost of the time committed to their training by paid staff, etc.

PAID STUDENT WORKERS

Paid student workers may be more committed, as they are receiving another source of benefit (pay) over and above what is received by volunteers. However, in the context of the comparatively large amounts of money that they are paying (or which is being paid on their behalf) to attend school, they may see little difference between the paid and unpaid positions in terms of any sense of commitment or longer-term responsibility. Pay, thus, may overcome (or assist in rectifying) lack of interest issues, but may not assist with semester-to-semester turnover issues or commitment in the face of other time draws.

INTERNS

Interns (in the case of non-university employers) may see a multi-faceted benefit which may cause particular (comparative) commitment. The intern may be earning credit for their participation, getting paid, gaining experience and gaining an opportunity to demonstrate their capabilities to a prospective employer. The foregoing (particularly if the intern sees the employer as a desirable place to seek post-graduation employment) may cause interns to place the internship amongst their highest priorities, overcoming most of the common (controllable) risk factors and creating a particularly high level of diligence. Interns may or may not have ongoing coursework during the internship period (the lack thereof reducing another set of risk factors). As a generally fixed-term period of employment, however, scheduled turnover is expected.

JUNIOR EMPLOYEES

Junior employees may see performance as critical to their future livelihood; however, this perception may not always be the case (even if it is accurate, it may not be perceived or employment may be perceived as an entitlement). While most will want to set their careers off on a 'good foot', others may find the change in structure (more or less control, different control structures and a need to be self-starting) problematic and not know how to function effectively under the changed structure. Employees may also be looking for new positions, if they take a position that is not of their liking simply to 'pay the bills' and may lack the professional discipline to continue to perform while in a job they dislike (or which they are not particularly excited about).

CONCLUSIONS AND FUTURE WORK

This paper has expanded the SQUIRM framework into a new SQUIRM-E version that adds additional assessment criteria related to student-specific risk types. It has presented an analytical framework for assessing risk factors, relevant to student and inexperienced workers quantitatively, and evaluating the value of the use of a student/inexperienced worker on a given project. A limited extrapolation to non-student workers has been discussed.

Future work will involve the enhancement of the quantitative models presented as well as the collection of a data set to begin to characterize these common risk areas for various classes of projects. It will also involve the development of a SQUIRM-E-based model for junior employees that replaces student-specific factors with those more appropriate to junior employees.

ACKNOWLEDGEMENTS

Small satellite development work at the University of North Dakota is, or has been, supported by the North Dakota Space Grant Consortium, North Dakota NASA EPSCoR, the University of North Dakota Faculty Research Seed Money Committee, North Dakota EPSCoR (NSF Grant # EPS-814442), the Department of Computer Science, the John D. Odegard School of Aerospace Sciences and the National Aeronautics and Space Administration.

The involvement of the numerous students from multiple disciplines in UND small satellite development is gratefully acknowledged. Also, thanks are given to the numerous faculty mentors who have helped make this possible.

Thanks is given to John Nordlie, who provided a subset of the information used for the examples. Thanks is also due to all of the faculty, students and staff who participated in the examples presented herein.

REFERENCES

Ayob, A., Majid, R.A., Hussain, A. and Mustaffa, M.M., 2012, "Creativity enhancement through experiential learning" Advances in Natural and Applied Science, Vol. 6, No 2, pp. 94-99.

Bauerle, T.L. and Park, T.D., 2012, "Experiential learning enhances student knowledge retention in the plant sciences", HortTechnology, Vol. 22, No 5, pp. 715-718.

Boholm, A., 1998, "Comparative studies of risk perception: A review of twenty years of research" Journal of Risk Research, Vol. 1, No 2, pp. 135-163. doi: 10.1080/136698798377231.

Botterill, L. and Mazur, N., 2004, "Risk and risk perception: A literature review", Project no.BRR-8A, Rural Industries Research and Development Corporation, Barton.

Brodeur, D.R., Young, P.W. and Blair, K.B., 2002, "Problem-based learning in aerospace engineering education", Proceedings of the 2002 American Society for Engineering Education Annual Conference and Exposition, Montreal, Canada, pp. 16-19.

Broman, D., Sandahl, K. and Baker, M., 2012, "The company approach to software engineering project courses", Education, IEEE Transactions On, Vol. 55, No 4, pp. 445-452. doi:10.1109/TE.2012.2187208.

Brumbaugh, K.M. and Lightsey, E.G., 2013, "Application of risk management to university CubeSat missions", Journal of Small Satellites, Vol. 2, No 1, pp. 147-160.

Brumbaugh, K. and Lightsey, E.G., 2014, "CubeSat mission risk survey – a call for response from the small satellite community", Journal of Small Satellites, Vol. 2, No 1, pp. 83-84.

Bütün, E., 2005, "Teaching genetic algorithms in electrical engineering education: A problem-based learning approach", International Journal of Electrical Engineering Education, Vol. 42, No 3, pp. 223-233. doi: 10.7227/IJEEE.42.3.2.

Carroll, J.S., Rudolph, J.W. and Hatakenaka, S., 2002, "Lessons learned from non-medical industries: Root cause analysis as culture change at a chemical plant", Quality & Safety in Health Care, Vol. 11, No 3, pp. 266-269. doi: 10.1136/qhc.11.3.266.

Cohn, L.D., Macfarlane, S., Yanez, C. and Imai, W.K., 1995, Risk-perception: Differences between adolescents and adults, Health Psychology, Vol. 14, No 3, pp. 217-222.

Coller, B.D. and Scott, M.J., 2009, "Effectiveness of using a video game to teach a course in mechanical engineering", Computers & Education, Vol. 53, No 3, pp. 900-912. doi: 10.1016/j.compedu.2009.05.012.

Correll, N., Wing, R. and Coleman, D., 2013, "A one-year introductory robotics curriculum for computer science upperclassmen" Education, IEEE Transactions On, Vol. 56, No 1, pp. 54-60. doi:10.1109/TE.2012.2220774.

Dahlgren, M. A. and Dahlgren, L.O., 2002, "Portraits of PBL: Students' experiences of the characteristics of problem-based learning in physiotherapy, computer engineering and psychology" Instructional Science, Vol. 30, No 2, pp. 111-127. doi: 10.1023/A:1014819418051.

Deery, H.A., 1999, "Hazard and risk perception among young novice drivers", Journal of Safety Research, Vol. 30, No 4, pp. 225-236. doi:10.1016/S0022-4375(99)00018-3.

DeJoy, D.M., 1992, "An examination of gender differences in traffic accident risk perception" Accident Analysis & Prevention, Vol. 24, No 3, pp. 237-246. doi:10.1016/0001-4575(92)90003-2.

Doggett, A.M., 2004, "A statistical comparison of three root cause analysis tools", Journal of Industrial Technology, Vol. 20, No 2, pp. 2-9.

Doppelt, Y., 2003, "Implementation and assessment of project-based learning in a flexible environment" International Journal of Technology and Design Education, Vol. 13, No 3, pp. 255-272. doi: 10.1023/A:1026125427344.

Duch, B.J., 1996, "Problem-based learning in physics: The power of students teaching students", Journal of College Science Teaching, Vol. 15, No 5, pp. 326-329.

Elbaum, B., 1989, "Why apprenticeship persisted in Britain but not in the United States" Journal of Economic History, Vol. 49, No 2, pp. 337-349. doi:10.1017/S002205070000797X.

Fasse, B., Hotaling, N., Forest, C., Bost, L. and Hermann, C., 2012, "The case for multi-disciplinary capstone design: A quantitative analysis of impact on job placement and product quality", BMES Conference, Retrieved from http://pbl.gatech.edu/wp-content/uploads/2013/05/fasse_bmes_2012.pdf.

Fortescue, P., Swinerd, G. and Stark, J., 2011, Spacecraft systems engineering, 4th ed., Wiley.

Gilmore, M., 2013, "Improvement of STEM education: Experiential learning is the key", Modern Chemistry & Applications, Vol. 1, No 3, pp. e109. doi:10.4172/2329-6798.1000e109.

Hall, S.R., Waitz, I., Brodeur, D.R., Soderholm, D.H. and Nasr, R., 2002, "Adoption of active learning in a lecture-based engineering class", Frontiers in Education, 2002. FIE 2002. 32nd Annual, Vol. 1, pp. T2A-9-T2A-15. doi:10.1109/FIE.2002.1157921.

Hayne, S., Pendergast, M. and Smith, C.A.P., 2012 "Team composition, knowledge and collaboration", SAIS 2012 Proceedings, Paper 17.

Hotaling, N., Fasse, B.B., Bost, L.F., Hermann, C.D. and Forest, C.R., 2012, "A quantitative analysis of the effects of a multidisciplinary engineering capstone design course", Journal of Engineering Education, Vol. 101, No 4, pp. 630-656. doi:10.1002/j.2168-9830.2012.tb01122.xledema.

R.A., Jorm, C., Long, D., Braithwaite, J., Travaglia, J. and Westbrook, M., 2006, "Turning the medical gaze in upon itself: Root cause analysis and the investigation of clinical error" Social Science & Medicine, Vol. 62,No 7, pp. 1605-1615. doi:10.1016/j.socscimed.2005.08.049.

Jackson, D. and Hancock, P., 2010, "Non-technical skills in undergraduate degrees in business: Development and transfer", Education Research and Perspectives, Vol. 37, No 1, pp. 52-84.

Jayaram, S., Boyer, L., George, J., Ravindra, K. and Mitchell, K., 2010, "Project-based introduction to aerospace engineering course: A model rocket". Acta Astronautica, Vol. 66, No 9-10, pp. 1525-1533. doi:10.1016/j.actaastro.2009.10.038.

Larsen, J.A. and Nielsen, J.D., 2011, "Development of cubesats in an educational context", Recent Advances in Space Technologies (RAST), 2011 5th International Conference On, pp. 777-782. doi:10.1109/RAST.2011.5966948.

Larsen, J.A., Nielsen, J.F.D. and Zhou, C., 2013, Motivating students to develop satellites in problem and project-based learning (PBL) environment, International Journal of Engineering Pedagogy (iJEP), Vol. 3, No 3, pp. 11-17.

Levinson, R.A., Jaccard, J. and Beamer, L., 1995, "Older adolescents' engagement in casual sex: Impact of risk perception and psychosocial motivations", Journal of Youth and Adolescence, Vol. 24, No 3, pp. 349-364. doi:10.1007/BF01537601.

Mathers, N., Goktogen, A., Rankin, J. and Anderson, M., 2012, "Robotic mission to mars: Hands-on, minds-on, web-based learning", Acta Astronautica, Vol. 80, pp. 124-131. doi:10.1016/j.actaastro.2012.06.003.

Mitchell, V.W., 1995, "Organizational risk perception and reduction: A literature review", British Journal of Management, Vol. 6, No 2, pp. 115-133. doi:10.1111/j.1467-8551.1995.tb00089.x.

Mountrakis, G. and Triantakonstantis, D., 2012, "Inquiry-based learning in remote sensing: A space balloon educational experiment" Journal of Geography in Higher Education, Vol. 00, No 0, pp.1-17. doi: 10.1080/03098265.2011.638707.

Nagda, B.A., Gregerman, S.R., Jonides, J., von Hippel, W. and Lerner, J.S., 1998, "Undergraduate student-faculty research partnerships affect student retention" The Review of Higher Education, Vol. 22, No 1, pp. 55-72. doi:10.1353/rhe.1998.0016.

Neily, J., Ogrinc, G., Mills, P., Williams, R., Stalhandske, E., Bagian, J. and Weeks, W.B., 2003, "Using aggregate root cause analysis to improve patient safety", Joint Commission Journal on Quality and Safety, Vol. 29, No 8, pp.434-439.

Nordlie, J. and Fevig, R., 2011, "Blending research and teaching through high-altitude balloon projects", Proceedings of the 2nd Annual Academic High Altitude Conference.

Okudan, G.E. and Rzasa, S.E., 2006, "A project-based approach to entrepreneurial leadership education", Technovation, Vol. 26, No 2, pp. 195-210. doi:10.1016/j.technovation.2004.10.012.

Pollard, C.E., 2012, "Lessons learned from client projects in an undergraduate project management course", Journal of Information Systems Education, Vol. 23, No 3, pp. 271-282.

Prince, M.J., Felder, R.M. and Brent, R., 2007, "Does faculty research improve undergraduate teaching? An analysis of existing and potential synergies", Journal of Engineering Education, Vol. 96, No 4, pp. 283-294. doi:10.1002/j.2168-9830.2007.tb00939.x.

Qidwai, U., 2011, "Fun to learn: Project-based learning in robotics for computer engineers", ACM Inroads, Vol. 2, No 1, pp. 42-45. doi:10.1145/1929887.1929904.

Renn, O., 1998, "The role of risk perception for risk management", Reliability Engineering & System Safety, Vol. 59, No 1, pp. 49-62. doi:10.1016/S0951-8320(97)00119-1.

Rippl, S., 2002, "Cultural theory and risk perception: A proposal for a better measurement", Journal of Risk Research, Vol. 5, No 2, pp. 147-165. doi:10.1080/13669870110042598.

Ribeiro, L.R.C, 2008, "Electrical engineering students evaluate problem-based learning (PBL)" International Journal of Electrical Engineering Education, Vol. 45, No 2, pp. 152-161. doi:10.7227/IJEEE.45.2.7.

Robson, N., Dalmis, I.S. and Trenev, V., 2012, "Discovery learning in mechanical engineering design: Case-based learning or learning by exploring?", 2012 ASEE Annual Conference, American Society for Engineering Education.

Roh, K.H., 2003, "Problem-based learning in mathematics", ERIC Clearing House for Science Mathematics and Environmental Education.

Ronan, K.R. and Johnston, D.M., 2001, "Correlates of hazard education programs for youth", Risk Analysis, Vol. 21, No 6, pp. 1055-1064. doi:10.1111/0272-4332.216174.

Rooney, J.J. and Heuvel, L.N.V., 2004, "Root cause analysis for beginners", Quality Progress, Vol. 37, No 7, pp. 45-56.

Saunders-Smits, G.N., Roling, P., Brügemann, V., Timmer, N. and Melkert, J., 2012, "Using the engineering design cycle to develop integrated project based learning in aerospace engineering", International Conference on Innovation, Practice and Research in Engineering Education.

Simons, L., Fehr, L., Blank, N., Connell, H., Georganas, D., Fernandez, D. and Peterson, V., 2012 "Lessons learned from experiential learning: What do students learn from a practicum/internship?", International Journal of Teaching and Learning in Higher Education, Vol. 24, No 3, pp. 325-334.

Sjöberg, L., 2000, "Factors in risk perception", Risk Analysis, Vol. 20, No 1, pp. 1-12.

Slovic, P., Fischhoff, B. and Lichtenstein, S., 1982, "Why study risk perception?", Risk Analysis, Vol. 2, No 2, pp. 83-93. doi:10.1111/j.1539-6924.1982.tb01369.x.

Snell, K.D., 1996, "The apprenticeship system in British history: The fragmentation of a cultural institution", History of Education, Vol. 25, No 4, pp. 303-321. doi:10.1080/0046760960250401.

Steinberg, L., 2004, "Risk taking in adolescence: What changes, and why?", Annals of the New York Academy of Sciences, Vol. 1021, No 1, pp. 51-58. doi:10.1196/annals.1308.005.

Straub, J., Berk, J., Nervold, A. and Whalen, D., 2013a, "OpenOrbiter: An interdisciplinary, student run space program", Advances in Education, Vol. 2, No 1, pp. 4-10.

Straub, J., Fevig, R., Casler, J. and Yadav, O., 2013b, "Risk analysis & management in student-centered spacecraft development projects", Proceedings of the 2013 Reliability and Maintainability Symposium, Orlando, FL, USA.

Straub, J., Whalen, D. and Marsh, R., (in press), "Assessing the value of the OpenOrbiter program's research experience for undergraduates, Sage OPEN.

Straub, J., 2013, "OpenOrbiter: Analysis of a student-run space program", Proceedings of the 64th International Astronautical Congress, pp. 2-10.

Straub, J. and Fevig, R., 2012, "Formalizing mission analysis and design techniques for high altitude ballooning", Proceedings of the 3rd Annual High Altitude Conference, Nashville, TN.

Straub, J. and Whalen, D., 2013, "An assessment of educational benefits from the OpenOrbiter space program", Education Sciences, Vol. 3, No 3, pp. 259-278. doi:10.3390/educsci3030259.

Sulaiman, S.A., Git, H.M. and Rani, A.M.A., 2010, "Engineering team project course for development of teamwork skills".

Swartwout, M., 2004, "University-class satellites: From marginal utility to 'disruptive' research platforms", Proceedings of the 18th Annual AIAA/USU Conference on Small Satellites, Logan, UT, USA.

Swartwout, M., 2011, "AC 2011-1151: Significance of student-built spacecraft design programs it's impact on spacecraft engineering education over the last ten years", Proceedings of the American Society for Engineering Education Annual Conference, Retrieved from http://www.asee.org/file_server/papers/attachment/file/0001/1307/paper-final.pdf.

Weber, E.U. and Milliman, R.A., 1997, "Perceived risk attitudes: Relating risk perception to risky choice", Management Science, Vol. 43, No 2, pp. 123-144.

Weidl, G., Madsen, A. and Israelson, S., 2005, "Applications of object-oriented Bayesian networks for condition monitoring, root cause analysis and decision support on operation of complex continuous processes", Computers & Chemical Engineering, Vol. 29, No 9, pp. 1996-2009. doi:10.1016/j.compchemeng.2005.05.005.

Weinstein, N.D., 1984, "Why it won't happen to me: Perceptions of risk factors and susceptibility", Health Psychology, Vol. 3, No 5, 431-457. doi:10.1037/0278-6133.3.5.431.

Wertz, J.R., Everett, D.F. and Puschell, J.J., 2011, "Space mission engineering: The new SMAD", Hawthorne, CA: Microcosm Press.

Wildavsky, A. and Dake, K., 1990, "Theories of risk perception: Who fears what and why?" Daedalus, Vol. 119, No 4, pp. 41-60.

Zydney, A.L., Bennett, J.S., Shahid, A. and Bauer, K.W., 2002a, "Impact of undergraduate research experience in engineering", Journal of Engineering Education, Vol. 91, No 2, pp. 151-157. doi:10.1002/j.2168-9830.2002.tb00687.x.

Zydney, A.L., Bennett, J.S., Shahid, A. and Bauer, K., 2002b, "Faculty perspectives regarding the undergraduate research experience in science and engineering", Journal of Engineering Education, Vol. 91, No 3, pp. 291-297. doi:10.1002/j.2168-9830.2002.tb00706.x.

7

Life Cicle Inventory for Lead Azide Manufacture

Erick Braga F. Galante[1], Assed Haddad[2], Dieter Boer[3], Danielle Bonifácio[4]

ABSTRACT: Like any other manufactured chemical compounds, explosives are produced using chemical reactants and other utilities (steam, heat, compressed air, feed water and electricity) and generate a set of environmental impacts (waste water, solid and water residue and waste heat, for example). On top of that, one can count the intrinsic hazard characteristic of explosives and the possibility of accidents involving these compounds. Within this framework, explosives present themselves as chemical compounds suitable for both LCI (Life Cycle Inventory) and LCA (Life Cycle Assessment). This LCI study takes into account all the raw materials, utilities and wastes taking place during the production process. In this particular article, lead azide has its processed mapped and inventoried under the scope of ISO 14040. ISO 14040:2006 describes the principles and framework for life cycle assessment (LCA), including: definition of the goal and scope of the LCA, the life cycle inventory (LCI) analysis phase, the life cycle impact assessment (LCIA) phase, the life cycle interpretation phase, reporting and critical review of the LCA, limitations of the LCA, the relationship between the LCA phases and conditions for use of value choices and optional elements. The Lead azide was chosen due its singular explosive characteristics (very sensitive, which makes lead azide the explosive of choice as a primer in several applications. The results and conclusions of this study are drawn from the review of the process, its analysis, as well as from the application of life cycle inventory methods upon the manufactory of lead azide, a highly sensitive primer explosive, providing solid ground for the further studies, such as a full LCA assessment. Furthermore, for explosives, most LCA research works aims towards disposal, not addressing manufacturing, which is the main strength of this work.

KEYWORDS: Lead azide, Manufactory process, Life Cycle Inventory, Life Cycle Impact Assessment, Life Cycle Assessment.

INTRODUCTION

Explosives are chemical compounds capable of undergoing a very rapid decomposition (detonation), which generates heat and produces gases, which can be converted in usefull work (thermodinamically defined). Overall, explosives have many applications, ranging from military to civilian. While the military applications of explosives are straightforward in its nature (bombs, mortars, missiles, payloads, for example), the quantities of explosives destined to the civil applications are significantly larger, due to its usage in infrastructure (mining operations, oil extraction and demolitions, for example).

Like all other manufactured chemical compounds, explosives are produced using chemical reactants and other utilities (steam, heat, compressed air, feed water, and electricity) and generate a set of environmental impacts (waste water, solid and water residue and waste heat, for example). On top of that, one can count the intrinsic hazardous characteristic of explosives and the possibility of accidents involving these compounds. Within this framework, explosives present themselves as chemical compounds suitable for LCI (Life Cycle Inventory) and LCA (Life Cycle Assessment) studies.

This study is a LCI and takes into account all the raw materials, utilities and wastes taking place during the production process. Hence, in this particular study, lead azide was the chosen explosive to have its manufactory process studied. Lead azide was chosen due to its explosive characteristics (very sensitive, which increase risks of detonations but makes it a mandatory compound in almost every explosive use), intense use and the presence of lead in the raw materials (in the form of lead nitrate) and residues.

1. Instituto Militar de Engenharia – Rio de Janeiro/RJ – Brazil **2.** Universidade Federal do Rio de Janeiro – Rio de Janeiro/RJ – Brazil **3.** Universitat Rovira i Virgili – Tarragona – Spain **4.** Indústria de Material Bélico do Brasil – Magé/RJ – Brazil

Author for correspondence: Erick Braga F. Galante | Instituto Militar de Engenharia – Departamento de Engenharia Química – SE/5 – Praça Marechal Tibúrcio 80 CEP 23070-200 – Rio de Janeiro/RJ – Brazil | Email: egalante@ime.eb.br

Therefore, for this article, the manufactory process of lead azide was mapped using the fundamentals of ISO 14044, qualitatively analysed and had its environmental impacts identified and categorized (midpoint and endpoint). ISO 14040:2006 describes the principles and framework for life cycle assessment (LCA) including: definition of the goal and scope of the LCA, the life cycle inventory (LCI) analysis phase, the life cycle impact assessment (LCIA) phase, the life cycle interpretation phase, reporting and critical review of the LCA, limitations of the LCA, the relationship between the LCA phases and conditions for use of value choices and optional elements. ISO 14040:2006 covers life cycle assessment (LCA) studies and life cycle inventory (LCI) studies. It does not describe the LCA technique in detail, nor does it specify methodologies for the individual phases of the LCA.

Furthermore, for explosives, several LCA works (Alverbro *et al.*, 2009; Hochschorner *et al.*, 2006; Hochschorner and Finnveden, 2006) aim towards disposal and applications, not addressing manufacturing, which is the focus of this study.

This study aims to disclose the life cycle inventory of a largely used primary explosive (lead azide), aiming to help other researches addressing the environmental impacts in manufacturing lead azide, as well as other lead-based explosives. Thus, this study aims to review one of the processes for manufacturing lead azide, as well as to build the life cycle inventory of the overall process as a preparation for modelling the life cycle of lead azide manufacture.

BACKGROUND INFORMATION

This section provides useful background regarding LCA, explosives, lead azide and its manufacture processes.

LIFE CYCLE ASSESSMENT (LCA)

Sonnemann and Castells (2004) presents a list of some analytical tools for environmental managing, in which LCA ranks first as a standardized method for product-oriented environmental impact assessment (ISO 14001; ISO 14040; ISO14040:2006; Zharen, 1998). Alverbro *et al.* (2009) defined Life cycle assessment (LCA) as a method for assessing the potential environmental impacts and resources used throughout a product's life from raw material acquisition, production, use and waste management (ISO 14040:2006). It is important to determine that, under ISO, the term 'product' can also include services such as waste management.

The purpose of the LCA study of a particular product is to identify the main environmental impacts caused by the product or service evaluated to better determine the actions required for mitigation of impacts. According to ISO 14.040:2006, the steps of LCA are: goal and scope definition, inventory analysis, impact assessment, and interpretation. Modern work (Clift *et al.*, 2000) stated that LCAs are commonly carried out to compare alternative ways of delivering some function. The basis for comparison, common between all alternatives, is termed the Functional Unit for the study.

LIFE CYCLE INVENTORY

Life cycle inventory is a part of larger initiatives towards environmental management (ISO 14001; Zharen, 1998) under the guidance of ISO standards such as ISO 14001, ISO 14040 and ISO 14040:2006. To perform a life cycle inventory, it is necessary to determine inputs and outputs within a given system, collect data as well as looking at environmental aspects. According to recent work (Haddad *et al.*, 2013), a LCI study must be carried out for each subsystem involved, materials and energy, generation of products and by-products, as well as pollutants. This inventory is known as LCI. Furthermore, as addressed by Passer *et al.* (2012) for an environmental construction, LCI is one step prior to the a full LCA. While the LCA aims to determine an indicator that assesses key environmental aspects, the LCI aims to obtain a condensed and reliable array of data.

EXPLOSIVES OVERVIEW

Explosives are substances of great importance in the human development, being applied in many key areas, such as in the oil industry (Galante *et al.*, 2013). In addition to its traditional use in the military industry, it is usage is noticeable in great engineering projects, such as the construction of tunnels and exploitation of natural resources, like mining and oil exploration. Explosives are chemical substances or mixtures of substances that react rapidly by heating or attrition, generating huge volumes of gas and heat. The general formula for an explosive is $C_xH_yN_wO_z$ (Kinney and Graham, 1985). Explosives can be classified by several standards; however, the most common way of classifying explosives is according to its detonation velocity and activation energy. Table 1 presents a classification matrix with some examples.

Table 1. Explosives Classification matrix (compiled by Authors).

		Classification due to detonation velocity	
		High explosive	Low explosive
Classification due to activation energy	Primers	Lead azide	Lead styphnate
	Boosters	RDX, TNT, HMX	Gun powder Black powder

In general, to allow a full detonation to occur, a set of different explosives is required. The simplest configuration starts with a primer, which is ignited by spark, friction, flame or impact and provides energy to ignite other more powerful explosives. Among the primers, lead azide can be counted as the most used both in military and civilian applications.

THE LEAD AZIDE - PRIMER EXPLOSIVE

The lead azide is an explosive initiator, being more efficient than mercury fulminate. It requires a higher temperature for spontaneous combustion and does not decompose during long storage periods at moderately elevated temperatures. When compressed into a capsule, lead azide ignites detonating or exploding the spark (Ramaswamy *et al.*, 1981), but not as immediately as mercury fulminate. For this reason, the main charge initiation of a blasting cap composed of $Pb(N_3)_2$ is generally mixed in defined proportions with lead styphnate or other sensitizers to cause the explosion more easily, though with a less violentflame, which serves to initiate the explosion of azide.

The sensitivity to shock and friction rapidly increases in proportion as the particle size increases. Crystals of 1 mm in length are subject to a spontaneous explosion due to the energy content that they possess. Lead azide should not have needle-shaped crystals with a length greater than 0.1 mm. Lead azide with dextrin may appear to be safely stored under water for a long time. However, there is a belief that crystallized azide becomes more sensitive when stored under water due to an increase in the size of the crystals. Lead azide is produced in two crystal forms: orthorhombic (α), density 4.71 and monoclinic (β), density 4.93. The formation of the orthorhombic form, very sensitive, prevented the mixture of a diluted solution of lead nitrate and sodium azide, with which lead azide is prepared (Allan *et al.*, 1963; Imbel, 2013; Leslie, 1966).

Lead azide with dextrin can be soluble in cold water to extent measure of 1% because of the presence of dextrin. Both forms are almost insoluble in ether and acetone. Ethanol has little cold solvent action on the product and lead azide can be stored when wetted with a mixture of equal volumes of water and ethanol. Lead azide is soluble in aqueous solutions with ammonium acetate 10%. When exposed to an atmosphere with 90% of relative humidity at 30°C, lead azide in crystal form is hygroscopic and coated with a measure of 0.03 and 0.8%, respectively, of dextrin (Akhavan, 2011; Allan *et al.*, 1963; Cooper, 1996; Holloway *et al.*, 1965; Leslie, 1966; Meyer *et al.*, 2007; Urbanski, 1984). Dry lead azide does not react with or corrode iron, steel, nickel, aluminium, lead, zinc, copper, tin or cadmium. Coatings of acid-proof black ink, shellac or iron surfaces, do not show any effect. In the presence of moisture, it corrodes copper and zinc and, in the latter case, it forms the dangerous and highly sensitive copper azide. For this reason, lead azide is not loaded or placed in contact with uncoated copper or their alloys (bronze, brass and other metals) (Silva and Iha, 2010; Urbanski, 1984).

Lead azide is obtained as a white precipitate, by reaction of a solution of sodium azide in a solution of lead acetate or lead nitrate. It is absolutely necessary that the process be conducted so that the precipitate formed creates very very small crystals. There are several approaches to the manufactory process. Lead azide manufactory is well known in technical literature (Leslie, 1966) and can be obtained with dextrin (Allan *et al.*, 1963; Holloway *et al.*, 1965) or without it. From Meyer *et al.* (2007) as well as other sources (Akhavan, 2011; Cooper, 1996; Date *et al.*, 2009), one can design the process for dextrinated lead azide as presented in Fig. 1.

A standard plant that is able to hold a production unit as described in Fig. 1 contains vessels for the preparation of solutions (mainly solutions of sodium azide and lead nitrate), jacked vessels for the precipitation of lead azide, as well dosing pumps, filtering and drying installations. Due to the standard reaction using lead nitrate in excess, the wastewater is full of lead, and requires treatment before the final disposal. This treatment can be carried out using precipitation or other methods (like ionic resin). Since the simpler method is precipitation using sulphate, this study considers it in the inventory.

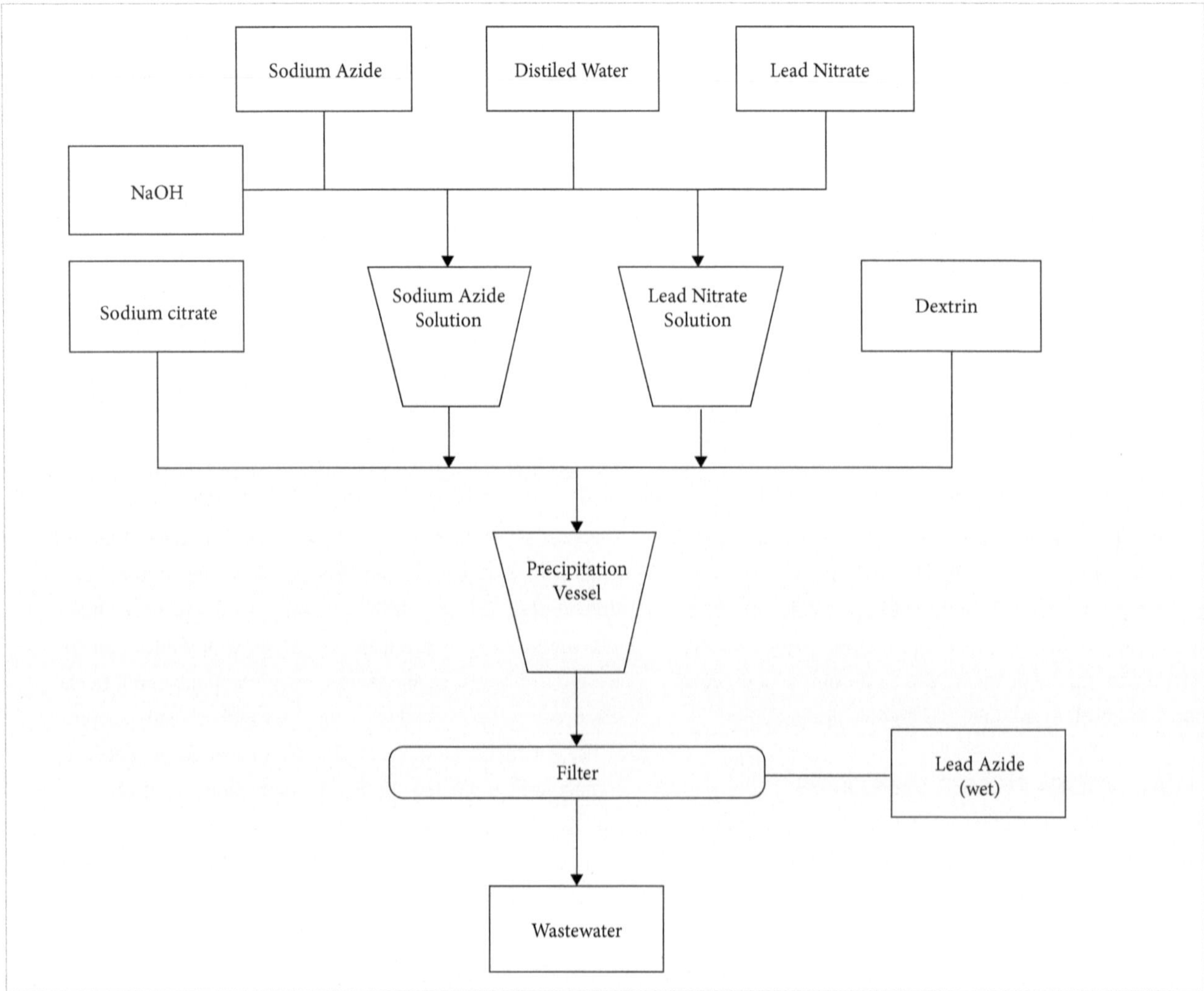

Figure 1. Manufactory process for lead azide (based on (Allan *et al.*, 1963; Holloway *et al.*, 1965; Imbel, 2013; Leslie, 1966)).

The steam generator consists of a set of equipment for producing water vapour. The water vapour produced is intended to distil the water from the solution, heating the reactive mixture and drying the lead azide produced.

LEAD AZIDE: LIFE CYCLE INVENTORY

The mapping starts with the identification of the manufacturing process and inventory stages of raw material inputs, energy inputs and service inputs themselves or other factors that may cause environmental impacts. With these data, enough information can be collected to understand the situation, qualitatively and quantitatively.

In the following sections of this chapter, the parameters of the four stages of the the manufacturing process life cycle of lead azide are defined, according to the goal and scope defined by ISO 14040:2006, from the inventory analysis of the life cycle assessment, life cycle impact and interpretation of the life cycle. This inventory was compiled from the processes described by (Allan *et al.*, 1963; Holloway *et al.*, 1965; Leslie, 1966) and used by (Imbel, 2013).

PRODUCTION SYSTEMS

The main part of the production system under study consists of both phases directly and indirectly linked to the production process of lead azide, as presented in Fig. 1. The operation and procedures taking place can be folded into subsystems. Each subsystem is described, quantified, analysed and summarized in Table 2.

Table 2. Subsystems for production of lead azide.

Subsystem	Description of the operation
Subsystem for the production of sodium azide	The production starts with 450 L of water (free of impurities and salts) being placed in a plastic container, in which are dissolved 4.9 kg of sodium azide with manual stirring. The pH of the solution is checked; if other than 10, 1.5 L of sodium hydroxide solution 14% are added. After this, it is linked to continuous stirring for several minutes, and then the solution is ready.
Subsystem for the production of lead nitrate	The production starts with 450 L of water (free of impurities and salts) being put into a plastic container, in which are dissolved in 33.7 kg of dried and crystallized lead nitrate. After that, the solution is stirred for several hours and allowed to sit for 8 h. The density of the solution should be 1.05 g/cm^3 and the pH should be between 5 and 6 before being transferred by gravity to the intermediate reservoirs.
Subsystem for the production of sodium hydroxide	For this solution, 1.4 kg of sodium hydroxide are dissolved in 10 L of water under agitation.
Subsystem for the production of dextrin aqueous solution	The coating agent (dextrin) is dissolved in filtered water at a proportion of 2 kg of agent for every 5 L of water. The solution is placed under heating (100°C) for 4 h with manual stirring every 30 min. After that, the solution must be cooled naturally and the density should be 1.154 g/cm^3.
Subsystem for the production of lead azide	The reactor is fed manually with 6 L of the lead nitrate solution prepared above; this is measured on a solution of sodium azide at a rate of of 5L/40min. During the reaction, it should be kept under constant stirring, heated to 80°C and 50mL should be added to the corn-starch dextrin solution. After the end of the dosage of sodium azide, the circulation of hot water is closed in the reactor jacket and the solution is allowed to sit for 1min. After this, the supernatant mother liquor is poured into a plastic container and the precipitate - lead azide - is filtered on a Büchner–Kitassato flask with the vacuum turned on. 500 ml of ethyl alcohol GL 96% are added to the ending process of the already filtered lead azide in order to optimize the drying process of the product.
Subsystem for effluent treatment	15 L of a sulphuric acid solution 50% are inicially placed in the tanks of the treatment plant. Then, the mother liquor effluent from each reaction is added and 1 L of a sodium nitrite solution 7% is added to the 2 cargo tank poured into the effluent. Sodium nitrite is explosive, and a distortion of the sulphuric acid precipitation is in the form of lead sulphate. The tank of the treatment plant may receive up to 12 loads of effluent, 6 L of sodium nitrite and 15 L of sulphuric acid. There is no stirring or heating. All of the loading and unloading of the solutions is done manually.

The following subsystems are indirectly linked to the cycle of the manufacturing process of lead azide, but were not included in this study:

- Production of fuel oil (boiler fuel);
- Production of electricity;
- Condensing of steam for use of process water;
- Production of steam;
- Manufacture of sodium azide;
- Port logistics (because the company model sodium azide is an imported product);
- Manufacture of lead nitrate;
- Manufacture of dextrin;
- Manufacture of sodium citrate;
- Manufacture of sodium hydroxide;
- Manufacture of ethyl alcohol 96% GL;
- Transportation: sodium azide, lead nitrate, corn-starch dextrin, sodium citrate, sodium hydroxide and ethanol 96% GL.

PRODUCT FUNCTION AND FUNCTIONAL UNIT

The product function is to allow the explosive lead azide to load fuses, a primer accessory used for initiation of the explosive train. Due to its high sensitivity, it is not necessary for bulk loading of the primer. The functional unit for the development of the study was fixed at 1.0 kg of lead azide produced. This value is the basis for the standardization of data in the preparation stage of the Life Cycle Inventory.

SYSTEM BORDERS

The product system is related with environmental aspects by the following steps: Consumption of natural resources (direct): crude oil and water; Energy consumption: electricity and fossil fuels; and Generation of waste: atmospheric emissions, liquid effluents and solid waste. A full set of borders and subsystems included in the study, as well as the excluded, are listed in Table 3.

SYSTEM BALANCES

From information, it is possible to consolidate inputs and outputs for each subsystem. These results are presented in Table 4.

Further analysing the subsystems (Table 4), it was necessary to determine the quantities of each raw material involved in the production of one kilogram of dry lead azide. This information is presented in Table 5, which was compiled using processes described by several sources (Allan *et al.*, 1963; Holloway *et al.*, 1965; Leslie, 1966) and under production in some industries (Imbel, 2013).

Furthermore, it is possible to determine that the two main raw materials are sodium azide and lead nitrate. Sodium azide, being a salt is not hygroscopic. It can be white or yellow. Its density ranges from 1.846 at 20°C in 100 parts of water to 40.7 at 15.2°C in parts of dissolved sodium azide. It is insoluble in ether and 100 g of absolute ethanol at 0°C can dissolve 0.22 g of sodium azide. The aqueous reaction is basic. When heated on a metallic plate, it does not melt, remaining unchanged up to 350°C, and detonates at higher temperature. When heated in vacuum, it decomposes into Na and N_2.

Table 3. Summary of the borders and limitations.

Subsystems considered	Subsystems outside this study
- Subsystem for the production of sodium azide - Subsystem for the production of lead nitrate - Subsystem for the production of sodium hydroxide - Subsystem for the production of dextrin aqueous solution - Subsystem for the production of lead azide - Subsystem for effluent treatment	- Production of fuel used for steam generation - Production of electric energy; - Production of steam - Manufactory of sodium azide - Manufactory of lead nitrate - Manufactory of dextrin - Manufactory of sodium citrate - Manufactory of sodium hydroxide - Manufactory of alcohol - Port logistics and transports

Lead nitrate is a salt of colourless crystals. It begins to decompose at 200°C, giving off gaseous nitrous oxide and lead. It is soluble in water. Lead nitrate is much less hygroscopic than the other nitrates, except potassium nitrate. While the

Table 4. Subsystems balances.

Subsystem	Input	Output	Product
Subsystem for the production of sodium azide	Sodium azide (powder) Process water Electricity	Heat	Sodium azide 99%
Subsystem for the production of lead nitrate	Lead nitrate (powder) Process water Electricity	Heat	Lead nitrate 98%
Subsystem for the production of sodium hydroxide	Sodium hydroxide (powder) Process water Electricity	Heat	Sodium hydroxide 93%
Subsystem for the production of dextrin aqueous solution	Dextrin Process water Electricity Heat	Heat	Dextrin (solution)
Subsystem for the production of lead azide	Ethanol 96% GL Sodium azide 99% Sodium citrate Sodium hydroxide 93% Dextrin Lead nitrate 98%	Heat Wasted water (with high concentration of lead)	Lead azide
Subsystem for effluent treatment	Waste water (with high concentration of lead) Sulphuric acid	Waste water (contaminated with diluted lead)	Lead sulphate

Table 5. Material balances.

Raw materials	Quantities
Ethanol 96% GL	0.84 L
Sodium azide 99%	0.51 Kg
Sodium citrate	trace
Sodium hydroxide 93%	0.01 kg
Dextrin	0.07 kg
Lead nitrate 98%	1.38 Kg

alkali nitrates and alkaline clay turn intoexplosion carbonates or oxides, lead nitrate's boiling point is at about 160°C. Lead nitrate, similarly to ammonium, has the property of being completely volatile, although only at high temperatures.

ENVIRONMENTAL IMPACTS

The environmental impacts of using lead azide occur from its manufacture to its disposal. The wastewater contains heavy metals and extremely low alkalinity, and there are also air emissions and solid waste generation. The environmental impacts arising from the manufacture of lead azide can be observed along all subsystems included in the study, in a greater or lesser extent, involved in the life cycle.

Energetic resource consumption

In the manufacture of lead azide, energy consumption comes from fossil fuels (used by the boiler to generate steam), and especially from electricity, which feeds the agitation, vacuum and product classification equipment, as well as the boiler. In the energy transportation steps, consumption comes from fossil fuel, used only by their respective transportation vehicles.

Atmospheric emissions

The main emissions of the system under study are mainly liquids and solids residues, due to the chemical reaction being a precipitation, taking place in water at temperatures below vaporization point. The majority of atmospheric emissions come from the steam production (CO_2, H_2O, NO_x), which varies depending on the fuel used.

Liquid effluents

For the manufacture of lead azide, excess lead nitrate is used to ensure complete consumption of sodium azide, themost expensive raw material in the manufacturing process. Thus, the mother liquor at the end of the reaction contains excess solubilized lead nitrate and traces of lead azide remaining in solution. This is necessary in the effluent treatment process, in which precipitation can lead to disfigurement and explosion. After addition of these components, a pH adjustment is made and the final effluent is discarded, considering pH and the legal emission limits of 0.01 mg/L for lead (Conama, 2000).

Solid residues

The solid residue is generated only in the effluent treatment step, in which sulphuric acid is later added to the mother liquor for the precipitation of lead sulphate. The recovered lead sulphate is a precipitated solid, which can be treated by co-processing (Curtin *et al.*, 2013; Garg *et al.*, 2013; Hasanogĺlu *et al.*, 2012), used in cement or other environmental ways of disposing solid residues.

CONCLUSION

From this study, it can be concluded that the Life Cycle Inventory for a standard lead azide manufactory unit was developed in full, allowing further analysis and a full LCA. Thus, considering the limitations and assumptions made during the study, it is understood that the objectives have been achieved with the desired quality. The raw material considered in this study did not have their manufactory and transportation taken into account. Regarding energy and utilities, this may vary according to the actual facility under study.

Regarding limitations, one of the most important of this study is the allocation and use of lead azide, due to the complexity of use of the explosive after being loaded. Given the wide range of environmental aspects raised in an LCI and LCA, Haddad *et al.* (2013) suggest the application of selection criteria that allow the separation of only the most significant environmental aspects from the total surveyed. From this study, it is possible to summarize the inputs and outputs of the lead azide manufactory process, which were presented in Table 6.

Table 6. Summary of inputs and outputs.

	Inputs	Outputs
Materials	Ethanol 96% GL Sodium azide 99% Sodium citrate Sodium hydroxide 93% Dextrin Lead nitrate 98%	Lead azide Wastewater Lead sulphate
Energy	Electricity	Heat
Utilities	Fuel (petrol based)	CO_2 H_2O NO_x Heat Steam Hot water

Furthermore, this study was successful in filling some of the gaps in the literature regarding the manufacturing of explosives, as well as their life cycle inventory. The environmental impacts were mapped and the process was inventoried, which allows researchers to use this study as background information on lead azide manufactory, as well as its LCI, in future researches.

REFERENCES

Akhavan, J., 2011, "The Chemistry of Explosives". Third Edit ed. Norfolk: Biddles Ltd., Kings Lynn, Norfolk.

Allan, G.B., Anders, M.L.L. and Stig, Y.E., 1963, "US 3095268 A - Process for the preparation of lead azide". No. United States.

Alverbro, K. *et al.*, 2009, "A life cycle assessment of destruction of ammunition". Journal of hazardous materials, Vol. 170, No. 2-3, pp. 1101–1109, doi:10.1016/j.jhazmat.2009.05.092.

Clift, R., Doig, A. and Finnveden, G., 2000, "The Application of Life Cycle Assessment to Integrated Solid Waste Management", Part 1ÐMethodology. Institution of Chemical Engineers, Vol. 78.

Conama, 2000, Res. Conama, No. 357.

Cooper, P., 1996, "Explosives Engineering" [S.l.]: Wiley-VCH, Inc. United States of America.

Curtin, V. *et al.*, 2013, "Reducing mechanical activation-induced amorphisation of salbutamol sulphate by co-processing with selected carboxylic acids". International Journal of Pharmaceutics, Vol. 456, No. 2, pp. 508–516.

Date, P. *et al.*, 2009, "Modelling the risk of failure in explosion protection installations". Journal of Loss Prevention in the Process Industries, Vol. 22, No. 4, pp. 492–498, doi:10.1016/j.jlp.2009.03.007.

Galante, E., Haddad, A. and Marques, N., 2013, "Application of Explosives in the Oil Industry". International Journal of Oil, Gas and Coal Engineering (OGCE), Vol. 1, No. 2, pp. 16–22, doi:10.11648/j.ogce.20130102.11.

Garg, N., Dureja, H. and Kaushik, D., 2013, "Co-processed excipients: A patent review". Recent Patents on Drug Delivery and Formulation, Vol. 7, No. 1, pp. 73–83.

Haddad, A.N. *et al.*, 2013, "Quality Indicators for Life Cycle Inventory: Real Cases Exploratory Application". Applied Mechanics and Materials, Vol. 431, pp. 350–355, doi:10.4028/www.scientific.net/AMM.431.350.

Hasanoglu, A. *et al.*, 2012, "Co-processing of Turkish high-sulfur coals and their blends with a petroleum asphalt. Part 1: In the absence of a catalyst". Energy and Fuels, Vol. 26, No. 12, pp. 7220–7229.

Hochschorner, E. *et al.*, 2006, "Environmental life cycle assessment of a pre-fragmented high explosive grenade". Journal of Chemical Technology and Biotechnology, Vol. 81, No. 3, pp. 461–475, doi:10.1002/jctb.1446.

Hochschorner, E. and Finnveden, G., 2006, "Life cycle approach in the procurement process: The case of defence materiel". International Journal of Life Cycle Assessment, Vol. 11, No. 3, pp. 200–208, doi:10.1065/lca2005.10.230.

Holloway, K.J. *et al.*, 1965, "US 3173818 A - Manufacture of dextrinated lead azide". No. USA, 1965.

IMBEL, "IMBEL", from www.imbel.gov.br. Accessed inb 17 jul. 2013.

Kinney, G. and Graham, K., 1985, "Explosive Shocks in Air", 2nd edition, USA, Springer-Verlag.

Leslie, J.P.M., 1966, "US 3264150 A - Explosive lead azide process". No. United States of America.

Meyer, R., Köhler, J. and Homburg, A., 2007, "Explosives ", 6th, Complete edition, Weinheim: Wiley-VCH Verlag GmbH, Weinheim.

Passer, A., Kreiner, H. and Maydl, P., 2012, "Assessment of the environmental performance of buildings: A critical evaluation of the influence of technical building equipment on residential buildings", International Journal of Life Cycle Assessment, Vol. 17, No. 9, pp. 1116–1130, doi:10.1007/s11367-012-0435-6.

Ramaswamy, C.P., Patwardhan, W.D. and Mukherjee, A., 1981, "EP 0070932 B1 - Initiatory explosive for detonators and method of preparing the same", No. [s.n.].

Silva, G.D. and Iha, K., 2010, "Polimorfismo: caracterização e estudo das propriedades de uma fase cristalina", Journal of Aerospace Technology and Management, Vol. 2, No. 3, pp. 331–338, doi:10.5028/jatm.2010.02038310.

Sonnemann, G. and Castellis, F., 2004, "Integrated Life-Cycle and Risk Assessment for Industrial Processes", [S.l.]: CRC Press, Lewis Publisher.

Urbanski, T., 1984, "Chemistry and tecnology of explosives, Vol 4", New York, USA, Pergamon Press.

Zharen, W.M., 1998, "ISO 14000 — Understanding the Environmental Standards", Rockville, MD.: Published by GaoLi.

Development and Optimization of a Catalytic Thruster for Hydrogen Peroxide Decomposition

Fernanda Francisca Maia[1], Leonardo Henrique Gouvea[1], Luis Gustavo Ferroni Pereira[1], Ricardo Vieira[1], Fernando de Souza Costa[1]

ABSTRACT: Hydrogen peroxide is a non toxic, low cost, green monopropellant with a significant potential for applications in satellite microthrusters. It presents a density-specific impulse similar to hydrazine. A new bulk mixed oxide catalyst for decomposition of hydrogen peroxide has been developed and tested in a 2 N thruster with modular parts, allowing the use of catalytic beds with different diameters and lengths. A method of experiment design with a star configuration was implemented to optimize the catalytic bed geometry, in order to yield complete peroxide decomposition, reduce thruster size and volume and provide maximum thrust.

KEYWORDS: Satellite propulsion, Hydrogen peroxide, Mixed oxide catalyst, Catalytic bed optimization.

INTRODUCTION

Nowadays, most satellites in Earth orbit utilize hydrazine catalytic decomposition within their propulsion systems. Hydrazine (N_2H_4) is an expensive and highly toxic (carcinogenic) monopropellant. In the last decade, due to growing concerns about the environment, there has been a significant interest for storable and nontoxic liquid propellants. Hydrogen peroxide (H_2O_2) is one of the most important candidates for application as a non-toxic, low cost, green-propellant in satellite propulsion systems (Wernimont, 2009).

The traditional catalyst used for H_2O_2 decomposition is made of superposed silver gauzes which form the catalytic bed. The use of solutions with high concentration of H_2O_2 leads to the increase of the adiabatic decomposition temperature of the monopropellant (632°C for 85%; 755°C for 90%; 953°C for 98% in mass), making not viable the application of pure silver or silver coated catalysts in long duration operations. The melting point of silver (962°C) is very close to the decomposition temperature of H_2O_2 and, besides, the formation of silver oxide does not allow the extended use of this type of catalyst (Kappenstein *et al.*, 2002). However, it is notorious that the H_2O_2 purity level, meaning the amount of stabilizers present in the monopropellant, has high influence in the efficiency of the catalytic system.

The reaction of H_2O_2 decomposition for propulsion applications is very fast, where the induction velocity is very dependent on the active sites exposed to the reaction. Since this

1.Instituto Nacional de Pesquisas Espaciais – São José dos Campos/SP – Brazil

Author for correspondence: Fernando de Souza Costa | Laboratório Associado de Combustão e Propulsão – INPE | Rodovia Presidente Dutra, Km 40 | CEP 12.630-000 Cachoeira Paulista/SP – Brazil | Email: fernando@lcp.inpe.br

reaction is extremely exothermic, a few miliseconds after the contact of the monopropellant with the catalyst, the catalytic bed reaches a temperature high enough to make the reaction almost entirely of thermal nature instead of catalytic.

A large number of transition metal oxides are catalytically active for H_2O_2 decomposition (Goldstein, 1974). Nevertheless, a catalyst frequently considered for H_2O_2 decomposition, is MnO_2. However if there is oxidation of this oxide, it will change to Mn_2O_3, that presents a low catalytic activity. On the other hand, higher temperatures lead to the formation of Mn_3O_4, which is as active as MnO_2 (Hasan *et al.*, 1999).

Rusek (1996) proposed manganese oxides, supported on alumina and silica, to replace the traditional silver catalysts. The results obtained suggest that the catalytic properties of manganese oxides are influenced by the type of support, hence the interest in new, supported catalyst, containing high concentration of surface active species, which can be stable under reaction conditions, typical of space propulsive applications.

Mixed oxides, containing cations of transition metals, are frequently used as bulk catalysts in different reactions. These oxides are generally prepared by thermal decomposition of different precursors, such as hydroxides, carbonates, nitrates and oxalates (Vaccari, 1998).

Taking into account these considerations and the results of preliminary tests, bulk catalysts were developed. Extrudates of cobalt/manganese based mixed oxides were prepared for H_2O_2 decomposition, employed in satellite microthrusters. The materials were tested in a modular 2 N microthruster, specially designed and built with catalytic beds of variable sizes (diameter and length), for catalytic decomposition of stabilized H_2O_2, concentrated to 90% in mass.

A method of experiment design, using a star configuration, was employed to optimize the catalytic bed dimensions, aiming to obtain complete decomposition of propellant, the reduction of thruster volume and mass, and to yield maximum thrust. The modeling of the process was made by the response surface method, adjusting quadratic models to the experimental results, obtained from a factorial planning. Through this modeling, the response sensitivity to the variables can be estimated, besides determining the levels where the response is optimal. The interest factors studied were diameter and length of the catalytic bed, and the response of interest was the thrust generated by the system.

EXPERIMENTAL METHODOLOGY

HYDROGEN PEROXIDE CONCENTRATION

Stabilized 50% H_2O_2, supplied by Solvay Chemicals, was concentrated to 90%, in mass, by a counter flow of hot dry air. A deionization treatment may be done before concentration, in order to remove the stabilizers present in the peroxide. Stabilizers can reduce catalytic stability, however, in this work, stabilized peroxide has been used to evaluate the new catalyst.

CATALYST PREPARATION

Extruded material was prepared from a bulk catalyst, based on manganese and cobalt oxides, capable to spontaneously decompose cold H_2O_2. The catalyst was prepared by co-precipitation of solutions of cobalt, manganese and aluminum nitrates (Sigma-Aldrich) in a Na_2CO_3 solution at room temperature, keeping a constant pH of 10, by addition of a NaOH solution. A mole ratio of 4:1:1 (Co:Mn:Al) was used. The product of this reaction was filtered, washed, dried and macerated. The powder was then peptized with HNO_3 solution and finally extruded, followed by stove drying and calcinations at 900°C temperature. The final product presented a cylindrical form with approximate dimensions of 2 mm diameter and 3 mm length.

MICROTHRUSTER

Performance parameters of the new mixed oxide bulk catalyst for H_2O_2 decomposition were obtained by tests with a 2 N thruster, manufactured with modules, including different chambers, nozzles and ring adapters built with 316 stainless steel (Fig.1). To vary the dimensions of the catalytic bed, three chamber modules were designed and manufactured, besides two nozzles and several rings of different lengths and thicknesses, to hold the catalytic beds in place.

The NASA CEA rocket performance code (McBride & Gordon, 1996) was used to calculate the thruster theoretical performance. Chamber pressure was chosen as 5 bar, and the nozzle expansion ratio was calculated as 1.45, to adapt the nozzle at 600 m altitude, where the tests were performed. Assuming frozen flow, with 90% H_2O_2, the calculated specific impulse was 1040 m/s, theoretical characteristic velocity was 940 m/s and ideal thrust coefficient was 1.106. Figure 2 shows a computer view of thruster and thrust balance. Pressure transducers and thermocouples were placed before and after injection and after the catalytic bed, as can be seen in Fig. 3. Thrust was measured by a 5 N load cell. A graphical user interface, written with LabView software, was used to monitor and control the tests.

TESTS

Before tests, the mass flow rate of propellant, as a function of differential pressure in the feeding line, was determined to estimate the peroxide tank pressure required to provide the desired mass flow rate during each test. After this procedure, the microthruster was assembled and mounted on the thrust balance, the load cell HBM 1-PW4C3/500G-1 (error±0.1g) was calibrated using known weights and, finally, continuous and pulsed test firings were performed.

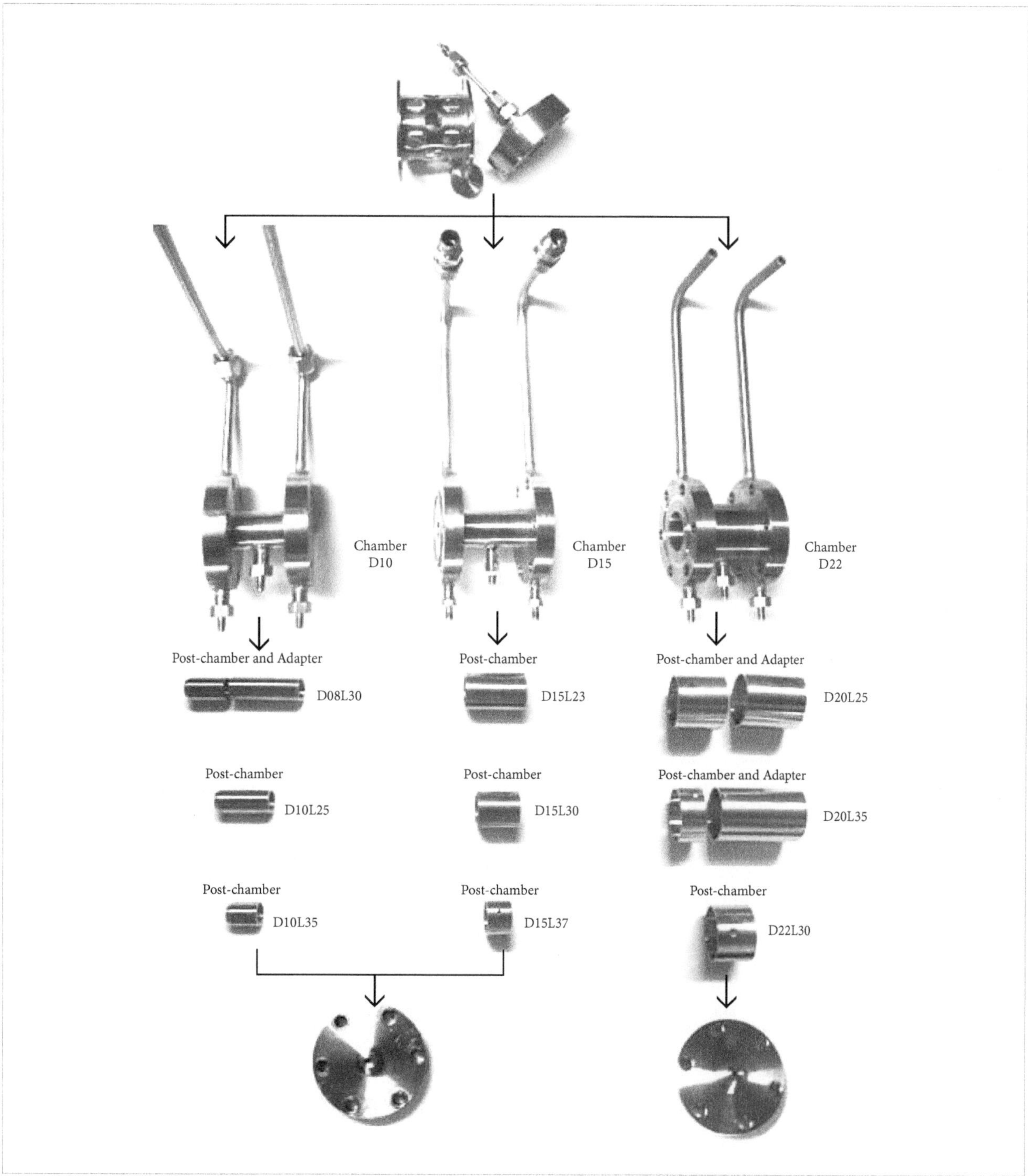

Figure 1. Photography of disassembled thruster parts.

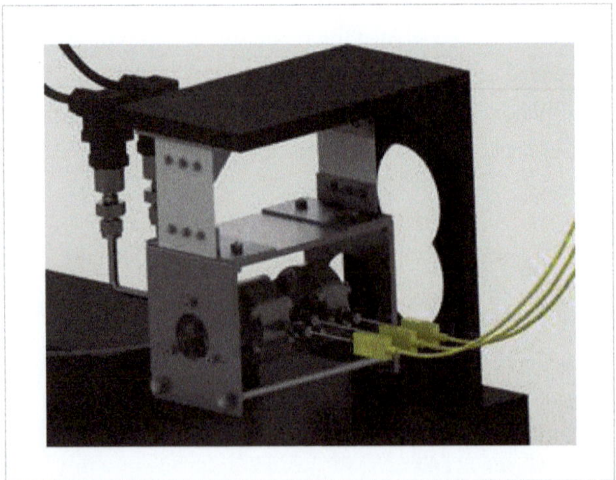

Figure 2. Computer view of thruster and thrust balance.

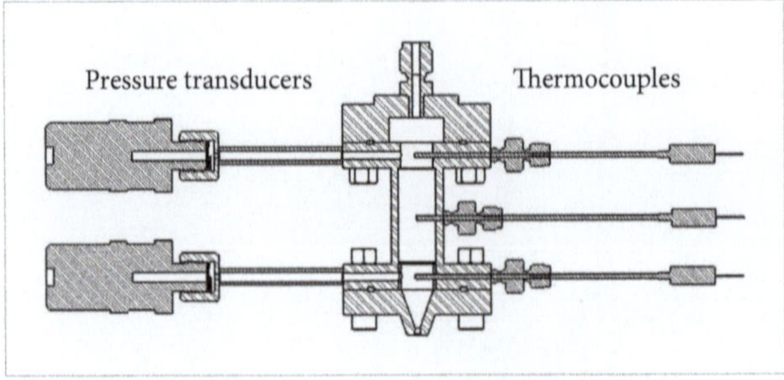

Figure 3. Cut view of thruster with thermocouples and pressure transducers.

RESULTS AND DISCUSSION

The mixed oxides bulk catalysts prepared in this work had high mechanical strength and low specific area (~6 m^2/g) due to the high temperatures in which they were calcined. Nevertheless, they showed no significant mass loss during tests in the microthruster.

The plan of experiments followed the methodology described by Barros-Neto *et al.* (1995). Table 1 presents the plan of experiments with a star configuration built from the factorial planning 2^3 (2 factors and 3 study levels), added to three replicates at the central point, totalling 11 tests.

A quadratic model was obtained using a factorial design, describing the influences of catalytic bed length (*L*) and catalytic bed diameter (*D*) on thrust (*F*):

$$F = 2.22 + 0.084L + 0.247D - 0.344L^2 + 0.0925LD - 0.573D^2$$

Table 1. Plan of experiments.

Test	Var 1	Var 2	*L* (mm)	*D* (mm)	*F* (N)
1	−1.00	−1.00	25	10	1.03
2	1.00	−1.00	35	10	1.06
3	−1.00	1.00	25	20	1.21
4	1.00	1.00	35	20	1.61
5	0.00	0.00	30	15	2.33
6	0.00	0.00	30	15	2.14
7	0.00	0.00	30	15	2.24
8	−1.41	0.00	23	15	1.52
9	1.41	0.00	37	15	1.69
10	0.00	−1.41	30	8	0.71
11	0.00	1.41	30	22	1.59

L = catalytic bed length; *D* = catalytic bed diameter; *F*= thrust.

From this model, the influences of the variables in the test conditions were represented by level lines of the response surface, as depicted in Fig. 4. The responses are indicated on *z*-axis (thrust), with a color bar, as function of the independent variables indicated in the *x* (diameter) and *y* (length) axes. Table 2 shows the regression analysis to fit the response function to experimental data.

As seen on Table 2, *L* and *LD* present lower coefficients than *D*, indicating that bed diameter presents a more significant influence than bed length in this experiment. The analysis of variance, depicted on Table 3, demonstrates that the model is significant,

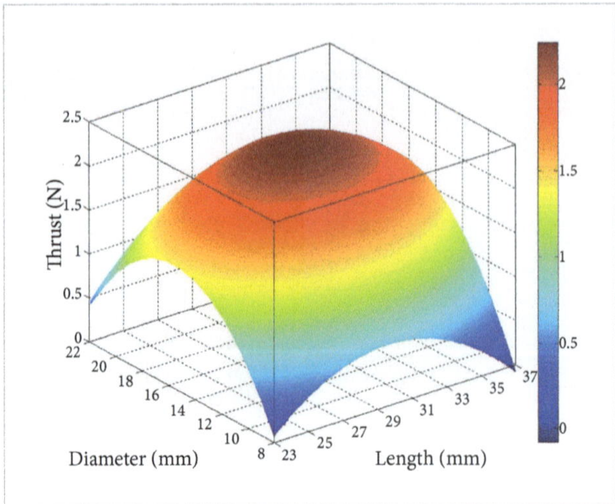

Figure 4. Contour graphic for the variable response studied: thrust *F* (N).

Table 2. Regression analysis using a 2^3 full factorial central composite design.

Predictor	Coefficient	Standard error coefficient	t-value	p-value
Constant	2.2172	0.0815	27.19	0.000
L	0.0839	0.0500	1.68	0.154
D	0.2471	0.0500	4.94	0.004
L^2	-0.3444	0.0597	-5.77	0.002
LD	0.0925	0.0706	1.31	0.247
D^2	-0.5732	0.0597	-9.61	0.000

Table 3. Analysis of variance.

Source	Degrees of freedom	SS	MS	F	P
Regression	5	2.6110	0.5222	26.17	0.001
Error	5	0.0997	0.0199		
Total	10	2.7107			

SS = sum of squares; MS = mean square; F and P = statistical parameters which indicate how far away a result is from the hypothesis

as evidenced from the *p*-value (0.001) obtained. The model showed a high determination coefficient, which explains 96% of the variability in the response.

According to the contour graphic (Fig. 4), monopropellant microthrusters using the new bulk mixed oxide catalyst for rocket grade stabilized H_2O_2 decomposition yield optimum thrust when the catalytic bed has length from 29 to 33 mm and diameter from 15 to 17 mm. Thrust, specific impulse and characteristic velocity curves obtained within these optimized dimensions (D = 15 mm and L = 30 mm), with chamber pressure P_c ~ 5.3 bar, are shown, respectively, in Figs. 5, 6 and 7.

The propellant mass flow rate ($\dot{m}$) was determined from injection pressure (ΔP) considering the following polynomial equation, obtained from a regression of 16 experimental data (up to 9 bar) using water injection, prior to fire tests in the thrust balance:

$$\dot{m} = 0.009\Delta P^2 + 0.613\Delta P + 0.766;\ R^2 = 0.999 \quad (1)$$

The thruster specific impulse (*Isp*) was then determined by:

$$Isp = \frac{F}{\dot{m} g_o} \quad (2)$$

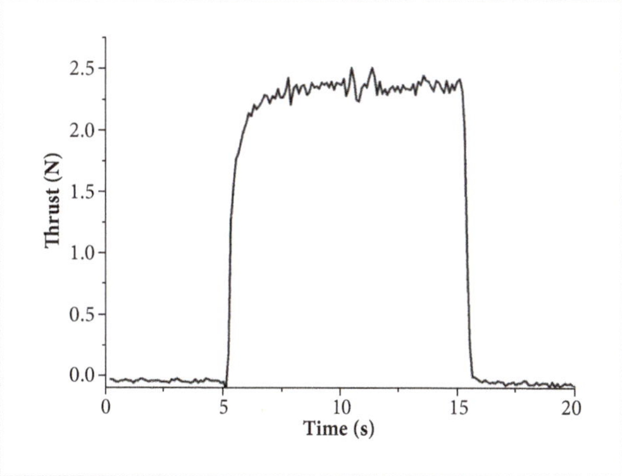

Figure 5. Thrust (N) obtained with CoMnAl based catalyst: D = 15 mm and L = 30 mm (P_c ~ 5.3 bar).

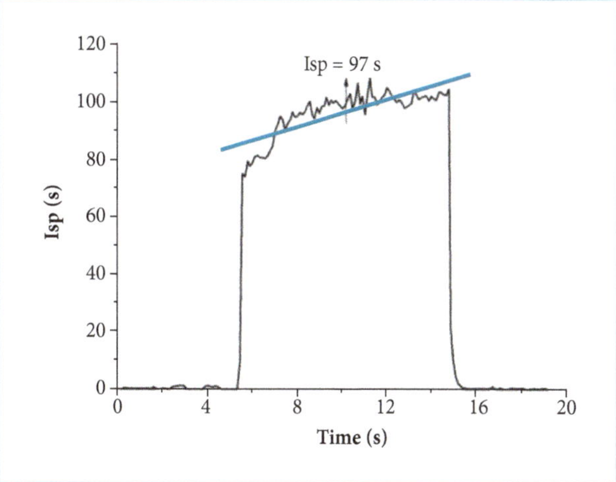

Figure 6. Specific impulse obtained with CoMnAl based catalyst: D = 15 mm and L = 30 mm (P_c ~ 5.3 bar).

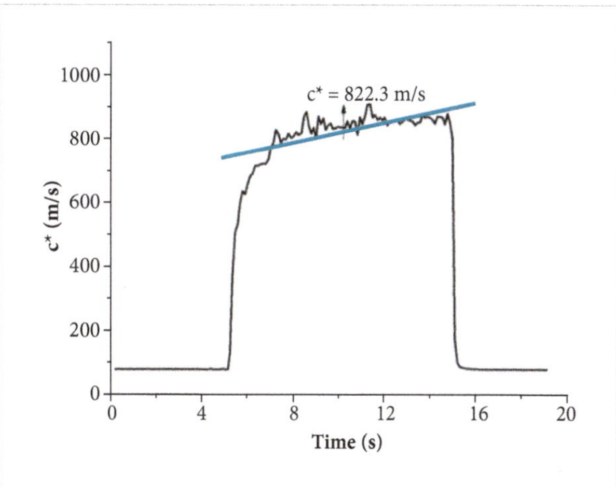

Figure 7. Characteristic velocity obtained with CoMnAl based catalyst: D =15 mm and L = 30 mm (P_c ~ 5.3 bar).

where F is the measured thrust and $g_0 = 9.8065$ m/s^2 is the standard gravity acceleration (Sutton and Biblarz, 2000). An average experimental specific impulse of 97 s was obtained (Fig. 6). This value is close to 106 s, the value calculated with the NASA CEA code (1996). Therefore, the experimental result is satisfactory, since various losses occur in the process.

The experimental characteristic velocity was obtained by:

$$c^* = \frac{A_t P_c}{\dot{m}} \tag{3}$$

where P_c is the measured chamber pressure and $A_t = 3.6$ mm^2 is the nozzle throat area.

Since the characteristic velocity is a function only of temperature and gas properties in the combustion chamber, one can use this parameter to examine the chamber independently of the nozzle. The theoretical value obtained was 940 m/s and the experimental average value was 822.3 m/s (Fig. 7), corresponding to a characteristic velocity efficiency of 87%, which indicates incomplete catalytic decomposition. However, this can be considered a high c^* efficiency for hydrogen peroxide decomposition, when compared to results of Pasini *et al.* (2007) as depicted on Table 4.

Finally, a continuous firing of 30 seconds (Fig. 8) and a sequence of 12 pulsed firings of 5 seconds, at intervals of 5 seconds (Fig. 9), were performed to check the stability of the CoMnAl catalyst, both tests adjusting $P_c \sim 5$ bar. There was no evidence of catalyst deactivation with respect to time, despite the use of stabilized hydrogen peroxide. No fragmentation of the catalyst grains was observed after tests. Pressure drops in the catalytic bed were about 0.5 bar and approximately constant (Fig. 10). However, long duration and shorter pulse tests, at various mass flow rates and pressures, will still be performed to evaluate thruster performance on a broader operational range.

Table 4. Comparison of c^* efficiencies.

c^* efficiency (%)	Catalyst
58	Platinum supported in alumina
80	Silver screen
87	Present work

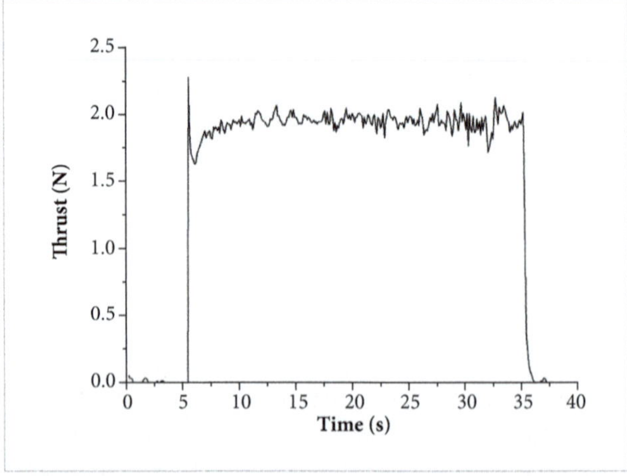

Figure 8. Thrust curve obtained during continuous 30-s test with CoMnAl based catalyst ($P_c \sim 5$ bar).

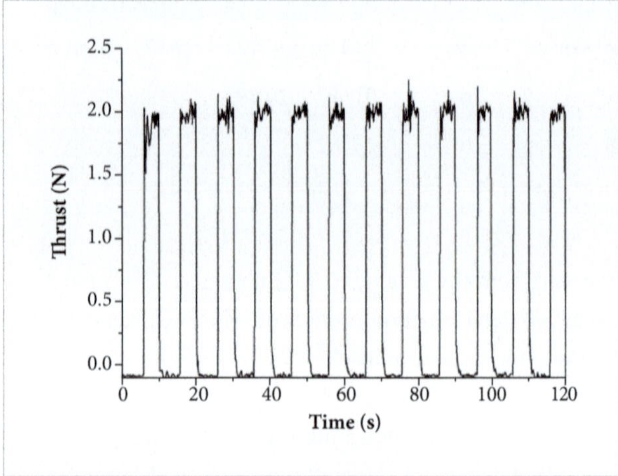

Figure 9. Thrust curve during pulsed test, 5s on/5s off, with CoMnAl based catalyst ($P_c \sim 5$ bar).

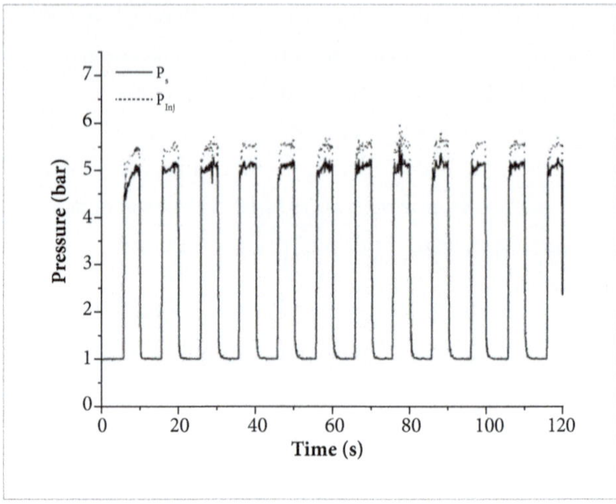

Figure 10. Pressure curves obtained during a pulse test, 5s on/5s off, with CoMnAl based catalyst.

CONCLUSIONS

Bulk catalysts based on mixed oxides, containing Co, Mn and Al, were prepared and showed capable of promoting various cold starts in the decomposition reaction of H_2O_2 in a 2 N microthruster. The catalyst showed high catalytic activity, despite its low specific surface. The catalytic bed sizes, for rocket grade H_2O_2 decomposition using the new CoMnAl based catalyst, were optimized to obtain maximum thrust in a modular microthruster. A design of experiments methodology allowed the determination of the optimum dimensions of the catalytic bed: length from 29 to 33 mm and diameter from 15 to 17 mm. For an optimized catalytic bed, the measured specific impulse was 97s, and the characteristic velocity efficiency was about 87%. Additional tests will be performed to evaluate thruster performance in a broader operational range.

ACKNOWLEDGMENTS

The authors acknowledge the Fundação de Amparo à Pesquisa do Estado de São Paulo, the Agência Espacial Brasileira and the Conselho Nacional de Desenvolvimento Científico e Tecnológico for supporting this research.

REFERENCES

Barros-Neto, B., Scarminio, I.S. and Bruns, R.E, 1995, Planejamento e Otimização de Experimentos, UNICAMP, Campinas, SP, Brazil (in Portuguese).

Goldstein, J.R. and Tseung, A.C.C., 1974, "The kinetics of hydrogen peroxide decomposition catalyzed by cobalt-iron oxides". Journal of Catalysis, Vol. 32, No. 3, pp. 452-465.

Hasan, M.A., Zaki., M.I., Pasupulety, L. and Kumari, K., 1999, "Promotion of the hydrogen peroxide decomposition activity of manganese oxide catalysts". Applied Catalysis A, Vol. 181, No. 1, pp. 171-179.

Kappenstein, C., Pirault-Roy, L., Guérin, M., Wahdan, T., Ali, A.A., Al-Sagheer, F.A. and Zaki., M.I., 2002, "Monopropellant decomposition catalysts: thermal decomposition and reduction of permanganates as models for the preparation of supported MnO_x catalysts". Applied Catalysis A, Vol. 234, No. 1, pp. 145-153.

McBride, B.J. and Gordon, S., 1996, "Computer program for calculation of complex chemical equilibrium compositions and applications, II-Users manual and program description", NASA-RP-1311.

Pasini, A., Torre, L., Romeo, L., Cervone, A., d'Agostino, L., Musker, A.J. and Saccoccia, G., 2007, "Experimental characterization of a 5 N hydrogen peroxide monopropellant thruster prototype", AIAA 2007-5465, 43rd AIAA/ASME/SAE/ASEE Joint Propulsion Conference & Exhibit, Cincinnati, OH, USA.

Rusek, J.J., 1996, "New decomposition catalysts and characterization techniques for rocket grade hydrogen peroxide". Journal of Propulsion and Power, Vol. 12, No. 3, pp. 574-579.

Sutton, G.P. and Biblarz, O., 2000, Rocket Propulsion Elements, 7a ed., John Wiley & Sons.

Vaccari, A., 1998, "Preparation and catalytic properties of cationic and anionic clays". Catalysis Today, Vol. 41, No. 1-3, pp. 53-71.

Wernimont, E. and Ventura, M., 2009, "Low temperature operation of hydrogen peroxide gas generators: verification testing and possible applications", AIAA Paper 09-4617.

High Power Laser Weapons and Operational Implications

Nelson Alex Roso[1], Romero da Costa Moreira[1], José Edimar Barbosa Oliveira[1]

ABSTRACT: The operational implications of high power laser weapons are constantly growing in countries with advanced military technological level. As well as on progress in integration with air, land and naval platforms, this paper discusses the necessary development and implementation of the operational concepts into Armed Forces, which should target orientation in the improvement process of the appropriate warfare material, i.e. the laser, as well as remodeling the existing combat's doctrine. Finally, we highlight some capabilities and limitations inherent in the technology of lasers and present some applications in defense and attack operations enabled by the implementation of laser weapons.

KEYWORDS: Weapons systems, Laser, Operational applications, Photonics.

INTRODUCTION

The experimental demonstration of the first laser was performed by Theodore Maiman on May 16, 1960, and preceded numerous scientific and technological areas, which have provided great advances in telecommunications, instrumentation and armaments industries, among others (Hecht, 2010).

In Brazil, it is noteworthy that the formation of human resources and the development of research on laser within the *Departamento de Ciência e Tecnologia Aeroespacial* (DCTA) — former *Centro Tecnológico da Aeroespacial* (CTA), are performed by the *Instituto Tecnológico de Aeronáutica* (ITA) and the *Instituto de Estudos Avançados* (IEAV) for over thirty years. In these institutes, significant results have been obtained in the areas of high power laser (HPL), optical sensors and communications links with optical fiber, among others (Monteiro *et al.*, 1994).

History reveals that, in areas of defense interest, conventional weapons have dominated warfare for centuries. Nowadays, the primacy of cartridge is replacing the projectiles for HPL, as technological advances in photonics introduce systems capable of destroying materials hardness, as well as protecting centers of Command and Control with early warning systems integrating lasers. Projectiles, e.g., mortars and missiles, are passive of interception and destruction by lasers that concentrate large amounts of electromagnetic energy (EME) in very small areas, destroying control systems of guided weapons or causing previous detonation.

The coexistence of conventional weapons systems and concentrated EME weapons is not nearly over, since these complement one another. Emerging possibilities of available technological resources require the fighters, scientists and commanders to have a comprehension of their implications,

1. Instituto Tecnológico de Aeronáutica – São José dos Campos/SP – Brazil

Author for correspondence: Nelson Alex Roso | Instituto Tecnológico de Aeronáutica | Praça Marechal Eduardo Gomes 50, São José dos Campos / SP | CEP: 12.228-901 – Brazil | Email: nelsonalexroso@gmail.com

since the war is won with the best job proficiency, e.g., the troops with greater efficiency will dictate the asymmetry of the conflict.

Within the Brazilian Air Force (FAB), the thematic "laser weapons" has received attention as reveal issues on the Air General Commander (COMGAR) Spectrum Magazine, published in 2000, 2001 and 2012.

This paper proves the success of graduate programs and research in the area of defense, provided by FAB for over ten years, making insight into operational implications of laser weapons. Initially, the growing capacity in developing these types of weapons, as well as progress in integration with air, land and naval platforms, is discussed. The need to develop operational concepts into the Armed Forces is also highlighted, mainly to guide the development of laser weapons; as well as a parallel is made between the capabilities and limitations of such weaponry. In addition to that, an approach of attack and defense enabled by the use of laser weapons is made; and finally, a retrospective completion of the main elements addressed is presented.

DEVELOPMENT AND INTEGRATION OF HIGH POWER LASER WEAPONS

Laser weapons are being created and tested throughout the world, specially in Armed Forces with notable weapons capabilities such as the United States, China, France and Russia, among others. The current example is the Tactical High Energy Laser (THEL), developed for use by the U.S. Army against tactical missiles with short and medium range, as well as the Anti-missile Airborne Laser (ABL) platform used by Boeing 747 U.S. Air Force (USAF) against intercontinental ballistic missiles (Dunn, 2005).

The major challenge in understanding the operational implications of the progress on laser weapons technology is the great variety of characteristics of such weaponry. Technologically, the type of laser that has been developed for military purposes was the chemical type employed in the THEL and ABL, which produces an intense beam of infrared radiation from chemical reactions between its solutes. Also, the solid state lasers (SSL), electrically energized, are less potent, and they depend on crystals which are responsible for originating their beams, but continue in constant evaluation, promising great benefits in light weapons used in the front battle. In a statement, the Department of Defense of the U.S. (DoD), through the high power solid-state laser development program — announced in December 2004 —, had already reached a 25 kW of power with SSL in the laboratory, although, according to the DoD, it is way far from the power required to be tactically effective, which should be greater than 100 kW (Dunn, 2005). The system's weight is also a problem, since it has been estimated at about 11,000 Kg for a laser of 100 kW. On the other hand, the American defense industry estimates that the technological development of SSL can reduce this weight to less than 4,000 Kg (Dunn, 2005), allowing the use of lighter and more agile platforms. However, power level is not the only feature to improve. An effective laser gun system must be created, a strong integration between subsystems for acquiring and tracking targets must be made, as well as the system base should monitor and provide enough energy power in several engagements at combat.

Another type of laser with potential applicability in weapons systems is the Free Electron Laser (FEL), whose base of operation is in free motion of a relativistic electron beam through a fixed magnetic structure (Federation of American Scientists, 2013). This type of laser has larger area than others and can be easily tuned to a specific wavelength bandwidth. The FEL technology is being evaluated by the U.S. Navy (U.S.N.) as a good candidate in the introduction of direct energy weapons against missiles and aircrafts. Significant progress has been achieved in increasing the power level of the FEL, and it is possible to consider, in the near future, the construction of compact weapons with this type of laser with megawatt's power. Recently, the North American Thomas Jefferson National Accelerator Facility (2013) — Jefferson Lab — showed that it has reached the power of 14 kW. With emphasis on this expectation, on June 9, 2009, the office of the Research Department of the U.S.N., responsible for promoting science and technology programs of the U.S.N. and U.S. Marine Corps, announced the contract closing with Raytheon Company, for the development of an experimental 100 kW FEL (Popular Mechanics Journal, 2009).

Researchers have also been made to improve the quality of the laser beam at high powers. The laser's feature called Beam Quality (BQ) essentially measures the scattering cross section of the laser beam at a given distance, measuring the ability to focus the laser on the target to form a point of intense light from the platform which employ its. In many cases, the intensity of the focused spot is proportional to $1/BQ^2$, where BQ equals 1, meaning a perfect focus. Therefore, a small increase in

BQ could result in a decrease of the intensity of the laser energy deposited on the target. Hence, the maintenance of a near BQ equals 1 is a very important requirement in the development of the laser system.

The integration of HPL on air, land and naval platforms must progress in order to suit the needs of the missions focus. As mentioned before, the ABL suited to the B-747 USAF platform, to execute strategic missions, was necessary multiple structural changes on the aircraft in order not to lose the features of the platform performance. Another example of adaptation was the modification on THEL, towards implementing it as a fighting system known as High Energy Laser Rockets and Mortars Artilhary (HELRAM), which may be used to destroy multiple types of threats, including artillery rockets and light mortars (Dunn, 2005).

Several other concepts using the laser from military platforms have been developed, but the limitations in space and weight still remain. Unmanned Aerial Vehicles (UAV), bombers, fighters, ships, submarines, satellites, tanks and many other weapon system platforms may involve HPL, serving as offensive and defensive weapons in the Operation Theatre (OT).

OPERATIONAL CONCEPT BASED ON CAPABILITIES AND LIMITATIONS

It is necessary that the Armed Forces inform the defense industry theirs requirements, in order to provide the best performance possible. The investment at HPL technologies must be made in order to promote research's development in technology institutes, as well as justifying budget on National Security. At FAB, according Oliveira (2002), the Operational Applications Postgraduate Program (PPGAO), that commissioned militaries from the Air Force, the Navy and the Army, develops operational research on these technological innovations, integrating operational and scientific communities, in order to clarify doubts and to maximize the conceptual and operational outcomes.

Acting as Electronic Protective Measures (Brasil, 2006), self-defense systems such as the Direct Infrared Countermeasure System (DIRCM) (Northrop Grumman, 2013), are capable to divert the spotlight of laser-guided missiles, confusing their trajectories.

Acting as Electronic Attack Measure (Brasil, 2006), lasers can be used in conjunction with weapon systems of AF to appoint or engage multiple targets, increasing the offensive capability. This function is extremely desirable against fast maneuverable targets such as UAV, surface-to-air missiles (SAM), air-to-air missiles (AAM), rockets and mortars. Notwithstanding this, the currently laser designation, carried out in the FAB by Litening System (Fig. 1), enables the use of "dumb bombs" (with Lizard control kit) with a high degree of accuracy (Northrop Grumman, 2013).

Compatible Energy consumption is proportional to the consumption of chemical fuel (COIL – Chemical Oxygen Iodine) or electricity (SSL) of the system, according to the type of laser. This is a tactical advantage, since the production of electric power at aircrafts, tanks and warships is constant, then the systems will be limited to components heating. The chemical lasers, for example, can operate according to the autonomy of its chemical cartridges, as is the case of HELRAM, which can be used between 10 and 20 times without being refilled, and furthermore, it is possible to operate with serials cartridges such as USAF B-747 ABL, improving the reload speed. In the SSL lasers, the number of shots is limited by the dissipate heat ability and batteries capacity on board. This limitation is called Duty Cycle Limit (Dunn, 2005).

The HPL has low cost per shot, improving the military availability and replacement when compared to guided missiles, that are expensive to manufacture because of the embedded technology involved (rocket motors, guidance systems, avionics, seekers and others).

The HPL has a cost directly proportional to the used energy, which can be produced without additional costs. For instance, the GBU-31 Joint Direct Attack Munition (JDAM) costs around U$25,000.00, one AGM-154C Joint Standoff

Figure 1. Litening Pod Designator of FAB AMX aircraft (adapted from aereo.jor.br, 2009).

Weapon (JSOW) costs around U$720,000.00, one AGM-158 Joint Air-to-Surface Standoff Missile (JASSM) approximately U$700,000.00, one AIM-120 Advanced Medium Range Air-to-Air Missile (AMRAAM) is estimated at around U$320,000.00 and the AGM-65 Maverick may reach the value of U$158,000.00 (depending on the variation) (Federation of American Scientists, 2013). In contrast to the values of the weapons above, the cost of the chemical fuel by shot in B-747 ABL, which uses the COIL as laser fuel, is around U$10,000.00 (Dunn, 2005), indicating operational feasibility of this weaponry.

The accuracy of these systems allows the operator to choose the impact place of the laser beam on the target and to adjust it according to the desired level of damage. This adjustment may be inserted in the beam Dwell Time, setting the duration of laser illumination, reducing the collateral damage.

The combat logistics for the maintenance of HPL is reduced in comparison with conventional systems. For example, guns and launchers need to be reloaded with projectiles and missiles after the shot, while laser weapons need only chemical fuel or electricity to generate energy for their systems.

The ability to adapt to several types of missions, controlling the Dwell Time of the laser beam, makes the HPL weapon platforms a Multi mission crafts, reaffirming the Air Force flexibility feature. However, the technical limitations of laser systems must also be understood by the operational community, some of them are listed below:

- Turbulence and atmospheric attenuation are factors that can affect the propagation of the laser beam, since dust particles, aerosols, water vapor and atmospheric instability lines, absorbing or scattering the laser energy, reducing operational range of application;
- Dependence of line-of-sight with targets, because the laser beam does not make ballistic trajectories;
- Due to the high concentration of energy in a point with a relatively small cross section, the lasers beams are more adjusted for engaging targets with small areas. For example, against combat vehicles, the laser will be effective in components such as antennas, control surfaces, sensors and fuel tanks;
- While chemical lasers generate wavelengths beams which are safe for human vision, the SSL are dangerous regarding these wavelengths. In case of HPL use against a target that does not fully absorb the energy issued, the spread energy by reflection can blind or damage the allied troops at OT if they are not properly protected.

Once the capabilities and limitations are known, the consequences of HPL weapons should be properly discussed.

OPERATIONAL IMPLICATIONS

The approach of operational implications in the use of HPL weapons does not underestimate the complex doctrinal study for development of military tactics and strategies, therefore, a synthetic implication in offensive and defensive operations on the ground, water, air and space OT's will be explored.

OFFENSIVE OPERATIONS

Offensive operations using HPL should consider that features of laser itself, such as accuracy, speed and the number of engagements, will be more important than destruction power itself. For example, the same laser that provides an active defense for an aircraft can be used against ground targets, providing a significant air-to-ground capability. Combining these capabilities with new aerial platforms and new sensors, it is possible to increase its potential use, increasing the airpower effectiveness in counterinsurgency and counterterrorism missions, as was done in Pakistan by the USAF. Moreover, the reduction of collateral damage is remarkable, because damage may decrease from many meters to some centimeters.

Strategic air targets such as radio, television and satellite antennas, power transformers, refrigeration and heating systems, which depend on these sites, may be blocked without complete destruction. It is also possible to assume that the SSL development, will change a lot of conventional weapons loaded in modern attack aircraft by HPL systems, which are capable of being used hundreds of times, setting a new limit on station combat, i.e., the human limit and the electromechanical system limits of air platforms. At this point it is possible to foresee the efficient union of SSL lasers and combat UAV.

On surface warfare context, the offensive ground operations should use the HPL weapons on precision missions that require high speed on engagement, adjustability and minimal collateral damage, such as missions against snipers. Laser weapons can be associated with optical and acoustic sensors to locate and to neutralize the sniper instantly, before they perform a shooting. Lasers do not have time of flight compared to the missiles and conventional weapons. However, the stealth features of lasers are also used in benefit of sniper weapons, since they are only visible through target damage.

The same limiting factors of laser weapons in ground operations will also limit the naval operations. A good example for the use of these weapons is the integrated laser on submarine periscope, which may engage surface targets such as ships, boats and aircrafts flying at low altitude with greater precision and less collateral damage, not revealing their positions when launching missiles and torpedoes.

In addition to conventional air, land and sea scenarios, there is also the space OT, a perfect environment for the lasers, because a space platform can engage, for example, a spy satellite or ground targets with the use of HPL; however, the reverse is true (Brasil, 2006). An example of this, according to Dunn (2005), occurred in 1984, when the Union of Soviet Socialist Republics (USSR) used a Ground Based HPL (GBL) against the U.S.' space shuttle, the Challenger, causing malfunctioning of the onboard equipment and stress on the crew.

DEFENSIVE OPERATIONS

In defensive operations, the HPL can provide self-defense measures for the air, land and naval platforms, as well as protect other platforms without protective systems. Lasers can be used against missiles, aircraft, satellites, UAV, bombs, artillery projectiles, rockets or mortars, neutralizing these threats before they reach their targets.

The first use of the laser for self-defense purposes occurred in 1973, when the USAF Weapons Laboratory (AFWL) hit a drone flying at 200 Kts, using a CO_2 laser based on ground. In 1983, the same laboratory carried out tests against the AIM-9B Sidewinder missile, which was kept at 2000 Kts (Dunn, 2005).

In air operations, lasers incorporated into self-defense systems are responsible for increasing the survivability of many aerial platforms, particularly the subsonic platforms, such as the E-8 Joint STARS (Joint Surveillance Target Attack Radar System), the E-3 Sentry AWACS (Airborne Warning and Control System) and the ABL B-747 (Federation of American Scientists, 2013). These aircrafts are considered strategic targets in OT and are called High Value Airborne Asset (HVAA). Furthermore, the use of laser systems for self-defense in attack aircraft against SAM and AAM missiles reduces the suppression of enemy air defense workload, making the OT less widespread. In other words, an air campaign can begin with interdiction actions before obtaining air superiority stage, reaching the enemy gravity centers in a shorter time, which also will result in lower costs and less collateral damage.

In land operations, HPL can increase the survival of the troops, as well as enable greater land mobility. These systems can be mounted on tanks to intercept missiles, artillery shells, rockets and mortars, increasing the OT self-defense. An example, previously mentioned in this paper, is the THEL type system, which has the ability to destroy rockets, mortars and other artillery ballistic projectiles.

Several features that benefit the air defense and land operations also provide naval defense operations, increasing the fleet flexibility. For instance, lasers can provide an effective shield from ballistic or cruise missiles, allowing the permanence of warships greater proximity to the enemy coastline within the firing range of tactical missiles. One of the biggest threats of naval forces is the Supersonic Sea-Skimming Missile Antiship (ASCM) (Federation of American Scientists, 2013), which may be undetected until seconds before impact. In this case, HPL can offer a high rate of engagement of these threats.

Strategically, as well as in offensive space operations with HPL, defensive operations by satellite based lasers are able to monitor and defend large areas on Earth's surface against ballistic missile much faster than land protection systems, because this type of threat can be engaged shortly after launch and at higher altitudes.

CONCLUSIONS

The presentation of the operational implications of HPL weapons is the first step to aware for operational and scientific community of the evolution in this area and it is used in order to alert the authorities about the military importance given on the nations with high technologies developed.

Initially, this paper described the capacity of HPL, as well as the progress in its integration with air, land and naval platforms. It also was presented the attractive costs and benefits ratio. Thereafter, it was explained the necessity of operational concepts development on Armed Forces, in order to guide appropriate manufacturing laser weapons, highlighting the HPL capabilities and limitations. Finally, it was presented an overview of offensive and defensive operations enabled by use of HPL weapons.

Therefore, according to the evidence of operational applications with laser weapons systems in many conflict situations, a new age in military and civilian community is possible, where postgraduate programs as the PPGAO can guide the development of concise operational concepts, joining the scientific knowledge with operative know-how, since one complements the other and vice-versa.

REFERENCES

Brasil, Ministério da Defesa do. Comando da Aeronáutica, 2006, "Comando-Geral de Operações Aéreas. NSCA 500-1: Sistema de Guerra Eletrônica da Aeronáutica", Brasília: COMGAR.

Dunn, R.J., 2005, "Operational Implications of Laser Weapons", Retrieved in June 22, 2013, from http://www.northropgrumman.com/AboutUs/AnalysisCenter/Documents/dfs/Operational_Implicationsof_La.pdf.

Federation of American Scientists, 2013, "Website about science and technology in the fields of biosafety, technology analysis and strategic security", Retrieved in June 20, 2013, from http://www.fas.org/.

Hecht, J., 2010, "A Short History of Laser Development", Optical Engineering, Vol. 49, No 9. Retrieved in June 25, 2013, from http://www.if.ufrgs.br/~jgallas/pub/FisAtoMol/hecht_short_-history_of_laser_development_OE2010.pdf.

Monteiro, E.C., Geraldo, J.A. and Oliveira, J.E.B., 1994, "Fotônica e Optoeletrônica: Pesquisas Desenvolvidas na Divisão de Engenharia Eletrônica do ITA", Revista ITA Engenharia, Vol. 1, No 1, pp. 13-16.

Northrop Grumman, 2013, "DIRCM, Directed Infrared Countermeasures", Retrieved in June 24, 2013, from http://www.northropgrumman.com/Capabilities/DIRCM/Pages/default.aspx.

Oliveira, J.E.B., 2002, "Ciência, Tecnologia e Inovação em Áreas de Interesse da Defesa", Retrieved in June 22, 2013, from https://www.defesa.gov.br/arquivos/pdf/ciencia_-tecnologia/palestras/ctidefesa.pdf.

Poder Aéreo, 2009, "Pod Litening no A-1 da FAB na Operação Laçador", Retrieved in June 25, 2013, from http://www.aereo.jor.br/2009/11/19/pod-litening-no-a-1-da-fab-na-operacao-lacador/.

Popular Mechanics Journal, 2009, "The Navy's New 100 kW laser weapons", Retrieved in June 25, 2013, from http://www.popularmechanics.com/technology/gadgets/4321422.

Thomas Jefferson Laboratory, 2013, "Free Electron Lasers", Retrieved in June 21, 2013, from https://www.jlab.org/FEL/.

Fire Resistant Aircraft Unit Load Devices and Fire Containment Covers: A New Development in the Global Air Cargo Industry

Glenn Baxter[1], Kyriakos Kourousis[1], Graham Wild[1]

ABSTRACT: Unit load devices (ULDs) are pallets and containers which are used to carry air cargo, mail and passenger baggage on wide-body aircraft. This paper analyses two important recent developments in the suppression of fires on freighter aircraft – fire resistant containers and fire containment covers for palletised cargo. In July 2013, United Parcel Service (UPS) placed an industry-first order for 1,821 fire-resistant shipping containers; this represented a major milestone in aviation history, offering unprecedented protection from intense fires on the airline's freighter aircraft. The new fire-resistant containers have been designed to withstand intense fires of four hours duration or longer, in order to provide pilots with sufficient time to land their aircraft during such an emergency. Fire containment covers have also been developed to enhance safety on freighter aircraft by isolating individual palletised cargo under special fire retardant covers that are capable of withstanding the intense heat and explosions associated with the ignition of dangerous goods.

KEYWORDS: Aircraft unit load devices, Air cargo, Fire containment covers, Fire resistant aircraft containers, Freighter aircraft, Innovation.

INTRODUCTION

Innovation initiatives are becoming exceptionally important for firms seeking greater competitiveness (Maldonado *et al.*, 2009). In the global air transport industry, an important area of innovation in recent times has been in aircraft unit load devices (ULDs). Unit load devices are pallets and containers which are used to load passengers' baggage, air cargo and mail onto a wide-body aircraft (Lu and Chen, 2011). The International Air Transport Association (IATA) estimates that there are currently around 700,000 ULDs in use by the world's airlines, with an estimated replacement value of close to approximately $USD1 billion (IATA, 2013a).

As a by-product of the aeronautical design, space is created underneath the wide-body aircraft's passenger deck, where passenger baggage, postal mail and air cargo can be carried (Dempsey and Gesell, 1997). With the introduction of wide bodied aircraft, such as the Boeing B747 and the McDonnell Douglas DC10 aircraft in the 1970s, a large volume of space was required to be filled in the lower deck belly holds of passenger flights, and a faster method of loading and unloading needed to be introduced (Morrell, 2011). Consequently, airlines utilise special containers (aircraft unit load devices) that are designed to fit the lower deck belly holds of their wide-body passenger aircraft, together with specially designed containers to fit the main deck of their freighter aircraft (Baxter, 2011; Morrell, 2011). The utilization of ULDs assists airlines in the standardization and unitization of loading and discharging passenger baggage and air cargo at airports (Lu and Chen, 2012).

1.RMIT University – Melbourne/VIC - Australia

Author for correspondence: Glenn Baxter | RMIT University | School of Aerospace, Mechanical and Manufacturing Engineering, Building 57, Level 3, 115 Queensberry Street - Carlton South/VIC - Australia 3053 | Email: glenn.baxter@rmit.edu.au

Nearly 60 per cent of world air cargo traffic is carried in freighter aircraft (Boeing, 2013). Wensveen (2011) notes that a freighter aircraft is an aircraft that is expressly designed or converted to carry air cargo, express, and so forth, rather than passengers. Most air cargo is carried in containers or ULDs, with the remainder transported loose in dedicated aircraft lower deck belly-hold compartments (Nobert and Roy, 1998).

Freighter aircraft are operated by the integrators, such as FedEx and United Parcel Service (UPS), dedicated all-cargo carrying airlines, for example, Luxembourg-based Cargolux, and combination passenger airlines that operate freighter aircraft in addition to their scheduled passenger services, for instance, Lufthansa or Korean Airlines. The integrators services are underpinned by the use of dedicated, efficient global multi-modal networks - they own and operate most of their own aircraft, smaller aircraft, trucks, and automated handling and storage facilities (Milenković, 2001). FedEx and United Parcel Service (UPS) are the world's two largest integrators (Bowen, 2012).

The main deck of a freighter aircraft has no amenities and utilises a roller deck to move cargo on and off the aircraft quickly so as to optimise the aircraft turnaround time. A roller deck is an aircraft main deck that is equipped with rollers on the floor, which enables palletized or containerised cargo/ULDs to be pushed or moved into position (Coyle *et al.*, 2011).

In November 2012, the United States National Transportation Safety Board (NTSB) announced recommendations to reduce the harm from fires aboard freighter aircraft. The NTSB recommendations followed three fire-related freighter aircraft accidents that had occurred around the world since 2006 (Jansen, 2012; Sperry, 2012). In response to these recommendations and to suppress fires on freighter aircraft, new innovative fire-resistant containers and fire containment covers for palletized cargo have been developed and introduced into service in the world air cargo industry.

THE ROLE OF AIRCRAFT UNIT LOAD DEVICES IN AIRLINE OPERATIONS

In addition to the ULDs used to carry passenger baggage, airlines use special ULDs for the carriage of air cargo and postal mail, such as containers and pallets. An aircraft pallet and its net and/or igloo, and/or aircraft container, enable individual pieces of air cargo and passengers baggage to be assembled into a standard size unit which expedites the loading and unloading of aircraft having compatible handling and restraint systems during the aircraft's ground turnaround time (Allaz, 2004).

Airlines can select customized types of ULDs for matching the inner contours of the main and lower decks of various wide-bodied aircraft sizes. Since ULDs are also compatible with a variety of wide-body aircrafts, airlines generally seek the benefit of commonality to purchase as many similar types of ULDs as possible (Lu and Chen, 2012).

The International Air Transport Association (IATA) recognizes a set of standardized ULDs in the form of open pallets, igloos and ULD containers (IATA, 2013b). These ULDs are designed to be compatible with a number of aircraft types and are compatible with cargo terminal handling systems, the airport apron, and the aircraft loading equipment (Ashford *et al.*, 2013).

The dimensions (and capacities) of the aircraft container ULDs used by the world's airlines depend on the aircraft flown and are often closed aluminium devices. Their size and shape depend on whether they are loaded onto lower or upper aircraft decks and whether they have been designed for use on narrow or wide-bodied aircraft (Reynolds-Feighan, 2013). The Airbus A319, A320 and A321 aircraft, for example, offer airlines a containerised baggage/cargo system that saves on the aircraft ground handling turnaround time (Airbus, 2013). Some airline ULDs are customised to fit specific aircraft types, whereas others as noted are compatible with multiple aircraft types (Nobert and Roy, 1998). Also, some containers/ULDs are refrigerated (Reynolds-Feighan, 2013). For example, LD8 containers can only be utilised on Boeing B767 aircraft (Nobert and Roy, 1998), whereas LD9s can be used on Airbus A330, A340 and A380 as well as Boeing B777, 747 and 787 passenger aircraft. LD9s can also be used on Boeing MD11, B777 and 747 freighter aircraft.

The major types of ULDs used by the world's airlines are as follows:

Containers: Rigid bodied containers are used by airlines to protect air cargo (and passenger baggage) and to ease the handling of multiple small, individual consignments of air cargo.

Pallets: Pallets are devices providing a rigid base, suitable for forklifting, on which air cargo consignments can be loaded. The load is secured by nets, and the complete load can be manhandled, forklifted, or moved mechanically as a unit (Ashford *et al.*, 2011: 465). Similar to container ULDs, pallet size and shape/contour will vary on the aircraft operated by the airline. Pallets can assist an airline to accommodate larger size consignments (Reynolds-Feighan, 2013). Wide-bodied aircraft can accommodate pallets with standard dimensions of 96 x 125 x 64 inches (244 x 318 x 163 cm) within their lower deck belly holds (Ashford *et al.*, 2011). Dedicated freighter aircraft can also accommodate pallets with a 96 x 125 inch base, but up to 96 inches (Q6 Contour) or 118 inches (Q7 Contour) (244 x 318 x 243 cm or 244 x 318 x 300 cm) in height on their main deck (CPC, 2012).

Igloos: Igloos are rigid bodied pallets used principally to prevent damage to air cargo or to the interior of the aircraft, where passenger cabins are converted to all-cargo use. A structural igloo is a fully enclosed shell constructed integrally with a pallet base to ensure that cargo conforms to the required aircraft contours. The shell and the pallet base form a single structural unit. A non-structural igloo is a bottomless shell that is designed to fit over an aircraft pallet and to provide shape to loaded cargo. The shell is used in conjunction with the aircraft pallet but does not provide any structural strength (Ashford *et al.*, 2011).

FIRE RESISTANT AIRCRAFT UNIT LOAD DEVICES

BACKGROUND

As noted earlier, in November 2012, the United States National Transportation Safety Board (NTSB) announced recommendations to reduce the harm from fires on board freighter aircraft. The NTSB recommended that the United States Federal Aviation Administration (FAA) require active fire-suppression systems in all cargo compartments of aircraft, the development of fire resistant containers/ULDs, and also to improve the early detection of fires within cargo containers (ULDs) and pallets (Jansen, 2012; Sperry, 2012).

The NTSB recommendations followed three fire-related aircraft accidents that had occurred around the world since 2006. In two of the accidents, fires on the aircraft resulted in the deaths of the flight crew and the loss of the aircraft (UPS Boeing B747-400F freighter aircraft in Dubai, UAE in 2010 and an Asian Cargo Boeing B747-400F freighter aircraft that crashed into the ocean off South Korea in 2011) (Jansen, 2012). The third accident occurred in a UPS aircraft at Philadelphia Airport in 2006, in which the crew was able to escape from the aircraft with some minor smoke-inhalation difficulties, but the aircraft was significantly damaged by the fire (Sperry, 2012).

The accident investigations revealed that, during its early stages, a fire burning inside a cargo container or ULD is often concealed. Subsequently, when a fire grows, it quickly burns through the container to threaten the aircraft and its crew (Jansen, 2012). In the three accidents, the fires started within the cargo containers aboard the aircraft, but by the time the aircraft's fire warning system alerted the pilots to the dangers of the fires, there was little time for them to react (Sperry, 2012).

In the United States, Federal regulations require the fire detection systems on cargo airline's aircraft to alert pilots within one minute of a fire starting. However, the NTSB's investigation discovered that current systems detected fire and smoke anywhere from two and half minutes to more than 18 minutes after the fire started on the aircraft (Sperry, 2012).

The NTSB concluded in its report that cargo containers/ ULDs made of flammable materials significantly increase the intensity of the on-board aircraft (Sperry 2012).

The aviation industry, responding to these requirements has taken several actions to design safer equipment. It should be noted that one of the first attempts to mitigate the fire safety issue has originated from the Pan Am flight 103 accident (Ushynskyi, 2009). In particular, in a 1993 FAA report focusing on design of explosion resistant containers, particular requirements for fire and explosion suppression covers were defined (FAA, 1993).

The International Organization for Standardization (ISO) TC20/SC9 'Air cargo and ground equipment' committee has initiated a project on defining a standard for fire safety of air cargo containers (FAA, 2013). Under this task the following standards are going to be developed and published in the coming period:

- ISO 14186:2013 specifies the minimum design and performance criteria and testing methods of fire containment covers (FCCs): either in those cargo compartments of civil transport aircraft where they constitute one means of complying with applicable airworthiness regulations, or secondly, on a voluntary basis, when deemed appropriate by operators to improve fire protection in aircraft cargo compartments where airworthiness regulations do not mandate their use (International Organization for Standardization, 2013b).

- SAE AS6453. This SAE Aerospace Standard (AS) is identical to the ISO 14186 standard
- ISO/SAE standard for Fire Resistant Containers (FRC). The FAA is going to develop a unique 'Technical Standards Order' (TSO) referencing this standard (FAA, 2013). The SAE AS6453 and ISO 14186:2013 for Fire Containment Covers (FCC) standards are identical in content, covering covers used in cargo compartments of civil transport aircraft. FAA will include this in the relevant Technical Standards Order (TSO) – TSO C90 (FAA, 2013).

BASICS OF AIRCRAFT FIRES

Types of fires

There are four types of fire, classified by the fuel source, which are identified by alphabetical labels.

- Class A: ordinary combustible materials, wood, cloth, paper, or upholstery materials, etc.
- Class B: flammable or combustible liquids, petroleum, greases, solvents, or paints, etc.
- Class C: energised electrical equipment.
- Class D: metals (Dornan 2008).

These are the US identifiers, Europe and Austrália use different letters. Different types of fire require the use of different types of fire extinguishers, specifically, the substance used to extinguish the fire, and whether it removes the fuel, oxygen, or the heat source (or a combination of these).

Fires zones

A fire zone is an area or region of an aircraft that is designed by the manufacturer to require specific fire protection. As with fire types, fire zones are also identified by alphabetical labels (Flynn, 2001; Tooley and Wyatt, 2009):

- Class A – is the cockpit, cabin, or cargo area, where the fire can be visually detected, reached and combated by crew,
- Class B – is a cargo area, where the fire can be reached and combated by crew, and is detected by a remote fire detection system,
- Class C – is an engine or cargo area, where automated systems are required for both the detecting and extinguishing of the fire,
- Class D – is a cargo area, on passenger aircraft, where the crew cannot reach fire, but the fire is confined within a minimally ventilated fire resistant area (no longer used),
- Class E – is a cargo area on a cargo only aircraft, where the crew cannot reach the fire and ventilation controls are required.

Fire Suppression Systems

While an aircraft has essential fire suppression systems for the engine and for the lavatories, of interest to this work are portable fire extinguishers and hold fire suppression systems. Portable fire extinguishers are used in Class A and B fire zones, where the crew can use them to actively combat the fire. As with their terrestrial kin, all types of portable fire extinguishers are red in colour, with a different colour label, and each is suited to different types of fires, having a specific effect.

- Red label: water, which is suitable for Class A fires, removing the heat.
- Blue label: dry powder, which is suitable for Class B fires, isolating the oxygen.
- Black label: CO_2, which is suitable for Class C fires, isolating the oxygen.
- Green label: Halon 1211, which is suitable for fires of all Classes, stopping the chemical reaction.

Hold fire suppression systems for Class C fire zones are based on Halon 1301 (the two version of Halon have different chemical compositions). Here, the Halon is used as a flooding agent to halt the chemical reaction that takes place in a fire. These suppression systems continue to supply Halon for a period of time after the fire has been extinguished to ensure the heat has dissipated such that re-ignition cannot occur. The specific requirements are an initial volume concentration greater than 5%, and a minimum concentration of 3% for 60 minutes or 180 minutes under ETOPS (Flynn, 2001).

DEVELOPMENT OF FIRE RESISTANT CONTAINERS AND FIRE CONTAINMENT COVERS

Fire resistant containers (FRC)

The SAE AS 6278 Standard *Fire Resistant Container - Design, Performance and Testing Requirements* "specifies the minimum design and performance criteria and testing methods of passive fire resistant containers (FRCs) for carriage on aircraft main deck, to be used: (a) either in those cargo compartments of civil transport aircraft where they constitute one means of complying with applicable airworthiness regulations, (b) or on a voluntary basis, when deemed appropriate by operators to improve fire protection in aircraft cargo compartments where airworthiness regulations do not currently mandate their use".

On June 19, 2013, the commercial aerospace and defence contractor, AAR, announced the development of a Fire Resistant Container (FRC) that has proved capable of containing an onboard aircraft fire for up to six hours of flight. The AAR subsidiary, Nordisk, partnered with DuPont, and was able to incorporate DuPont™Kevlar® brand fibre and Nomex® XF flame barrier into the innovation, making it the lightest weight FRC on the market. The new FRC is up to 30 per cent lighter than aluminium containers/ULDs (AAR, 2013).

The new FRC is a main deck container designated with the airline industry code "AAD". The code (AAD) refers to its size and shape, which is over 485 cubic feet. The testing of the new FRC strictly measured fire containment, not blast resistance. Depending upon the aircraft's size, anywhere from 24 to 30 FRCs could be accommodated inside the aircraft (AAR, 2013).

On July 22, 2013, United Parcel Service (UPS), one of the world's major integrators, placed an industry-first order for 1,821 fire-resistant shipping containers. The development and introduction of these unit load devices (ULDs) represented a major milestone in aviation history, offering unprecedented protection from intense fires (UPS, 2013b). UPS took delivery of the first of these ULDs in October 2013 and will receive the final ULDs by Spring 2014 (Putzger, 2013).

The new fire-resistant ULDs containers are built with a revolutionary new panel material, *MACROLite*, that are stretched across an aluminium frame (Croft, 2013) (Putzger, 2013). According to United Parcel Service (2013b), "MACROLite is a fibre-reinforced plastic composite similar to the material used in ballistic body armour". Burn testing conducted by UPS and the United States Federal Aviation Administration (FAA), and witnessed by the National Transport Safety Board (NTSB), has revealed that a ULD constructed of *MACROLite* panels can contain a fire with a peak temperature of 1,200 degrees Fahrenheit (649°C) for more than four hours (United Parcel Service, 2013b). Therefore, the new fire-resistant containers have been designed to withstand intense fires of four hours or longer, in order to provide pilots with sufficient time to land their aircraft during such an emergency. The units work by depriving any fires that may break out of oxygen (Putzger, 2013).

Fire containment covers (FCC)

Air cargo is also carried on the main and lower-deck freighter aircraft belly holds, on aircraft pallets. The relative volumes of cargo carried on aircraft pallets is of significance since fire mitigation measures needed to be developed to suppress fires that break out when cargo is carried on these units. As a result, fire containment covers have been developed to enhance safety on freighter aircraft by isolating individual pallets under a special fire retardant cover (Waldron, 2011).

In recent times, several firms, such as AmSafe and Newtex Industries, have responded to the industry requirement for fire containment covers. The *AmSafe* fire containment cover provides air cargo airlines with a safer flying environment by mitigating the risk of serious fire caused by undeclared dangerous goods. The fire containment cover operates as a passive system which keeps a fire isolated from other cargo being carried on an aircraft. The *AmSafe* fire containment cover has been designed for palletised loads of cargo and the cover is comprised of a patented fire retardant fabric with a detachable *QuickZip* pallet net. The manufacturer reports that the cover "can contain a fire with temperatures of up to 1500°F (815°C) for up to four hours". This combination of net and fabric cover provides a two-way fire barrier that effectively isolates each cargo position on the aircraft and prevents any fire from spreading and escalating. The fire containment cover suppresses the localised threat of the fire on the aircraft through oxygen starvation (AmSafe, 2013).

The *AmSafe* fire containment cover has many applications but it has been principally designed for palletized loads being transported in the presently unprotected Class E compartments found on the main deck of the majority of freighter aircraft. These covers are also intended to be used to protect the remaining Class D[4] lower lobe aircraft belly-holds that are still in operation (AmSafe, 2013).

Another firm that has been developing a fire containment cover is United States-based Newtex Industries, Inc. Since 2007, Newtex has been working closely with aviation experts and airline safety officials to develop a structure capable of withstanding the intense heat and explosions associated with the ignition of dangerous goods, including lithium-ion batteries. The Newtex *Z-Block*™ Fire Containment Cover (Fig. 1) has been developed following the outcome of this research and development, and the firm notes that its product is the most advanced protection available for containing deadly smoke and fire in commercial aircraft cargo bays. The system is made of

Z-Block™, a polymer coated fabric treated with a proprietary formulation of fire retardants that is able withstand temperatures up to 1800°F (980°C) and offers exceptional resistance to fire, smoke, and weather conditions (Newtex , 2013a). Similar to all of Newtex Industries products, Z-Block™ is formulated with inorganic materials and will not produce hazardous outgassing (Newtex, 2013b).

Z-Block™ coating and substrate fabrics were originally designed to channel smoke and prevent the spread of fire in aircraft hangars, picture theatres, and large public gathering spaces. The firm has subsequently optimized the fabrics for use in cargo operations by creating a multi-layer system for increased strength and puncture resistance and by adding an interior membrane that contains weather and mildew resistant properties (Newtex, 2013a).

The Z-Block™ fabric has been tested to the following standards:

- ASTM D6413 - Vertical Flame Resistance
- ASTM E-84 - Surface Flame Spread & Smoke Density
- FAR 25, Appendix F, Part III and IV - Flame Penetration Resistance & Smoke Density
- BSS 7239 - Toxicity of Products Combustion (Newtex, 2013a).

Photo courtesy of Newtex Industries, Inc.

Figure 1. Airline air cargo pallet fire containment cover.

UPS Airlines, for example, has purchased 575 fire containment covers that are able to withstand fires with a peak temperature of 650° C / 1,200° F for at least four hours. The airline is deploying these fire containment covers on palletized cargo, and on routes where there are large volumes on electronics being shipped (UPS, 2013a).

CONCLUSIONS

The carriage of air cargo has become increasingly standardised with the vast majority of world scheduled air cargo traffic now being carried in aircraft unit load devices (ULDs) (Reynolds-Feighan, 2013). Following the investigation into three fire-related freighter accidents that occurred between 2006 and 2012, the NTSB recommended, in October 2012, that the United States Federal Aviation Administration (FAA) require active fire-suppression systems in all cargo compartments of aircraft, the development of fire resistant ULDs, and also to improve the early detection of fires within cargo containers (ULDs) and pallets.

The industry has responded to this critical aviation safety issue. This paper has examined two recent innovations in response to the mitigation and suppression of fires on board freighter aircraft -fire resistant containers and fire containment covers for palletised cargo. Using the latest technologies, this equipment has been designed to suppress fires in the aircraft for around 4 hours thereby providing the pilots with sufficient time to land their aircraft during such an emergency.

ACKNOWLEDGMENTS

The authors would like to express their sincere gratitude to Newtex Industries, Inc. for permission to use the image of the firm's fire containment cover in this paper.

REFERENCES

AAR, 2013, "AAR's Nordisk announces lightest weight cargo container now has advanced fire resistance: Providing up to 6 hours of protection, innovation could set a new safety standard", Retrieved in January 3, 2014, from http://www.aarcorp.com/news/AAR_Cargo_Container_061913.htm.

Airbus, 2013, "A320 family: the market leader", Retrieved in January 2 2014, from http://www.airbus.com/fileadmin/media_gallery/files/brochures_publications/aircraft_families/A320_Family_market_leader-leaflet.pdf.

Allaz, C., 2004, "History of air cargo and mail from the 18th century". Christopher Foyle Publishing in association with The International Air Cargo Association, London.

AmSafe, 2013, "Fire containment cover", Retrieved in December 7 2013, from http://www.amsafe.com/products-services/commercial-aviation/cargo-airframe-products/fire-containment-cover/.

Ashford, N. J., Mumayiz, S. A. and Wright, P. H., 2011, "Airport engineering: planning, design, and development of 21st century airports", 4th ed. , John Wiley & Sons, Hoboken, NJ.

Ashford, N. J., Stanton, H. P. M.and Moore, C. A. et al. 2013, "Airport operations", 3rd ed., McGraw-Hill, New York.

Baxter, G. S., 2011, "Restructuring air freight chains: strategic options for competitive advantage", Ph.D. Thesis, Griffith University, Brisbane, Australia.

Boeing, 2013, "World air cargo forecast 2012-2013", Retrieved December 23 2013, from http://www.boeing.com/assets/pdf/commercial/cargo/wacf.pdf.

Bowen, J. T., 2012, "A spatial analysis of FedEx and UPS: hubs, spokes, and network structure", Journal of Transport Geography, Vol. 24, pp. 419-431.

CPC, 2012, "ULD specifications: pallet", Retrieved in December 23 2013, from http://www.cathaypacificcargo.com/en-us/helpsupport/uldspecifications/pallet.aspx?code=PMC.%20PQP.%20P6P.

Coyle, J. J., Novak, R. A., Gibson, B. J., Bardi J. Edward, 2011, "Transportation: a supply chain perspective", 7th ed, South-Western Cengage Publishing, Mason, OH.

Croft, J., 2013, "UPS takes proactive steps on fire protection", Retrieved in January 3 2014, from http://www.aviationweek.com/Article.aspx?id=/article-xml/awx_07_23_2013_p0-599998.xml.

Dempsey, P. S. and Gesell, L. E., 1997, "Air transportation: foundations for the 21st century", Coast Aire Publications, Chandler, AZ.

Dornan, S. 2008, "Industrial Fire Brigade: Principles and Practice", Jones and Bartlett Publishers, Sudbury, MA.

Flynn, T, 1999 "Rule Change: Cargo-Compartment Smoke Detection & Fire Suppression", AERO Magazine, No. 6.

IATA, 2013a, "ULD update", Retrieved in December 5 2013, from http://www.iata.org/publications/tracker/jan-2013/Pages/ULDR.aspx.

IATA, 2013b, "IATA ULD regulations", IATA, Montreal-Geneva.

Jansen, B., 2012, "NTSB urges fire-suppression upgrades for air cargo", USA Today, November 28, Retrieved in December 4 2013, from http://www.usatoday.com/story/todayinthesky/2012/11/28/ntsb-urges-fire-suppression-upgrades-for-air-cargo/1731897/.

Lu, H. A. and Chen, C. Y., 2011, "A time-space network model for unit load device stock planning in international airline services", Journal of Air Transport Management, Vol. 17, No. 2, pp. 94-100.

Lu, H. A. and Chen, C. Y., 2012, "Safety stock estimation of unit load devices for international airline operations", Journal of Marine Science and Technology, Vol. 20, No. 4, pp. 431-440.

Maldonado, M. U., Dias, N. and Varvakis, G., 2009, "Managing innovation in small high-technology firms: a case study in Brazil", Journal of Technology Management and Innovation, Vol. 4, No. 3, pp. 130-142.

Milenković, G., 2001, "Early warning of organizational crises: a research project from the international air express industry", Journal of Communications Management, Vol. 5, No.4, pp. 360-373.

Morrell, P. S., 2011, "Moving boxes by air: the economics of international air cargo", Ashgate Publishing, Farnham, UK.

Newtex, 2013a, "Cargo fire containment covers", Retrieved in December 6 2013, from http://www.newtex.com/EngineeredSystems/FireContainmentCovers/.

Newtex 2013b, "Advanced Fire & Smoke Barrier", Retrieved in December 6 2013, from http://www.newtex.com/Data/Documents/ZBlock.pdf.

Nobert, Y. and Roy, J., 1998, "Freight handling personnel scheduling at air cargo terminals", Transportation Science, Vol. 32, No.2, pp. 295-301.

Putzger, I., 2013, "Fire-proof ULDs for UPS sparks heated discussion", Retrieved in December 6 2013, from http://www.aircargonews.net/news/single-view/news/fire-proof-ulds-for-ups-sparks-heated-discussion.html.

Reynolds-Feighan, A., 2013, "Comparative analysis of air freight networks in regional markets around the globe", Ed. J. H. Bookbinder, Handbook of global logistics: transportation in international supply chains, Springer Science+Business Media, New York, pp. 324-366.

Sperry, T., 2012, "Fire protection aboard freight aircraft unsafe, NTSB says", Retrieved in December 5 2013, from http://edition.cnn.com/2012/11/28/us/cargo-plane-fires/index.html.

Tooley, M. and Wyatt, D., 2009, "Aircraft electrical and electronic systems: principles, maintenance and operation", Butterworth-Heinemann, Jordan Hill, UK.

UPS, 2013a, "UPS Airlines fire safety enhancements", October 31, Retrieved in December 6 2013, from http://www.icao.int/safety/DangerousGoods/DGP%2024%20Working%20Papers/UPSPresentation.pdf.

UPS, 2013b, "UPS pioneers aviation safety, implements new fire resistant shipping containers", Retrieved in December 6 2013, from http://www.pressroom.ups.com/Press+Releases/Archive/2013/Q3/UPS+Pioneers+Aviation+Safety,+Implements+New+Fire-Resistant+Shipping+Containers.

FAA, 1993, "DOT/FAA/CT-93/18, Hardened Luggage Container Design Survey", Retrieved in December 11 2013, from http://www.dtic.mil/dtic/tr/fulltext/u2/a291485.pdf.

FAA, 2013, "International Aircraft Systems Fire Protection Working Group Meeting, May 22-23 2013", Retrieved in December 11 2013, from http://www.fire.tc.faa.gov/pdf/systems/May13Meeting/May2013SystemsMin.pdf.

Ushynskyi, S., 2009, "Pan Am flight 103 investigation and lessons learned", Aviation, Vol. 13, No. 3, pp. 78-86.

Waldron, G., 2011, "AA11: Cargo majors keen on Amsafe fire containment covers", Retrieved in December 7 2013, from http://www.flightglobal.com/news/articles/aa11-cargo-majors-keen-on-amsafe-fire-containment-covers-354052/.

Wensveen, J.G., 2011, "Air transportation: a management perspective", 7th ed., Ashgate Publishing, Farnham, UK.

Investigations on Directed Infrared Countermeasures Risks to Fighter Aircraft Pilots

Lester de Abreu Faria[1], Luciano Barbosa Magalhães[2], Roberto d'Amore[1]

ABSTRACT: Directed Infrared Countermeasures (DIRCM) are used to protect aircrafts against missiles with infrared (IR) guidance. They are employed by military and civilian aircrafts, drawing away the guidance system of a missile as it attempts to lock onto the IR signature of the turbines. Unfortunately, the protection provided by these devices generates risks and challenges that must be overcome. In this paper, investigations on DIRCM risks to fighter aircraft pilots are carried out. Different kinds of lasers employed in actual DIRCMs are analyzed and the results show that, depending on their frequency (wavelength), damages can occur up to a distance of 4.8 km. The transmittance through the canopy of an F-5 fighter aircraft is evaluated and its effects on the IR propagation are predicted by the use of software called Counter-Measurements in PYTHON (CMePy). Results show that, even when there are interfaces between the pilot and the source of radiation, damages can occur, showing the importance of this investigation to the right understanding of this subject and future mitigations.

KEYWORDS: Aerospace, Safety, Defense systems, Laser, DIRCM, Infrared.

CONTEXTUALIZATION

Since September 11th, 2001, the entire world, but specially the USA, focuses its attention on the use of man portable air defense system (MANPADS) as the main terrorist's weapon for striking civilian and military aircrafts. Today, there are approximately 500,000 shoulder-fired missiles in military arsenals around the world, while there are from 5,000 to 150,000 in the hands of up to 30 non-state organizations, according to a report by the Congressional Research Service (CRS) (Bolkcom *et al.*, 2003). An analysis provided by the CRS indicates that, in the last 30 years, there were, at least, 35 civilian aircraft attacks through the use of MANPADS, of which 24 were shot down (Laurenzo, 2005; Nakagawara and Montgomery, 2008).

Since then, there has been a trend not only for military but also for civilian aircrafts to be equipped with some kind of countermeasure that can draw away the guidance system of a missile as it attempts to lock onto the infrared (IR) signature of the turbines of the aircraft.

Flare-based C-MANPAD systems were a main solution for military aircrafts, but the fear of collateral damage from these kinds of countermeasures, should the flares be deployed by mistake, makes it non-optimum. Therefore, some alternative countermeasure systems were developed to be employed as infrared countermeasures (IRCM). These systems disrupt the guidance systems of surface-to-air missiles (SAM), using multiple-wavelength lasers that emit radiation in the IR portion of the electromagnetic spectrum. Unfortunately, new

1.Instituto Tecnológico de Aeronáutica – São José dos Campos/SP – Brazil 2.Comando-Geral de Operações Aéreas – Rio de Janeiro/RJ – Brazil
Author for correspondence: Lester de Abreu Faria | Praça Marechal Eduardo Gomes, 50 – Vila das Acácias | CEP 12.228-900 São José dos Campos/SP – Brazil | Email: lester@ita.br

systems generate new risks and new challenges to scientific and medical areas that must be overcome.

IR emissions from these laser systems can be hazardous to ocular tissues and skin under certain circumstances. Depending on factors like energy delivered per unit area, exposure time, and wavelength, several damages can occur.

The major danger in IR emissions to the human vision is the lack of response of the ocular system to this type of radiation. The human eye can detect only wavelengths in the visible optical range and, in the case of the IR radiation, it is only noticed when some kind of damage has already happened (Nakagawara and Montgomery, 2008).

Figure 1 indicates which wavelengths are mainly absorbed in the different parts of the eye: the cornea, the lens, or the retina (Brunetaud and Hill, 2012).

Concerning the IR wavelengths, if the source is within the nominal ocular hazard distance (NOHD) of the pilot, near-IR (NIR) (780–1400 nm) laser radiation may damage the retina, while middle-IR (MIR) and far-IR laser radiations (>1400 nm) can injure the cornea and, to a lesser extent, the crystalline lens.

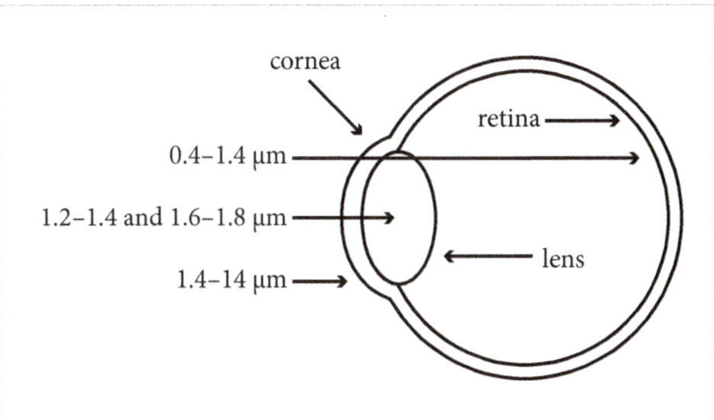

Figure 1. Parts of the eye at risk from different wavelengths (adapted from Brunetaud and Hill, 2012).

On the other hand, although skin burns are not generally very serious if a person is able to feel the damage and leave the radiation area, it becomes much more serious if people cannot react to the exposure, as the situation where a pilot is engaged in a combat arena (Brunetaud and Hill, 2012).

Table 1 demonstrates the potential injuries resulting from the excessive exposure to visible and IR radiation.

Excessive exposure to optical radiation in different ranges is currently a concern to industrial hygienists, safety engineers, and public health officials in many developed countries around the world for their potential hazard to health and safety (Nakagawara and Montgomery, 2008). In the literature, a lot of information and works can be found about dangers associated with exposure to excessive levels of visible and UV radiation in the USA National Airspace System (NAS) (Diffey and Roscoe, 1990; Caidin, 1992; West *et al.*, 1998) and other airspace systems (NASA, 1998; Wisegeek, 1999; Setlow, 2003). There are, as well, many studies on the transmittance of polycarbonate in different ranges of IR. However, only a few works conjugate both information and deal with the potential hazards to human health from the exposure to high levels of IR radiation, in free space or through an additional interface (Nakagawara and Montgomery, 2008). Above all, such kind of research has not yet been performed in Brazil, leading to a lack of information on professional diseases and impacting the performance of the pilots. The same results found in this work can be expandable to the "occupational health and safety", for all of those who work with laser emissions.

This paper shows a study made by the Electronic Warfare Laboratory (LAB-GE), at Technological Institute of Aeronautics (ITA), Brazil, concerning to the IR and visible radiation risks to the pilot's health and the protection provided by aircraft canopies. Transmittance properties of a fighter aircraft canopy from the visible to the IR range

Table 1. Potential injuries from excessive exposure to visible and infrared radiation.

<table>
<tr><th>Spectral band</th><th>Visible</th><th>NIR</th><th>MIR</th><th>FIR</th></tr>
<tr><td>Wavelenght (nm)</td><td>380–780</td><td>780–1400</td><td>1400–3000</td><td>3000–10^6</td></tr>
<tr><td rowspan="4">Potential injury</td><td colspan="2">Retinal burns</td><td colspan="2">Corneal burns</td></tr>
<tr><td colspan="3">Cataracts</td><td></td></tr>
<tr><td>Color and night vision degradation</td><td></td><td></td><td></td></tr>
<tr><td colspan="4">Thermal skin burns</td></tr>
</table>

(670–12000 nm) are presented. Further discussion on the electromagnetic spectrum leads to a "big picture" of the environment where the pilot operates and the risks to which he is submitted.

DIRECTED INFRARED COUNTERMEASURES

Directed Infrared Countermeasures (DIRCM) is used to protect aircraft against missiles with IR guidance, especially the portable ones that can be shoulder-launched (also known as MANPADS). Its operation is given in conjunction with missile approach warning system (MAWS), which is responsible for the detection of an approaching missile, determining if it is a threat to the aircraft, and providing the approaching direction of threat to DIRCM. This signal activates the laser transmitter, which will engage the threat and transmit energy modulated beams to deflect it from the aircraft.

Those equipments can be classified in two kinds of systems: directional and non-directional. Considering this work, we will focus on the directional DIRCM, once they are the most advanced and the most efficient ones recently in use.

The directional DIRCM operates by emitting radiation beams with small angular apertures, as lasers. This feature leads to low operation power and low response time. Besides that, they show spectral and spatial power densities much higher than any other typical source, due to the low divergence found in the optical laser beam (order of milliradians). Its main disadvantage is the need of using turrets to direct the beam emission against the missile, requiring high pointing accuracy and integrated communication with the MAWS. An example of this kind of equipment is AN/AAQ-24(V) NEMESIS, from Northrop-Grumman (GlobalSecurity.org. 2011), which is shown in Fig. 2.

There are also multi-spectral DIRCMs, which can handle multiple threats fired from any direction, and are adaptable to threat changes. These equipments are able to operate in different IR bands (Elbit Systems of America , 1999).

In order to analyze the emission of a directional DIRCM, it must be considered the main characteristics of a laser beam, such as: high spectral and spatial power density, high directivity, low operating power, low bandwidth beam, and pulse operation.

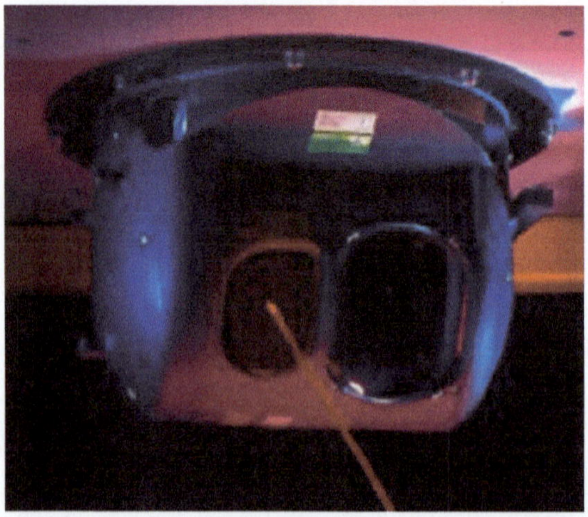

Figure 2. AN/AAQ-24(V) NEMESIS, from Northrop-Grumman.

GENERAL CONSIDERATIONS

Given the possibility of physiological tissue damage, a study was performed considering the potential effects that IR countermeasure equipment radiation could cause not only to human vision but also to human skin. Therefore, initially, some actual DIRCM data were collected from the literature and their potential risks to human health were dimensioned. After that, the transmittance of the canopy of an F-5 fighter aircraft was measured from visible to IR band, since this structure can be perceived as a filter, or a first barrier, between the emitting source and the pilot. The measurements were performed in LAB-GE/ITA. In order to predict the potential risks before and after the insertion of interfaces (canopy or any others), it was developed and used a software called Counter-**Me**asurements in **PY**THON (CMePy) , which calculates the optical power that reaches the eye, or the skin, of a pilot, helping us to dimension the problem.

An aircraft canopy is the transparent enclosure over the cockpit. Its main function is to provide a weatherproof and reasonably quiet environment for the aircraft's occupants, protecting them during flight. The canopy must be as aerodynamically shaped as possible to minimize drag and as transparent as possible to improve the outside visibility (AS 2211, 1991; Nakagawara and Montgomery, 2008). A high-performance aircraft's canopy is vital not only for the previous features but also for the enhancing and protecting of the pilot's vision. High-performance aircrafts from the Brazilian Air Force use polycarbonate (plastic) as canopy compound. The transmittance of a canopy compound can affect pilot's visual performance while providing protection from harmful electromagnetic radiation.

On the other hand, transmittance may be defined as the ratio of the total radiant flux that is transmitted, to that incident on the surface of the aircraft canopy. Higher is the ratio, higher is the transmittance.

MEASUREMENTS AND EXPERIMENTAL RESULTS

Once it was understood how the energy generated by the DIRCM can interact with the human eye, it was necessary to perform some transmittance measurements of aircraft canopies, in order to figure out how radiation is transmitted through these interfaces.

To perform this, it was collected random samples of the material used in the canopy of the F-5 fighter aircraft. The measurements were performed using a Spectrum 400 spectrometer, manufactured by PerkinElmer. This is a device that relies on the concepts of the Michelson interferometer to perform the measurement of transmittance of the sample at the wavelengths of interest. It allows measurements in the band from 0.67 to 18 μm.

The measurements were conducted in a controlled environment in terms of temperature (23±0.1°C) and relative humidity (65±1%), for wavelengths from 0.67 to 12 μm. The samples were placed inside the spectrometer using a support, which was set an incidence angle of 90°. Figure 3 shows a sample being measured in the spectrometer.

The measurement results are calculated, as a function of wavelength, using the ratio between the intensity of radiation that reaches the detector and the one emitted by the source. A correction factor is applied considering the background emission. The actual result of the measurement performed is shown in Fig. 4.

As seen in Fig. 4, the canopy presents a high transmittance, from the initial wavelength (0.67 μm) up to 1.6 μm with some absorption valleys around 1.19 and 1.4 μm.

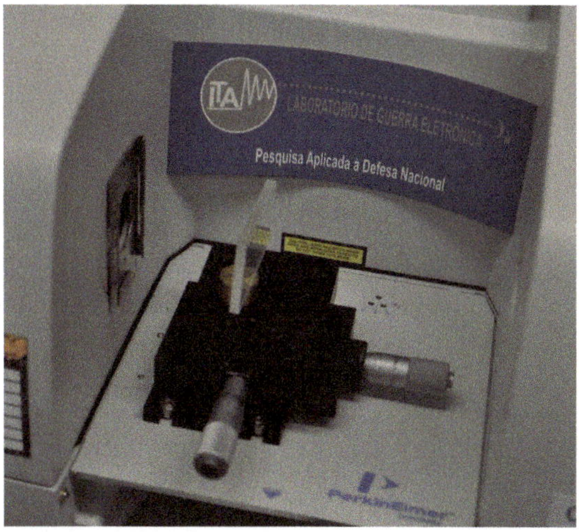

Figure 3. Canopy sample inside the spectrometer in order to perform the measurements.

From 1.6 μm on, transmittance drops sharply close to zero, presenting peaks around 1.8 and 2 μm, showing a transmittance level around 40%. Thereafter, the measured values are almost zero, up to the limit of the measurement, 12 μm, although the graph is presented only up to 3 μm. We must highlight the presence of valleys in 1.19, 1.4, and 1.9 μm, which correspond to the wavelengths where a high absorption by the water molecules in the air occur (Hudson, 1969). On the other hand, it is estimated that the observed absorption at 1.7 μm refers to the polycarbonate, once, synthesized from hydrocarbons, it will present absorption bands around the wavelengths of 1.68, 1.70, 1.72, and 1.78 μm (Araújo and Kawano, 2001).

The results presented in Fig. 4 show that there is no transmission through the canopy for emission whose wavelengths are above 2.2 μm (except for a small region around 2.6 μm), because these wavelengths are completely absorbed or reflected by the material. This result agrees with other ones found in the literature (Nakagawara and Montgomery, 2008). Therefore, the canopy shows high effectiveness against directional DIRCM MIR emissions. However, if the DIRCM is multi-spectral and presents spectral components in the NIR, the canopy does not serve as protection, since the transmittance is around 90%, and only 10% of the incident power is reflected or absorbed by it. Based on these results, a theoretical analysis was performed, focusing on possible damages to the pilot's health.

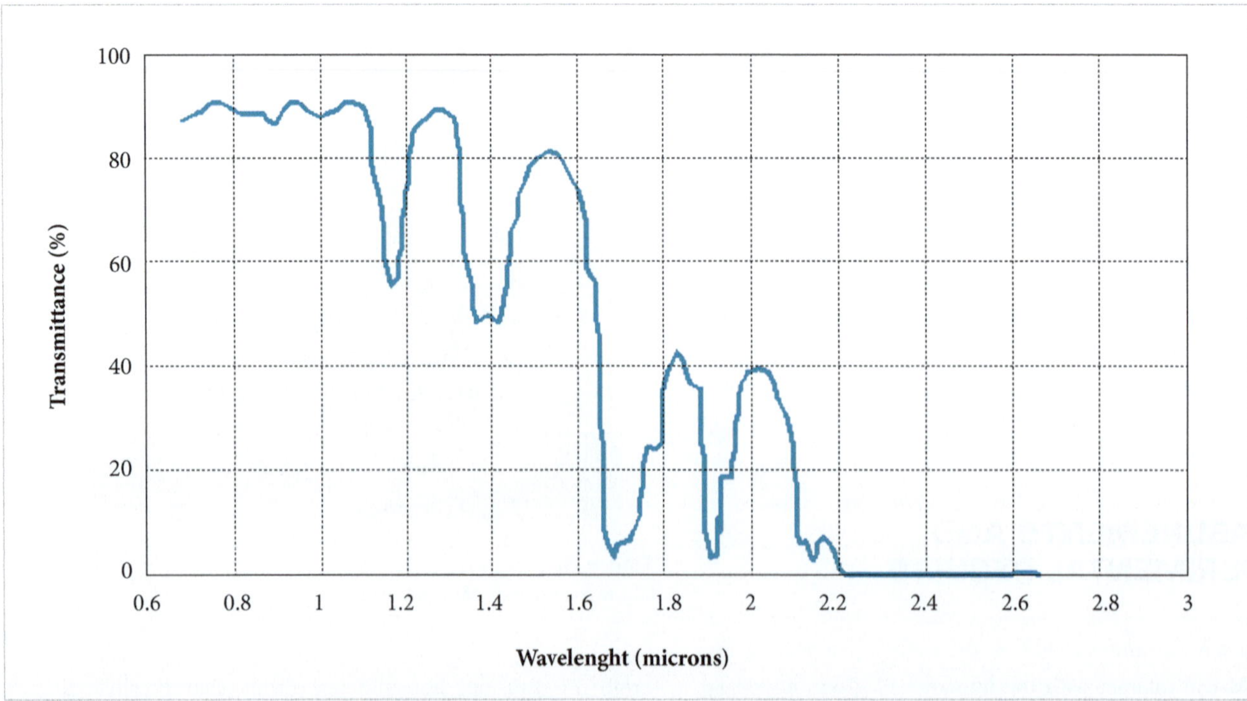

Figure 4. Experimental results of visible, near-infrared, and middle-infrared transmittance.

THEORETICAL ANALYSIS

Due to the high transmissivity of the canopy in the NIR range, it was made a theoretical analysis of possible damages to the pilot's vision, given the main features of well-known lasers and safety parameters for human vision. Therefore, some definitions are necessary, as shown in AS 2211:

- **Maximum permissible exposure (MPE):** *maximum level of radiation to which human tissue can be exposed without suffering injury or damage immediately after exposure or after some time;*
- **Nominal ocular hazard distance (NOHD):** *minimum safety distance where the radiant intensity of the ocular tissue (retina) is considered to be below the level defined as MPE; and*
- **Nominal skin hazard distance (NSHD):** *minimum safety distance at which the radiant intensity on the skin is below the level defined as MPE.*

Although the aim of this work is to analyze the possible damage to vision, some calculations on the potential damage to the human skin were included, once the procedure for performing both calculations is similar and the used software (CMePy) provides the complete information on them (with appropriate changes in reference tables).

The analysis starts by using the equation that calculates the intensity of a laser beam (W/m^2) at a distance z from the laser source, which is given by (Lelek, 2007):

$$I = \frac{4P_0 exp(-z\mu)}{\pi(2w + z\theta)^2} \quad (1)$$

In Eq. 1, P_0 is the power of the source (in Watts), or eventually the total energy carried by one pulse (in Joules), considering a Gaussian beam; w is the waist of the Gaussian beam (m) at a distance where the beam intensity is $1/e^2$ of P_0; θ is the divergence of the beam (rad); μ is the atmospheric absorption (usually neglected) (Nakagawara and Montgomery, 2008); and z is the distance between the source and the target.

Since, for safety reasons, it is necessary to calculate the NOHD of a laser with respect to the retina, variable z should be replaced by NOHD in Eq. 1; while the variable I must be replaced by MPE. Thus, considering the transmittance of any interfaces (such as the canopy) as τ_m and a multiplicative factor of 2.5 for non-Gaussian beams, one arrives to (AS 2211; Lelek, 2007):

$$NOHD = \frac{\sqrt{\frac{4P_0\tau_m 2.5}{\pi MPE}} - 2w}{\theta} \quad (2)$$

Equation 2 was applied on data of Table 2, achieving the results presented in Table 3. The calculation of the MPE is required for each one of the lasers. It was obtained by following the instructions and tables provided in AS 2211; Nakagawara and Montgomery (2008), Northwestern University (2012), and Fred Seeper (2012), considering always the most restrictive value of this parameter, as established by the AS 2211. In general, the procedures for the calculation of the MPE can be summarized as follows:

- For a continuous wave (CW) laser source, the wavelength of the CW laser beam, the duration time, and the size of the laser source beam must be determined. By selecting the appropriate MPE table, use the laser wavelength (nm) and exposure time (*t*) to locate the MPE value or formula on the table and calculate it.
- For a pulsed laser source, the pulse length, the pulse repetition rate, the laser wavelength, the exposition time, and source beam size must be determined. Based on these data, (1) calculate the MPE for a single laser pulse, using the pulse length as the duration time. Then, (2) calculate the MPE for a repeated laser pulse by dividing the MPE for a single pulse by $N^{0.25}$, where "*N*" is the actual number of pulses. Finally, (3) calculate the single pulse MPE again, using the duration time and divide this result by the number of pulses (*N*) to determine the average MPE for the pulsed laser source. Choose the smallest of the three values as the MPE for the pulsed laser source.

Besides considering the parameters shown in Table 2 for the calculations of each one of the lasers, the following assumptions were made (Nakagawara and Montgomery, 2008):

Table 2. Lasers technical specifications used in simulations.

Laser*	Wavelength (nm)	FRP (Hz)	P_0 (W)	Beam divergence (mrad)	Pulse width (ns)	Beam waist (mm)	Exposition time (s)	Gaussian beam?
GaAs	840	15	300×10^{-3}	1	12	0.4	10	No
Nd:YAG	1064	15	300×10^{-3}	1	12	0.4	10	No
Nd:YAG	1330	15	300×10^{-3}	1	12	0.4	10	No
Cr^{2+}:CdSe	2600	15	300×10^{-3}	1	12	0.4	10	No
HeNe	3390	15	300×10^{-3}	1	12	0.4	10	No
CO_2	10600	15	300×10^{-3}	1	12	0.4	10	No

*Nakagawara and Montgomery (2008)

Table 3. Safety distances for the lasers under test.

Laser*	Wavelength (nm)	MPE eye (J/m²)	MPE skin (J/m²)	NOHD (m)	NSHD (m)
GaAs	840	2.72×10^{-3}	108.89	4834.98	23.38
Nd:YAG	1064	14.29×10^{-3}	285.74	2110.10	14.13
Nd:YAG	1330	114.30×10^{-3}	285.74	745.51	14.13
Cr^{2+}:CdSe	2600	28.57	28.57	46.40	46.40
HeNe	3390	28.57	28.57	46.40	46.40
CO_2	10600	28.57	28.57	46.40	46.40

*Nakagawara and Montgomery (2008)

- Negligible atmospheric attenuation (equals to zero);
- Transmittance of the optical interfaces equal to 100%;
- The beam is stationary during each one of the analysis; and
- Only one wavelength is emitted.

The previous assumptions were established in order to simplify the analysis. Although being slightly different from the ones found in an actual environment, they serve as a basis for the conclusions to which we want to reach.

The results presented in Table 3 were calculated with the aid of a program called CMePy developed specifically for this purpose, in a PYTHON environment. It is able to calculate both the MPE and NOHD for eye and skin in atmosphere transmission, with or without the presence of any interfaces, considering that the spectral transmissivity of that interface is known.

The analysis of Table 3 shows that lasers operating in the NIR (as may be the case for multi-spectral DIRCMs) present a high danger to the pilot's eye, since the calculated NOHD shows that, even at distances of 4834.98 m (GaAs, 840 nm), the laser beam irradiance is sufficient to cause damage to the retina. For higher wavelengths, the danger is reduced, since the NOHD and NSHD are much smaller and highly improbable to occur in actual situations. Although being improbable in an actual flight situation, these data must be considered if a ground test is being conducted. This conclusion leads to the necessity of using specific personal protection equipment (PPE) during the tests.

It is important to mention that the procedure followed for the calculation and analysis of results is the same of that one published by the Laser Institute of America (LIA) under number ANSI Z136.

Once the NOHD and NSHD are known, in free space, it is important to calculate these parameters considering the influence of the canopy, according to Fig. 4, since it works as a filter. Therefore, the values of transmittance shown in Fig. 4 were inserted in Eq. 2 as the variable τ_m. The results are shown in Table 4.

The results show that the IR attenuation provided by the canopy reduces NOHD of the three different types of lasers operating in the region of danger to the retina (<1.4 μm) in only 10%, in average. On the other hand, for wavelengths above 2.2 μm, in regions of MIR and FIR, the canopy acts as a perfect filter, protecting the pilot completely, except laser Cr^{2+}:CdSe (2600 nm), which, despite of suffering a strong attenuation, still presents some danger, depending on the distance of emission. The same conclusions can be extended to NSHD measurements. However, in this last case, the calculated distances are very low, showing to be highly improbable to occur in practice.

CONCLUSIONS

In this paper, we presented an investigation on the DIRCM risks to fighter aircraft pilots. Through experimental results and using a computational tool, it was showed that the F-5 fighter aircraft canopy provides only partial protection against IR radiation.

Measurements performed using a spectrometer showed that the average transmittance for canopies in the NIR range (0.67–1.3 μm) was around 90% (around 10% of absorption and reflection of canopy), presenting valleys at 1.19 and 1.4 μm.

Table 4. Safety distances for lasers when interfaced by the canopy.

Laser*	Wavelength (nm)	NOHD (m)	NSHD (m)	Transmittance
GaAs	840	4535.56	21.88	0.88
Nd:YAG	1064	2013.06	13.44	0.91
Nd:YAG	1330	570.09	10.61	0.60
Cr^{2+}:CdSe	2600	5.87	5.87	0.02
HeNe	3390	-0.80[1]	-0.80[1]	0.00
CO_2	10600	-0.80[1]	-0.80[1]	0.00

[1]Negative values do not have physical meaning and must be considered "0". For these mathematical results, laser is shown to be safe at any distance (AS 2211)
*Nakagawara and Montgomery (2008)

From 1.5μm on, transmittance drops sharply close to zero, presenting two peaks at 1.8 and 2 μm (around 40%). For wavelengths higher than 2.2 μm, transmittance show to be almost zero.

Due to the high transmissivity in NIR, additional theoretical calculations were performed, which showed that the safety distances for certain lasers operating in this range can reach up to 4.8 km, depending on some parameters of operation. Even with the canopy as an interface, this distance does not seem to be considerably reduced (≈4.5 km).

These results demonstrate that the F-5 canopy offers only limited protection against IR emissions, especially against DIRCMs operating at NIR wavelengths, to which the NOHD can reach more than 4.5 km. In order to complement this work, it is suggested that the same measurements are performed to visors of helmets used by pilots (mandatory item in fighter aircraft), as well as on other kinds of canopies, in order to have a more comprehensive picture of the problem, since different materials with different thicknesses will infer different spectral transmittances to the canopies

REFERENCES

Araújo, S.C. and Kawano, Y., 2001, "Espectro vibracional no infravermelho próximo dos polímeros poliestireno, Poli (Metacrilato de Metila) e Policarbonato", Polímeros Vol.11, No. 4, pp. 213-221.

Bolkcom, C., Elias, B. and Feickert, A., 2003, "Homeland Security: Protecting airlines from terrorist missiles", Congressional Research Service Report for Congress, The Library of Congress, USA, 27 p.

Brunetaud J.M. and Hill, S., 2012, "Risks from lasers". Retrieved in July 20, 2012, from http://www3.univlille2.fr/safelase/english/haza _en.html

Caidin, M., 1992, "Growing threat of UVR", Aviation Safety, Vol. 12, No. 3, pp.1-15.

Diffey, B. L. and Roscoe, A. H., 1990, "Exposure to solar ultraviolet radiation in flight", Aviation Space and Environmental Medicine, Vol. 61, No. 11, pp. 1032-1035.

Elbyt Systems, 1999, "Sensor and electro-optics solutions. MUSIC". Retrieved in July 04, 2013, from http://www.elbitsystems-us.com/sensor-electro-optics-solutions/airborne-maritime/survivability/music

GlobalSecurity.org, 2011, "AN-AAQ-24 Directional Infrared Countermeasures (DIRCM)". Retrieved in May 31, 2012, from http://www.globalsecurity.org/military /systems/aircraft/systems/an-aaq-24.htm

Hudson, R.D., 1969, "Infrared system engineering", New York: John Wiley & Sons, USA, 642 p.

Laurenzo, R., 2005, "Antimissile systems for airliners". Aerospace America: American Institute of Aeronautics and Astronautics, Vol. 43, Ed. 3, pp. 33-37.

Lelek, M., 2007, "Laser and non-linear optics – Laser safety". Retrieved in January 15, 2012, from http://www.optique-ingenieur.org/en/courses/OPI_ang_M01_C02

Nakagawara, V. B. and Montgomery, R. W., 2008, "Infrared radiation transmittance and pilot vision through civilian aircraft windscreens", Federal Aviation Administration, (DOT/FAA/AM-08/15), 18 pp.

NASA, 1998, "Ask an Astrophysicist". Retrieved in July 20, 2012, from http://imagine.gsfc.nasa.gov/docs/ask_astro/answers/980119b.html

Northwestern University, 2011, "Laser Safety Handbook", 20pp. Retrieved in July 19, 2012, from http://www.research.northwestern.edu/ors/forms/laser-safety-handbook.pdf

Seeber, F., 2012, "Fundamentals of Photonics", Module 1.2 - Light Sources and Laser Safety, Retrieved in July 19, 2012, from http://spie.org/Documents/Publications/00%20STEP%20Module%2002.pdf

Setlow, R.B., 2003, "The hazards of space travel". Retrieved in July 20, 2012, from http://www.nature.com/embor/journal/v4/n11/full/embor7400016.html

West, S.K., Duncan, D.D., Munoz, B., Rubin, G.S., Fried, L.P., Bandeen-Roche, K. and Schein, O.D.,1998, "Sunlight exposure and risk of lens opacities in a population-based study: the Salisbury Eye Evaluation project". The Journal of the American Medical Association, Vol. 280, No. 8, pp. 714-718.

Wisegeek, 1999, "How Dangerous is UV Radiation?". Retrieved in July 20, 2012, from http://www.wisegeek.com/how-dangerous-is-uv-radiation.htm

The Prediction of the Man-Hour in Aircraft Assembly Based on Support Vector Machine Particle Swarm Optimization

Tingting Yu[1], Hongxia Cai[1]

ABSTRACT: As the representative of manufacturing industry, aircraft assembly lacks of effective method to forecast man-hour. The forecasting accuracy of existing methods is universally pretty low. On the basis of full analysis of aircraft assembly's feature, this study proposes a forecasting model based on support vector machine (SVM), which is optimized by particle swarm optimization. It can carry out quantitative prediction of the process' man-hour during aircraft's assembly. Firstly, we decompose aircraft's assembly work by the concept of work breakdown structure. Further, the process parameters related to man-hour were listed and we made necessary correlation analysis of these historical data. Parameters with high contribution are then used as input of forecasting model. A new forecasting model utilizing SVM is proposed, which carries out the process as the minimum research granularity. Its performance is compared with back propagation neural network. The process of automatic drilling & riveting is adopted as an example in order to present and validate the model. Experimental results reflect that SVM has high forecast precision and good fitness, so that it is suitable for small sample prediction. Through the optimization, it can effectively predict man-hour of assembly work in a short time while maintaining sufficient accuracy.

KEYWORDS: SVM, PSO, Aircraft assembly, Man-hour prediction, Predictive model.

INTRODUCTION

Due to the large size, complex shape and numerous parts, the amount of aircraft assembly accounts for more than a half of total aircraft manufacturing (Enming, 2005). Aircraft assembly is the process of assembling a large number of aircraft parts according to a certain order and gradually putting them into components, forge pieces and sub-units. Finally, the components are butted into an entire aircraft. Therefore, aircraft assembly, as a very important part in the industry of large passenger aircraft, need be carried out strictly in accordance with the production plan, which includes not only the distribution of production site, equipment and resources, but also the arrangement of man-hour. Man-hour quota not only directly affects the working time, the utilization rate of the equipment, but also is the basic unit for calculating cost (Bin and Zuhua, 2006). At the same time, it is widely used as a toll for cost management in manufacturing enterprises. Therefore, man-hour prediction as the core of man-hour quota management is directly related to economic accounting, production schedule control, resource optimization, production cycles shortening, cost control and product quotation. Besides, it ultimately promotes the improvement of labor productivity of enterprises and enhances their market competitiveness (Chao and Danchen, 2010).

Just realizing this point, advanced international aircraft manufacturing companies have changed the condition that arranges production plan passively into predicting man-hours actively. Through the analysis, they can determine the production planning and production scheduling. Since the civil aircraft industry in China started late, there is a big gap in man-hour

1.Shanghai University – Shanghai – China.

Author for correspondence: Hongxia Cai | Shanghai Key Laboratory of Intelligent Manufacturing and Robotics | Shanghai University | Yanchang Road 149, Zhabei District | 200072 – Shanghai – China | Email: hxcai@staff.shu.edu.cn

prediction between China and international level. The overall utilization of workshop equipment is less than 40% (Changqing *et al.*, 2011). At present, man-hour prediction approach is mainly estimated through workers' experience, analogy and other methods based on relevant technical files and it has to calculate the production hours according to the detailed production process step by step. This method has the disadvantages of slow speed, low efficiency, big error and high dependence on personal experience, which causes the uncertainties in production planning and scheduling (Yajie *et al.*, 2013). As a result, it is unable to adapt to the needs of modernization, scientific and fine production management, seriously affecting the production schedule and production efficiency.

Different industries including computer, ships, computer numerical control (CNC) machining and aerospace have devoted a lot to develop a series of man-hour management software with good performance. Man-hour prediction has been treated as an important method to improve production efficiency, optimize resource management and shorten the manufacturing cycle. Among this series, the most representative are Timesheet, AceTeamwork, Appfarm, G TimeSheet and Replicon Web TimeSheet. However, most of these programs are integrated with enterprise resource planning (ERP) system, manufacturing execution system (MES) and computer aided process planning (CAPP). Their core function is to manage and share the document and the information during production. Thereby, man-hour management is deficient. So far, the time forecasting methods in manufacturing industry mainly includes standard data method, simulation method by numerical control (NC) program and artificial intelligence method. Boothroyd (1994) used an injection molding as standard and estimated other injection moldings rely on the relative value. Wang *et al.* (2002) used a triangular fuzzy number to estimate ready time and solved the ready time scheduling problem. Siller (2006) predicted the man-hour by simulation on NC program and specially forecasted the cycle of high-speed milling. Coelho (2010) also presented a practical mechanistic method for milling time estimation when machining free-form geometries by process simulation method. Bustillo and Correa (2012) presented a predictive model by using artificial intelligence to optimize deep drilling operations under high speed conditions for the manufacture of steel components. A feed-forward neural network and support vector machine (SVM) are used for surface's roughness prediction by Çaydaş and Ekici (2012). Two soft computing techniques, namely, neuro fuzzy logic technique and support vector regression technique were used for the assessment of the remaining useful life (RUL) of cutting tools (Gokulachandran and Mohandas, 2013). Li *et al.* (2012) tested various possible linear regression methods by using six dependent variables to enhance the production yield in the color filter (CF) manufacturing process. In his study, the Support Vector Machine for Regression (SVR) model was found to be the most appropriate for forecasting manufacturing performance. In these methods, SVM owns few adjustable parameters, quick computing speed, accurate rate, good robustness and good generalization ability (Cao and Tay, 2003). In the absence of more background information, it can also achieve high prediction accuracy (Kotsiantis, 2007). It solves the non-linear, high-dimension and local minima and some other issues effectively.

On the basis of extensive study of time prediction methods and the present situation in manufacturing industry, aircraft assembly work is broken down according to the concept of work breakdown structure (WBS). Then, we listed various related process parameters. Using process as the smallest granularity, the predictive model based on SVM is established. Typically, one of the important processes — automatic drilling & riveting process — is introduced as an example to verify the model. Moreover, the predictive performance is also compared with the back propagation (BP) neural network. Eventually, considering that the grid algorithm is time-consuming, we choose to use PSO algorithm and genetic algorithm (GA) to optimize the predictive model.

PROBLEM AND ALGORITHM DESCRIPTION

Man-hour has different meanings under different situations. In this paper, from the perspective of work assignment, it refers to a combined unit of human labor and time required to carry out individual task needed in the process of production (Hur *et al.*, 2013). For instance, a task involves cutting steel plate with the size of 1,000 x 800 x 5 mm and it needs two persons to complete this work. If it costs 10 minutes for each person, then the man-hour of the task is 20 (2 x 10) minutes. The man-hour of each task is related to comprehensive factors. Therefore, we have to predict the man-hour by taking into account various features of the task's properties, the environment and the requirements of the target. Because the process includes basic man-hour data and is the foundation for time management and prediction, the minimum unit of our research is the process.

A complex project can be decomposed into a series of clearly detailed subprojects by WBS, which groups project elements through deliverables-oriented ideas. These elements organize and define the work scope of a whole project. The subprojects decomposed by WBS cover labor and non-labor resources, including man-hour information which provide an important foundation for schedule planning, resource requirements, cost budget, purchasing plan etc. Top-down approach is applied in this paper to complete the WBS decomposition. Namely, aircraft assembly is decomposed by the order of project-sortie-workstation-station-process. Then, we can predict the man-hour of each process. The total time of a project can be obtained by accumulating all the process' man-hour.

SUPPORT VECTOR MACHINES

SVM is established on the basis of statistical learning theory (Cristianini and Shawe-Taylor, 2000). It is a kind of machine learning method which applies the principle of Vapnik-Chervonenkis (VC) dimension and structural risk minimization. It extensively overcomes the problems of "dimension disaster" and "too much learning" that exist in traditional machine learning methods such as neural network (Peng and Wang, 2009; Choi, 2009). It shows a lot of unique advantages in solving small sample, non-linear and high dimensional recognition problems. Especially, it is widely used in pattern recognition, regression analysis, function estimation, time series forecasting and other fields (Ge *et al.*, 2004; Guo and Li, 2003).

We establish a training set containing n training samples as $\{(x_i, y_i), i = 1, 2, ..., n\}$. Among them, x_i ($x_i \in R^d$) is the ith input column vector of training samples. In this study, they are a series of variables associated with the man-hour; $x_i = [x_i^1, x_i^2, ..., x_i^d]^T$, $y_i \in R$, is the corresponding output value, namely, process' man-hour, and d is the number of rows. It is clear that man-hour forecasting belongs to the typical non-linear problem. The regression function in high dimensional space, which represents the relationship between process' man-hour and input variables, can be expressed as:

$$f(x) = \omega\varphi(x) + b \qquad (1)$$

where $\varphi(x)$ represents the high dimensional feature space. It is a non-linear mapping function; ω and b are parameters to be estimated.

ζ limits the error of the regression function and slack variables, ξ_i, ξ_i^* ($\xi_i \geq 0, \xi_i^* \geq 0$)are introduced. Namely,

$$\begin{cases} \min \frac{1}{2}\|\omega\|^2 + C\sum_i^k(\xi_i + \xi_i^*) \\ s.t.\begin{cases} y_i - \omega\varphi(x_i) - b \leq \varepsilon + \xi_i \quad ,i = 1,2,...k \\ -y_i + \omega\varphi(x_i) + b \leq \varepsilon + \xi_i^* \\ \xi_i \geq 0, \xi_i^* \geq 0 \end{cases} \end{cases} \qquad (2)$$

where C is the penalty factor.

In order to simplify the learning process of SVM, the problem can be changed as follows by using Lagrange function:

$$\begin{cases} \text{Max}\left[-\frac{1}{2}\sum_{i=1}^{k}\sum_{j=1}^{k}(\alpha_i - \alpha_i^*)(\alpha_j - \alpha_j^*)K(x_i, x_j) - \sum_{i=1}^{k}(\alpha_i + \alpha_i^*)\varepsilon + \sum_{i=1}^{k}(\alpha_i - \alpha_i^*)y_i\right] \\ s.t. \begin{cases} \sum_{i=1}^{k}(\alpha_i - \alpha_i^*) = 0 \\ 0 \leq \alpha_i \leq C, \quad i = 1,2,...,n \\ 0 \leq \alpha_i^* \leq C, \quad i = 1,2,...,n \end{cases} \end{cases} \qquad (3)$$

where $K(x_i, x_j) = \phi(x_i)\,\phi(x_j)$ (the inner product of x_i and x_j) is kernel function.

We applied radial basis function (RBF) kernel function $K(x,y) = \exp(-\frac{|x-y|^2}{\sigma^2})$ in our study. Assuming that the optimal solution of Eq. 3 is , $\alpha = [\alpha_1, \alpha_2, ..., \alpha_k]$, $\alpha^* = [\alpha^*_1, \alpha^*_2, ..., \alpha^*_k]$, then we get the regression function:

$$f(x) = \Sigma^k_{i=1}(\alpha_i - \alpha^*_i)\,\phi(x_i)\,\phi(x) + b = \Sigma^k_{i=1}(\alpha_i - \alpha^*_i)\,K(x_i - x_j) + b \qquad (4)$$

In SVM regression problems, the parameter σ controls the flexibility of RBF kernel function. It directly affects the generalization performance of models. The penalty factor C plays the role of balancing decision function and error classification samples, which influences the generalization ability of the model.

PARTICLE SWARM OPTIMIZATION

Particle swarm optimization (PSO) was firstly proposed Kennedy and Eberhart (1995) and it is based on the research of group behavior of birds. PSO looks each individual in n-dimension as a particle without weight and volume. It flies in the search space with a certain speed. The flight speed is adjusted dynamically by individual flight experience and group flying experience.

In a n-dimensional space, the ith particle's position and speed can be expressed respectively as: $X_1 = (X_{i1}, X_{i2}, ..., X_{in})$, $V_i = (V_{i1}, V_{i2}, ..., V_{in})$. The optimal position that each particle has experienced is $P_i = (P_{i1}, P_{i2}, ..., P_{in})$. The global optimal position is $P_{gbest} = (P_{g1}, P_{g2}, ..., P_{gn})$. $P_g(t)$ means the best location of a population containing s particles.

$$P_g(t) \in \{P_0(t), ... P_g(t)\} | f(P_g(t)) \min \{f(P_0(t)), ... f(P_s(t))\} \quad (5)$$

The updating formula of the ith particle's velocity and position are:

a) velocity updating formula:

$$V_i(t+1) = wV_i(t) + c_1 r_1 (P_i(t) - x_i(t) + c_2 r_2 (P_g(t) - X_i(t)) \quad (6)$$

b) position updating formula:

$$X_i(t+1) = X_i(t) + V_i(t+1) \quad (7)$$

where w is the inertia coefficient whose value is also in the scope of (0,1); c_1 is the cognitive coefficient; c_2 is the social learning coefficient. The values of c_1 and c_2 are usually in the scope of (1, 2); r_1, r_2, are the random number in the scope of (0,1).

The parameters C and g should be considered as a particle in space. Under this assumption, the problem of parameter optimization is substantially to find the global optimal position of all particles. Then, the SVM model can be optimized by utilizing the advantages of fast global optimization possessed by PSO.

THE PREDICTIVE MODEL BASED ON SUPPORT VECTOR MACHINES-PARTICLE SWARM OPTIMIZATION MODEL CONSTRUCTION

Historical data is a comprehensive reflection of the internal mechanism of a system's changes (Bing, 2014). The number of historical data shows the mechanism of the changes in an extent. Process' man-hour is an important basis for reasonable plan and arrangement of production. Man-hour related data are dependent on the specific process. Therefore, process is the basic unit for our study. We analyzed the technological characteristics of a specific process to extract influence factors. Besides, intrinsic relationships between these factors and the process' man-hour were further mined through machine learning, so as to achieve the purpose of man-hour prediction.

The steps of establishing the predictive model of SVM are described as follows:

Influence factors analysis

The total man-hour of aircraft assembly is affected by many factors including product parameters, process parameters and some other kind of parameters. Besides, the relationship among these factors is complex, leading the forecasting work to a typical non-linear problem. As input parameters of the predictive model, man-hour driving factors directly affect the predictive accuracy of the model. So, we need to analyze this influence from aspects of the contribution and relevance of the factors. This step is to make correlation analysis of related factors by using statistical product and service solutions (SPSS), an IBM software widely used in statistical analysis, data mining, prediction analysis and decision support.

Sample data acquisition and processing

The data's changing scope of sample set imposes an important influence on training model. The sample set should reflect the characteristics of a process as far as possible. Then the data were observed and collected, and invalid data were excluded. Due to the big differences in magnitude and unit, the data should be normalized before establishing the regression model.

Model selection and parameter determination

When building SVM, the selection of kernel function is directly related to the performance of the model. We choose to use radial basis function (RBF) which is widely used owning to its best performance. The penalty factor C and the kernel parameter g directly influence the accuracy of the predictive model. The role of C is to adjust the learning machine's confidence interval range in a certain data subspace. Kernel parameter g affects the distribution complexity of sample data's subspace and determines the minimum error (Keerthi and Lin, 2003). In this study, cross validation is firstly adopted to determine the parameters C and g. Subsequently, we apply PSO to make further optimization and we get the best parameter combinations of C and g.

Sample training and model establishment

Generally, for a given time prediction problem, sample data $\{x_1, x_2, x_3, ..., x_r\}$ are usually divided into 2 parts. The front m data is set as the training sample to build predictive model and the latter n-m data is for prediction test. Model building is actually the process of solving regression function (Eq. 2) by SVM with its good performance.

Application and forecast

Put the input data that need to be forecast in accordance with the specified format. Then, predict them and make error analysis through the predictive model achieved in step "Sample training and model establishment". For regression problems, the measures to value the performance of a model are mean square error (MSE or E) and determination coefficient (R^2):

$$E = \frac{1}{n}\sum_{i=1}^{n}(y'_i - y_i)^2 \quad (8)$$

$$R^2 = \frac{(n\sum_{i=1}^{n} y'_i . y_i - \sum_{i=1}^{n} y'_i . \sum_{i=1}^{n} y_i)^2}{(n\sum_{i=1}^{n} y_i'^2 - (\sum_{i=1}^{n} y'_i)^2)(n\sum_{i=1}^{n} y_i^2 - (\sum_{i=1}^{n} y_i)^2)} \quad (9)$$

where n is the number of test samples; y_1(i = 1, 2, …, n) is the real value of the ith sample; y'_1(i = 1, 2, …, n) is the predictive value of the ith sample. The smaller MSE means the higher prediction accuracy. R^2 determines the related degree and the value more closer to 1 indicates the greater fitness.

Build the man-hour predictive model based on SVM-PSO as previously mentioned (Fig. 1).

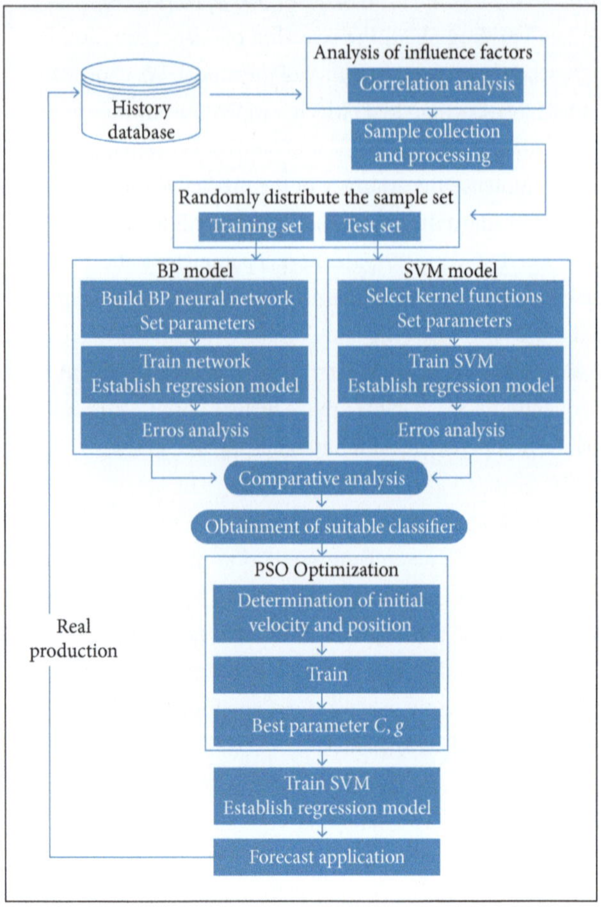

Figure 1. The man-hour predictive model based on SVM-PSO.

MAN-HOUR PREDICTION OF AUTOMATIC DRILLING & RIVETING BY SVM-PSO

WBS decomposition demonstrates that the aircraft assembly involves multiple workstations and stations. Assembly work eventually composes a series of assembly process. There is a variety of connections during the course of aircraft assembly, such as riveting, bolting, welding, gluing etc. According to statistics, 70% of the aircraft accidents due to fatigue failure were attributed to the joint connection. Moreover, 80% of the fatigue cracks occurred in hole connection (Liming and Chongneng, 2008). It is visible that connection quality greatly affects the life of an aircraft. Riveting occupies a very important position among different kinds of connections. It is estimated that the amount of assembly labor accounts for about half or even more of the entire aircraft manufacturing labor. And riveting accounts for 30%. Owning to the good fatigue resistance and high reliability, automatic drilling & riveting is widely used in the assembly process of large plane and wing panel. It is an important process for aircraft assembly. Therefore, this paper takes automatic drilling & riveting process as an example to conduct the research of man-hour prediction.

ANALYSIS OF THE INFLUENCE FACTORS OF PROCESS' MAN-HOUR

According to statistics, headless riveting accounts for more than 80% in wing manufacturing process of aircraft ARJ21. Riveting & drilling parameters are the key-factors that affect the efficiency and quality of assembly (Lianxi *et al.*, 2013). Generally, we analyze the driving factors of the automatic drilling & riveting process from the aspects of general features of assembly (I_F), processing technic (I_p), equipment parameters (I_E) and performance indexes (I_I). The function relation between man-hour and various factors can be expressed as:

$$T = f(I_F, I_p, I_E, I_I) \quad (10)$$

General features of assembly includes assembly parts (F_p), material's composition (F_i) and thickness of the material (F_t). At present, the material's airframe (including fuselage, wings, tail etc.) mainly adopts the materials of aluminum alloy, titanium alloy, steel and composite materials in the world. However, aluminum alloy material is used in more than 90% of the cases due to the restrictions of China's civil aircraft development. So, material composition is removed. Then, general features of assembly can be expressed as:

$$I_F = f_1(F_p, F_t) \quad (11)$$

The processes can be achieved by automatic drilling & riveting, which mainly includes: drilling, countersinking, glue, nail feeding, fastener installation or completing one or more of these operations. We need to use the combination of these process. So, it is unnecessary to consider process technic when analyzing the influence factors.

Equipment parameters associated with the man-hour are mainly related to drilling parameters including spindle speed (E_r), clamping force (E_c), feed rate (E_d), riveting force (E_f), riveting residence time (E_t) and lubrication pressure (E_p). The relation between equipment parameters and man-hour can be expressed as:

$$I_E = f_2(E_r, E_c, E_d, E_f, E_t, E_p) \tag{12}$$

The indexes of main performance refer to scratch length of hole (I_l), surface roughness (I_{Ra}) and burr height (I_h).

$$I_l = f_a(I_l, I_{Ra}, I_h) \tag{13}$$

After these considerations, Eq. 10 can be expressed as:

$$T = f(F_p, F_t, E_r, E_c, E_d, E_f, E_t, E_p, I_l, I_{Ra}, I_h) \tag{14}$$

It is found that the major driving factors are extremely related to some other detail parameters. Besides, these parameters may influence each other or some just have weak influence. In the current modeling process, we should pay enough attention to the pre-condition that improves the accuracy of the model with least driving factors to reduce the complexity and the noisy interference of the model (Wang *et al.*, 2012).This requires further analysis of the factors' correlation and contribution. This study used Statistical Package for the Social Sciences (SPSS) to analyze the correlation of 99 sample data extracted from historical database. Firstly, the sample data was saved as in the format file.sav, which can be recognized by SPSS. Then, the analysis was made to judge the correlation between F_p and process' man-hour, including Pearson's correlation and conspicuousness. Similarly, we obtained the correlation between each influence factor and process' man-hour through various analyses, respectively. Finally, the results are demonstrated in Table 1.

In the mentioned table, Pearson's correlation coefficient describes the relationship between variables as well as their statistical correlation degree. Pearson's correlation coefficients bigger than 0.4 mean good correlation, bigger than 0.5 mean strong correlation and bigger than 0.6 represent very strong correlation. Conspicuousness can be understood as probability, which is used to determine the overall conspicuousness difference between reality and hypothesis. It requires conspicuousness levels below 0.05, because 5% can be considered as a small probability event. It is easy to find from Table 1 that driving factors which influence the man-hour of automatic drilling & riveting process are:

$$T = f(F_t, E_r, E_c, E_d, E_f, E_t) \tag{15}$$

SAMPLE DATA ACQUISITION AND PROCESSING

Firstly, we have to screen the sample data extracted from the historical database to eliminate invalid data of unconventional or high reusable degree. Then, 99 valid data were obtained and they were separated into test set (19 data) and training set (80 data) when building the model. The training set and the test set were randomly selected in order to ensure the accuracy of the model. Because the units of F_t, E_r, ..., E_t are different and the magnitude of them differs tremendously, we made the normalization processing of these samples by the following formula:

$$y = \frac{(y_{max} - y_{min}) + (x - x_{min})}{x_{max} - x_{min}} + y_{min}, (y_{max} = 1, y_{min} = -1) \tag{16}$$

The data were normalized to the range of (-1,1) and 80 training data after normalization are shown in Table 2.

Table 1. Correlation comparison of different influence factors.

	F_p	F_t	E_r	E_c	E_d	E_f	E_t	E_p	I_l	I_{Ra}	I_h
Pearson's correlation	-0.013	0.485	-0.441	-0.468	-0.337	-0.447	0.930	-0.078	-0.209	0.161	0.179
Conspicuousness (bilateral)	0.927	0.066	0.000	0.000	0.001	0.000	0.001	0.444	0.146	0.265	0.213

Table 2. Sample set after normalization.

	F_t	E_r	E_c	E_d	E_f	E_t	T
1	-0.3333	0.1667	-0.1538	-0.1765	-0.2598	0.5	0.2992
2	0.3333	0.8333	0.8462	1	-1	0.5	0.2356
3	-0.3333	0.1667	-0.1538	-0.1765	-0.2598	0.5	0.2992
4	0.3333	0	-0.2308	-0.2941	-0.2598	-0.3	-0.1811
5	0.6667	-0.6667	-0.6154	-0.5882	-0.4173	0.5	0.6252
6	0.3333	-0.5000	-0.6923	-0.2941	-0.5748	1	0.8425
7	-0.6667	0.3333	-0.0769	-0.1765	-0.1024	0.5	0.2402
8	-1	-0.8333	-0.8462	-0.8824	-0.7638	0.5	0.3228
9	-1	0.1667	-0.3846	-0.1765	-0.4173	0.5	0.1811
10	1	0.5000	0.2308	0.2941	0.2126	-1.1102	0.0236
11	-0.5000	0.4167	0.0769	-0.0588	0.1339	-1.1102	-0.1431
12	1	0.5000	0.2308	0.2941	0.2126	-1.1102	0.0236
…	…	…	…	…	…	…	…
80	-0.3333	0.8333	0.8462	0.8824	0.6850	0.6850	-0.6157

REGRESSED PREDICTION BASED ON SUPPORT VECTOR MACHINES

We will begin to train the sample data and establish predictive model in this part after completing the mentioned preparations. Firstly, use cross validation ($v = 5$) to find the best parameters of C and g. Secondly, put the normalized sample data into predictive model and complete sample learning and prediction. The prediction results of training set and test set are shown in Figs. 2 and 3.

Both training set and test set were randomly generated in the experiment, so it could ensure the universality and accuracy of the result. Nevertheless, it is particularly necessary to conduct comparative analyses on other methods. In order to further validate the performance of the model, we also presented BP neural network to make predict experiment. The forecasting result is shown in Fig. 4.

Table 3 assesses the comparison of the forecasting results by SVM and BP neural network.After comparison, it is obvious that the prediction result of SVM is much better than BP neural network. MSE of BP neural network is much higher than SVM and R^2 of SVM is closer to 1. Thus, it proves that SVM can be effectively applied to the prediction problem of small sample while maintaining sufficient accuracy.

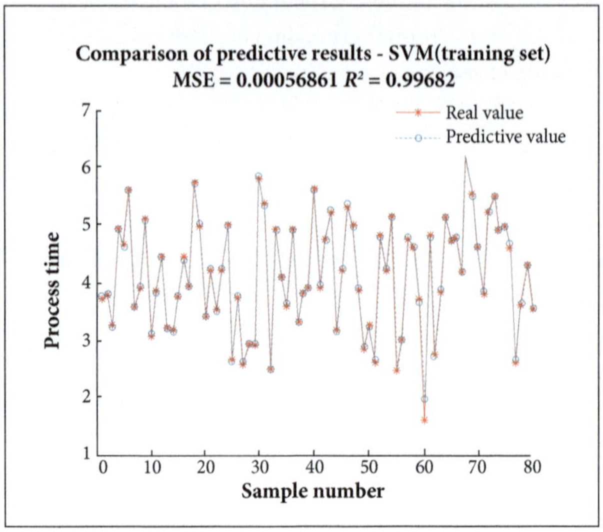

Figure 2. The prediction result of training set.

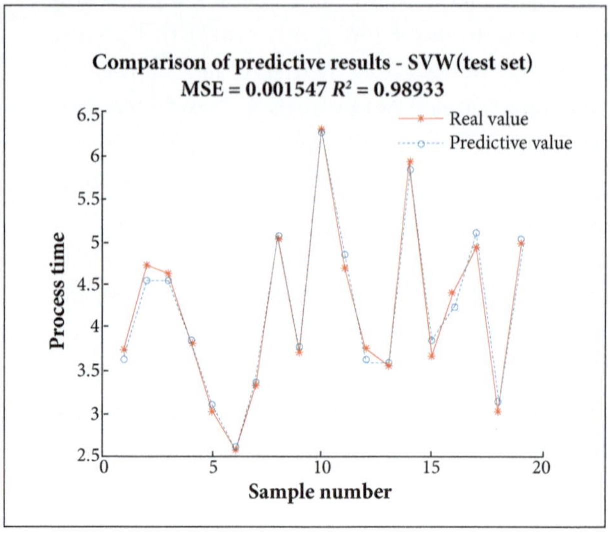

Figure 3. The prediction result of test set.

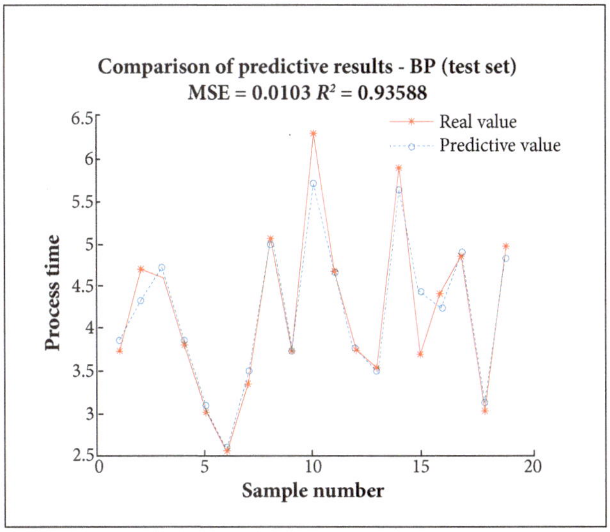

Figure 4. Prediction result of the test set by BP neural network.

Table 3. The comparison of the forecasting results.

Real man-hour (s)	Prediction value by SVM	Prediction value by BP neural network
3.7143	3.6083	3.8610
4.6923	4.5447	4.3332
4.6000	4.5341	4.7431
3.8000	3.8413	3.8626
3.0286	3.0816	3.0772
2.5760	2.5947	2.5792
3.3500	3.3457	3.4554
5.0500	5.0346	5.0085
3.7200	3.7602	3.7193
6.2800	6.2320	5.7196
4.6750	4.8345	4.6614
3.7385	3.6001	3.7587
3.5360	3.5565	3.4862
5.9200	5.8023	5.6485
3.6462	3.8205	4.4489
4.3840	4.2309	4.2387
4.9000	5.0474	4.8978
3.0286	3.1098	3.1021
4.9474	4.9944	4.8441
MSE	0.001547	0.0103
R2	0.98933	0.93588

OPTIMIZATION OF MAN-HOUR PREDICTIVE MODEL

The parameters of SVM decide its study performance and generalization ability. In practical application, in order to reduce the dependence of initial sample and improve the accuracy of SVM model, it is necessary to optimize the parameters of *C* and *g*. It means that SVM optimization is actually the optimization of the relevant parameters, namely, quickly find optimal SVM parameters *C* and *g*. Grid search can find the highest accuracy of classification, namely, the global optimal solution. However, if you want to find the best parameters of *C* and *g* in greater scope, it will be very time-consuming. With the decrease of the grid density, time grows exponentially. Heuristic algorithm can find the global optimal solution without traversing all the points in the grid. Currently, the widely used heuristic algorithms for finding optimal parameters are PSO algorithm, genetic algorithm, and simulated annealing algorithm.

Considering that the principle of PSO and GA is simple and they are easy to implement with high efficiency, this study tried to imply PSO and GA to optimize SVM prediction results obtained above. Figures 5 to 7 show the results of optimization using the PSO and their corresponding prediction results.

Figures 8 to 12 show the results of optimization using the PSO and their corresponding prediction results.

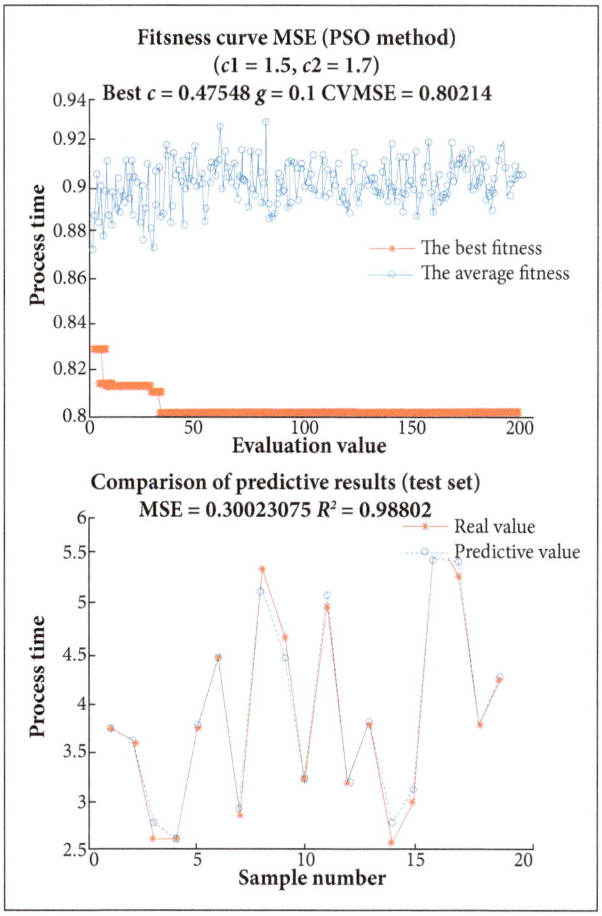

Figure 5. The optimization by PSO when maxgen = 200; pop = 20.

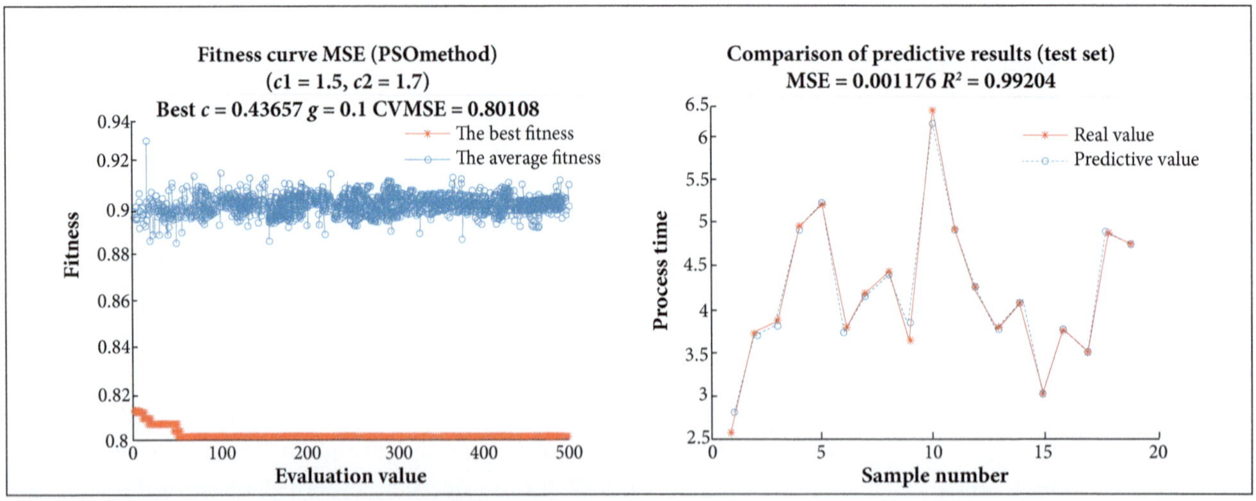

Figure 6. The optimization by PSO when maxgen = 500; pop = 50.

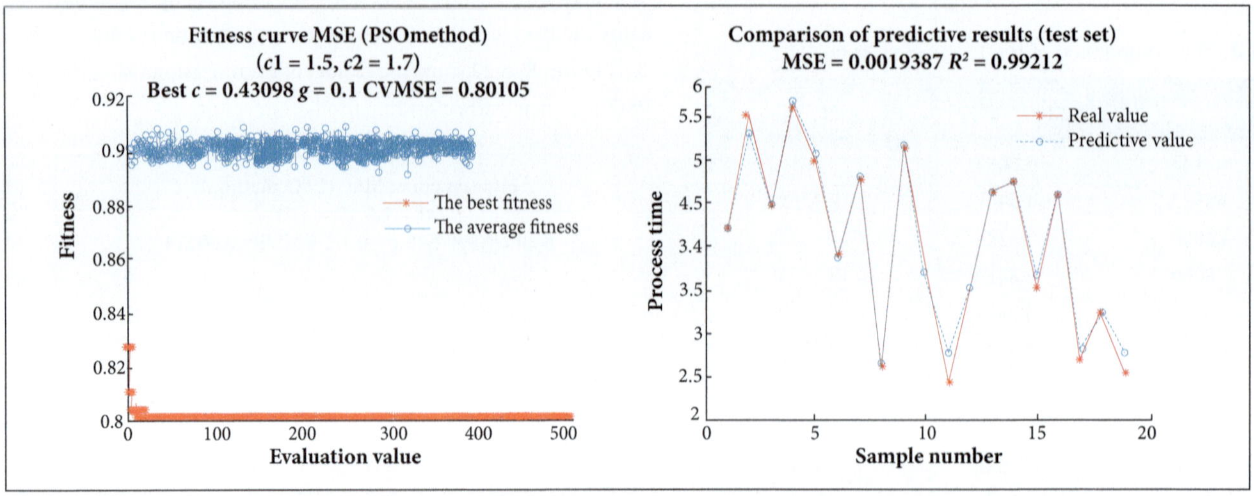

Figure 7. The optimization by PSO when maxgen = 500; pop = 100.

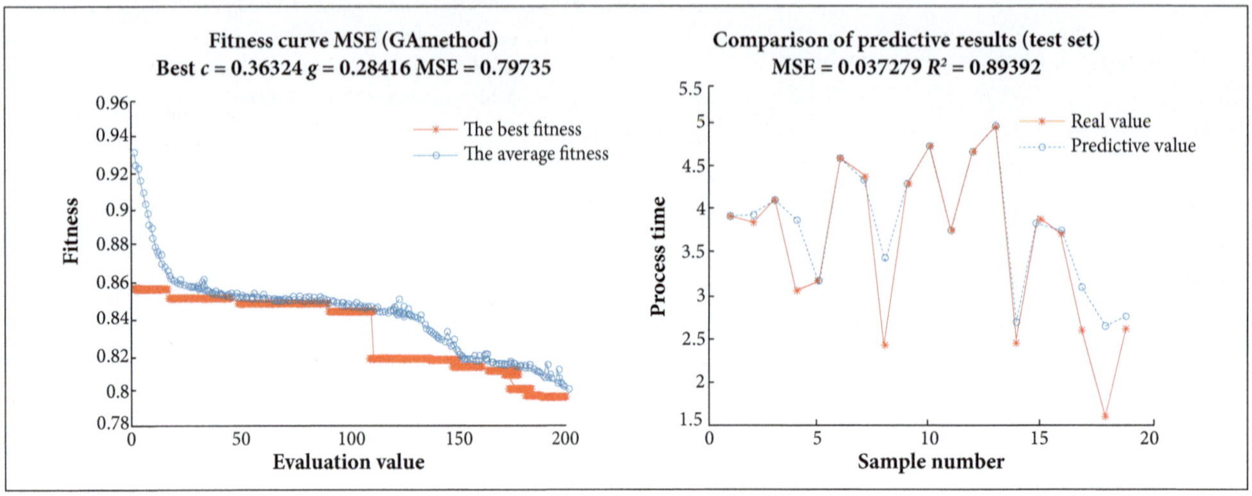

Figure 8. The optimization by GA when maxgen = 200; pop = 20.

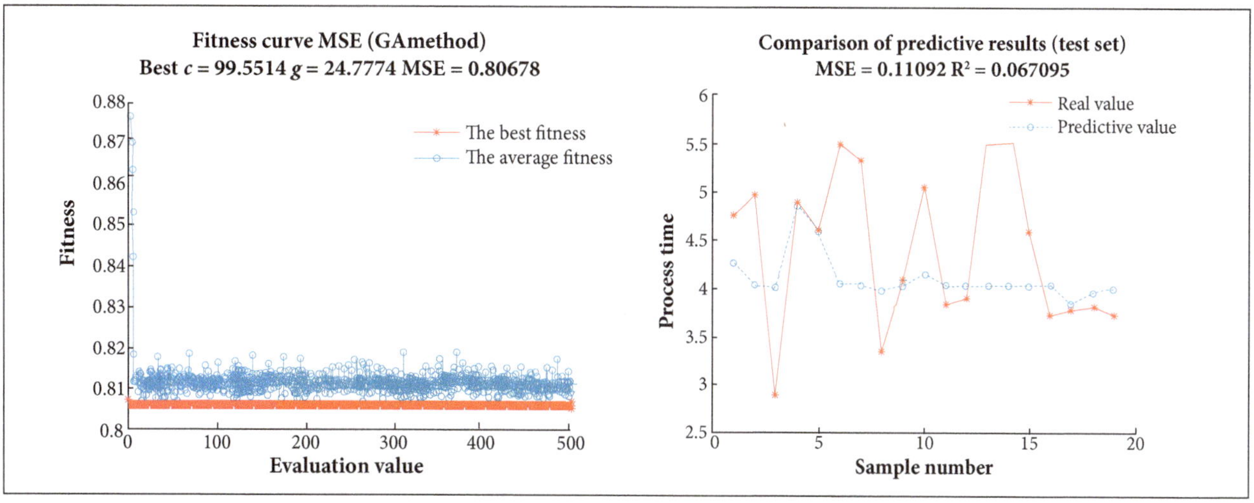

Figure 9. The optimization by GA when maxgen = 500; pop = 50.

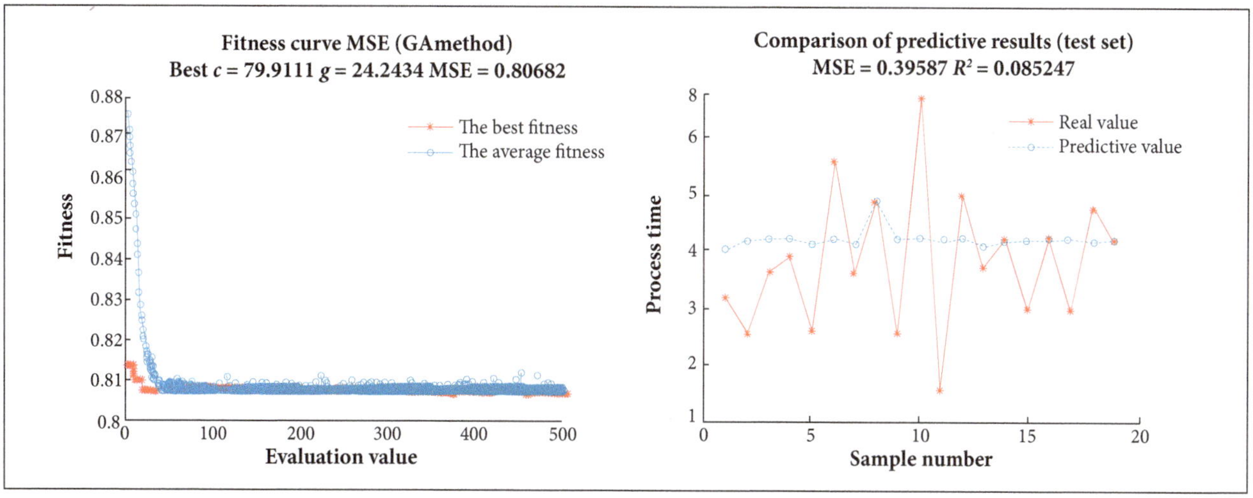

Figure 10. The optimization by GA when maxgen = 500; pop = 100.

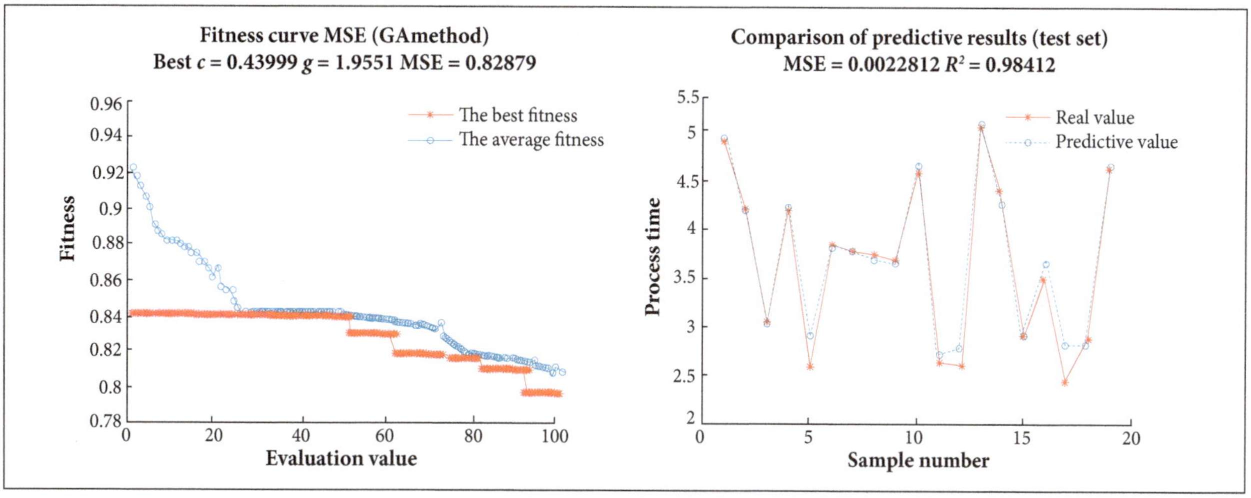

Figure 11. The optimization by GA when maxgen = 100; pop = 20.

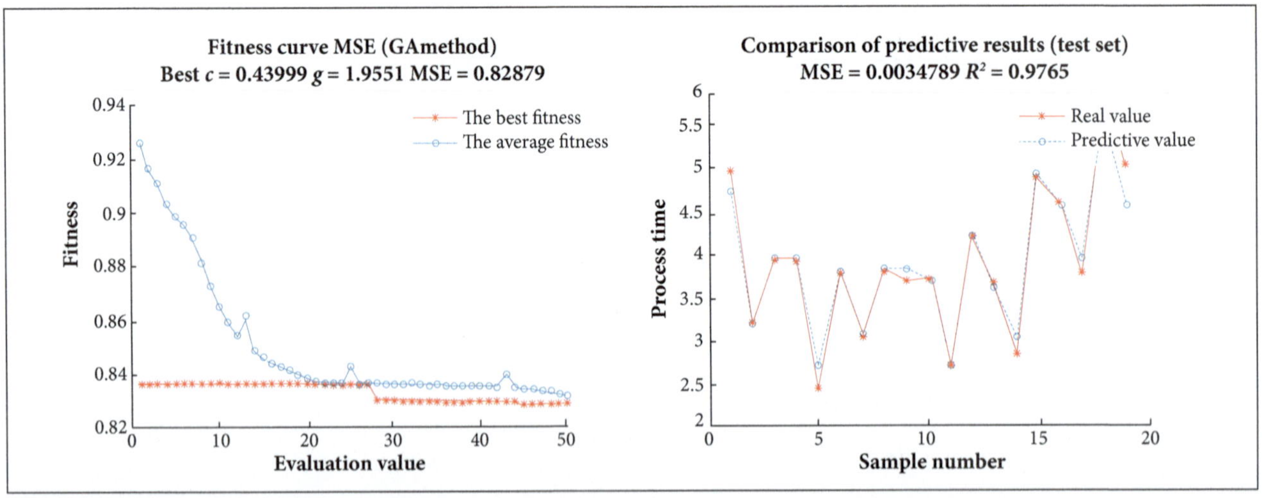

Figure 12. The optimization by GA when maxgen = 50; pop = 20.

From the optimization experiments of PSO and GA, we could find that the optimal prediction result is obtained when the maximum number of evolution (maxgen) = 500; population (pop) = 100 of PSO (MSE = 0.0019387; R^2 = 0.99212). And the optimal prediction result is obtained when maxgen = 100; pop = 20 of GA (MSE = 0.0022812; R^2 = 0.98412). As a result, PSO optimization is much better than GA.

CONCLUSION

Aircraft assembly is crucial for its production. It requires effective prediction of man-hour at the beginning of a project, in order to arrange the resources and make production plan reasonably. So far, time prediction mainly relies on work experience. It has a lot of problems, such as slow speed, low efficiency, big error and serious dependence on personal experience. Because the aircraft assembly work is complex and be related to various process' parameters, this research decomposed the assembly work through the concept of WBS. Then, the parameters that affect process were extracted by analyzing process' features. Parameters with high correlation are useful. Further, we selected the parameters that highly contributed to predicting man-hour as driving factors. Finally, man-hour predictive model based on SVM was established. Moreover, we made a comparison with BP neural network. As the parameters of C and g extremely determine the model's performance, this study proposed PSO and GA to optimize the predictive model. Various experiments were made with different parameters to evaluate MSE and R^2. The results indicate that PSO is more accurate with better fitness. The process of automatic drilling & riveting is used as an example to validate the model with the comparing result of MSE = 0.0019387; R^2 = 0.99212. It means that the MSE between predictive value and the real value is very small and the coefficient of determination reached 0.99212. Namely, the goodness of fit is close to 1, which can well reflect the relation between the dependent variable (man-hour) and the independent variable (F_t, E_r, E_c etc.).

The predictive model based on SVM PSO can effectively forecast the man-hour of different processes according to their driving factors. It is an important approach for enterprises to realize scientific management of the man-hour. In the next step, we will continue to do more studies and experiments so as to make the prediction more precise.

ACKNOWLEDGEMENTS

This work was supported by Shanghai Science and Technology Committee, under grant No. 13521103604. It was also supported by Shanghai Key Laboratory of Intelligent Manufacturing and Robotics. The authors are grateful for the financial support and also would like to thank the anonymous reviewers and the editor for their comments and suggestions.

REFERENCES

Bin, L. and Zuhua, J., 2006, "The Intelligent Man-Hour Estimate Technique of Assembly for Shipbuilding" (In Chinese), Journal of Shanghai Jiaotong University, Vol. 39, No. 12, pp. 1979-1983.

Bing, D., 2014, "Reliability Analysis for Aviation Airline Network Based on Complex Network", Journal of Aerospace Technology and Management, Vol. 6, No. 2, pp. 193-201. doi: 10.5028/jatm.v6i2.295

Boothroyd, G., 1994, "Product Design for Manufacture and Assembly", Computer-Aided Design, Vol. 26, No. 7, pp. 505-520.

Bustillo, A. and Correa, M., 2012, "Using artificial intelligence to predict surface roughness in deep drilling of steel components", Journal of Intelligent Manufacturing, Vol. 23, No. 5, pp. 1893-1902. doi: 10.1007/s10845-011-0506-8

Cao, L.J., and Tay, F.E.H., 2003, "Support Vector Machine with Adaptive Parameters in Financial Time Series Forecasting", IEEE Transactions on Neural Networks, Vol. 14, No. 6, pp. 1506-1518. doi: 10.1109/TNN.2003.820556

Çaydaş, U. and Ekici, S., 2012, "Support Vector Machines Models for Surface Roughness Prediction in CNC Turning of AISI 304 Austenitic Stainless Steel", Journal of Intelligent Manufacturing, Vol. 23, No. 3, pp. 639-650. doi: 10.1007/s10845-010-0415-2

Changqing, L., Yingguang, L., Wei, W. and Yong, L., 2011, "Feature Based Man-Hour Forecasting Model for Aircraft Structure Parts NC Machining", Computer Integrated Manufacturing Systems, Vol. 17, No. 10, pp. 2156-2162. doi:10.13196/j.cims.2011.10.66.liuzhq.026

Chao, G. and Danchen, Z., 2010, "Man-Hour Quota System Based on Genetic Neural Network" (In Chinese), Computer Applications and Software, Vol. 27, No. 8, pp. 205-208.

Choi, Y.S., 2009, "Least Squares One-Class Support Vector Machine", Pattern Recognition Letters, Vol. 30, No. 130, pp. 1236-1240.

Coelho, R.T., de Souza, A.F., Roger, A.R., Rigatti, A.M.Y., Ribeiro, A.A.L., 2010, "Mechanistic Approach to Predict Real Machining Time for Milling Free-Form Geometries Applying High Feed Rate", The International Journal of Advanced Manufacturing Technology, Vol. 46, No. 9-12, pp. 1103-1111. doi: 10.1007/s00170-009-2183-8

Cristianini, N. and Shawe-Taylor, J., 2000, "An Introduction to Support Vector Machines and Other Kernel-Based Learning Methods", Cambridge University Press, Cambridge, UK. 189p.

Enming, G., 2005, "Flexible Assembly Technology of Foreign Aircraft" (In Chinese), Aeronautical Manufacturing Technology, No. 9, pp. 28-33.

Ge, M., Du, R., Zhang, G. and Xu, Y., 2004, "Fault Diagnosis Using Support Vector Machine with an Application in Sheet Metal Stamping Operations", Mechanical Systems and Signal Processing, Vol. 18, No. 1, pp. 143-159. doi:10.1016/S0888-3270(03)00071-2

Gokulachandran, J. and Mohandas, K., 2013, "Comparative Study of Two Soft Computing Techniques for the Prediction of Remaining Useful Life of Cutting Tools", Journal of Intelligent Manufacturing, pp.1-14. doi: 10.1007/s10845-013-0778-2

Guo, G.D. and Li, S.Z., 2003, "Content-Based Audio Classification and Retrieval by Support Vector Machines", IEEE Trans on Neural Network, Vol. 14, No. 1, pp. 209-215. doi: 10.1109/TNN.2002.806626

Hur, M., Lee, S.K., Kim, B., Cho, S., Lee, D. and Lee, D., 2013, "A Study on the Man-Hour Prediction System for Shipbuilding", Journal of Intelligent Manufacturing, pp.1-13. doi: 10.1007/s10845-013-0858-3

Keerthi, S.S. and Lin, C.J., 2003, "Asymptotic Behaviors of Support Vector Machines with Gaussian Kernel", Neural Computation, Vol. 15, No. 7, pp. 1667-1689. doi:10.1162/089976603321891855

Kennedy, J. and Eberhart, R., 1995, "Particle Swarm Optimization", IEEE International Conference on Neural Networks, Vol. 4, pp. 1942-1948. doi: 10.1109/ICNN.1995.488968

Kotsiantis, S.B., 2007, "Supervised Machine Learning: a Review of Classification Techniques", Informatica, Vol. 31, No. 3, pp. 249-268.

Li, D.C., Chen, W.C., Liu, C.W. and Lin, Y.S., 2012, "A Non-Linear Quality Improvement Model Using SVR for Manufacturing TFT-LCDs", Journal of Intelligent Manufacturing, Vol. 23, No. 3, pp. 835-844. doi: 10.1007/s10845-010-0440-1

Lianxi, L., Xining, L., Zhongqi, W. and Weiping, W., 2013, "Headless Rivet Test of Automatic Drilling Riveting Process" (In Chinese), Journal of Northwestern Polytechnical University, Vol. 31, No. 1, pp. 77-82.

Liming, W. and Chongneng, F., 2008, "Application of Digital Automatic Drill-Riveting Technology in Aircraft Manufacture" (In Chinese), Aeronautical Manufacturing Technology, Vol. 11, pp. 42-45.

Peng, X. and Wang, Y., 2009, "A Normal Least Squares Support Vector Machine (NLS-SVM) and its Learning Algorithm", Neurocomputing, Vol. 72, No. 16-18, pp. 3734-3741. doi:10.1016/j.neucom.2009.06.005

Siller, H., Rodriguez, C. A. and Ahuett, H., 2006, "Cycle Time Prediction in High-Speed Milling Operations for Sculptured Surface Finishing", Journal of Materials Processing Technology, Vol. 174, No. 1-3, pp. 355-362. doi:10.1016/j.jmatprotec.2006.02.008

Wang, C., Wang, D., Ip, W.H. and Yuen, D.W., 2002, "The Single Machine Ready Time Scheduling Problem with Fuzzy Processing Times", Fuzzy Sets and Systems, Vol. 127, No. 2, pp. 117-129. doi:10.1016/S0165-0114(01)00084-7

Wang, M.Z., Deng, Y.D. and Chang, X.J., 2012, "Estimation Model of Plane Man-Hours in Manufacture Based on Partial Least-Squares Regression", Acta Aeronautica et Aeronautica Sinica, Vol. 33, No. 8, pp. 1448-1454.

Yajie, L., Weiping, H., Rong, D. and Yanli, H., 2013, "Man-hour Forecasting and Evolution Based on MES Data Collection", Computer Integrated Manufacturing Systems, Vol. 19, No. 11, pp. 2810-2818. doi:10.13196/j.cims.2013.11.liyajie.2810.9.20131118

13

A Multidisciplinary Design Optimization Tool for Spacecraft Equipment Layout Conception

Valentino Lau[1], Fabiano Luis de Sousa[1], Roberto Luiz Galski[1], Evandro Marconi Rocco[1], José Carlos Becceneri[1], Walter Abrahão dos Santos[1], Sandra Aparecida Sandri[1]

ABSTRACT: One issue the design team has to face in the process of building a new spacecraft, is to define its mechanical and electrical architecture. The choice of where to place the spacecraft´s electronic equipment is a complex task, since it involves simultaneously many factors, such as the spacecraft´s required position of center of mass, moments of inertia, equipment heat dissipation, integration and servicing issues, among others. Since this is a multidisciplinary task, the early positioning of the spacecraft´s equipment is usually done "manually" by a group of system engineers, heavily based on their experience. It is an interactive process that takes time and hence, as soon as a feasible design is found, it becomes the baseline. This precludes a broader exploration of the design space, which may lead to a suboptimal solution, or worse to a design that will have to be modified later. Recently, it has been shown the potential benefits of automating the process of spacecraft´s equipment layout using optimization techniques. In this paper, a prototype of an Excel® based tool for multidisciplinary spacecraft equipment layout conception is described. Provided the geometric dimensions, mass and heat dissipation of the equipment, and the available positioning area, the tool can automatically generate many possible trade-off solutions for the layout. It allows the user to set specific equipment to specific areas of positioning, and different combinations of objective functions can be used to drive the design. The features of the tool are shown in a simplified three dimensional problem.

KEYWORDS: Layout optimization, Spacecraft, Conceptual design, Electronic equipment.

INTRODUCTION

In the conceptual phase of the development of a new spacecraft, different candidate solutions for its electrical and mechanical architectures are assessed, in a search for one which would fit the spacecraft mission, within the constraints of cost and schedule. It is in this phase that the main features of its subsystems are defined, and where the systemic and multidisciplinary character of the design process becomes more relevant to the definition of its cost and performance.

The assessment of different solutions for the mechanical and electrical architecture includes the positioning of the spacecraft's equipment over its structure panels, aiming at satisfying mechanical and electrical requirements or constraints. A target position for the system's mass center, preference of moment of inertia in a given direction, minimization of electromagnetic interference, avoidance of high heat dissipation due to equipment being positioned close to another, and minimization of cabling are examples of such concerns. The early positioning of the spacecraft's equipment is usually done "manually" by a group of system engineers, heavily based on their experience. Coupled to an analysis stage, where the system's performance and constraints are verified, the spacecraft's equipment layout definition is an interactive process that takes time and hence, as soon as a good feasible design is found, it becomes the baseline. This reduces the exploration of the design space, and increases the probability that better designs are missed. Thus, increasing the creation of candidate solutions by numeric automatization of the

1. Instituto Nacional de Pesquisas Espaciais – São José dos Campos/SP – Brazil.

Author for correspondence: Valentino Lau | Instituto Nacional de Pesquisas Espaciais | Avenida dos Astronautas, 1.758, CEP: 12.227-010 – São José dos Campos/SP – Brazil | Email: valentino.lau@inpe.br

search through the conceptual design space would increase the possibilities that better designs are found.

The works of Ferebbe Jr. and Powers (1987) and Ferebbe Jr. and Allen (1991) are probably the firsts to propose numerical optimization methods for automating the process of determining the layout of equipment during the conceptual phase of spacecraft design. In a series of works, Teng *et al.* (2001), Sun and Teng (2003), Zhang *et al.* (2008) and Teng *et al.* (2010), studied the efficacy of the approach when applied to a spinning telecommunication satellite, considering also the influence of the application of different optimization methods. These works have in common the focus on placing the equipment driven by the system's mass properties (position of mass center and magnitude and direction of principal axis of inertia) requirements, subject to geometric constraints. In Jackson and Norgard (2002), thermal issues and minimization of wiring between equipment were introduced as objectives to be considered in the search for candidate solutions in the design space. Thermal requirements are in fact one of the main drivers of the spacecraft layout design, and in the context of conceptual layout optimization they have been treated either by trying to meet requirements of equipment heat dissipation uniformity over the spacecraft's structural panels (Jackson and Norgard, 2002; Hengeveld *et al.*, 2011) or target temperatures on them (De Sousa *et al.*, 2007). In the later work the problem was treated as fully multi-objective, that is, opposed to the usual approach of transforming it in mono-objective before optimization is performed, a set of trade-off solutions is the objective of the search. This provides more information about the design space, leaving for a posteriori analysis the choice of which solution will be implemented.

Coupling optimization algorithms with Computer Aided Design (CAD) and engineering analysis packages, provides an efficient way to tackle the spacecraft equipment layout problem, as highlighted in the works of Baier and Pühlhofer (2003), Pühlhofer *et al.* (2004) and Cuco (2011). In the later one, a new methodology was proposed to address the problem. The Cuco's methodology (Cuco, 2011; Cuco *et al.*, 2014) considers the main drivers commonly used to define the equipment layout during the spacecraft's conceptual design:

- The position of the system's center of mass;
- The alignment and strength of the system's main axis of inertia;
- Avoidance of concentration of high heat dissipation equipment over the satellite panels; and
- Equipment functional requirements.

The methodology of Cuco (2011), or different versions of it, may be implemented in different ways using commercial or custom made software. Cuco (2011) used modeFrontier® to couple Solidworks®, Matlab®, Excel® and an executable written in C, for such purpose. The advantage of using a software such as modeFrontier® as the core tool to implement the methodology is that, since it was specially developed to tackle optimization problems and act as an integrator of other CAD or Computer Aided Engineering (CAE) tools, it has readily available on its internal features different optimization algorithms and techniques to be used on the problem, and provides a user-friendly interface to integrate other tools and analyze the results. On the other hand, the user has limited or no access to changes on the workings of these tools, what may affect his/her ability to explore new ways of addressing the problem. In the context of a research tool for the exploration of different concepts and algorithms to address the spacecraft equipment layout optimization problem, Excel® would provide a convenient alternative, since it can be used as a platform where new optimization algorithms can be embodied, as a calculator for engineering analysis, data storage, visualization of results, as well as an integrator of CAD and CAE tools. It also has the advantage of being known and be available largely in the engineering community.

In the present paper, an Excel® based tool for spacecraft equipment layout is presented. Built using Cuco's (2011) methodology as the optimization framework, it can provide the spacecraft design team an efficient and easy way to explore the layout conceptual design space. Excel® was coupled to SolidWorks®, which is used to calculate design parameters and as a graphical interface, where candidate layout configurations can be visualized. The tool's concept and early application tests were first presented in the 22th International Congress of Mechanical Engineering (COBEM 2013) (De Sousa *et al.*, 2013). The present paper is an updated version of the former. It introduces new features such as integrated decision making criteria for selection of solutions on the approximate Pareto frontier, additional objective functions and three-dimensional (3D) capability.

In the Sections that follow, the general spacecraft equipment layout optimization problem is formulated, a prototype of the layout tool is presented and a simplified 3D example of application is shown.

SPACECRAFT EQUIPMENT LAYOUT PUT AS AN OPTIMIZATION PROBLEM

The spacecraft's layout problem can be tackled as a multidisciplinary multiobjective optimization problem and can be generally stated as:

Minimize

$$f_i(x)\ ;\ i = 1 \text{ to } I \tag{1}$$

Subject to:

$$g_k(x) \leq 0\ ;\ k = 1 \text{ to } K, \tag{2}$$

$$h_l(x) = 0\ ;\ l = 1 \text{ to } L, \tag{3}$$

$$x_{jmin} \leq x_j \leq x_{jmax}\ ,\ ,\ j = 1 \text{ to } J \tag{4}$$

where f_i is a vector of I objective functions, x_j is a vector of J design variables, g_k and h_l are vectors of K and L inequality and equality constraints, respectively, and x_{jmin} and x_{jmax} are the bottom and upper boundary constraints on the design variables.

The objective functions encode the design requirements for the spacecraft, such as a target position for its mass center, whereas the constraints define the viable design space. For example, there must be no mechanical interference among the equipment. In the simplified case study showed further in the text these points will be made clear.

The approach for the spacecraft equipment layout problem proposed by Cuco (2011) is used as the general framework to build the tool presented herein. The design variables are defined considering the faces of the panels where the equipment would be positioned and, over a given panel face, the local coordinate position of the equipment mass center being positioned on that panel. For a box-shaped equipment, the angle formed between the box edge and the panel axis is also a design variable. Hence, each equipment has 4 design variables: one index indicating in which panel's face it is allocated, two local coordinates x and y of the position of the center of mass projected on the panel, and one orientation angle. For example, if there are 8 box-shaped equipment to be positioned and 2 panels available for positioning, there are 32 design variables for optimization. Design requirements for the position of the system's mass center, grouping of sets of equipment, avoidance of "hot spots" over the panels and, the alignment of the principal axis of inertia and the proportion of the principal moments of inertia are tackled by five objective functions, while geometric and functional requirements are take into account as constraints. In Cuco's methodology, the result of the optimization is a set of candidate non-dominated solutions for the layout and its respective approximate Pareto frontier. The decision of which solution, or solutions, would be subject to a further analysis to become the baseline layout is left for the engineering team responsible for the layout design. The Cuco's methodology framework embodies the basic aspects to be considered by any computational environment aimed at providing the system´s engineering team, a tool for the spacecraft´s equipment layout, during its conceptual design phase. It is flexible enough to accommodate design goals being treated either as objective functions or constraints.

Adding to Cuco´s work, the tool presented herein also incorporates decision making criteria to help the design team choose one or more candidate solutions for further evaluation, after the approximates Pareto set and Pareto frontier are returned.

For the general three-dimensional spacecraft layout problem, a practical tool must allow the exploration of any combination of equipment and positioning panel, as well as the possibility for the user to set a specific combination of equipment/panel. For example, it may be desirable that some equipment is positioned over a panel of the spacecraft with the least incidence of Solar thermal radiation. Moreover, functional aspects may require that some equipment is fixed in a given position, or that a set of them are positioned close to each other. All these features are implemented in the tool presented here.

The main aspects considered for the conceptual spacecraft equipment layout are taken into account in the tool by 5 objective functions, which may be activated by the user independently. Mechanical interference between equipment is taken into account by constraints penalizing the objective functions, while parameterization of the design variables assures that the equipment remains inside the available positioning areas. The general optimization framework used for development of the tool is presented in Fig. 1.

In the present version of the layout tool, the multi-objective optimization problem was posed as:

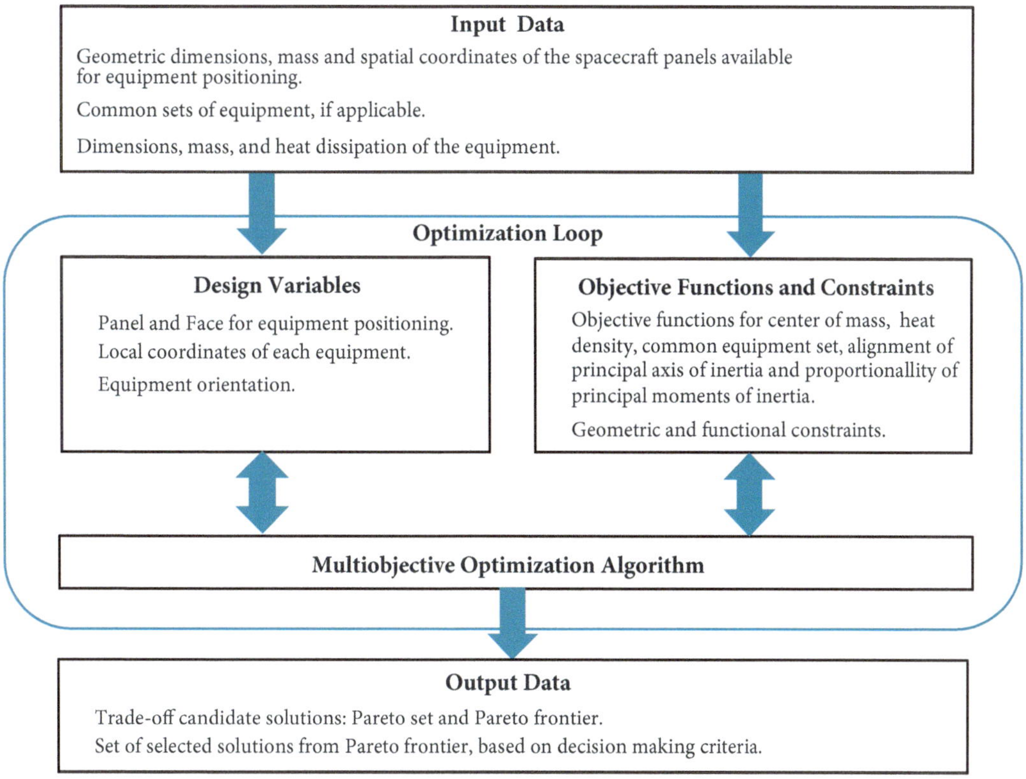

Figure 1. General optimization framework for the spacecraft equipment layout tool.

Minimize:

$$f_1 = \sqrt{\sum_{i=1}^{3} \lambda_i \cdot (x_{i\_CM\_sis} - x_{i\_CM\_target})^2} \tag{5}$$

$$f_2 = f_2^{Global} + \frac{r_{\min}^2}{A_{Total}} \cdot \sum_{p=1}^{Npanel} f_{2,p}^{Local} \tag{6}$$

$$f_2^{Global} = \sqrt{\frac{\sum_{p=1}^{N_{panel}} \left(\sum_{i=1}^{N_{equi,p}} \frac{P_i}{A_p} - \frac{P_{Total}}{A_{Total}} \right)^2}{N_{panel}}} \tag{6.1}$$

$$f_{2,p}^{Local} = \sqrt{\frac{\sum_{k=1}^{N_{cell,p}} \left(\sum_{i=1}^{N_{equi,p}} \frac{P_i}{r_{i,k}^2} - \frac{\sum_{j=1}^{N_{cell,p}} \sum_{i=1}^{N_{equi,p}} \frac{P_i}{r_{i,j}^2}}{N_{cell,p}} \right)^2}{N_{cell,p}}} \tag{6.2}$$

$$f_3 = \sum_{k=1}^{N_{set}} \sum_{i=1}^{N_{equip,k}-1} \sum_{j=i+1}^{N_{equip,k}} d_{i,j,k} \tag{7}$$

$$f_4 = \sqrt{\sum_{i=1}^{3} \rho_i \cdot (\alpha_i - \alpha_{i\_target})^2} \tag{8}$$

$$f_5 = \frac{\left\| \vec{v}_{inertia} \times \vec{v}_{target} \right\|}{\left\| \vec{v}_{inertia} \right\| \cdot \left\| \vec{v}_{target} \right\|} \tag{9}$$

Subject to:

$$I_i^{face} \in [1,\ 2,\ \ldots,\ N_{face,i}] \tag{10}$$

$$0 \le s_{i,1}, s_{i,2} \le 1 \tag{11}$$

$$I_i^{\theta} \in [1,\ 2,\ \ldots,\ N_{division,i}] \tag{12}$$

$$V_{inter} = 0 \tag{13}$$

f_1 represents the goal of having the center of mass (CM) of the system, $x_{i\_CM\_sys}$, as close as possible to a given target center of mass, $x_{i\_CM\_target}$. The parameters λ_i, which may assume zero or one value, are used to disable or enable the CM coordinate components. For example, if the CM longitudinal component of an spacecraft is less constrained than the other components, then only the lateral components could be enabled to drive optimization.

f_2 is an object function devised to approximate the heat density over the spacecraft's panels. This objective function is composed of two components. The first one, f_2^{Global}, measures how far the layout is from an ideal condition of uniformly heat distribution over the entire spacecraft. N_{panel} is the number of panels; $N_{equi,p}$ is the number of equipment installed in panel p; P_i represents the heat dissipated by equipment i; A_p is the projected area of panel p; P_{Total} is the total heat dissipated; and A_{Total} is the total projected areas of the panels. The second component, $f_{2,p}^{Local}$, evaluates the heat dissipated by the equipment installed in panel p over discrete regions of this panel. The panel is divided in $N_{cell,p}$ rectangular cells, each side of them with the size of r_{min}, which is the size of the smallest dimension, in contact with the panels, of all equipment divided by 2. $r_{i,j}$ is the distance between the center of the ith equipment to the center of jth cell, as seen in Fig. 2.

Minimizing $f_{2,p}^{Local}$ means that the standard deviation of the quantity $\sum_{i=1}^{N_{equi,p}} P_i / r_{i,j}^2$ is minimized, that is, the combined influence of all equipment over each panel cell would be the same. This would avoid "hot spots" over the panel. The ratio r_{min}^2 / A_{Total} is applied for scaling compatibility of these two components.

f_3 represents the goal of minimizing the distance between equipment belonging to the same common set. We define here a common set, as a group of equipment that should be positioned near each other. $d_{i,j,k}$ is the Euclidian distance between the geometric centers of equipments i and j belonging to a common set k, $N_{equi,k}$ is the number of equipment in common set k and N_{set} is the number of common sets.

f_4 measures the alignment of the principal axis of inertia to the spacecraft global coordinate system; α_i are the angles formed between the i-axis of the principal inertia and global coordinate systems, as shown in Fig. 3; and α_{i_target} is a given target angle. Analogously to λ_i in f_1, the parameters ρ_i, which are set to zero or one, are used to disable or enable angle components.

The goal of f_5 is to achieve a given proportionality between the principal moments of inertia. $v_{inertia}$ is a vector which components are the tree principal moments of inertia, and v_{target} is a given target vector, which components are positive values, that keep a desired proportion. For example, in a spacecraft controlled by spin, the longitudinal moment of inertia should be larger than lateral ones, say n times, while lateral moments of inertia could be of the same order. Setting the longitudinal component of v_{target} to n and the lateral components to 1 would represent this proportion. The vector norm of the cross product gives the area of the parallelogram formed by these two vectors. If they

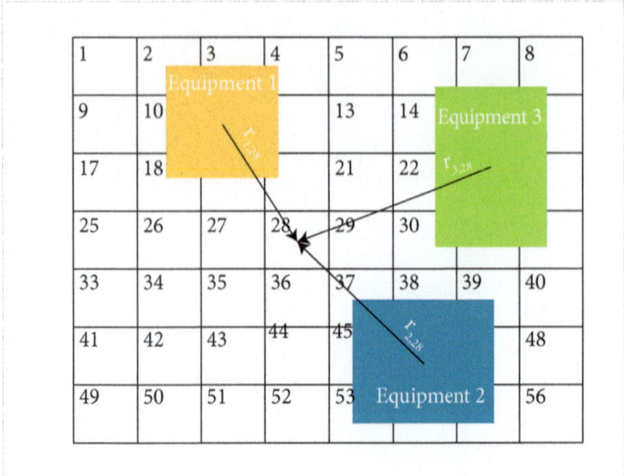

Figure 2. Representation of how the distance between the equipment and the panel's cells is considered in the heuristic used to calculate f_2^{Local}. Example with three equipment and 56 cells. Only distances for cell 28 are shown in the example, but all cells are considered when calculating the value of f_2^{Local}.

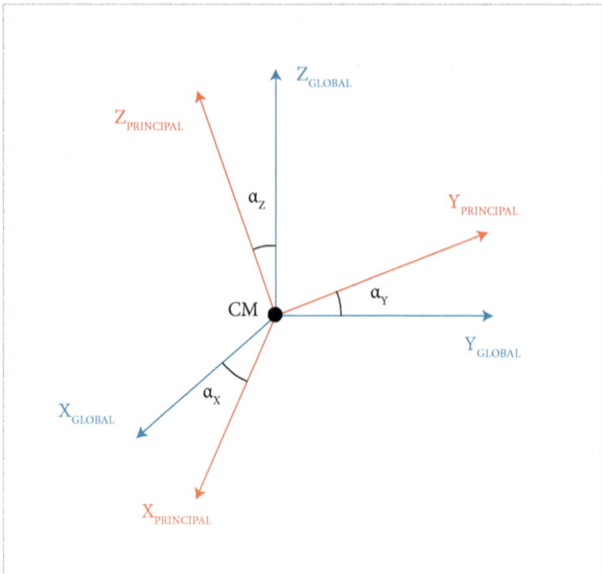

Figure 3. Angles between principal axis of inertia and global axis.

are aligned, what means that v_{inertia} has the same proportion of v_{target}, this area vanishes. On the other hand, if they are not aligned, a positive value is obtained. Normalization is used in order to keep f_5 in the range of [0,1].

T_i^{face}, $s_{i,1}$, $s_{i,2}$ and I_i^{θ} are the design variables.

The first defines the panel and the face where equipment i is installed. This is an integer variable corresponding to the index of an element in a list that contains all available panel faces, coded as the panel ID number with a signal, positive for top face and negative for bottom face. The number of available faces, $N_{\text{face},i}$, can vary for each equipment *i*, since constraints may be applied to restrict equipment installation.

The next two variables define the parameterized position of the geometric center of an equipment i over the panel. *B*, *L* and *H* are the equipment dimensions, while L_1 and L_2 are the panel dimensions, and θ is the equipment rotation angle, as shown in Fig. 4. The distances e_1 and e_2 are defined in Eq.14, and the relationship between the parameterized variables and the local coordinates D_1 and D_2 are presented in Eq.15. The values of the parametric variables $s_{i,1}$ and $s_{i,2}$ can vary in the range [0, 1]. This parameterization guaranties that the boundaries of the equipment always lies inside the area of the panel.

$$e_1 = B \cdot \|\cos\theta\| + L \cdot \|\sin\theta\| \;; \qquad e_2 = B \cdot \|\sin\theta\| + L \cdot \|\cos\theta\| \tag{14}$$

$$D_1 = \frac{e_1}{2} + s_{1,i} \cdot (L_1 - e_1) \;; \qquad D_2 = \frac{e_2}{2} + s_{2,i} \cdot (L_2 - e_2) \tag{15}$$

I_i^{θ} is an integer variable used to evaluate the rotation angle θ as indicated in Eq.16. The number of increments $N_{\text{division},i}$ is defined by the user. The angle θ varies in the range 0°≤θ≤180°.

$$\theta = \frac{\left(I_i^{\theta} - 1\right)}{N_{\text{division},i}} \times 180^{\circ} \tag{16}$$

V_{inter} is the total volume of mechanical interference among equipment and structure. The equality constraint (Eq. 13) is treated as a penalty for the objective functions when it is violated, using an exterior penalty method (Vanderplaats, 2007) approach.

DESCRIPTION OF THE SPACECRAFT EQUIPMENT LAYOUT CONCEPTUAL DESIGN TOOL

The main components of the optimal layout tool are presented in Fig. 5.

The Excel workbook consists of 7 worksheets and 3 main macros. From the Read Me worksheet, a description of all parameters used in this tool is presented. In the Control worksheet, parameters used to define and activate

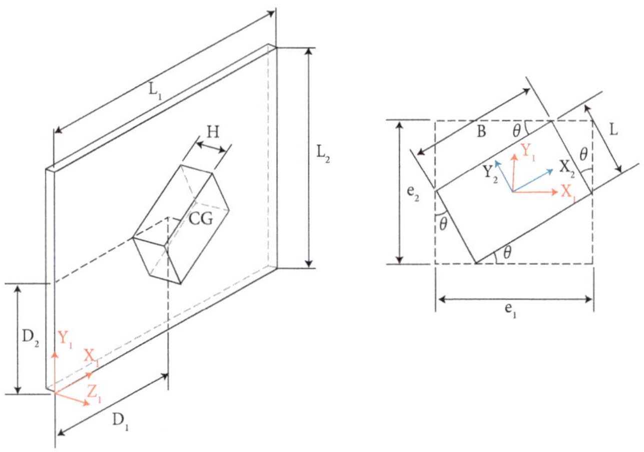

Figure 4. Equipment position over the panel.

objective functions and constraints are entered. A specific geometric configuration, defined in the Panels and Equipment worksheets, is built in SolidWorks®, which is launched by clicking a macro button inside the Control worksheet. All design parameters calculated inside SolidWorks®, are returned to the Control worksheet. In the Panels worksheet the geometric characteristics of each panel available for equipment positioning is entered. In the present version of the layout tool only rectangular panels are modeled. In the Equipment worksheet, the mechanical and thermal characteristics of the equipment are entered, together with the information of what subsystem they belong. In the current version of the layout tool, rectangular, cylinder and sphere solid shapes can be used to simulate the equipment. The solids may be assigned with different colors. In this worksheet, it can also be entered values for the design variables. In the Problem Description worksheet, a brief description of the objective functions, constraints and design variables being considered in the optimization problem is provided. In the Choice of Optimizer worksheet, the optimization algorithm to be used is chosen and information concerning its operational, such as parameters and stopping criteria, is entered. The optimization process is initialized from this worksheet, by clicking a macro button representing an available optimization algorithm. This calls a routine that embodies the algorithm and links it to other routines that launch and control SolidWorks®. Finally, in the Results worksheet, the approximate Pareto set and Pareto frontier obtained during the search are presented. Different types of graphs available in Excel® may be used in order to show the approximate Pareto frontier. For example, for problems with three objective functions, bubble or surface graphs may be used. In Fig. 6, screen prints of the seven worksheets are presented for illustration purposes.

The macros for the optimization algorithms, objective functions and routines that link Excel® to SolidWorks® are built using the VBA editor, in a modular approach, such that new optimization algorithms or objective functions can be added or removed from the tool, as desired. In its present version, only a real coded implementation of the M-GEO optimization algorithm (Galski, 2006), was incorporated to the layout tool. In Fig. 7 screen prints of the VBA editor and the SolidWorks® environment are shown.

The layout optimization process embodied in the layout tool just described is fully automatic. That is, once the "button" linked to an optimization algorithm is clicked in the Choice of Optimizer worksheet (for example, Play M-GEO in Fig. 3), the information on the Panels and Equipment worksheets are accessed, SolidWorks® is launched and linked to Excel®, the optimization performed and the results sent to the Results worksheet. The graph that plots the approximate Pareto frontier is also automatically updated. After the approximate Pareto frontier is retrieved, a particular layout solution may be visualized in SolidWorks® by selecting a solution ID and clicking in a macro button available in Result worksheet.

The automatic selection of particular solutions from the approximate Pareto frontier, based on dedicated decision making criteria that will be described ahead, is also calculated and listed in the Result worksheet.

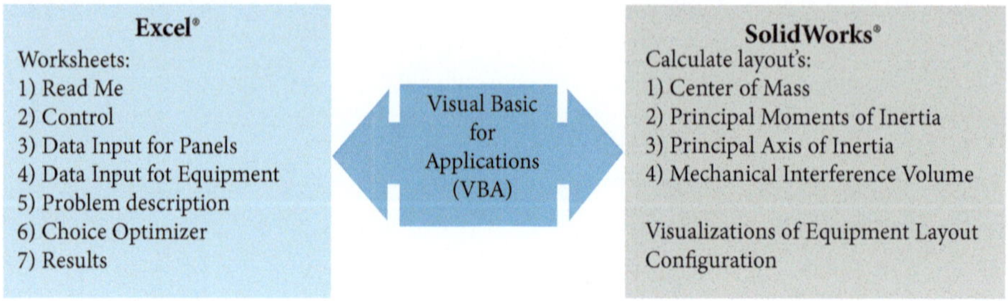

Figure 5. Main components of the spacecraft equipment layout tool.

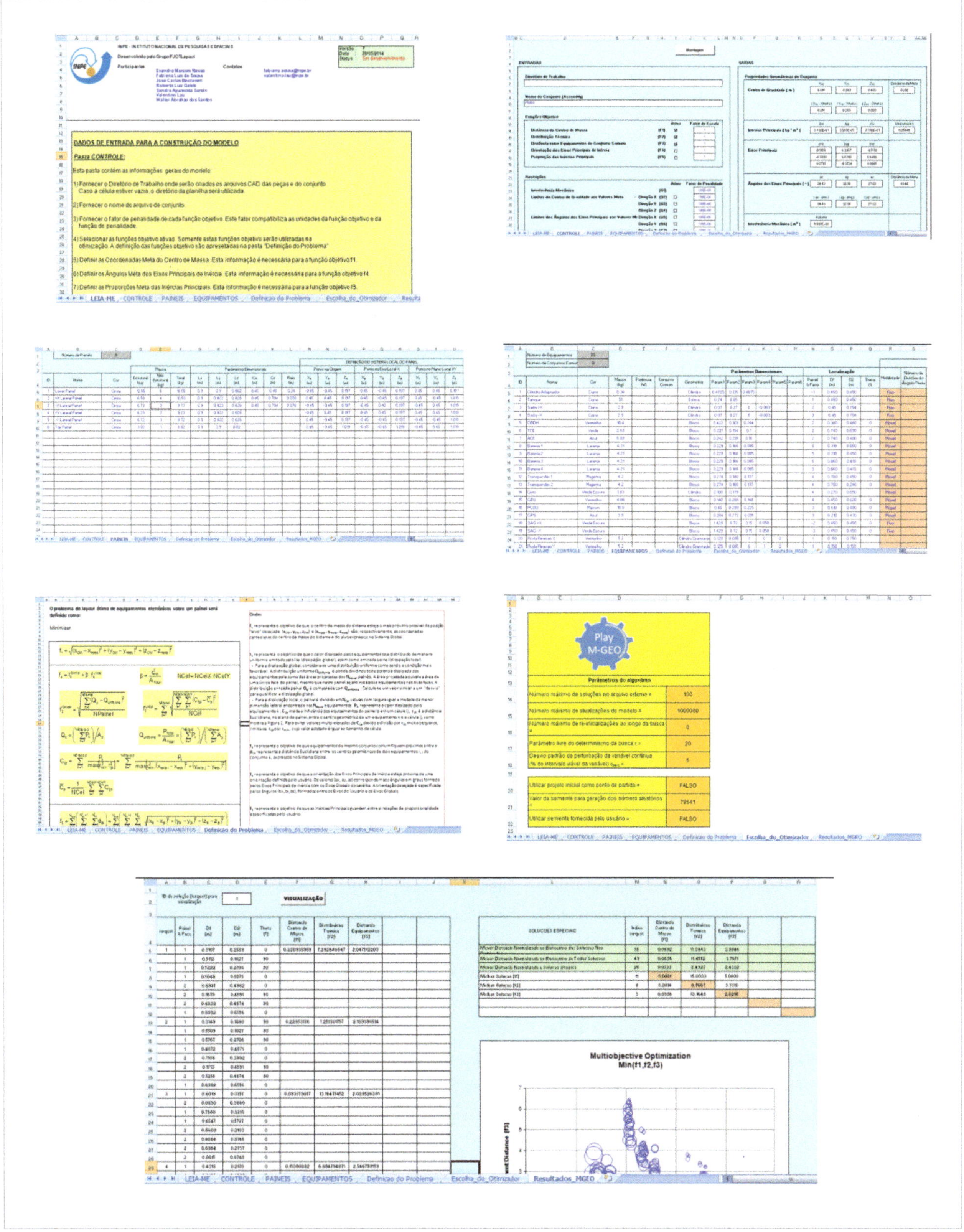

Figure 6. Screenshots of the layout tool Excel® worksheets.

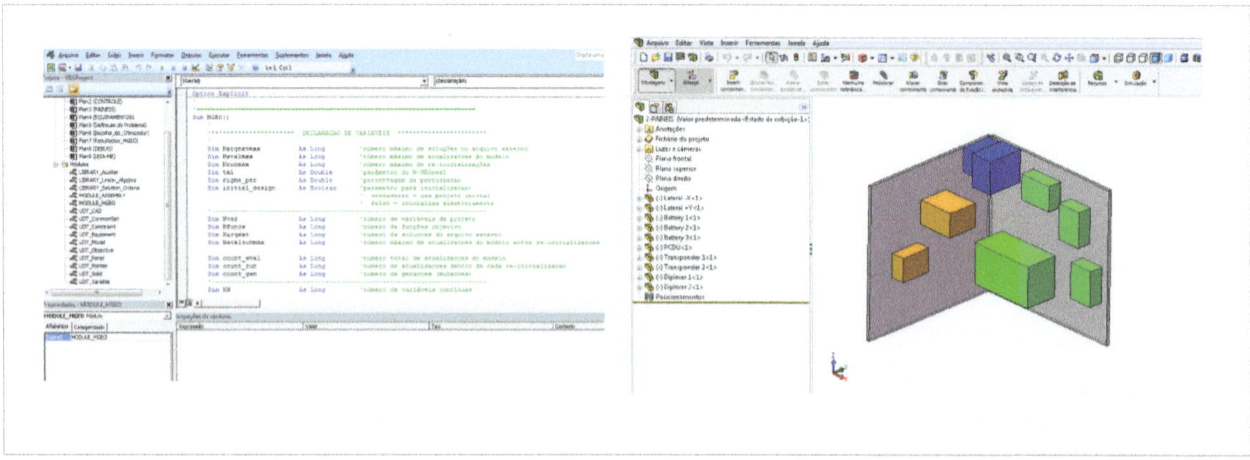

Figure 7. Screenshots of the VBA editor showing the M-GEO macro (left view) and SolidWorks® environment (right view).

THE OPTIMIZATION TOOL

The optimization algorithm implemented in the layout tool so far is based in M-GEO (Galski, 2006), a multiobjective version of GEO evolutionary algorithm (De Sousa *et al.*, 2003; De Sousa, 2002). In an early version of the layout tool, the canonical M-GEO was used (De Sousa *et al.*, 2013). As in the original GEO, in the canonical M-GEO the design variables are codified in binary strings. However, it has been shown that for problems where the design variables are continuous, a real coded GEO may perform better than its canonical version (Mainenti-Lopes *et al.*, 2008), what was also verified with real coded versions of M-GEO (Mainenti-Lopes *et al.*, 2012, Mainenti-Lopes, 2013). It has also been showed that GEO can work successfully treating discrete variables directly (De Sousa and Takahashi, 2005). Because in the spacecraft equipment layout problem there is a mix of discrete ($I_i^{face}$and I_i^{θ}) and continuous ($s_{i,1}$, $s_{i,2}$) design variables, was decided for the present version of the layout tool, to implement the M-GEO using the variables directly, instead of codifying them in binary strings. The main steps of the M-GEO algorithm as implemented in this work is described in Fig. 8.

Because the number of non-dominated solutions found during the optimization search can become very large, the user of the layout tool can set the maximum number of non-dominated solutions desired to be stored in the computer's memory and retrieved at the end of the search. Each time this number is exceeded, the "crowded distance" strategy proposed by (Deb *et al.*, 2000) is used to select the point on the approximate Pareto frontier that is on its most crowded region, and it is removed from the solution set to be retrieved. For problems with a large number of non-dominated solutions, this approach helps the user to keep the approximate Pareto set within a size more manageable for decision making analysis, while keeping on the solution set representative solutions of the entire approximate Pareto frontier.

SELECTING CANDIDATE SOLUTIONS ON THE APPROXIMATE PARETO FRONTIER

Though a multiobjective problem may be considered formally solved when the approximate Pareto set is found, from the practical point of view it is not over, since at least one of the non-dominated solutions has still to be choose to be implemented, or further investigated. Hence, some decision making criteria were included in the layout tool to help the designer in choosing solutions on the approximate Pareto Frontier (PF). Following The Smallest Loss Criterion, defined by Rocco *et al.* (2003) and used by Venditti *et al.* (2010) and Rocco *et al.* (2013), the solutions on the approximate Pareto frontier closest to its barycenter, calculated either considering all solutions on the frontier or only its edge values, and the utopian solution (the coordinates on the objective space that represents the optimal solution of each objective isolated), are used as references to choose solutions on the PF, as shown in Fig. 9. Since the edge solutions on the PF are the best solutions for each objective function, they are also candidate solutions to be further examined. In a problem with two objective functions, such as the hypothetical one shown in Fig. 9, there may be up to 5 solutions on the PF chosen by the criteria just outlined. It must be pointed out that the final choice of which solutions would be subject of further analysis and eventual implementation is always up to the designer. Automatic decision making strategies, such as the ones described above, should be used to help the decision making process and not as a substitute for the decision maker.

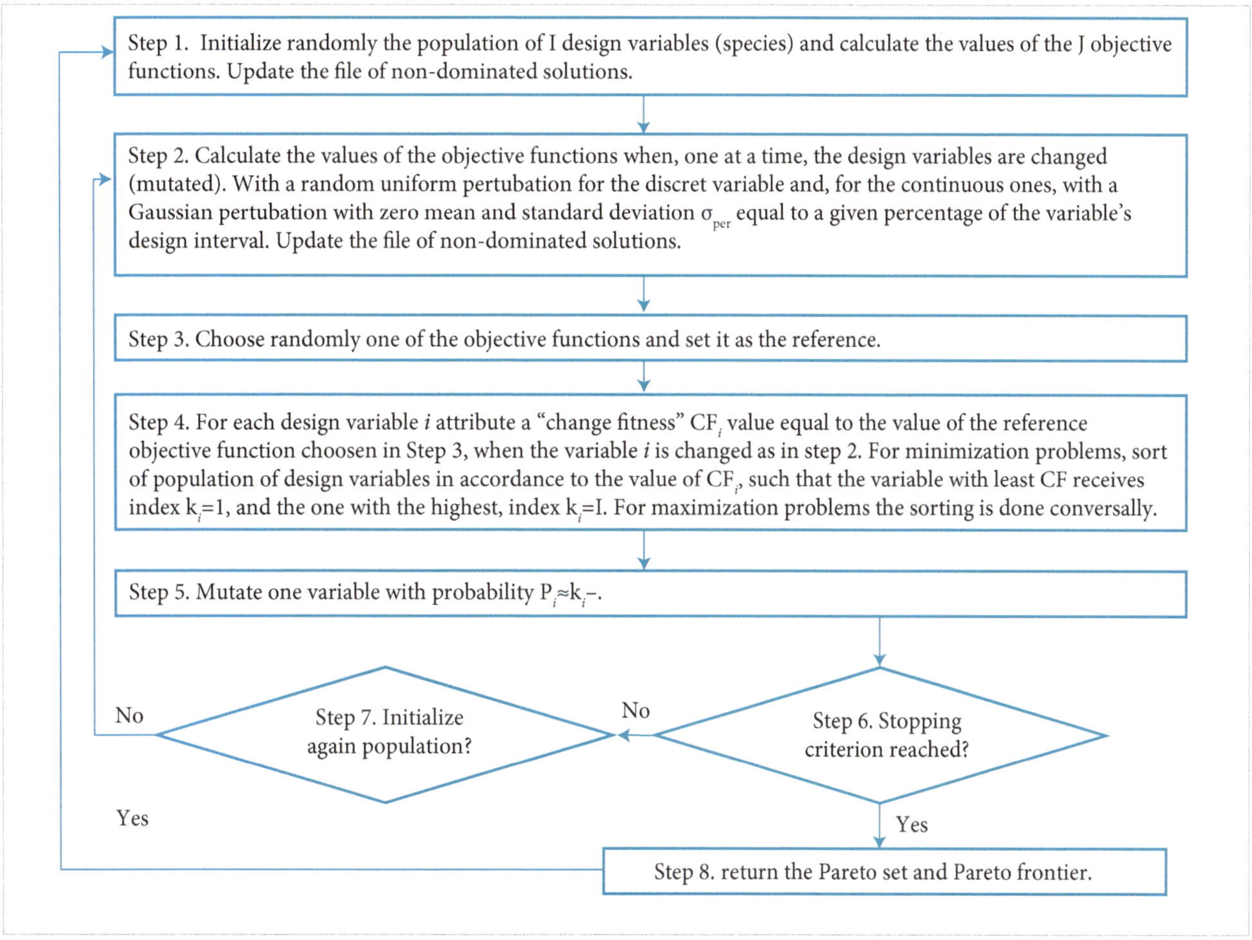

Figure 8. Main steps of M-GEO multiobjective optimization algorithm as implemented in this work.

EXAMPLE OF APPLICATION

A simplified three dimensional (3D) example is used for illustration of the tools features. It consists of placing 8 typical spacecraft equipment belonging to three different "common sets", over two squared panels, each one with an area of 1 m^2. Only the panels' top faces were selected as available for equipment installation. The equipment positions were defined using a total of 32 design variables. Optimization was performed using two different sets of objective functions. In the first run, the selected objective functions are the heat density (f_2) and the common set distance (f_3). In the second run, the same objective functions previously used are selected, and one additional objective function, the center of mass (f_1), is included. The chosen target center of mass is located 0.3 m far from the panels' top faces, with a height of 0.5 m from the lower edge of these panels. Therefore, two approximate Pareto

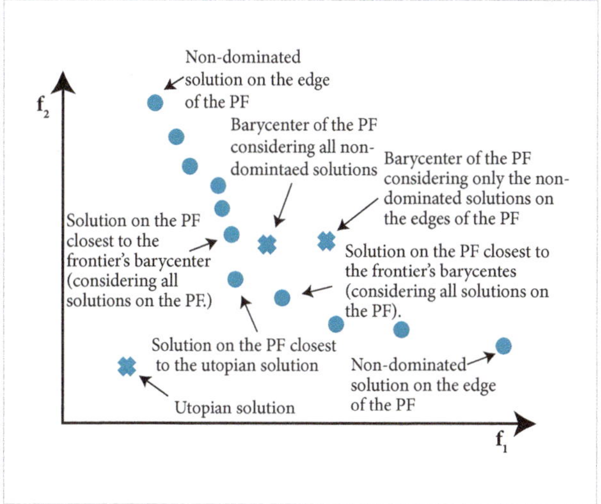

Figure 9. Some criteria to select solutions on the approximate Pareto Frontier (PF) for further analysis. A hypothetical example with two objective functions is presented here. Circles are non-dominated solutions. Crosses are reference marks based on the criteria (see text).

frontiers are calculated, one with two objective functions and other with three objective functions.

RESULTS OF SIMPLIFIED 3D CASE STUDY

The values used for the geometric dimensions, mass, heat dissipation and common set for each equipment is shown in Table 1.

The M-GEO algorithm was used for the optimizations, starting with initial configurations randomly generated. The number of model updates was selected as the stopping criterion. All the selected objective functions are evaluated in each model update. In both runs, a total of 500,000 model updates were evaluated, corresponding to 15,624 generations in M-GEO. The deterministic parameter τ was set to 20, the variable standard perturbation parameter σ_{perc} was set to 5%, and 5 re-initializations were used during optimizations. The stored non-dominated solutions were limited to a maximum of 100 solutions.

In the first run, with two objective functions (f_2,f_3), 89 non-dominated feasible solutions were recovered at the end of this search. The obtained approximate Pareto frontier is shown in Fig. 10. The large number of non-dominated solutions makes evident the convenience of using decision making techniques to help the design team select candidate solutions for further analysis. Using the Smallest Loss criterion described above and the edge solutions, five non-dominated solutions were

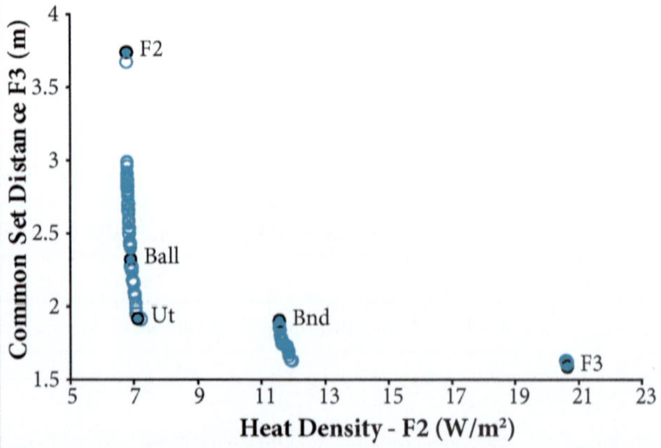

Figure 10. Optimization with two objective functions (F2, F3) Approximate Pareto frontier found using M-GEO.

Table 1. Geometric, mass, power and common set data of the equipment.

Equipment	Mass (kg)	Heat Dissipation (W)	Dimensions			Common Set
			LX (m)	LY (m)	LZ (m)	
Battery 1	4.000	4.71	0.166	0.229	0.095	1
Battery 2	4.000	4.71	0.166	0.229	0.095	1
Battery 3	4.000	4.71	0.166	0.229	0.095	1
Diplexer 1	0.750	1.30	0.156	0.21	0.025	2
Diplexer 2	0.750	1.30	0.156	0.21	0.025	3
PCDU	18.800	44.43	0.45	0.265	0.225	1
Transponder 1	2.800	29.20	0.156	0.21	0.094	2
Transponder 2	2.800	29.20	0.156	0.21	0.094	3

Table 2. Optimization using two objective functions - Selected solutions on the approximate Pareto frontier.

Selection Criterion		Index of solution on the approximate Pareto frontier	f_2 (W/m²)	f_3 (m)
Best value obtained for the thermal uniformity (f_2) objective function.	F_2	24	6.7826	3.7427
Best value obtained for the distance between equipment of the same common set (f_3) objective function.	F_3	6	20.6329	1.5931
Non-dominated solution closest to the barycenter calculated only considering the edges of the approximate Pareto frontier.	B_{nd}	3	11.5459	1.9071
Non-dominated solution closest to the barycenter obtained considering all solutions on the approximate Pareto frontier.	B_{all}	74	6.9032	2.3249
Non-dominated solution closest to the utopian solution.	U_t	31	7.1172	1.9220

picked from the frontier. They are presented in Table 2 and shown in Fig. 10.

In Fig. 11, colors were used to distinguish equipment of common sets: green for set 1, blue to set 2, and orange for set 3.

In the second run, with three objective functions (f_1, f_2, f_3), 100 non-dominated feasible solutions were recovered. Figure 12 shows a plot of all these solutions. A colored scale was used to represent the f1 objective function. Table 3 presents the six selected solutions and Fig. 13 shows their layout configurations.

The results shown on Figs. 10 and 12 shows clearly the capacity of the layout tool generate a great number of feasible non-dominated solutions for a 3D problem, starting from a completely random configuration. The two objective functions in the first optimization run are naturally competitive: the heat density f_2 drives the layout to a spread equipment configuration to avoid "hot" spots, while the common set distance f_3 drives

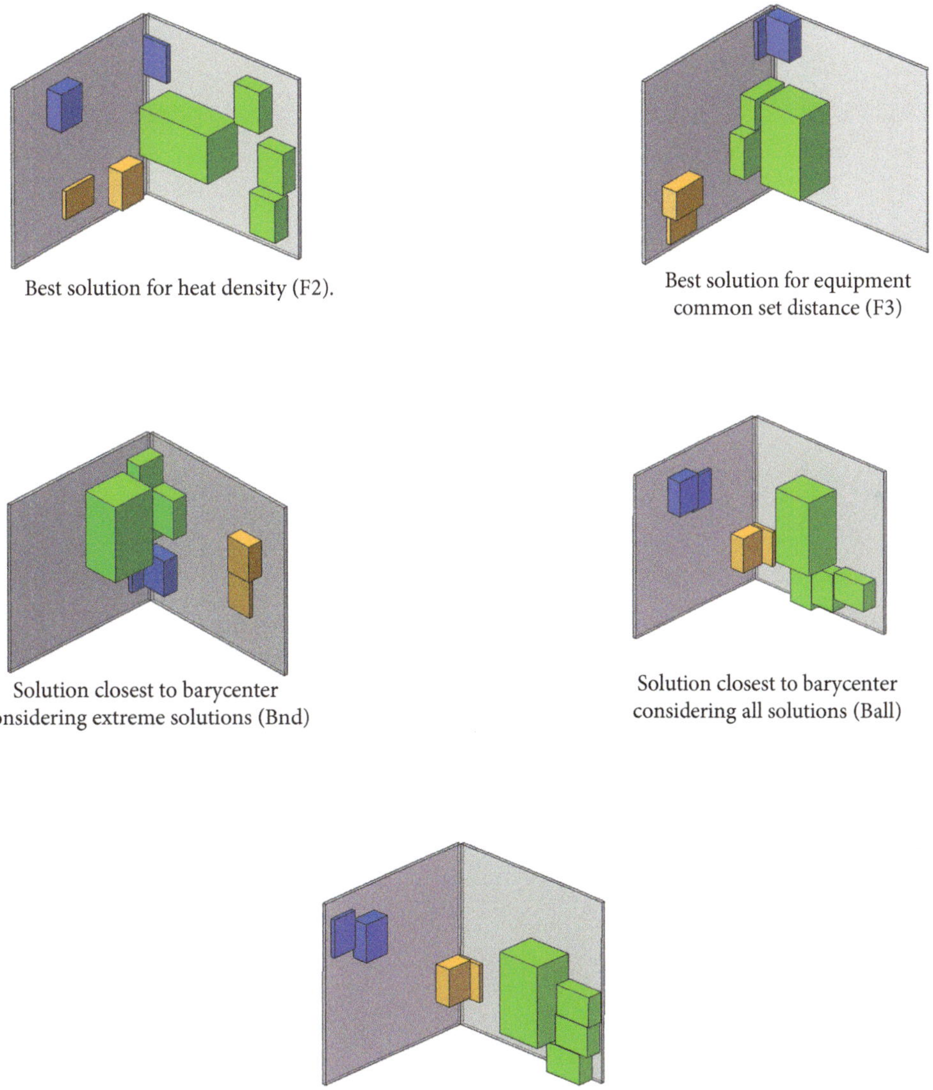

Best solution for heat density (F2).

Best solution for equipment common set distance (F3)

Solution closest to barycenter considering extreme solutions (Bnd)

Solution closest to barycenter considering all solutions (Ball)

Solution closest to utopian solution (Ut).

Figure 11. Optimization with Two Objective Functions - Layout configurations for selected solutions.

equipment to dense clusters in order to reduce the distances between equipment. Plotting the results with these two functions as coordinate axis highlights this competitive behavior. While the mass center objective function may not conflict with heat density and the equipment common set ones, it drives the search towards equipment configuration which mass center position is close to the desired one. Examining, in Figs. 11 and 13, the solutions selected using the automatic decision making criteria, it can be seen clearly that the edge criteria generate very different layout solutions, due to the fact that they represent trade-off solutions that privileges one of the objective functions. On the other hand, the solutions chosen using the barycenter

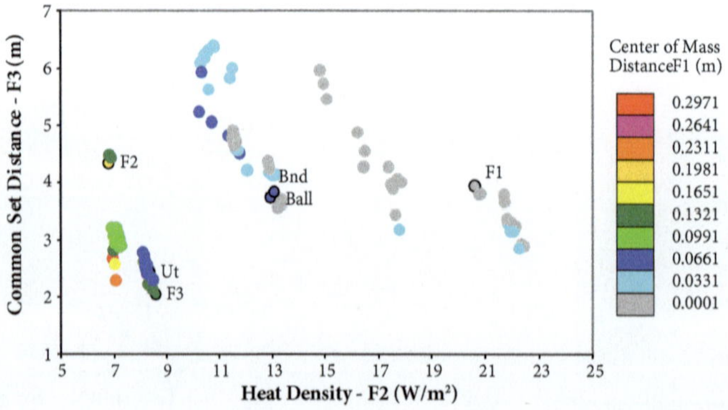

Figure 12. Optimization with Three Objective Functions - Approximate Pareto frontier found using M-GEO.

Table 3. Optimization with Three Objective Functions - Selected solutions on the approximate Pareto frontier.

Selection Criterion		Index of solution on the approximate Pareto frontier	f_1 (m)	f_2 (W/m²)	f_3 (m)
Best value obtained for the mass center (f_1) objective function.	F_1	5	0.0001	20.6003	3.9522
Best value for the thermal uniformity (f_2) objective function.	F_2	16	0.1981	6.7834	4.3506
Best value for the distance between equipment of the same common set (f_3) objective function.	F_3	2	0.1457	8.5488	2.0540
Non-dominated solution closest to the barycenter calculated only considering the edges of the approximate Pareto frontier.	B_{nd}	57	0.0705	12.9192	3.7523
Non-dominated solution closest to the barycenter obtained considering all solutions on the approximate Pareto frontier.	B_{all}	58	0.0553	13.0616	3.8427
Non-dominated solution closest to the utopian solution.	U_t	67	0.0738	8.3236	2.4376

or utopian approach, are less dissimilar, but still provide a lot of information on alternative design solutions. Confirming what was observed previously for the two dimensional test example (De Sousa *et al.*, 2013), it is noteworthy how the layout tool can provide potentially significant design gains. For example, in the present 3D application with three objective functions, a 38 % reduction on the value of objective function f_3 is obtained if solution Ut is chosen instead of solution F_1. In a real design application this would mean a significant reduction on the cabling connecting the equipment, which can lead, for example, to cost savings and mitigation of integration problems.

The processing time spent for running each optimization was approximately 3 hours and 32 minutes, in a PC with a Core i5 CPU, 2.5 GHz of clock and 4 GB of RAM memory.

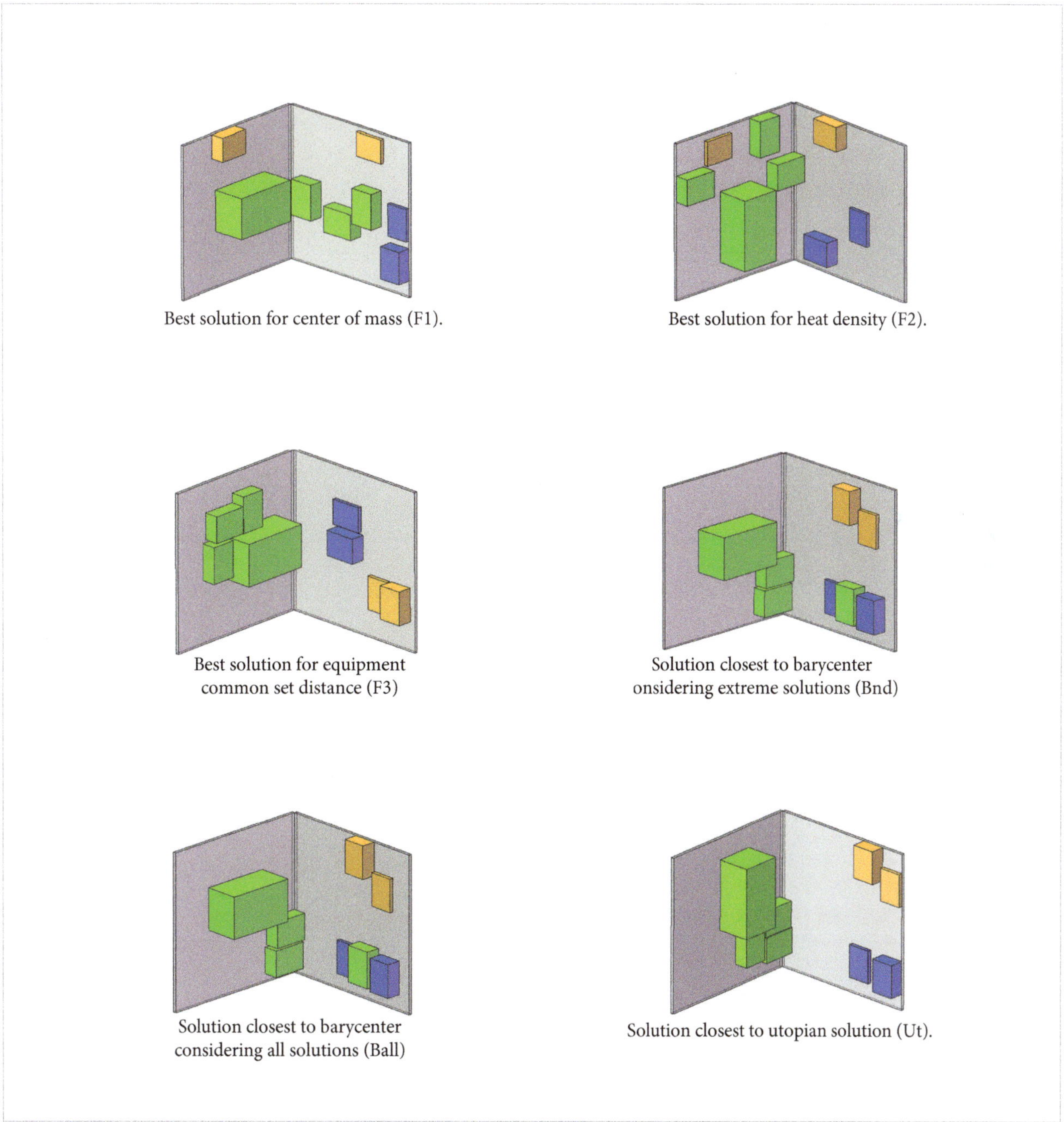

Figure 13. Optimization with Three Objective Functions - Layout configurations for selected solutions.

CONCLUSIONS

In this paper a tool for three dimensional multidisciplinary design conception of spacecraft equipment layout was presented. It is an evolution of an early prototype with 2D capability, which main features where presented in COBEM 2013 (De Sousa *et al.*, 2013). The tool can be used either as a research bed for testing different candidate methodologies and optimization algorithms to the problem, as well as an operational tool to be used by an engineering design team. The choice of using Excel® as the main software platform over which the optimization tool is built, was based on the convenience of having a readily available and broadly known software, which could be easily used for data input, numerical calculations, output of results and integrator of CAD or CAE software.

The tool uses Cuco's multiobjective methodology (Cuco, 2011; Cuco *et al.*, 2014) as the main framework for the layout optimization, which is performed by a customized implementation of the M-GEO (Galski, 2006) algorithm. The search for the optimal solutions, the approximate Pareto set, is performed from an initial completely random layout configuration. The user can select up to 5 different objective functions to guide the search. The user can also set which spacecraft panel's faces are available for positioning a given set of equipment. Excel® was coupled to SolidWorks®, which is used to calculate design parameters and as a graphical interface, where candidate layout configurations can be visualized. Results are automatically retrieved to a dedicated Excel® worksheet, becoming available to be further analyzed, either graphically or using internal Excel® features. The tool also embodies an automatic decision making procedure to select solutions on the approximate Pareto frontier, which, for a frontier with many non-dominated solutions, may help the user to decide which of them are more suitable to be further investigated. All these characteristics were exercised in a simplified three dimensional application example, which highlighted the potential benefits such a tool can provide.

The Excel® based spacecraft equipment layout tool presented in this paper can be considered the first "operational" version of a tool which preliminary results were presented at COBEM 2013 (De Sousa *et al.*, 2013). It was conceived to be continuously improved with new features, and short term goals in its development are the inclusion of new optimization algorithms and new objective functions to address additional engineering issues, as well as its application to a full real spacecraft layout problem. This would imply in a much larger design problem. For instance, for the service module of a middle size satellite of 500 kg, such as the MMP (Multi-Mission Platform) developed currently at INPE, the software would have to deal with around 88 design variables (for example, 22 equipment each one with four design variables). Moreover, there may be an increase in the number of constraints, depending on requirements posed on the positioning of some equipment.

REFERENCES

Baier, H. and Pühlhofer, T., 2003, "Approaches for further rationalization in mechanical architecture and structural design of satellites", In Proceedings of 54th International Astronautical Congress, Bremen, Germany.

Cuco, A.P.C., 2011, "Development of a Multiobjective methodology for layout optimization of equipment in artificial satellites" (in Portuguese), Master dissertation, Postgraduate Course in Space Technology and Engineering, National Institute for Space Research (INPE).

Cuco, A.P.C, De Sousa, F.L. and Silva Neto, A.J., 2014, "A multi-objective methodology for spacecraft equipment layouts", Optimization and Engineering. doi:10.1007/s11081-014-9252.

De Sousa, F.L., Muraoka, I. and Galski, R.L., 2007, "On the optimal positioning of electronic equipment in space platforms", In Proceedings of the 19th International Congress of Mechanical Engineering, Brasilia, Brasil.

De Sousa, F.L., 2002, "Otimização extrema generalizada: um novo algoritmo estocástico para o projeto ótimo", (INPE-9564-TDI/836), Ph.D. Thesis in Computação Aplicada, Instituto Nacional de Pesquisas Espaciais, 142p.

De Sousa, F.L., Ramos, F.M., Paglione, P. and Girardi, R.M., 2003, "New stochastic algorithm for design optimization", AIAA Journal, Vol. 41, No. 9, pp. 1808-1818.

De Sousa, F.L. and Takahashi, W.K., 2005, "Generalized Extremal Optimization Applied to Three-Dimensional Truss Design", Proceedings of the 18th International Congress of Mechanical Engineering (COBEM2005), CDROM, Ouro Preto, Brasil.

De Sousa, F.L., Galski, R.L., Rocco, E.M., Becceneri, J.C., Santos, W.A. and Sandri, S.A., 2013, "A toll for multidisciplinary design conception of spacecraft equipment layout", 22nd International Congress of Mechanical Engineering (COBEM, 2013), Ribeirão Preto, SP, Brazil.

Deb, K., Agrawal, S., Pratap, A. and Meyarivan, T., 2000, "A Fast Elitist Nondominated Sorting Genetic Algorithm for multi-objective optimization: NSGA-II". In: International Conference on Parallel Problem Solving From Nature, 6., Paris, France. Proceedings... Paris, France: Springer, pp. 849–858.

Galski, R. L., 2006, "Desenvolvimento de versões aprimoradas híbridas, paralela e multiobjetivo do método da otimização extrema generalizada e sua aplicação no projeto de sistemas espaciais", (INPE-14795-TDI/1238), Ph.D. Thesis in Computação Aplicada, Instituto Nacional de Pesquisas Espaciais, São José dos Campos, Brazil, 279p.

Hengeveld, D.W., Braun, J.E., Eckhard, A.G. and Williams, A.D., 2011, "Optimal placement of electronic components to mininize heat flux nonuniformities", Journal of Spacecraft and Rockets, Vol. 48, No. 4, pp. 556-563. doi: 10.2514/1.47507.

Jackson, B. and Norgard, J., 2002, "A stochastic optimization for determining spacecraft avionics box placement", IEEE Aerospace Conference, Vol. 5, pp. 2373-2382.

Ferebee Jr., M.J. and Powers, R.B., 1987 "Optimization of payload mass placement in a dual keel space station", NASA Technical Memorandum 89051, March.

Ferebee Jr. M.J. and Allen C.L., 1991. "Optimization of payload placement on arbitrary spacecraft", Journal of Spacecraft and Rockets, Vol. 28, No. 5, pp. 612-614. doi: 10.2514/3.26288.

Mainenti-Lopes, I., De Sousa, F.L. and Souza, L.C.G., 2008, "The Generalized Extremal Optimization With Real Codification". Proceedings of International Conference on Engineering Optimization - EngOpt2008, Rio de Janeiro, pp. 01-05.

Mainenti-Lopes, I, Souza, L.C.G. and De Sousa, F.L., 2012, "Design of a nonlinear controller for a rigid-flexible satellite using multi-objective Generalized Extremal Optimization with real codification", Shock and Vibration, Vol. 19, No. 5, pp. 947-956. doi: 10.3233/SAV-2012-0702.

Mainenti-Lopes, I., 2013, "A Multiobjective Approach to the Optimization of Solar Sail Trajectories" (In Portuguese), Ph.D. Thesis, Pós-graduação em Engenharia e Tecnologia Espaciais, área Mecânica Espacial e Controle, INPE.

Pühlhofer, T., Langer, H., Baier, H. and Huber, M., 2004, "Multicriteria and discrete configuration and design optimization with applications for satellites", In Proceedings of 10th AIAA/ISSMO Multidisciplinary Analysis and Optimization Conference, Albany.

Rocco, E.M., Souza, M.L.O. and Prado, A.F.B.A., 2003, "Multi-Objective Optimization Applied to Satellite Constellations I: Formulation of the Smallest Loss Criterion", Proceedings of the 54st International Astronautical Congress (IAC'03), Bremen, Germany.

Rocco, E.M., Souza, M.L.O. and Prado, A.F.B.A., 2013, "Station Keeping of Costellations Using Multiobjective Strategies", Mathematical Problems in Engineering, Hindawi Publishing Corporation, Vol. 2013, pp. 15. doi:10.1155/2013/476451.

Sun, Z-G and Teng, H-F., 2003, "Optimal layout design of a satellite module", Engineering Optimization, Vol. 35, No. 5, pp. 513-529.

Teng, H-F, Sun, S-L, Liu, D-Q and Li, Y-Z., 2001, "Layout optimization for the objects located within a rotating vessel – a three-dimensional packing problem with behavioral constraints", Computers and Operations Research, Vol. 28, pp. 521-535.

Teng, H-F., Chen, Y., Zeng, W., Shi, Y-J. and Hu, Q-H., 2010, "A Dual-system variable grain cooperative coevolutionary algorithm: satellite-module layout design", IEEE Transactions on Evolutionary Computation, Vol. 14, No. 3, pp. 438-455.

Vanderplaats, G., 2007, "Multidiscipline Design Optimization", Vanderplaats Research and Development Inc. ISBN 0-944956-04-1.

Venditti, F.C.F., Rocco, E.M., Prado, A.F.B.A. and Suhkanov, A., 2010, "Gravity-assisted maneuvers applied in the multi-objective optimization of interplanetary trajectories", Acta Astronautica, Elsevier Ltd., Vol. 67, No. 9–10, pp. 1255–1271. doi.:10.1016/j.actaastro.2010.06.022.

Zhang B., Teng H-F. and Shi Y-J., 2008, "Layout optimization of satellite module using soft computing techniques", Applied Soft Computing, Vol. 8, pp. 507-521.

The Open Innovation Journey in Emerging Economies: An Analysis of the Brazilian Aerospace Industry

Fabiano Armellini[1], Paulo Carlos Kaminski[2], Catherine Beaudry[1]

ABSTRACT: Open innovation is the systematic integration of collaborative, sourcing and revealing practices into a firm's business strategy. Its implementation does not happen at once, but through a journey. This paper investigates this subject in an emergent economy context, the Brazilian aerospace industry, presenting the critical analysis of a questionnaire-based survey performed by means of personal in-company interviews in 22 firms. It comprises a wide range of practices associated with open innovation, connected to a conceptual model. We find open innovation elements in the sample, with no open business strategy behind them, though. Deficiencies regarding funding, R&D maturity and intellectual property protection prevent the cluster from being fully open innovator. Nevertheless, the culture company in the sample is very prone towards openness. From that, we conclude that open innovation in the cluster is still "unfreezing", but with great potential to emerge, once these problems are solved.

KEYWORDS: Open innovation, Aerospace Industry, Brazil, Innovation management.

INTRODUCTION

This work is about the applicability and importance of open innovation to the Brazilian aerospace industry.

Open innovation is a term coined by Chesbrough (2003a) to designate a new mindset within industrial organization in which companies make use of "purposive inflows and outflows of knowledge to accelerate internal innovation, and expand the markets for external use of innovation, respectively" (Chesbrough, 2006).

The subject of collaboration, networking and use of external sources for research and development (R&D) is not a novelty in the academic and specialized literature about innovation management. Freeman (1991), in a review paper in the early 1990's, has shown evidence that the use of formal and informal R&D networks and other kinds of collaborative arrangements are in order since before World War II. Freeman also states that the growth of structured networks of innovators, claimed by Chesbrough (2003b) to be at the root of the open innovation era, dates from the 1980's. Also within new product development (NPD) literature, review papers on the subject prior to the rise of open innovation (e.g. Brown and Eisenhardt, 1995; Liyanage *et al.*, 1999; Krishnan and Ulrich, 2001) identify a clear outward tendency in their time.

In spite of all that, the coining of the term "open innovation" has brought to the literature a binding perspective on a number of existing practices, and the necessity to structure such practices into firms' strategies. According to the OECD (2008), the novelty of the open innovation approach lies on the systematic

1.Polytechnique Montréal – Montréal – Canada 2.Universidade de São Paulo – São Paulo/SP – Brazil

Author for correspondence: Fabiano Armellini | Department of Mathematical and Industrial Engineering - Polytechnique Montréal | 2900 Boulevard Edouard-Montpetit-Montréal/QC H3T 1J4 | Canada | Email: fabiano.armellini@polymtl.ca

integration of such practices into firms' strategy, as well as on the exploitation of the outbound (inside-out) process by firms.

It is evident that the adoption of open innovation is different from firm to firm, and from industry to industry. So it should be, since the "optimum" level of openness depends on variables such as technology intensity, value chain position, and product development average lead times, to name a few (Chesbrough, 2003b; West *et al.*, 2006). According to Chesbrough and Appleyard (2007), a good open business strategy "balances the tenets of traditional business strategy with the promise of open innovation". In other words, it is about the decision of what shall and what shall not be disclosed.

The adoption of open innovation does not happen at once, though. Enkel *et al.* (2011), for instance, developed a maturity level framework, which identifies five stages of adoption of open innovation: (1) Initial/arbitrary; (2) Repeatable; (3) Defined; (4) Managed; and (5) Optimizing.

Chiaroni *et al.* (2011), in their turn, classify the level of adoption of open innovation by means of a three-axed framework, which they name "the open innovation journey".

In the first axe, the process of adoption of open innovation is split into three stages: the first is the "unfreezing" stage, which implies in establishment of a sense of urgency of change and a cultural shift towards openness, although not yet put in practice. The second, called the "moving" stage, concerns the actual implementation of change through the establishment of new procedures and patterns of behavior consistent to the new vision derived from the cultural shift. Finally, the third one is the "institutionalizing" stage, which the company achieves when open innovation is incorporated to its formal procedures and internal business process maps, with its own metrics and subjected to continuous improvement procedures.

The second axe is with respect to the direction of openness, and adopts the terminology of Enkel *et al.* (2009): outside-in and inside-out, the first usually preceding the latter. The third axe regards the structuring of managerial levers for open innovation, namely:

- Networking;
- Organizational structures;
- Evaluation processes; and
- Knowledge management systems.

From these two papers (Enkel *et al.*, 2011; Chiaroni *et al.*, 2011), one realizes that it is not just a matter of determining whether open innovation is adopted or not, but also of establishing at which step one sits. This question can also be posed at the industry level.

Another issue identified in the emerging open innovation literature is that although the concept has been presented as a general trend (sometimes even as a paradigm change), what is observed in practice is that open innovation publications are clustered into a few "open-dominated industry segments" (Chesbrough and Appleyard, 2007), where the evidences of openness are more clearly found. Another limitation is that most of the studies published so far analyze open innovation in the context of developed economies (especially the U.S. and the European innovation systems) as first noticed by West *et al.* (2006), which persists to date.

That being said, this paper has, therefore, a double-edged purpose: the first one is to present the descriptive analysis of the results of a survey whose goal was to search for evidences of open innovation trends within the Brazilian aerospace cluster. The second goal is to answer, in an explorative basis, and based on the data from the same survey, the following two research questions:

- Does open innovation makes sense for companies of the aerospace industry in Brazil?
- How mature are companies in the cluster towards open innovation?

In other words, the goal behind this work is to verify whether collaborative and outsourcing activities within this industry are indeed part of an open business strategy or not, and where companies sit within the open innovation journey framework. This is an important issue to investigate, since most open innovation publications ignore the importance of business models to date (West and Bogers, 2013).

The relevance of this work lies on the previously mentioned lack of literature about open innovation in developing contexts. Moreover, there is a worldwide interest in Brazil since 2003, when Goldman Sachs issued a paper about the BRIC (Brazil-Russia-India-China) economies (Wilson and Purushothaman, 2003). One finds a number of publications which show evidences of openness within developing economy contexts, but there is no conclusive answer so far whether this issue should be managed in the same manner they are in developed countries.

Moreover, the open innovation literature lacks studies about more traditional high-technology segments, since most of the evidence found is from industries where innovation dynamics is more intense, such as the information and communications technology (ICT) and the pharmaceutical sectors. The aerospace industry is an example of such an industry for which one finds few studies discussing open innovation.

In order to fill both gaps, this paper presents the result of a research project which took place between 2011 and 2013 to investigate Brazilian aerospace companies in search of open innovation patterns within their R&D and NPD strategies. This work's results and findings shall add to the current discussion concerning the dynamics of these "new entrants" in the global market.

The structure of the remainder of the article will present an overview of the Brazilian aerospace industry and its journey towards openness. Afterwards, the research methodology employed to investigate the subject is presented, which includes a conceptual model and the structuring of a survey followed by the main results and a descriptive and critical analysis of the data obtained from the survey will be presented. Finally, some conclusions will summarize this paper's contributions and makes suggestions for future investigation on the subject.

THE BRAZILIAN AEROSPACE INDUSTRY AND ITS PATH TOWARDS R&D OPENNESS

A striking paradox in the Brazilian aerospace industry is that, although the country is proud to have Alberto Santos Dumont (1873-1932), a Brazilian-born inventor, among the pioneers of flight, the establishment of the industry has happened much later and did not occur spontaneously, but by means of government intervention. The creation of the Brazilian aerospace cluster in the city of São José dos Campos, strategically located in between São Paulo and Rio de Janeiro, the two largest metropolises in the country, is the cornerstone of the national aerospace. It was accomplished by the establishment, in the mid-twentieth century, of the aeronautic research institute *Centro Técnológico de Aeronáutica*, nowadays called *Departamento de Ciência e Tecnologia Aeroespacial* (DCTA), and the Air-Forces engineering college *Instituto Tecnológico de Aeronáutica* (ITA) in the city (Ferreira, 2009; Gomes, 2012). Later on, in 1971, the establishment in the same city of the Brazilian space research institute (INPE – *Instituto Nacional de Pesquisas Espaciais*) consolidated São José dos Campos and surroundings as the heart of the aerospace cluster in Brazil. The history of most Brazilian-owned aerospace companies is connected to these public institutes, as they are either spin-offs of these institutes, or have their founders coming from ITA, DCTA or INPE. That is the case of Embraer, Mectron and Avibras, to name a few.

Aerospace comprises three segments: aeronautics, defense and space (IMAP, 2011). Undoubtedly, the main segment within the Brazilian aerospace in terms of revenues is aeronautics, as shown in Fig. 1. This is mostly due to Embraer, a global leader in the regional jets segment. In 2007, for instance, the aeronautic segment was responsible for 92.9% of the total revenues in the Brazilian aerospace market, of which 84.5% is attributable to Embraer (Ferreira, 2009).

Not neglecting Embraer's importance to the cluster, one may not reduce the Brazilian aerospace industry to one single company. There are several other niche markets exploited by a number of different companies acting in the country that must be taken into account. One such niche market is the helicopter segment exploited by Helibras, a subsidiary of EADS Eurocopter, the only helicopter manufacturer in Latin America. There are also a number of smaller aircraft manufacturers that develop and produce smaller vessels for agriculture, flight training and leisure.

The defense and space segments in Brazil are highly segmented, with small and medium enterprises acting in very specific niche markets, related either to civil and military air control infrastructure, or to the modest but relevant Brazilian space program or yet to other initiatives that have the government as the demander and purchaser.

In all three segments, though, the role of the government is central for funding and supporting product development projects, beyond its role as purchaser and final user. Local authorities acknowledge this situation by considering the aerospace industry as a strategic segment for the national development plan (Gomes, 2012). Although strategic, throughout their history, aerospace companies often find a hard time regarding public support to develop their own technologies and products. This is partly

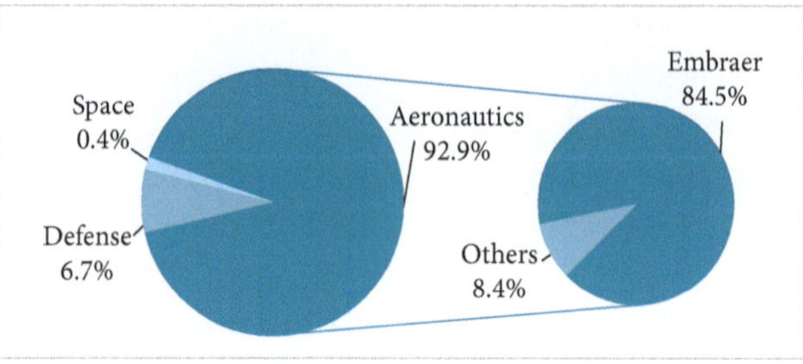

Figure 1. Brazilian aerospace market share in 2007 (Ferreira, 2009).

due to the novelty of the regulatory framework for innovation in Brazil, whose cornerstone is the so-called 'innovation law' (federal law 10.973/04), issued only in 2004.

As a result, despite the country having a distinguished competence in this industry globally, the cluster is very dependent on foreign knowledge, even for the prime contractor Embraer. When analysing Embraer's supply chain, one realizes that 93% of its suppliers are located abroad due to the lack of qualification of local suppliers (Figueiredo *et al.*, 2008). According to Cassiolato *et al.* (2002), the risk-sharing partnership model adopted by the company to develop its aircrafts since the mid-1990's was the solution Embraer found to gain access to the technology within the systems provided by its suppliers.

As for Helibras, since its establishment in the late 1970's, its business model was basically based on the manufacture and sale of aircrafts developed by its controller Eurocopter. More recently, though, the company is engaged with R&D, thanks to some technology transfer agreements from Eurocopter, that put the Brazilian subsidiary in charge for the development of a couple of platforms. This R&D is being performed in collaboration with local universities and public research institutes (Caiafa, 2012).

Finally, regarding the defense and space segments, the very structure of the market leads companies to collaboration, since their very small economies of scale and scope hinders them to invest in self-dependent R&D infrastructures. Collaboration is one of the few alternatives for survival in this scenario. One such example was the Atech-Omnisys joint venture for the development of the first 100% national S-band weather radar, which resulted in the spin-off Atmos Systems (Silveira, 2005). Another example was Mectron's collaborative arrangement with Brazilian and South-African enterprises for the development of the fifth-generation missile A-Darter (Silveira, 2009).

Recently, these segments are attracting the interest of new and bigger entrants, such as the French company Thales, which controls the Brazilian firm Omnisys, since the acquisition of 51% of the company in 2005 (Mileski, 2011). Likewise, the Swedish company Saab started operations in Brazil, through the launch of R&D facilities in 2011 (Pedroso, 2011), anticipating its victory in the fighter jet bid from the Brazilian Air Forces (Soto and Winter, 2013). Another important change in the segment's panorama was Mectron's acquisition by the Odebrecht group, the largest Brazilian company in the construction sector (Rolli, 2011).

All these recent changes motivated us to put forth the research project, which takes place within this dynamic environment, of alliances, partnerships, mergers and acquisitions, also characterized by secrecy and concealing due to the involvement of the military and matters of national sovereignty.

METHOD

The literature exploring the adoption of open innovation practices in aerospace is rather sparse and close to inexistent. Therefore, this work aims at performing an extensive investigation of open innovation concepts, tools, practices and strategies to verify what is in order for aerospace product development. To that purpose, we performed a questionnaire-based survey covering all relevant aspects related to an open innovation strategy.

After gathering some information about the firm itself, the questionnaire inquires about technology innovation management in general (product and process innovation), using the OECD's Oslo Manual framework (OECD and EUROSTAT, 2005). The most extensive part of the questionnaire, though, is about open innovation issues. Departing from the tripartite division of open innovation proposed by Enkel *et al.* (2009), we elaborated a list of the pertinent issues associated to each core processes, presented in Table 1.

In order to analyze 'when' and 'how' open innovation occurs within product development process, we propose a conceptual model, based on the three-phased R&D framework found in the Frascati manual (OECD, 2002). This model, presented in Fig. 2, identifies the internal "products" within the three activities of the R&D framework, that is, "knowledge" as basic research's main output, "technology" for applied research and "products" for development. The combination of such assets are the ideas, here defined as creative impulses, that allow the combination of existing data, information, knowledge and technologies into new knowledge, technologies, products and/or processes.

This conceptual model, along with the issues identified in Table 1, serve as a guide for the elaboration of an extensive survey questionnaire that structured the interviews performed during data collection.

Moreover, in order to cover for the open innovation journey, the questionnaire also inquires about open innovation culture (to evidence the "unfreezing" phase), formal business processes (for the "moving" phase) and open innovation metrics (for the "institutionalising" phase).

Table 1. Open innovation issues.

Core process	Issues associated
Outside-in	External knowledge/technology sourcing
	Integration of the customer and/or user in the innovation process
	Integration of the supplier in the innovation process
	Licensing in
	Spin-ins, mergers and acquisitions (M&A)
Inside-out	IP portfolio activity
	Licensing out
	Spin-offs and divestments
	R&D outsourcing
Coupled	Co-development with other companies
	Collaboration with universities and other science and technology institutes (STI)
	Venture capital

The result is a very comprehensive questionnaire of 71 questions, which took in average one hour and a quarter to be filled out during interviews.

RESULTS AND ANALYSES

The survey sample comprises a total of 22 Brazilian aerospace companies that presented R&D activities and high degree of maturity in the product development process (Oliveira and Kaminski, 2012). These companies provided detailed information between the years 2007 and 2011. Data collection was performed through personal interviews with R&D managers or directors responsible for the innovation process management within the company. All interviews took place inside the participating companies.

The sample is representative of the population being studied. The official Brazilian aerospace catalogue CESAER (DCTA, 2011) comprises a list of 270 companies. However, when analyzing the catalogue, one realizes that this list includes companies that, although being part of the aerospace supply chain, are not indeed aerospace companies. After filtering out these

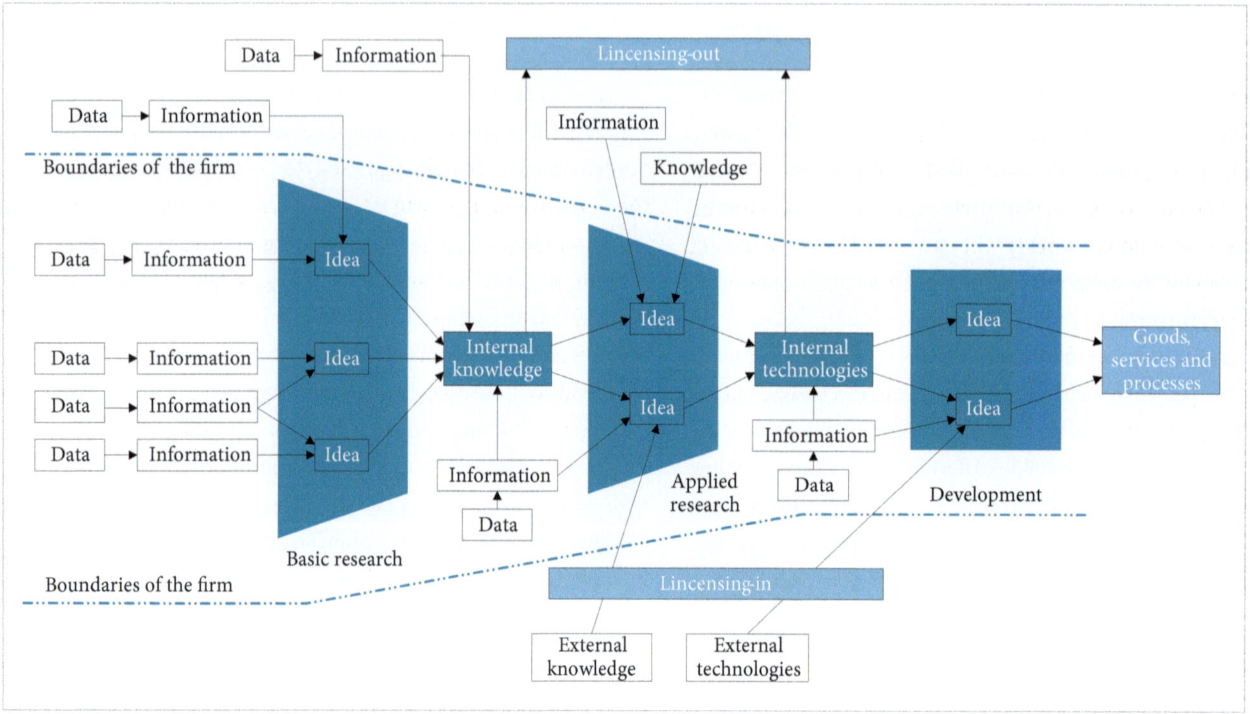

Figure 2. Open product-development conceptual model.

companies, one finds 150 aerospace companies. The catalogue includes a brief description of each company's activities, which allows one to perform a second filtering, retaining only those engaged on R&D activities. By doing so, only 59 firms are left in the population. Therefore, the sample obtained represents about 37% (22/59) of the population of interest for this study.

The aerospace industry comprises companies with varying technical expertise, since aerospace products embrace many different technologies. Looking up the distribution within the dataset, the largest proportion of firms lies in the electronic and avionic systems, followed by aircraft manufacturers/integrators and technical consulting firms. The others are specialized in satellite components, defense equipment, ICT/software, simulation equipment and mechanical machining. The small size of the population prevents us from disclosing the exact numbers of each category.

One can also cluster the according to the firm value chain position, among four possible options: prime contractors, equipment manufacturers, subcontractors and final users. This research does not target final users. Regarding the other three positions, Fig. 3 shows the distribution considering three scenarios:

- The entire aerospace industry (150 companies);
- R&D-engaged firms (59 firms); and
- This survey's sample (22 firms).

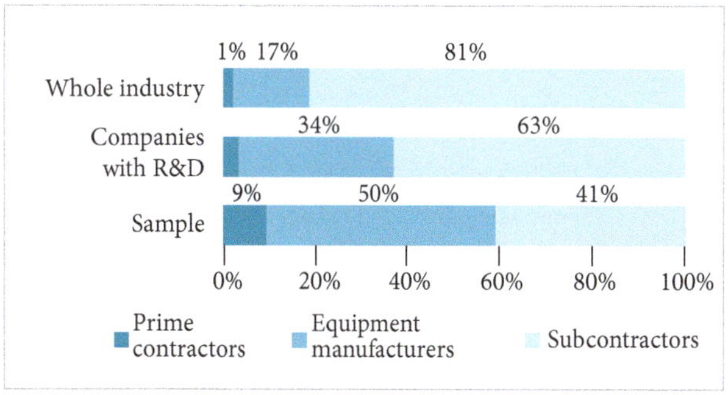

Figure 3. Firm distribution with respect to value chain position.

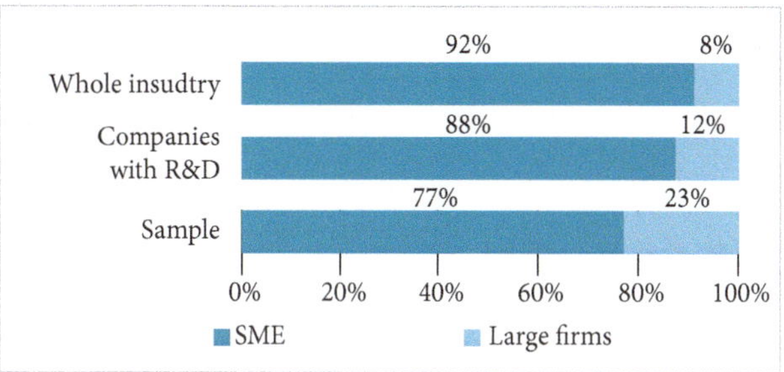

Figure 4. Distribution according to firm size.

One realizes that the survey's subset oversamples the final links of the chain (prime contractors and equipment manufacturers), even when compared with the R&D-engaged subset.

Turning now to size representativeness, in this study a firm is considered a Small and Medium Enterprise (SME) if it has 500 employees or less worldwide (that is, accounting all plants around the world). The distributions are shown in Fig. 4. Following the trend observed in the value-chain position analysis, the sample studied in the survey focuses on larger firms, when compared both to the whole industry population or to the share engaged with R&D. Nevertheless, more than 3/4 of the sample is composed of SMEs, which means that their position is undoubtedly captured in the analyses that follow.

INNOVATION MANAGEMENT

The first section after the introductory questions to the characterization of the sample focuses on traditional innovation management indicators, as established by the Oslo Manual (OECD and EUROSTAT, 2005). Although this manual distinguishes four different types of innovation (product, process, organizational and marketing), this survey concentrates only on technology product and process (TPP) innovation. All data covers the period ranging from 2007 to 2011.

As one can infer from Fig. 5, no single company claimed to be engaged only in process innovation, which indicates that innovation within this industry (or at least within the sample under analysis) is product-oriented. Figure 6 confirms this finding, by showing that the impact of product innovation tends to be higher when compared with that of process innovation. This corresponds exactly to what Pavitt (1984) defines as a science-based industry in the taxonomy he proposed in the 1980's.

Besides product and process innovation indicators, a common metric for innovation performance are intellectual property (IP) protection tools. Table 2 shows the percentage of companies within the survey sample that used IP protection tools in the 2007-2011 period: the proportion of firms using formal methods (e.g. patents) is low while strategic methods, such as secrecy and complexity of design, are more frequent in the sample. This result follows a Brazilian tendency, partly due to well-known shortcomings in patent registration in the country (Cruz and Mello, 2006; Gosain, 2013), in addition to a world-level low-patent tendency observed in the aerospace industry globally.

Moreover, Brazilian aerospace companies are in general not quite structured to deal with IP issues. As Fig. 7 shows, more

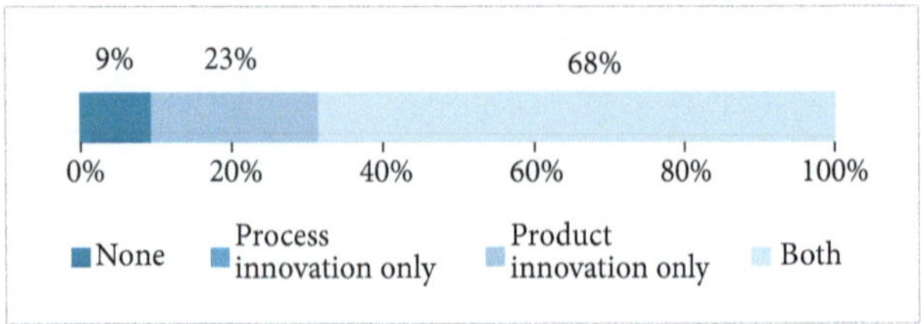

Figure 5. Types of innovation.

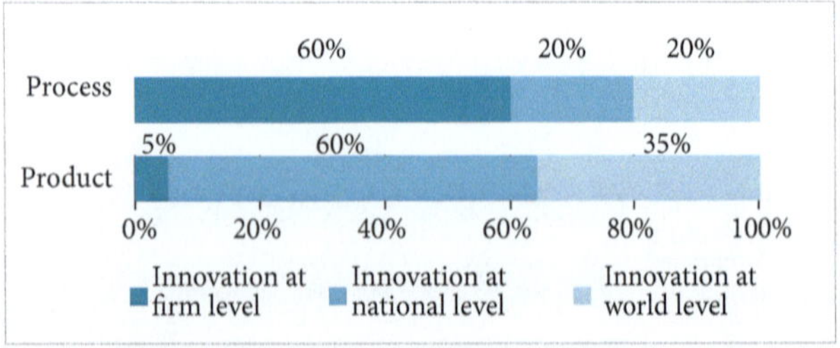

Figure 6. Highest innovation impact.

Table 2. IP protection methods.

Type of protection	IP protection method	Sample
Formal methods	Patents and utility models	18%
	Trademarks	23%
	Registration of industrial designs	5%
Strategic methods	Secrecy	55%
	Complexity of design	45%
	Lead-time advantage	18%

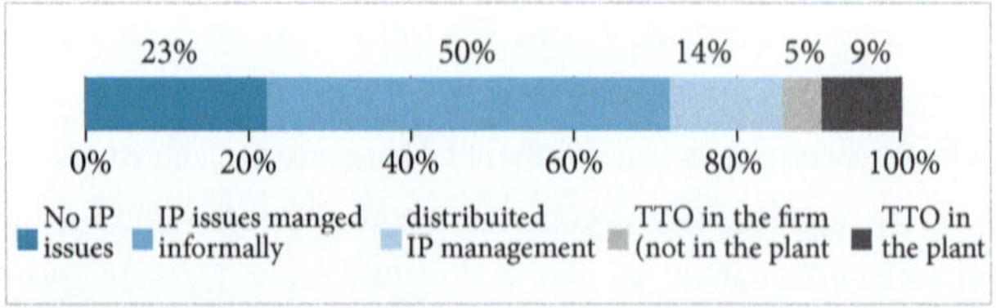

Figure 7. IP management.

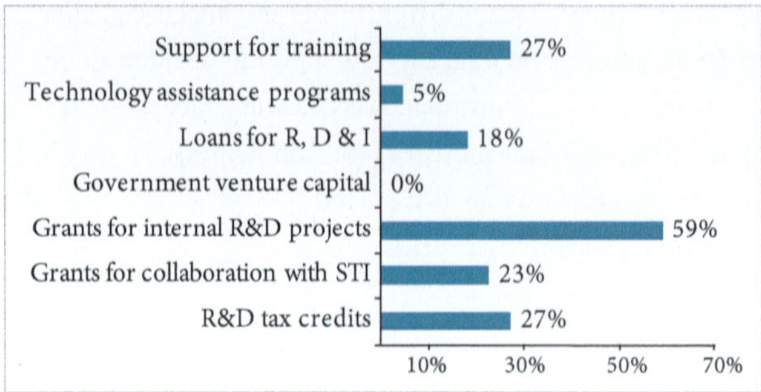

Figure 8. Use of innovation public policies.

than 3/4 of the sample claimed either not to have any IP issues or to deal with them informally, when they occasionally occur. Only 28% of the respondents claimed to a formal structure to manage IP, and only half of it (14%) within a Technology Transfer Office (TTO) or equivalent (9% in the plant where the interview took place and 5% in the firm, at a different plant).

Regarding at last the use of governmental support policies for innovation, Fig. 8 summarizes the adoption of a number of policies in the survey sample. The information that stands out is that more than half of the companies in the sample are users of non-refundable resources (grants) for innovation projects. Federal innovation agency FINEP and São Paulo state research agency FAPESP are the main culprits for this high rate, due to their well-established innovation support programs: "*SubvençãoEconômica*" from FINEP and "*PIPE – Pesquisa Inovadora na Pequena Empresa*" from FAPESP, the latter addressed to small enterprises (with less than 250 employees).

As for the other indicators, the percentage is relatively low. This is partly due to the novelty of most programs, since they were in general established after the Brazilian innovation law, issued in 2004.

PRODUCT DEVELOPMENT AND OPEN INNOVATION

Recalling the conceptual model shown in Fig. 1, the questionnaire begins the open innovation section of the survey by inquiring where companies act across the R&D spectrum. As expected, all companies in the survey claimed to be committed to development activities. About 2/3 of the sample claimed to perform applied research internally. As for basic research, only 9% of the companies claimed to perform such activities in Brazil and other 9% claimed to perform it globally (in others plants) but not in Brazil.

In order to perform this R&D, firms often find inspiration from others sources of knowledge. Figure 9 draws a radar diagram of the average importance attributed to a number of types of players that often contribute to R&D. Since the number of samples that claimed to be engaged to basic research was very low, the radar comprises and compares only applied research with development. The importance was given through a seven-point Likert scale with no central point.

One can take interesting insights from the comparison between the two remaining shapes drawn in Fig. 9:

- Internal R&D personnel are the most important, regardless of the R&D stage;

- Firms from different industrial sectors are barely used as sources of knowledge for aerospace companies, regardless of the phase;
- Science and Technology Institutes (STI), such as universities and research labs are more important during research activities than for development;
- Industry players, especially clients and suppliers are of great importance during development, but less important during the research phase.

Among the same list, companies could name up to two players as the most useful for their respective R&D. Figure 10 shows the frequency at which each player was mentioned.

Three players stand-out in this analysis: clients (59%), universities (50%) and suppliers (36%). Interestingly, the option "firms from other industries" is a choice that no respondent made, which corroborates the low importance of cross-industry collaboration for the segment.

OPEN INNOVATION STRATEGY

As mentioned in the introductory part of this paper, a recent review paper on open innovation research (West and Bogers, 2013) observed that there is a tendency on open innovation research to neglect the importance of open business models and strategies, in spite of their importance to distinguish open innovation from earlier research on interorganizational collaboration in innovation. Opposing such tendency, this research dedicates a full set of questions to investigate how openness is connected to the firm strategy and their respective business models.

According to Dahlander and Gann (2010), there are two types of strategy for open innovation: pecuniary and non-pecuniary. Pecuniary strategies consist of external practices directly related to acquiring or selling companies, the first an inbound activity and the latter outbound. Figure 11 shows that a very low percentage of companies claimed to be engaged in this kind of practices in the 2007-2011 period, in both directions.

The low percentages may be attributed to the low importance of such practices inside aerospace companies' strategies, or may be due to the small time span of analysis (5 years). Merger and Acquisition (M&A), spin-offs and divestments in most cases are not part of the ordinary agenda of companies, and their lead-time is often greater than other open innovation practices. In order to cover for that, the survey also inquired about former involvement of the plant in M&A, divestment or spin-off processes, whose result is found in Fig. 12.

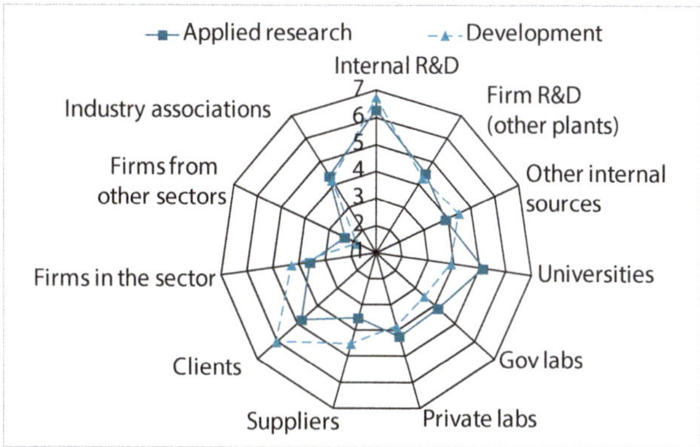

Figure 9. Importance of players as sources of knowledge for R&D activities.

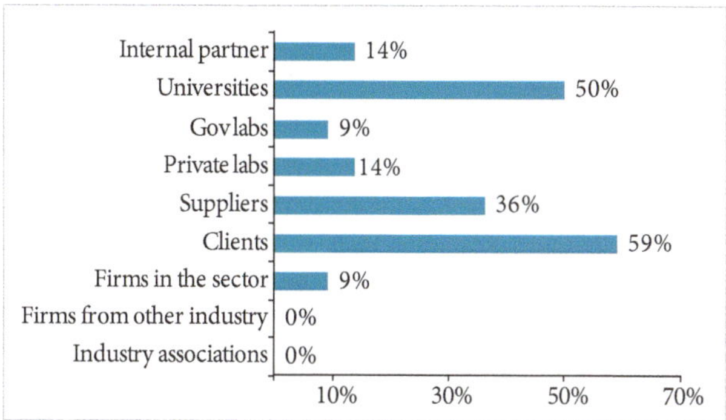

Figure 10. Top-of-mind partners for R&D.

With that approach, we found a larger M&A incidence within the sample. Therefore, even though 64% of the companies in the sample claimed they have never been involved with these matters, this result shows that this subject should not be disregarded when analyzing the Brazilian aerospace cluster, as 36% of the firms surveyed resulted from divestments, spin-offs or were merged or acquired by another company.

Turning now to non-pecuniary strategies, these are connected to sourcing (inbound) and revealing (outbound) practices (Dahlanderand and Gann, 2010). One first example of such strategy is licensing. Figure 13 shows the percentage of companies in the sample that claimed to have performed licensing during the 2007-2011 period.

The information that stands out is that not a single company in the sample claimed to have out-licensed their internal IP in this five-year period. In the inward direction, there were positive responses, but most of them are for the acquisition of specific development software tools or embedded software for the company's own products. Figure 14 indeed shows that a high percentage of licenses (70% out of the 45%) is due to software development firms.

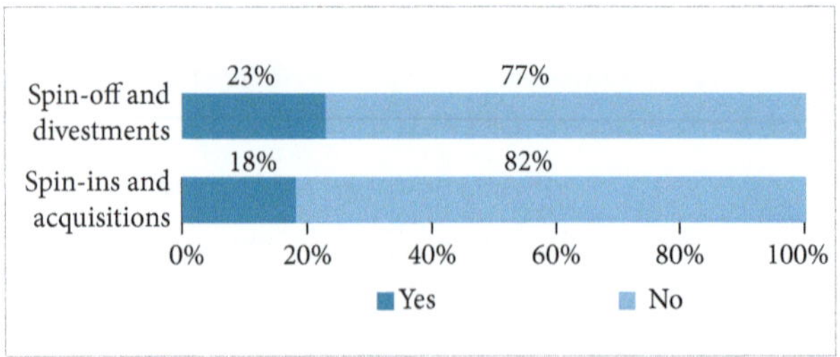

Figure 11. Strategic pecuniary practices related to the acquisition or selling of firms.

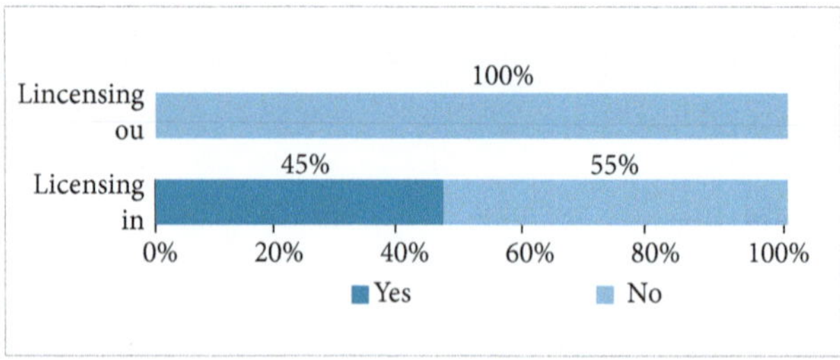

Figure 13. Licensing in and out.

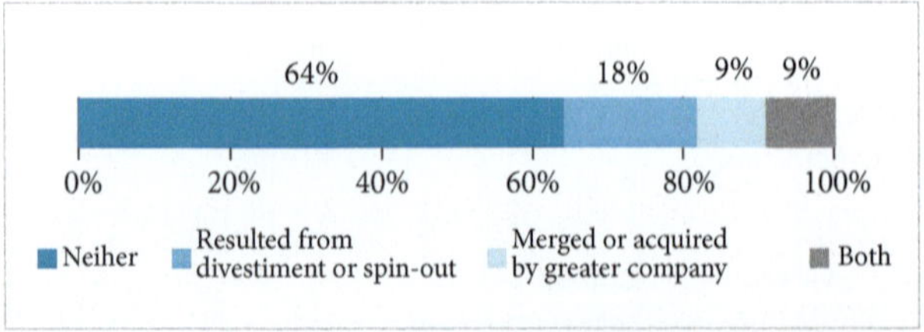

Figure 12. Plants previously involved in M&A or divestment processes in the past.

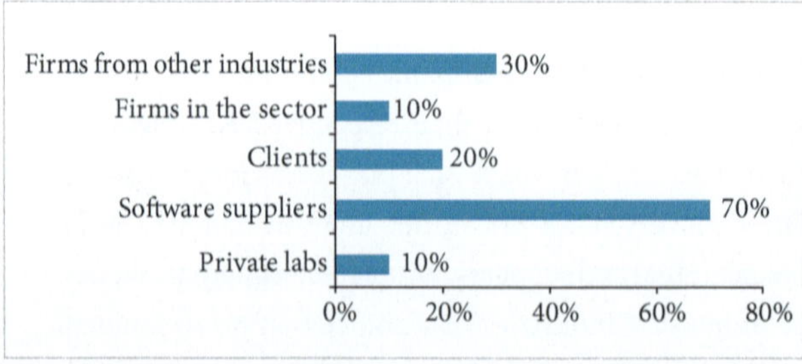

Figure 14. Type of organizations from which IP is licensed.

When inquired about the importance of licensing activities for the company strategy, on a Likert-scale from 1 to 7 (where 1 means 'not important' and 7 means 'very important'), the average remained below 2. Therefore, one concludes that firms in the sample do not adopt this kind of open strategy.

Another non-pecuniary revealing strategy is the provision of R&D services to third-parties. Now this is a practice that is found to be of great importance among surveyed firms. As Fig. 15 shows, roughly 3/4 (73%) of the firms claimed to perform this kind of activity, and close to half (41%) of them stated to perform this practice with great intensity.

The clients of these R&D services (Fig. 16) are mostly local, or located within the country, with low international incidence, which indicates that collaboration takes place with firms' own networking links. Moreover, only 1/3 of such services are provided within collaborative arrangements in a strict sense, the remaining 2/3 are direct-contracted services for the development of parts for third-parties, in almost all cases, public institutes such as INPE, the DCTA or one of the many divisions of the Brazilian Air-Force Command (COMAER).

This kind of practice, performed through direct-contracted development services, show a poor adherence of the open innovation model to the modus operandi of these companies. First, they are not performing R&D to develop their own products, but providing R&D services for a living, by outsourcing their own technical expertises to third parties. Besides, this kind of collaboration cannot be considered a partnership in the strict sense of the word; it is more akin to a supplier-customer relationship. In addition, there is not an open business strategy behind the decision of balancing the disclosure and revealing of internal knowledge and technologies.

On the other hand, in line with Tranekjer and Knudsen (2012), the very provision of technical services to their customers grants these companies updated technical knowledge and privileged information about the market, which is crucial to their survival in the market, but also represents a source of opportunities to these companies to develop their own products. As a matter of fact, 68% of the R&D service providers in the sample claimed to use federal or state grants to finance the transformation of internal expertise into own products.

What differs from the findings of Tranekjer and Knudsen (2012) in Denmark is that Brazilian aerospace R&D-service providers are not better product innovators in comparison to non-providers, since these firms often fail to accomplish the successful launching of such product innovations in the market. According to three R&D-service providers in the sample, Brazilian authorities lack uniformity in their purchasing policies of aerospace technologies, which makes it difficult for local companies for profiting from their own NPD initiatives, since opportunities identified today may not be confirmed as a purchase order when the product is ready for deployment.

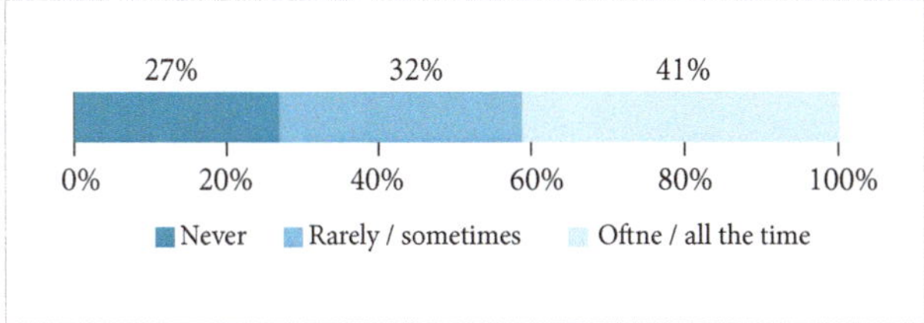

Figure 15. Proportion of firms providing R&D services by frequency.

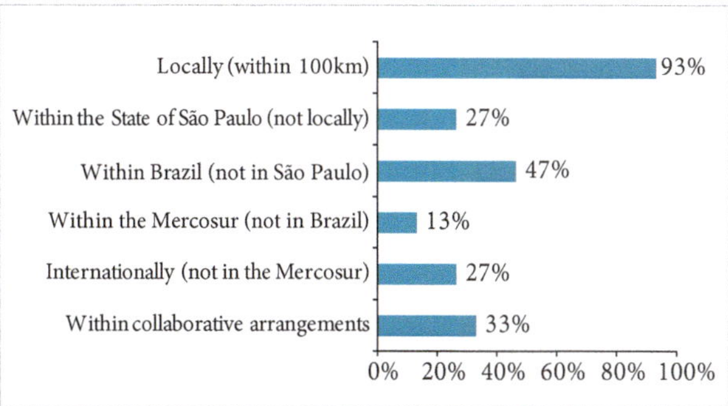

Figure 16. Location of R&D service clients.

OPEN INNOVATION CULTURE

Even though open innovation may not be a reality in many aspects of Brazilian aerospace firms' behavior or strategy, its principles may yet be present in their internal culture. This is what Chiaroni *et al.* (2011) refer to as the "unfreezing" phase, which is addressed in the last part of the survey questionnaire.

By means of a series of questions about the importance the company accords to a number of aspects, the survey is able to identify which open innovation practices are more related to the company routine, and therefore to its culture. The cultural facet is an aspect that open innovation research papers often lack, according to West and Bogers (2013).

In Fig. 17, the level of importance attributed to each practice is shown in a 1-to-7 scale, since all questions were graded through a seven-point Likert scale (with no central point), and grouped by the three core processes (Enkel *et al.*, 2009). As hinted from the results previously presented, there is a higher predominance of outside-in practices, followed by coupled activities and inside-out practices scored the lowest, with the exception of R&D services, as previously discussed.

Performing a simple average of the importance attributed to each practice within the three processes shown in Fig. 17, one obtains overall indices of 3.3 for outside-in, 2.6 for coupled and 2.0 for inside-out practices. Not surprisingly, outside-in scored highest, followed by coupled, and inside-out with the lowest score, which is consistent to a number of open innovation publications (OECD, 2008; Gassmann *et al.*, 2010; West and Bogers, 2013) that assert that there is a dominance of the outside-in process over the others.

The final question to pose at this point is whether this scenario is consistent to a journey towards innovation openness. Since averages could be deceitful, one alternative is the creation of dummy variables to determine whether companies adopt each one of the three core-processes. To that effect, all three indices from all surveyed firms were put together, and the median value (found to be 2.7) was set as the threshold for determining whether a particular open innovation process is part of the culture for each participating firm. Afterwards, firms were grouped according to the processes that were found to be present at their respective cultures.

By doing so, it was found that there is a pattern of adoption of open innovation processes. A group of companies (about 28% of the sample) was found not to adopt any of the three processes, which were labeled as "closed innovators". A second group of companies comprises those with one single process found within its culture (outside-in). This group corresponds to 5% of the sample. Among dual-process open innovators, two situations were found: outside-in is present in both cases, while the adoption of the second process differ: 14% of the sample adopts inside-out, while 43% of the companies, the largest group in this classification system, are coupled innovators. Finally, there is the full open innovators group that accounts for 10% of the sample, which adopt all three core processes. This classification system allows one the sketch the open innovation path illustrated in Fig. 18.

Outside-in is the first process to be incorporated into a firm's culture; pure inside-out or coupled innovators are inexistent in the sample. Companies only embody these core processes after they have mastered outside-in innovation. This finding is consistent to what Chiaroni *et al.* (2011) found in Italy.

With respect to the process of adoption of openness within the open innovation journey framework, this survey's questionnaire investigates the existence of a department formally responsible for open innovation processes and about the reference to open innovation activities in formal business process procedures of the firm. In both cases the incidence of positive answers was low (36%), even though the questions were structured in order to find the slightest evidence of these rather than fully operational open innovation management.

The questionnaire also includes a few questions intended to verify whether "not-invented-here" (NIH) and the "not-sold-here" (NSH) syndromes (Chesbrough, 2003a) were present in the firm's culture. According to the responses provided during the interviews, we found a very low presence of these syndromes in Brazilian aerospace companies: NIH affects only 4.5% of the companies in the sample, and NSH 9.1%. These answers clearly indicate a willingness to perform more collaboration and outsourcing than these firms actually do.

All these results allow one to conclude that open innovation is indeed present within these firms' culture, although not fully implemented in their strategies, internal procedures and market activity. It places the cluster in general in the "unfreezing" phase of the open innovation journey.

CONCLUSIONS

This paper aims to identify the level of integration of open innovation within the product development process. Open innovation is an emergent mindset observed in some technology segments, the so-called open-dominated segments (Chesbrough and Appleyard, 2007). Within this new mindset, firms become increasingly aware of both external knowledge for using in internal technologies and markets, as well as external opportunities for the use of internal knowledge in different markets. The body of knowledge built in the specialized literature, though, is mostly based on the case studies within these open-dominated segments, often within developed economies.

With that in mind, this research paper analyzed the applicability and relevance of open innovation within a different product development context in the Brazilian aerospace industry, a traditional high-tech industry within an emerging country. This industry is characterized by one large player (Embraer) and a large network of smaller companies, mostly concentrated in the cluster located in São José dos Campos and surroundings.

Through an extensive survey, this research was able to raise some interesting data to add to the open innovation

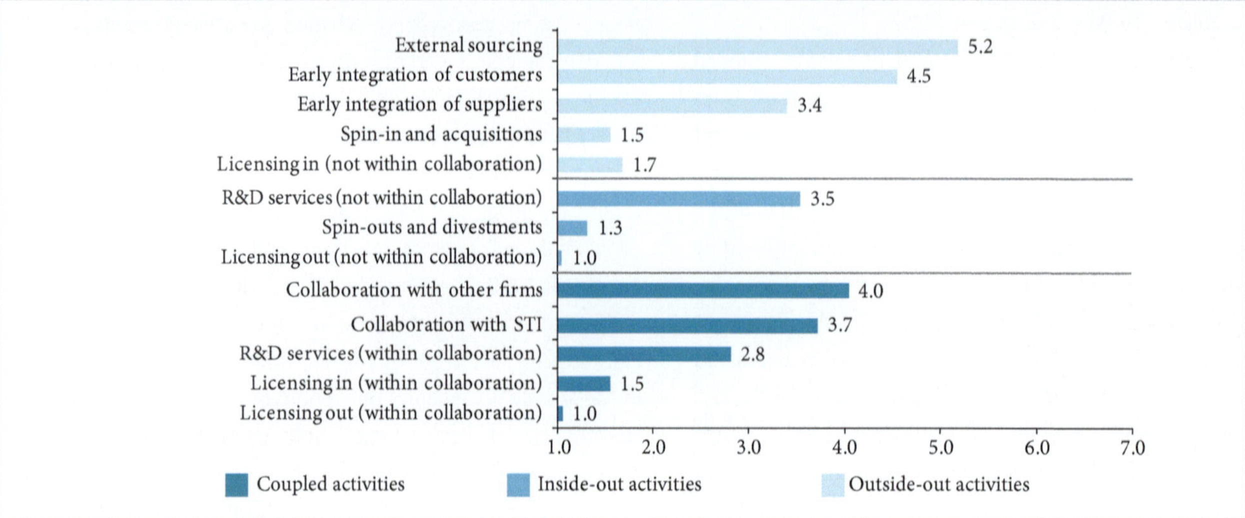

Figure 17. Open innovation practices as part of the day-by-day of Brazilian aerospace firms.

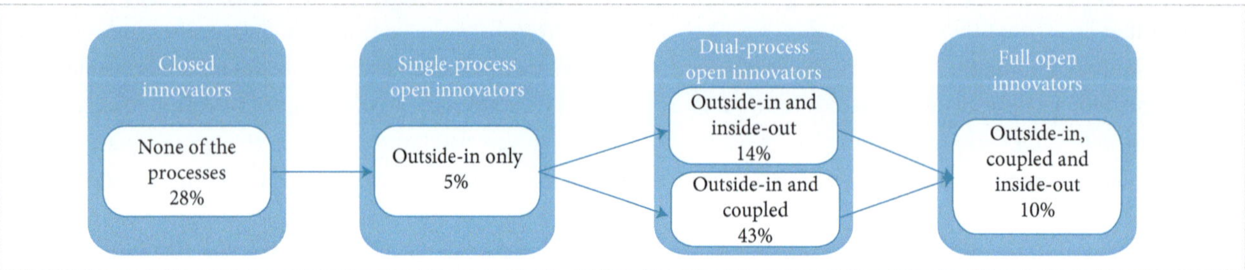

Figure 18. From closed to open: the journey of Brazilian aerospace firms.

literature. This survey was performed by means of personal in-company interviews with 22 aerospace companies in the cluster, based on a questionnaire elaborated after the conceiving of a conceptual model to integrate open innovation to NPD, and a profound literature review for raising all pertinent issues within an open business strategy.

By means of traditional innovation metrics, based on the Oslo Manual (OECD and EUROSTAT, 2005), we have found in general a product-oriented low to medium innovative performance, with low IP management structuring, low-adoption of public innovation policies and local impact. The prominent relation is the supplier-customer relationship, followed by the collaboration with STI. The latter, though, is much more relevant during research activities in comparison to development phases.

Regarding, collaboration trends, we found that collaborative links are much more intense locally, and mostly with other aerospace companies. Besides, we found an awkward rate of R&D services provided through direct contract, which shows that many interviewed companies do collaborate, but not as a strategy for enhancing their own technologies and incomes, as advocates the open business model (Chesbrough, 2006; Chesbrough and Appleyard, 2007).

In short, these results seem to lead to the conclusion that collaboration in this industry follows traditional standards, and that it is not fully adherent to an open business strategy, as defined by Chesbrough and Appleyard (2007).

Analyzing the internal culture of the companies in the sample, though, we found that companies are willing to collaborate more and also to adopt more complex models for innovation and for sourcing and revealing outwards their own boundaries. However, institutional gaps in the Brazilian cluster prevent this willingness to become a reality. Among these gaps, the insight from some respondents point towards the lack of uniformity of the government in its role of purchaser and the prevention of companies from performing R&D for the development of their own products, which in part is due to the first gap.

Comparing our findings to the open innovation journey framework proposed by Chiaroni *et al.* (2011), we found that the Brazilian aerospace cluster is indeed undergoing an openness process, but still in the "unfreezing" phase. In these stages, the sense of urgency for opening has been established, but institutional gaps prevents the clear adoption of an open business strategy. Through the analysis of the survey data, we can provide answers, on an exploratory basis, to the research questions formulated in the beginning of this paper. On the one hand, yes, open innovation does make sense in the Brazilian aerospace cluster; however, firms in the cluster do not seem to be very mature in dealing with open innovation, as they are in general still framed in the unfreezing phase of adoption of the new mindset.

Although this research focuses only one industrial segment (aerospace), from one single emerging economy country (Brazil), its results contribute for the general understanding of emerging economies in general, from an inductive perspective. The general panorama described in the data presented in this paper shows an industry with some very strong points from the knowledge and absorptive capacity perspective, but with structural gaps that prevents this potential to be fully developed. To change this scenario, effective government measurements are in order, to allow local companies to be able to develop their own technologies and products. For such, more than just offering funding alternatives, the government should revise its role as purchaser of aerospace technologies, and grant local companies a stable demand in the medium term that provides a stable ground for them to grow. On the other hand, there is also a step local companies should take, in order to organize further their R&D structures, so that they reach higher NPD maturity, with less informalities and more consistent strategies. Currently, the industry sits on a vicious cycle, where companies do not invest in R&D more because of lack of government support and the government do not support local companies because they lack R&D structure.

As a final remark: one should be aware that all these results are based on a simple average of the results obtained from the interviews, accounted equally regardless of firms' income. If we performed weighted averages with respect to income, we would find a much different scenario, because one player (Embraer), if present in the sample, would overshadow the whole sample, which is a problem normally found in studies that analyze the Brazilian aerospace industry. It is a common mistake to assume that the Brazilian aerospace industry resumes to Embraer alone.The focus intended for this paper is precisely the opposite. Regardless of having Embraer in the sample or not, what we desired to analyze here was the reality of the industry as a whole, including the needs of all those small players normally neglected in business sectors analyses.

REFERENCES

Brown, S.L. and Eisenhardt, K.M., 1995, "Product development: past research, present findings and future directions", Academy of Management Review, Vol. 20, No. 2, pp. 343-78. doi:10.2307/258850.

Caiafa, R., 2012, "R$200 milhões em investimento: Centro Tecnológico de Helicópteros em Itajubá", Revista Tecnologia e Defesa, No. 130, São Paulo.

Cassiolato, J. E., Bernardes, R. and Lastres, H., 2002, "Innovation systems in the South: a case study of Embraer in Brazil", Paper prepared for UNCTAD-DITE investment policy and capacity-building branch, New York and Geneva, United Nations.

Chesbrough, H.W., 2003a, "Open innovation: the new imperative for creating and profiting from technology", Harvard Business School Press.

Chesbrough, H.W., 2003b, "The era of open innovation", MIT Sloan Management Review, Spring 2003, pp.35-41.

Chesbrough, H.W., 2006, "Open business models: how to thrive in the new innovation landscape", Harvard Business School Press.

Chesbrough, H.W. and Appleyard, M.M., 2007, "Open innovation and strategy", California Management Review, Vol. 50. No.1, pp. 57–76.

Chiaroni D., Chiesa, V. and Frattini, F., 2011, "The open innovation journey: how firms dynamically implement the emerging innovation management paradigm", Technovation, Vol.31, No. 1, pp. 34-43. doi: 10.1016/j.technovation.2009.08.007.

Cruz, C.H.B. and Mello, L., 2006, "Boosting innovation performance in Brazil", Economics Department Working Paper No. 532, Organization for Economic Co-operation and Development.

Dahlander, L. and Gann, D.M., 2010, "How open is innovation?", Research Policy, Vol. 39, No. 6, pp. 699–709. doi:10.1016/j.respol.2010.01.013.

DCTA, 2011, "CESAER: Catálogo de empresas do setor aeroespacial", Departamento de Ciência e Tecnologia Aeroespacial.

Enkel, E., Bell, J. and Hogenkamp. H., 2011, "Open innovation maturity framework", International Journal of Innovation Management, Vol. 15, No. 6, pp. 1161-89.

Enkel, E., Gassmann, O. and Chesbrough, H.W., 2009, "Open R&D and open innovation: exploring the phenomenon", R&D Management, Vol. 39, No. 4, pp. 311-16. doi: 10.1111/j.1467-9310.2009.00570.x.

Ferreira, M.J.B., 2009, "Perspectivas do investimento em ciência – documento setorial: aeroespacial & defesa", Vol. 12, Projeto PIB, Universidade Federal do Rio de Janeiro (UFRJ)/ Universidade Estadual de Campinas (UNICAMP).

Figueiredo, P., Silveira, G. and Sbragia, R., 2008, "Risk sharing partnerships with suppliers: the case of Embraer", Journal of Technology Management and Innovation, Vol. 3, No. 1, pp. 27–27.

Freeman, C., 1991, "Networks of innovators: a synthesis of research issues", Research Policy, Vol. 20, No. 5, pp. 499–514. doi: 10.1016/0048-7333(91)90072-X.

Gassmann, O., Enkel, E. and Chesbrough, H.W., 2010, "The future of open innovation", R&D Management, Vol. 40, No.3, pp.213-221.

Gomes, S.B.V., 2012, "A indústria aeronáutica no Brasil: evolução recente e perspectivas", in: Lage, F. (org.), "BNDES 60 anos: perspectivas setoriais", Banco Nacional do Desenvolvimento, Vol. 1, Ch. 4, pp. 139-185.

Gosain, R., 2013, "On the right track: the INPI ups its game", World Intellectual Property Review, May-June 2013, pp. 56-57.

IMAP, 2011, "Aerospace global report 2011", International Network of M&A Partners.

Krishnan, V. and Ulrich, K.T., 2001, "Product development decisions: a review of the literature", Management Science, Vol. 47, No. 1, pp. 1-21. doi:10.1287/mnsc.47.1.1.10668.

Liyanage, S., Greenfield, P. and Don, R., 1999, "Towards a fourth generation R&D management model – research networks in knowledge management", International Journal of Technology Management, Vol. 18, Nos. 3-4, pp.372-93.

Mileski, A.M., 2011, "O bom momento da Thales", Revista Tecnologia e Defesa, No. 123, São Paulo.

OECD, 2002, "Frascati manual – proposed standard practice for surveys on research and experimental development", 6th edition, Organization for Economic Co-operation and Development.

OECD, 2008, "Open innovation in global networks", Organization for Economic Co-operation and Development.

OECD and EUROSTAT, 2005, "Oslo manual: guidelines for collecting and interpreting innovation data", 3rd edition, Organization for Economic Co-operation and Development.

Oliveira, A.C. and Kaminski, P.C., 2012, "A reference model to determine the degree of maturity in the product development process of industrial SMEs", Technovation, Vol. 32, No. 12, pp. 671-80. doi:10.1016/j.technovation.2012.08.001.

Pavitt, K., 1984, "Patterns of technical change: towards a taxonomy and a theory", Research Policy, Vol. 13, No. 6, pp.343-73. doi:10.1016/0048-7333(84)90018-0.

Pedroso, R., 2011, "Governo quer transformar ABC paulista em polo da indústria de defesa", Valor Econômico, October 20th.

Rolli, C., 2011, "Odebrecht adquire controle da fabricante de mísseis Mectron", Folha de São Paulo, 25 March.

Silveira, V., 2005, "Atech constrói para SIVAM primeiro radar 100% nacional", Gazeta Mercantil, 12 September, p.A-6, São Paulo.

Silveira, V., 2009, "Mectron fornece mísseis para o Paquistão", Valor Econômico, 17 April, p.B-9, São Paulo.

Soto, B. and Winter, B., 2013, "Saab wins Brazil jet deal after NSA spying sours Boeing bid", Reuters.

Tranekjer, T.L. and Knudsen, M.P., 2012, "The (unknown) providers to other firms' new product development: what's in it for them?", Journal of Product Innovation Management, Vol. 29, No. 6, pp. 986-99. doi:10.1111/j.1540-5885.2012.00974.x.

West, J. and Bogers, M., 2013, "Leveraging external sources of innovation: a review of research on open innovation", Journal of Product Innovation Management, Vol. 31, No. 4, pp. 814-831. doi:10.1111/jpim.12125.

West, J., Vanhaverbeke, W. and Chesbrough, H.W., 2006, "Open innovation: a research agenda", in: Chesbrough, H.W., Vanhaverbeke, W. and West, J. (eds.), 2006, "Open innovation: researching a new paradigm", Oxford University Press, Ch.14, pp. 285-307.

Wilson, D. and Purushothaman, R., 2003, "Dreaming with BRICs: the path to 2050", Goldman Sachs Global Economics Paper No: 99.

Identification of Design Considerations for Small Satellite Remote Sensing Systems in Low Earth Orbit

Juan Sebastián Triana[1], Sebastian Bautista[1], Freddy Alexander Díaz González[1]

ABSTRACT: The Sergio Arboleda University is preparing its next satellite mission, named Libertad 2. It will carry out a system of image acquisition as a working tool for researches based on the obtained data. Complete methodologies are often used in the development of satellite missions for planning, execution and deployment, for example, the standards of the European Cooperation for Space Standardization; these methodologies, however, do not include technical specifications or requirements for the development of nano-satellites nor for their subsystems. For this reason, this article focuses on the identification of the characteristics, requirements and restrictions, which must be considered in the design of a remote sensing system for satellites under the CubeSat standard, in order to serve as a starting point for the development of the main payload of the Libertad 2 mission.

KEYWORDS: Remote sensing, PCB, Embedded systems, CMOS APS, Low earth orbit, CubeSat.

INTRODUCTION

The CubeSat satellites have won recognition as possible platforms for future scientific missions (Woellert *et al.*, 2011; Saundau, 2008). Previous studies have determined the capabilities of these satellites and their functionality in the field of Earth observation. Daniel Selva, from Massachusetts Institute of Technology (MIT), conducted a feasibility study of remote sensing (RS) system technologies on-board small satellites, where he concluded which are the types of support technologies that can be implemented in satellite systems under the restrictions contained in the CubeSat standard. Additionally, he proposed technologies to perform spectral estimations of the state of vegetation and its biomass, particularly to determine the normalized differential vegetation index (NDVI) (Greenland and Clark, 2010).

Taking into account studies such as Selva's research, Sergio Arboleda University is carrying out its second satellite mission, called Libertad 2, which consists of the development and deployment of a nano-satellite of 3 units designed under the standard of CubeSat. This mission aims wto validate the development of a prototype of RS system, which will open the field for academic research in Colombia, based on images that it could get; these investigations can focus on the analysis of use of the land for agriculture, water resources, planning of urban growth, among others. This article focuses on the identification of the characteristics, requirements and restrictions, which must be considered in the design of an RS system for CubeSat satellites, in order to serve as a starting point for the development of the main payload of the Libertad 2 mission.

1.Universidad Sergio Arboleda – Bogotá – Colombia
Author for correspondence: Juan Sebastián Triana | Universidad Sergio Arboleda | Calle 74 14-14 14-14 | Bogotá – Colombia | Email: juans.triana@correo.usa.edu.co

The main goal of this study was the identification of constraints and considerations for the technical design of RS systems with operation in Low Earth Orbits (LEO). In order to provide a starting point, the paper includes a brief account of CubeSat missions previously carried out, which have contained payload for Earth observation and whose developers have available information of their camera model and/or their development process. Then, the considerations at the level of hardware and software, which must be considered in the development of a CubeSat satellite system, will be established, emphasizing the characteristics of an Earth observation system. As an example, the development of the first prototype of RS system for the Libertad 2 mission is included, applying the specifications previously identified. Finally, there are the results' analysis, conclusions and future work.

CUBESATS AS REMOTE SENSING SYSTEMS

Several prototypes of the visible spectrum RS systems have been designed and tested in academic CubeSat missions. It was found that 176 CubeSats have been launched until 2013; 68 of them have been deployed with RS systems, of which only 8 met information from their development process. The first of these was the CubeSat XI-IV, developed by the University of Tokyo to validate the effective use of commercial of the shelf (COTS) components in the development of RS payloads (Funase *et al.*, 2006; Enokuchi *et al.*, 2015); this was based on the commercial module C3188A of Quasar Electronics that employs an OV7620 sensor. In the same way, the picosatellite ITU-pSAT-1 and the Compass-1 employed this same integrated camera module (Khurshid *et al.*, 2013; Scholz *et al.*, 2010). However, none of the three missions has published detailed information on the process of acquisition and storage of RS's images, for subsequent download to the ground station.

The AAU-CubeSat uses a KAC1310 manufacturer Kodak sensor as payload; this sensor makes use of direct memory access (DMA) module driver, which allows reaching high speed flow of data between the sensor and the main memory. Images were stored in 5 RAM memories of 512 KB and incorporated hamming-based algorithms of error detection and correction code (Alminde *et al.*, 2003). However, they do not specify with which hardware's features the system was developed. CANX-1 satellite employed 2 complementary metal-oxide-semiconductor (CMOS) sensors, a multispectral one (Hotel-2020) and a monochrome (ADCS-2120), but they do not give specifications of the developed control systems or the mode in which the images were transmitted (Stras *et al.*, 2003).

The ESTCube-1 employs a CMOS MT9V011 along with a microcontroller STM32F217ZG of ARM Cortex M3 architecture that implements a dedicated interface for image sensors and additionally has a flexible static memory controller (FSMC) which allows easy connection to an external static random access memory (SRAM) (Kuuste, 2012). On the other hand, the University of Tartu does not provide information about the control software.

Despite the information provided by the developers of CubeSats with RS systems, no document was found that recorded the process of development of the actions and the considerations to be taken into account for the design, manufacture and assembly.

CONSIDERATIONS FOR THE DESIGN OF REMOTE SENSING SYSTEMS OF DEPLOYMENT IN THE LEO

The total system is designed based on a bus architecture, in which there is only one master subsystem called C&DH and the rest of the subsystems are slaves, including the RS system, as show in Fig. 1. All the satellite subsystems are defined in the next paragraphs.

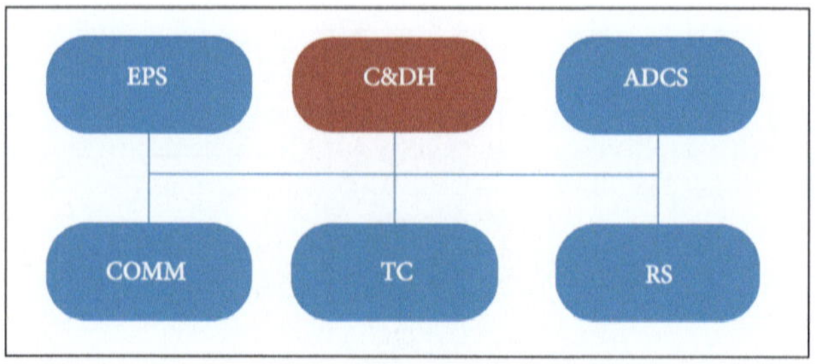

Figure 1. CubeSat subsystem's architecture (Schor *et al.*, 2009).

To meet the goals of a satelital mission, on certain occasions, it is necessary to control the orientation of the ship, in order to perform a specific task like provide effective communication with the ground station, capture images, load the batteries with solar panels or any action required by the mission that involves movement of the satellite. All of this is possible with the attitude determination and control system (ADCS) for CubeSat.

Another fundamental factor for the fulfillment of the objectives of the mission, decisive for its lifetime, is the energy.

Due to this, electrical power system (EPS) has autonomy in terms of energy management of the satellite; the C&DH is able to acquire information related to the battery condition, review of the voltages, and consultation of the battery conditions.

The communication subsystem (COMM) allows the satellite be capable to communicate with Earth. For this, the C&DH organizes the ship information into data packets to deliver them to the radio. In the same way, this subsystem receives the commands, which are related to the operations of the satellite, specific telemetry application and application data from the payload of the mission.

In the case of the RS subsystem, this would be controlled by C&DH, which can activate the RS system at the right time to execute the programmed actions.

In the thermal control (TC), the function of the C&DH will focus on the reading of sensors that indicate the temperature at certain nodes of the satellite, usually obtaining data of subsystems or modules in which sudden changes in temperature may result in permanent damage to physical level or at the level of data integrity.

RS systems serve as data acquisition instruments for experiments in different fields, either of basic science or advanced research, from studies of the Earth's surface or of the deep space. These RS systems are not limited to capturing images in the visible spectrum, but they also allow to capture representations in other bands of the electromagnetic spectrum as near infrared (NIR) or infrared (IR) (Rønning, 2012). The RS system's structure (Fig. 2) can be separated into the following subsystems: a lens that focus on the incoming light to the sensor, which could include light filters that only allow the passing of desired frequency bands according to the study; a mount as a mechanical support that holds the lens and electronic systems in place; an image sensor that allows the system to scan the scene; a control system that manages the sensor information and a thermal system that regulates the heat on the structure given to direct exposure to sunlight due to its satellite environment; this system can be passive as a heat sink, or active as it is the case of the cooled infrared camera refrigeration systems.

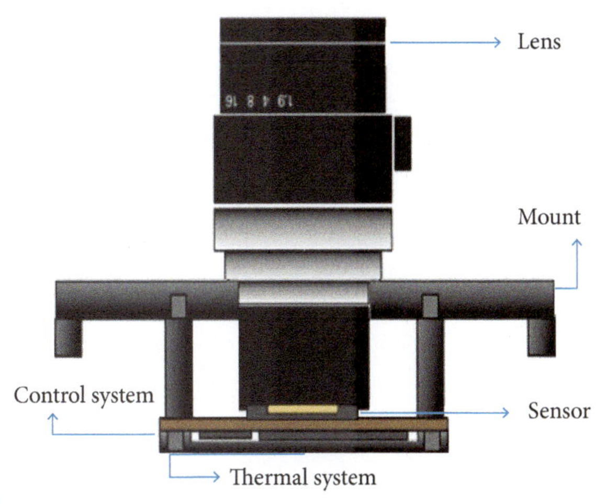

Figure 2. Image capture system structure (GomSpace, 2001).

OPERATING ENVIRONMENT IN LOW EARTH ORBITS AND ITS IMPLICATIONS ON THE HARDWARE OF A CUBESAT

The operation environment of these satellite systems limits the actions of support and maintenance that can be performed on it, which is why their development process must ensure robustness and fault tolerance, characteristic of a critical system (Knight, 2002). In addition, satellites in LEO, located at approximately 700 km, turn a full Earth in approximately 95 minutes (Montenbruck *et al.*, 2005), which means that the point on Earth to which the satellite is pointed will move with respect to the objective at 7 km per second; the pointing process is governed by the ADCS that guarantees the RS system pointing target for a short time period. Both the pointing process and the satellite speed leave a very precise time window in which the system is pointing to the target and the RS system is in a position ready to capture the image of the desired area. This is why the process of shooting must be synchronized with the sighting moment of the satellite to the objective, i.e. to be monitored and to ensure the exact time of execution of the image's capture process, which means it has to behave as a real time system to ensure deterministic times.

Also the restrictions surrounding a CubeSat satellite must be considered, such as the limited volume and mass and energy resources, which is why the implementation of the RS system of Libertad 2 satellite must be done on an embedded system and its software should optimize the available machine resources, including high-reliability quality attributes.

For the design of high-reliability systems for space applications, it is necessary to identify the factors that affect their behavior, in order to anticipate the failures that put at risk the goal of the mission; then, there are the most determining factors which were identified and which will be taken into account in the design of the RS system of the Libertad 2 nanosatellite.

Space Radiation in Low Earth Orbits

Space radiation is composed of different particles such as protons, ions, electrons, among others. It comes from sources as galactic cosmic rays or solar flares. The radiation mainly induces two types of effects: the total ionizing dose (TID) and the single event effect (SEE), which affect the correct performance of the electronic devices. The TID is a cumulative energy in the integrate circuits (IC) that changes parameters like leakage current or threshold voltage, causing permanent damages. The SEE refers to particular events and occurs when highly charged particles hit the circuit. There are two types of SEE errors: the soft ones — like a single event upset (SEU), that can be corrected by software — and the hard ones, like single event latchup (SEL) and single event bur (SEB), that cause definite changes in the way the device works or even destroy it (LaBel *et al.*, 1996).

The satellite systems in the LEO are a natural shielding by the magnetic field of the magnetosphere; however, they will be exposed to a radiation dose of about 100 rad/year (Petkov, 2003). One way to obtain additional shielding is adding an aluminum coat of 1.5 mm that might prevent permanent errors in the devices (Burlyaev and van Leuken, 2014). Additionally, SEL and SEU's effects can be prevented by software through the use of Cyclic Redundancy Check (CRC) algorithms in subsystems communication, as well as through storing backup copies of each subsystem software in the master system and the use of checksum fields in stored data.

Low Earth Orbits' thermal stress

Satellites in LEO must support extreme thermal cycles due to their intermittent exposure to direct sunlight. These thermal cycles cause the outer satellite structure's temperature to oscillate from -150°C to 150°C at its orbital period (Tennyson, 1995). In order to minimize effects on the hardware due to the thermal stress, usually a thermal control system is implemented. It regulates the heat throughout the structure in order to not let the peaks affect the components. Thermal systems can be active (heaters and coolers) or passive (thermal and insulating coatings). However, in the case of CubeSat satellite systems, an active thermal control is impractical, because the energy to maintain a stable temperature range within the structure may exceed the energy budget of the aircraft (Selva and Krejci, 2012). Due to this, the electronic components of the satellite must have a high thermal tolerance, to reduce the possibility of damage due to radiation.

For thermal management of the printed circuit board (PCB) parts, the thermal resistance must be calculated, which is the ability of a material to resist the heat flow. In the case of electronic devices operating in vacuum, it is necessary to contemplate the fact that there is not thermal resistance with the air, so there is only thermal resistance between the PCB and each device (Seo *et al.*, 2012).

An example of calculating the thermal resistance is presented using the ADP124 (8-lead MSOP) voltage regulator, whose tolerance to external temperature should not exceed 125°C. Equation 1 shows the thermal junction between the PCB and the device:

$$T_J = T_B + (P_D \,.\, \Theta_{JB}) \tag{1}$$

with

$$P_D = [V_{IN} - V_{OUT}]\, I_{LOAD} \tag{2}$$

where T_B is the PCB's temperature and P_D is the power loss. Assuming the maximum load current (I_{LOAD}) as 0.75 A; V_{IN} = 3.3 V; and V_{OUT} = 1.8 to P_D would be:

$$P_D = [3.3 - 1.8]\ 0.75 = 1.125\ \text{W} \tag{3}$$

Table 1 shows the thermal resistance coefficients between the device and the PCB according to the size of the copper zone assigned to the device die. It can be seen that the greater the area, the smaller will be the junction temperature for a given power loss. Supposing the case in which the PCB temperature reaches 75°C with a power loss of 1.125 W, consider the device ADP124 should not exceed 125°C to operate reliably. Replacing these values into Eq. 1:

$$\begin{gathered} T_J = T_B + (P_D \,.\, \Theta_{JB}) \\ 125^{\circ}\text{C} \geq 75^{\circ}\text{C} + (1.125\ \text{W} \,.\, \Theta_{JB}) \end{gathered} \tag{4}$$

Isolating Θ_{JB}:

$$44.44\ \frac{^{\circ}\text{C}}{\text{W}} \geq \Theta_{JB} \tag{5}$$

The operation of the ADP124 within the parameters of tolerated temperature can be guaranteed with Θ_{JB}= 44.44°C/W or

Table 1. Thermal junction versus copper size.

θ (°C/W)	Copper size (mm²)
108.6	25
75.5	100
42.5	500
34.7	1,000
26.1	6,400

less, which means that the copper prints must be at least with a size of 500 mm^2. Additionally due to the asymmetry between the surfaces, it is recommended to fill the joints between the copper and the device with thermal grease and to apply thermal pads, to reduce the thermal contact resistance and the gaps as well.

Low Pressures in Low Earth Orbits

In the LEO orbit, internally, the satellites have a close to absolute vacuum pressure (Grossman and Gouzman, 2003). Some materials at low pressures tend to have effects of degassing, where particles of the material are released in the form of gas and may end up condensing optics and solar cells, affecting the proper functioning of the satellite's system. In order to mitigate the effect of degassing, materials must comply with the requirements of the ECSS-Q-ST-70-02-C standard or the ASTM E595 standard, laying down the evidence these materials must submit and their standards of acceptance for deployment in vacuum (Fayazbakhsh and Abedian, 2010).

DESIGN FOR HIGH RELIABILITY OF ELECTRONIC SUBSYSTEMS FOR AEROSPACE APPLICATION

It is common to use redundant hardware for high-reliability systems, but it is not viable for CubeSat pace applications due to the increase in energy consumption and volume limitations. The reliability of hardware lies completely in the choice of components, circuit protection and in the design, manufacture and assembly of the PCB, which is the electromechanical hardware support. That is why international standards should be used for the electronic system design. These standards define PCB's development details, such as the Association Connecting Electronics Industries (IPC) standards and the European Cooperation for Space Standardization (ECSS), which provide industrial criteria to evaluate the quality of the designs under a single frame of reference. As following, some guidelines are recommended to be applied for the design of high-quality and reliability systems in space environments:

- Design Standard for Rigid Printed Circuit Boards and Assemblies (MSFC-STD-3425). In this standard, two documents of the IPC are named, the IPC-2221A and IPC-2222; the first talks about the general requirements for the design of PCBs and the second sets specific details over the designs for the installation and connection of components, assets and liabilities, using the 3-class performance for high-reliability electronic products.
- Design Rules for Printed Circuit Boards (Q-ST-70-12C) is an ECSS document that specifies rules for the design of printed circuit boards.

Another important issue to consider is that of ruptures of welding joints, whether by mechanical vibrations or difference between the coefficient of thermal expansion (CTE) of the components and the PCB. To decrease the rupture probability, the gap between the thermal expansion coefficients of the components and the board must be reduced, so that both will behave in a similar way at strong temperature changes.

To give greater rigidity to the devices against mechanical vibrations, the "footprints" of the components should be over dimensioned with respect to the physical size of the component, in order to apply greater volume of welding and give better mechanical support. The dimensions are given by the standard Naming Convention for Standard SMT Land Patterns (IPC 7351B), where the design parameters of the components footprints are established according to the application. In this case, it is used Level A for "Most Material Condition" for high-shock and vibration applications. In addition, physical supports should be added, which hold the components during periods of extreme vibration; the standard Class 3 High-Reliability Applications (IPC-2221) - section 8.1.9.2 mentions the criteria for selection of these supports.

Commercial off the shelf (COTS) devices used in CubeSat systems are not designed to operate in hostile environments such as space; that is why it is recommended to use a protective covering or conformal coating, to give extra protection against extreme changes in temperature and degassing (Hardy, 2011). According to military specification MIL-I-46058C, there are several types of conformal coatings which highlight the XY type, also called paraxylylenes, that offer protection to extreme temperature ranging from -200°C and 150°C and high tolerance to radiation. In addition, its implementation by vacuum-deposition process reduces the possibility of bubbles trapped under the overlay, which, at low pressures, may compromise the uniformity of the cover. The use of the paraxylylen Parylene D coating for CubeSat applications is recommended, since it maintains their physical and electrical properties at high temperatures. According to the standard MIL-I-46058C, a layer of 0.6 mm of thickness of this coating fully covering all the components can offer a substantial increase in tolerance, without adding considerable weight.

INTERNATIONAL TRAFFIC IN ARMS REGULATIONS RESTRICTIONS

In order to increase the reliability and damage tolerance of the satellite's hardware, ideally it must use components designed to operate in harsh environments, as the high radiation tolerance components also known as Rad Hard; however, the exportation of much of these components is controlled by the International Traffic in Arms Regulations (ITAR) (22 CFR 120 through 130), making it difficult to purchase in the market.

COMPONENT SELECTION FOR A CUBESAT REMOTE SENSING SYSTEM

SELECTION OF OPTICAL SENSOR FOR IMAGE CAPTURE

Image sensors are an array of light sensitive photosites, which quantify the amount of received light energy. An image is made up by organizing the values collected from each photosite on a matrix of pixels. There are currently two available technologies for image sensors, charge-coupled device (CCD) and CMOS.

The CCD technology consists of a matrix of metal-oxide-semiconductor (MOS) capacitors doped p-type, in which each capacitor builds up an electrical charge proportional to the perceived light intensity; then, an internal circuit transfers its charge to a neighbor in the same row in only one direction, the last pixel of the line delivery that burdens an operational amplifier, which is responsible for converting it to voltage. This results in an analog signal from each line of the matrix which is then sampled to digitize it (Litwiller, 2009).

A CMOS sensor is an active pixel sensor (APS) sensitive to visible light. It makes use of a digital signal processing (DSP) in an embedded system, to process the data of intensity that perceives each cell of the pixel matrix, recreating an image. This kind of sensor is known as CMOS since it uses technology based on metal-oxide-semiconductor field-effect transistors (MOSFET); each of its photosites contains an amplifier for electrical signal and usually includes a digital converter in the own chip (Litwiller, 2009).

To capture multispectral color images, usually an array of Bayer RGB filters is placed between the lens and the photodiode, to prevent the passage of all frequencies of light except for the bands of red, blue and green; with them, the RGB color model can be calculated and a good approximation of the visible scene to the human eye can be composed (Litwiller, 2009).

Table 2 shows a comparative analysis between the image sensor technologies, the CMOS and CCD, showing relevant characteristics useful for the images' capture process.

Table 2. CCD versus CMOS

Characteristics	CCD	CMOS
Dynamic range	Yes	No
Response capability	No	Yes
Reliability	Yes	Yes
Energy consumed	No	Yes
Size	No	Yes
Capture time	No	Yes
Blooming	No	Yes
Windowing	No	Yes
Temperature	No	Yes

The dynamic range is the ratio between the saturation level and noise in an image; since the CCD does not integrate the digitalization of the signal into the sensor, this process is carried out in specialized systems, which can get a lower level of noise through the use of DSP.

The response capability is the parameter that measures the ratio between the output voltage of the sensor and a certain incident light energy density; this is expressed in electric potential (V) by energy (J) in the unit of area (cm^2) — ($V/\mu Jcm^2$). CMOS technology has amplifiers that increase the profit by an integrated circuit, improving sensitivity to light in each pixel. In terms of reliability, both technologies have similar behaviors, so one cannot be considered as better than the other.

In the case of the power and size, CMOS has a clear advantage by integrating the required embedded system on a single chip to obtain the images, while CCD requires extra components. Additionally, the polarization voltage increases considerably for CCD technologies, as it uses voltages around 12 to 15 V, while CMOS uses voltages from 2.8 to 5 V. In terms of the speed of frame capture or capture time, it is similar in both technologies; nevertheless, CMOS presents the advantage of integrating all the capture process on a single chip, so power and data signals travel a shorter distance, improving propagation times and reducing the inductance.

A phenomenon to be considered in RS systems is blooming, that is an effect of overflowing of charge in a pixel due to the excess of incident light, resulting in a contribution of charge to neighboring pixels. CMOS technology presents an independent design, where each photosite delivers digital data separately to avoid blooming effect, while CCD sensors do not have this feature, making them more sensitive to this effect. The windowing is a feature added to CMOS sensors, where one can choose to capture only a fraction of the frame (picture capture), saving time and energy. This functionality is not implemented in the CCD sensors. Finally, the operating temperature represents a big problem in the CCD as it increases their sensitivity to noise, to the point of doubling it every 7°C. Therefore, they require to be refrigerated for space applications (Litwiller, 2009; Eastman Kodak Company, 2009).

SELECTION OF THE CONTROLLER FOR THE IMAGE SENSOR CONTROL SYSTEM

The control system must have communication interfaces with the CMOS sensor and with external memories in order to capture and storage the images. Different microcontroller systems that fill these requirements are presented in Table 3.

From the analyzed microprocessors, it was selected the STM32F407 of architecture ARM Cortex-M4, which possesses the digital camera interface (DCMI) for control and data transmission from the image sensor and flexible static memory controller (FSMC) module; it also has a DMA module to transfer data at high speed between peripherals and RAM memories. All these modules are implemented in hardware, which is configured from the CPU and runs concurrently to this, leaving the processor free for other operations. Additionally, this architecture of microcontrollers handle floating point logic, which optimizes it for DSP analysis. It adds special instructions that facilitate the handling and processing of images (ARM, 2013). Figure 3 shows the block diagram that describes the process for the image's capture and storage.

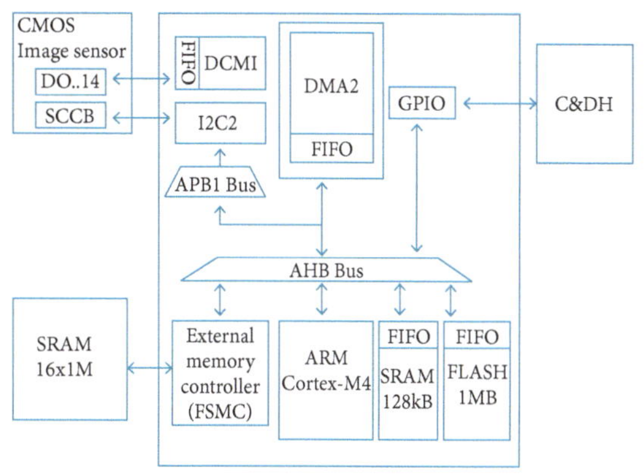

Source: Gulzar (2009).

Figure 3. Block diagram of the controller of the image acquisition system.

Table 3. Comparison of microcontrollers.

DSP	Maximum frequency (MHz)	Manufacturer	FSMC	DCMI	DMA	I/O interface	Program memory (KB)
Tms320dm643	700	TI	Si	Si	Si	I2C McBsp SPI CAN PCI Ethernet	No integrated memory
Tms320vc5509a	200	TI	Si	Si	Si	USB Full-Speed McBSP	64
OMAP3515	600	TI	Si	Si	Si	MMC/SD POP I2C UART USB	32
ADSP-BF561	600	Analog devices	Si	Si	Si	IrDA UART Ethernet	328
STM32F407	168	STMicro	Si	Si	Si	I2C CAN USART USB MMC/SD	Flash: 192 SRAM: 64

SELECTION OF THE MEMORIES FOR STORING IMAGES

Table 4 presents a comparative study between different types of volatile and non-volatile memories, in order to identify which of them is best suited for RS system, taking into account main operations, as the image capture and storage, and considering fundamental requirements as a low energy consumption and the available volume.

Due to the amount of data and the flow speed of information that the sensor delivers, the selection criteria of the memory are its speed and capacity. In order to avoid overloading in the internal RAM of the microcontroller, the system needs a fast enough memory and its capacity for data storage must be enough to allow staying in it a burst of images. Then, commercially available memory options are evaluated in order to select which is the one that is better suited to a RS system on-board a CubeSat.

SRAM memories are used as cache memory in personal computers since they offer faster speeds of reading and writing (8 ns) than all other memories; the SRAM, however, can store information when the system is energized. Once the power source is removed, the information will be lost. A dynamic random access memory (DRAM) is superior to the SRAM in terms of density, i.e. it has greater volume of data. However, it increases the time of reading/writing. SRAM memories, being the fastest but volatile, require a secondary storage to allow the download of images at any time.

After obtaining the images, RS system enters a wait state until it receives the command to send an image. In order to minimize the cost of energy in this state, the system should enter low consumption by disabling the peripherals, so it is necessary to store the images in a non-volatile memory for later transmission. There are three types of non-volatile memory technologies: Flash memory, ferro-electric random access memory (FRAM) and magnetic random access memory (MRAM). Flash memory is considered more suitable for satellite data transmission's needs for the following reasons:

- Flash memories can achieve the greatest amount of information per unit area, since a flash cell is made up of just one transistor, while the FRAM and MRAM cells need a transistor, a capacitor, or an union of magnetic tunnel, which implies greater physical space for the same amount of information (Semiconductor Industry Association, 2014; Kim *et al.*, 2002).
- Flash memories have the property of saved multi-bit per cell, which increases the density of the memory, reducing the cost per bit (King, 1999).
- The manufacturing process of the flash memory is compatible with current CMOS processes and is a sustainable solution for embedded system applications. A flash memory cell is a MOSFET; the rest of non-volatile memories require the integration of other materials that are not compatible with conventional CMOS processes, which makes them less reliable for integration with other analog or logical devices (She, 2003).

IMAGE CAPTURE PROCESS

Three stages for the acquisition of images are defined within embedded software: the first is the moment of image capture, followed by in-memory main storage and, finally, a stage of images saved in non-volatile memory.

At the moment of image's capture, the sensor loads their cells according to the intensity of light that receives and delivers this information as digital values organized as a Bayer matrix to the DSP. This data is processed to generate the delivery format (RAW-RGB, RGB or YUV) image, which is made by hardware and is called exposure time. CMOS image sensors are progressive scanners, i.e. they determine the amount of incident light at

Table 4. Memory technologies comparison.

Memory	DRAM	SRAM	Flash-NOR	Flash-NAND	FRAM	MRAM
Volatile/Non volatile	Volatile	Volatile	Non volatile	Non volatile	Non volatile	Non volatile
Reading method	Destructive	Partially destructive	No destructive	No destructive	Destructive	Destructive
Reading voltage	~ 1 V	~ 1 V	2 V	2 V	1.5 V	3.3 V
Writing voltage	~ 1 V	~ 1 V	10 V	18 V	1.5 V	3.3 V
Reading speed	8 ns	8 ns	60 ns	60 ns	80 ns	30 ns
Writing speed	50 ns	8 ns	1 us	1 ms	80 ns	50 ns
Biggest vector built			2 Gb	512 Gb	64 Mb	16 Mb

Source: She (2003).

discrete moments of time and discharge their voltage values whenever they deliver data; it is for this reason that CMOS sensors do not require a shutter to regulate the exposure time, since this is already determined by the manufacturer because it depends on the size of the frame and the speed of the internal DSP. This value is measured in frames per second (fps) and commercially it is between 580 and 3 fps.

The second stage consists on the image storage in the main memory. The DCMI, which is a synchronous parallel interface capable of receiving streams of data at high speeds from up to 14-bit CMOS camera module, is used to receive an image from the embedded system.

Figure 4 shows the block diagram of the DCMI interface. Signals DCMI_PIXCLK, DCMI_D[0:13], DCMI_HSYNC and DCMI_VSYNC are generated by the sensor; these serve as the control of transmission of the image. The data is stored in a 32-bit buffer that when it gets filled up, it sends a request to the DMA, which flushes the data from the DCMI buffer to a previously configured memory location. The DMA is used to provide data transfer at high speed between peripherals and memories, without action of the CPU. This module combines a dual master bus (AHB) with a separate first input-first output (FIFO) buffer architecture that optimizes the system bandwidth and improves response times.

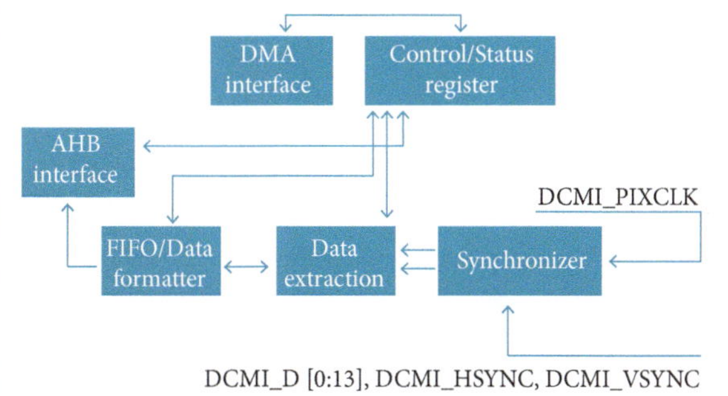

Source: ARM (2013).

Figure 4. DCMI block diagram.

The DMA manages requests for peripherals and starts a transfer whenever one is received; a DMA transfer is the flush of data from the peripheral directly to a space in memory, where it will remain while the power supply is not interrupted. High-speed memories, as the SRAM access, require the use of the flexible static memory control (FSMC) controller that acts as an interface for synchronous and asynchronous memories, translating the AHB transactions to the appropriate external reading or writing protocol (ARM, 2013).

The last stage consists of the non-volatile memory image storage, where it will remain until being required by the C&DH. The access to non-volatile memory is also performed by the module DMA, using memory-to-memory transfer mode. To counteract the effects of SEE, a checksum field is recommended to save the image to verify the consistency of the data when the C&DH receives a transmission request from Earth.

THE SATELLITE IMAGE SENSOR DATA PROCESSING

As mentioned previously, the CMOS sensor delivers information as a Bayer array with raw data and the size of n x m. The Bayer array consists of three color components: green, red and blue; it applies to this information an algorithm of color interpolation or rebuilding, which is a digital process used to reconstruct an image to the RGB color model from samples acquired from the multispectral CMOS image sensor (Adams *et al.*, 1998).

There are several rebuilding algorithms to calculate RGB color model from the Bayer array; among the most prominent ones, there are: Pixel Binning, Nearest Neighbor, Bilinear, Smooth Hue Transition and Edge-sensing Bilinear (Elizondo and Maestre, 2005). However, when using a Bayer to RGB conversion algorithm, the dimensions of the image are increased from (n x m) to 3 matrices of the same size, that is (n x m x 3). This is why it is recommended that the information delivered by the CMOS sensor be transmitted without being processed, so power consumption is reduced by having less amount of information in the queue.

DEVELOPMENT OF A REMOTE SENSING SYSTEM FOR A 3U CUBESAT

Taking into account the issues due to the radiation of high-energy particles, low atmospheric pressure, thermal shocks, vibration during the launch and low processing capacity — and according to the recommendations presented in the previous sections —, the software and hardware design of a RS system for CubeSat space applications are shown as follows.

DEVELOPMENT OF THE REMOTE SENSING SYSTEM CONTROL SOFTWARE

The available functionalities must be specified for the design of the RS system control software, which is made through the use of a case diagram that describes the driver in terms of the processes that runs each actor. At the same time, the actors represent subsystems or external devices that interact with the software. At first, the software operation environment must be established, identifying internal and external subsystems that will be in contact with the RS system. In this case, the master will be the central computer of the satellite (C&DH) and the payload driver is represented as an intern actor who executed the orders received from the master.

Figure 5 shows the minimum capabilities that the C&DH should carry out in the RS system; among them, there is the option to configure the camera in low power consumption when no actions are required. The other use cases are directly related to the process of image acquisition. In this case, one for ordering a capture and another for asking for the image transmission. The C&DH should be able to define the parameters to take the photo, like frame size, subsampling of the image, the gain and the delivery format. Another function that must be available is to be able to request the state of the camera's memories in order to verify if there is space left to store more images.

Next, the deployment of processes is explained, which defines the behavior of the control software. The activity diagram in Fig. 6 represents the time for execution of each task involved in the image's capture.

The "Command_Handler" task that runs in the background is a persistent and high priority task, which is on charge of reception and administration of the commands sent by the C&DH. When a capture command is received, it will set a timer to run the capture task at the proper time; this sequence corresponds to the steps to capture an image. During this task, the DMA module is activated, so that it will be attentive to the requests the DCMI peripheral could send, thus activating the DCMI and, as soon as it reaches a start of frame indicated by the VSYNC signal, the DMA will start the flush of data to the principal RAM memory.

As soon as the DCMI and DMA modules are activated, the sensor data will be flushed by hardware. The Capture_Finished method must set off the DCMI, so that the received data may remain static while it is copied to the external flash memory; the DCMI, however, is not disabled until it ends receiving the frame. So, in order to disable the DMA, a recurrent query must take place, asking if it has already finished the image capture, otherwise it can cause image damages.

After the DMA and DCMI modules are off, the software executes the routine of Persistent_Save. In this function, the DMA is used to flush the data from main memory (SRAM) to non-volatile memory (flash). Additionally, it adds the image metadata to the delivered package; for example, the values of gain and the average luminance. Finally, it turns a flag indicating

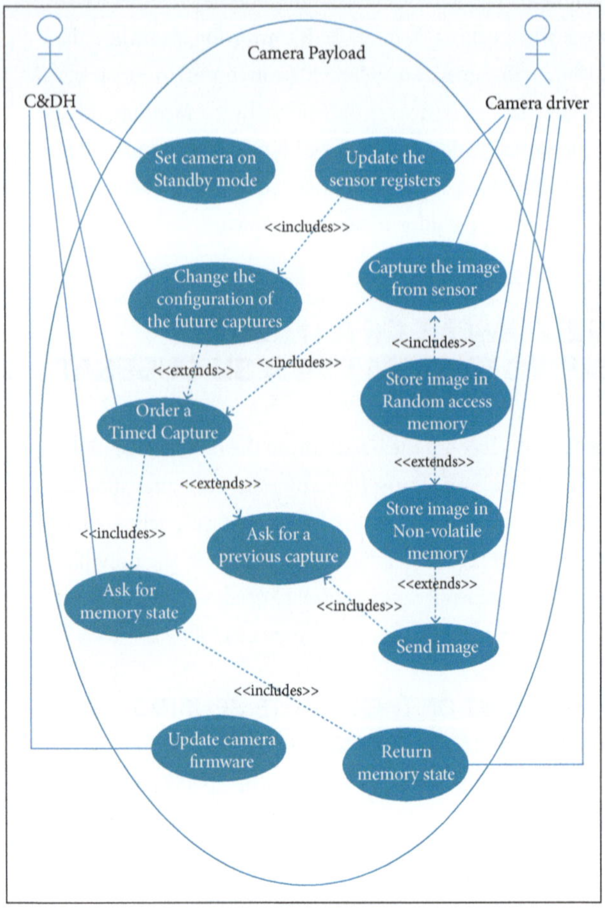

Figure 5. Use case diagram of the RS system software.

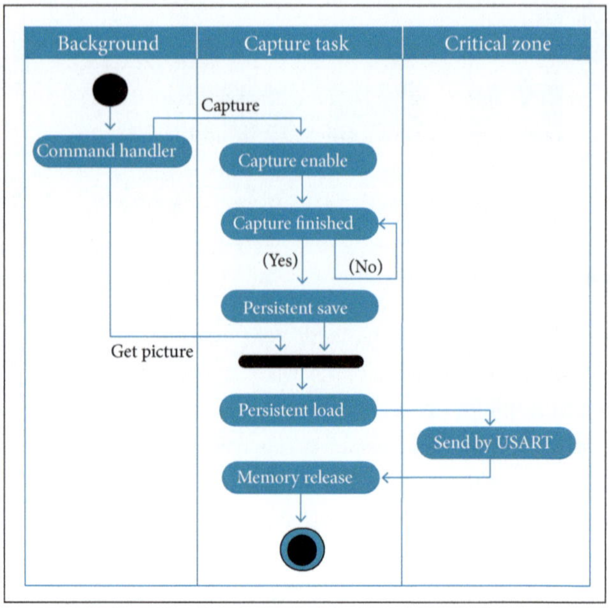

Figure 6. Activity diagram for tasks

that an image has been taken correctly and it is waiting to be transmitted.

Once an image is captured and stored, C&DH may submit the request for that image. When this command is received, the first flush of data will be initiated from the flash memory to the main memory through DMA, which must be reconfigured in the Persistent_Load method.

In order to send the image through serial port, the system interruptions should be disabled. For this, one must run the routine in a safe area of the code, to ensure the integrity of the data. While the interruptions are disabled, the system loses the ability to listen to the commands sent by the C&DH, so the use of this safe area should be limited to the transmission of images through the USART interface, with the Send_By_USART method. Many of the tasks of the camera driver have to run at specific times or even at the same time; this is why it is recommended to use a real time operating system (RTOS).

Radiation and other sources of external interference can cause critical errors in the software, such as corruption of the code, change in the data and unscheduled resets. Code and data corruption can be permanent or temporary. For permanent errors, the only solution available is to update the system's control software. Therefore, it is recommended to store versions of all the firmware of the subsystems in the C&DH.

All captured images must be transferred to a special memory region in the SRAM that can contain a complete image of 8 bits per pixel; the other part of the memory will store image compressions, cuts and other useful data for later use. Images are saved in flexible image transport system (FITS) format, that includes the following metadata in the header: exposure, capture time, temperature sensor, firmware version, and commentary on the current camera mode and some settings such as resolution and gain parameters (Pence, 2010).

DEVELOPMENT OF THE RS SYSTEM HARDWARE

Some considerations for the manufacture and assembly of RS system hardware are presented as following, implementing the design recommendations proposed in the section: Considerations for the design of remote sensing systems of deployment in the Low Earth Orbits.

Device's Footprints for High-Reliability Applications

For the design of the land pattern, it is recommended a specialized software tool like IPC-7351 LP Wizard® Mentor Graphics or "PCB Library Expert®", which has the ability to design land patterns with the specifications dictated by the standard IPC-7351B level A "Very Robust Solder Joint" for high reliability class 3 applications.

In the case of ball grid array (BGA) devices, the solder mask in the area around the copper pad must be removed to decrease the probability of the appearance of cracks into solder joints (Fig. 7a), due to the high mechanical stress that can cause contact between the mask and solder. This reduction of the mask allows the solder to stick around the pad, improving contact with the surface and increasing union's reliability (Fig. 7b).

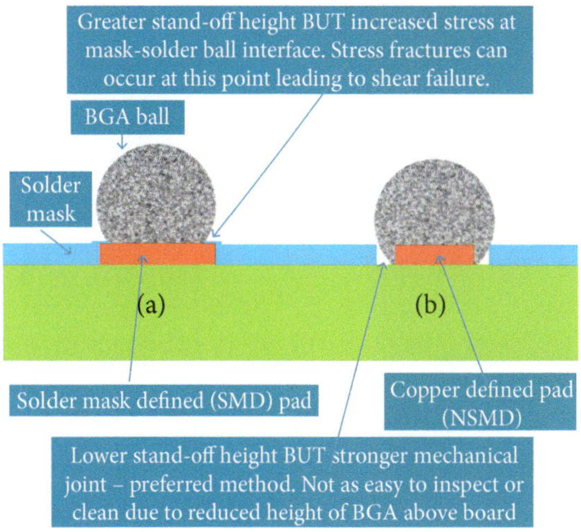

SMD: solder mask defined; NSMD: non-solder mask defined.

Figure 7. Definition of the BGA pad (Johnson, 2014). (a) BGA copper pad with solder mask around; (b) BGA copper pad without solder mask around.

Electromagnetic Compatibility in the Design

One of the most important aspects in the electronics equipment is the property of operating without causing interference to neighboring subsystems; there are different considerations that can be applied in the design of the PCB for better performance in electromagnetic compatibility (EMC). One must be sure to minimize the area of the loop, not to place circuits between connectors, control the transition time from digital signals and provide return signal plane (Hubing, 2003).

Another aspect to consider is electromagnetic interference (EMI) shielding signals. This can be done routing them between the power and ground planes. So, in the PCB design, it must consider the use of multiple layers to provide extra protection to the internal planes and make signals less likely to interact with external interference sources.

In the PCB design, the aspect ratio must not exceed 8:1 to prevent break in any via due to a mechanical stress (Marshall Space Flight Center, 2006). In addition to axial-leaded components with a weight equal or greater than 5 g per lead (leg), they must be secured to the PCB with mechanical stands and not only solder joints. It is advisable to assemble the PCB under the standards IPC j-STD-001 class 3 and IPC-A610 class 3.

Solder Joints for Space Applications

For space applications, the use of lead-based solder is allowed, which can prevent problems as delamination, rupture of internal paths and blistering, due to its lower melting point in comparison with the lead-free solder (Kessel, 2005). In the space applications electronic hardware addendum to the IPC-J-STD-001D standard described by IPC, titled as "Requirements for Soldered Electrical and Electronic Assemblies", it is specified that the solder alloys that can be used in space applications should be one of the following: Sn60Pb40, Sn62Pb36Ag2, Sn63Pb37 or Sn96.3Ag3.7 (Association Connecting Electronics Industries, 2008).

Table 5 shows the properties of the different solder alloys mentioned by IPC-J-STD-001D, whereas all solders have a similar behavior in the majority of its properties; it is used melting point and tensile strength parameters as criteria for solder selection. As seen in Table 4, the Sn62Pb36Ag2 presents the lower melting point, so it is less likely to damage the components by heat excess at the assembly time. It has, however, less resistance to mechanical stress. Moreover, Sn96.3Ag3.7 alloy presents the greater tensile strength. This indicates the maximum stress that can withstand before breaking the solder, which makes it more reliable to vibration, but with regard to its melting point, high value could affect components in their assembly process. For these reasons, it is recommended to use Sn60Pb40 solder or Sn63Pb37, since they offer the best relation between resistance to mechanical stress and fusion temperature, in addition to GomSpace used in the development of its prototype RS system for CubeSats, the NANOCAM-1U (GomSpace, 2001).

Printed Circuit Board Material for Vacuum Applications

The material selected must comply with the ASTM-E-595 standard, which specifies that the materials must not exceed 1% in "Total Mass Loss" (TML) and 0.1% in "Collected Volatile Condensable Materials" (CVCM) in vacuum testing. Table 6 shows a list of materials and their results in the vacuum test conducted by NASA, collected in the database "Outgassing Data for Selecting Spacecraft Materials" of this entity which meet the above mentioned requirements.

The type of material that one chooses affects the reliability of the system. The standard IPC-4101 describes the specifications that must comply with the materials to be used in PCB's rigid multilayer; some of the most important properties to be considered in the choice of the PCB materials are the dielectric constant for a better signal integrity, a greater transition glasses (Tg) and a lower CTE, allowing thermal stability against temperature cycles.

According to the standard ECSS-Q-ST-70-12c, the most used materials in the PCB's manufacture for space applications are the

Table 5. Solder alloys (Association Connecting Electronics Industries, 2006).

Solder alloy	Melting point (°C) solid/liquid	Density (g/cm³)	Electrical resistivity (ohm-m)	Thermal conductivity (W/m-K)	Tensile strength at break (Kgf/cm²)	Tensile elongation at break (%)	Brinell hardness (HB)
Sn60Pb40	183/191	8.50	0.153	49	535	40	16
Sn63Pb37	183/183	8.40	0.145	50	525	37	17
Sn96.3Ag3.7	221/221	7.37	0.123	55	580	35	15
Sn62Pb36Ag2	179/179	8.41	0.145	50	490	--	16

Table 6. Results of outgassing tests % TML, % CVCM for PCB materials.

Material	% TML	% CVCM
CER-10 Glass/PTFE/Ceramic laminate material	0.02	0.01
FR406, Woven e-glass, Copper foil	0.43	0.00
Alumina, Gold, PD/AG	0.02	0.00
Circuit board composite Nema G10 Mica/CE1155	0.48	0.01
G10 Fiberglass per MIL-I-24768/27	0.30	0.00
Laminate 602 Teflon/Fiberglass W/O Copper	0.01	0.00

FR4 and polyimide. The last one could have a Tg greater than 250°C and it would be more stable to heat than the FR4, which has a Tg up to 180°C. Among the commercial polyimides, the 35N, 85N and 85NT of Arlon are recommended, which present good behavior regarding the dielectric constant, coefficient of thermal expansion and Tg.

CONCLUSIONS

In this paper, the critical elements related to the operating environment were identified, which are decisive in the development of an embedded system to capture and storage satellite images, such as: thermal radiation, particle radiation, outgassing, and some ways to mitigate these effects.

A software/hardware architecture was proposed for the implementation of a system for capturing satellite images, and the capturing and storing process of an image taken with a CMOS sensor was established.

A software controller in real time was structured, with features of critical system for the control and management of the information of a CMOS sensor, detailing the capture sequence and storage techniques in memory arrays.

Finally, some parameters, materials and coatings were identified, which could be used in the PCBs manufacture for aerospace operation.

FUTURE WORK

It is expected the design of the optical RS system, according to the restrictions of the CubeSat standard, with emphasis on lenses and filters for an aerospace operation environment.

In addition, it is expected to include compression algorithms with and without loss in the proposed software controller, in order to transmit a mosaic to the ground station and then select the best for download in original format.

Besides, it is planned to implement and perform protoflight tests in laboratories specialized in aerospace systems, to verify the correct operation of the developed prototype, including a LEO thermal cycle emulation, as well as the observation of the material's behavior in vacuum.

REFERENCES

Adams, J., Parulski, K. and Spaulding, K., 1998, "Color processing in digital cameras", IEEE Micro, Vol. 18, No. 6, pp. 20-30. doi: 10.1109/40.743681

Alminde, L., Bisgaard, M., Vinther, D., Viscor, T. and Ostergard, K., 2003, "Educational Value and Lessons Learned from the AAU-CubeSat project", Proceedings of the International Conference Recent Advances in Space Technologies, Istanbul, Turkey.

ARM, 2013, "ARM® Cortex™-M4 Processor Technical Reference Manual", ARM Limited, USA.

Association Connecting Electronics Industries (IPC), 2008, "Requirements for soldered electrical and electronic assemblies - IPC J-STD-001D", Bannockburn, Illinois, USA.

Burlyaev. D. and van Leuken, R., 2014, "Systemfault-Tolerance Analysis of COTS-Based Satellite On-Board Computers", Microelectronics Journal, Vol. 45, No. 10, pp. 1335-1341. doi:10.1016/j.mejo.2014.01.007

Eastman Kodak Company, 2009, "Solid State Image Sensors Terminology", Microelectronics Technology Division, Rochester, USA.

Elizondo, J.E. and Maestre, L.P., 2005, "Fundamentos de Procesamiento de Imágenes", Universidad Autónoma de Baja California, México.

Enokuchi, A., Nagai, M., Funase, R., Nakamura, Y. and Nakasuka, S., 2015, "Remote Sensing by University of Tokyo's Pico-Satellite Project 'PRISM'", Department of Aeronautics and Astronautics, University of Tokyo, Retrieved in Jan 06, 2015, from http://www.dlr.de/Portaldata/49/Resources/dokumente/archiv5/0605_Enockuchi.pdf

Fayazbakhsh, K. and Abedian, A., 2010, "Materials Selection for Applications in Space Environment Considering Outgassing Phenomenon", Advances in Space Research, Vol. 45, No. 6, pp. 741-749. doi:10.1016/j.asr.2009.11.017

Funase, R., Takei, E., Nakamura, Y., Nagai, M, Enokuchi, A., Yuliang, C., Nakada, K., Noriji, Y., Sasaki, F., Funane, T., Eishima, T. and Nakasuka, S., 2006, "Technology Demonstration on University of Tokyo's Pico-Satellite 'XI-V' and its Effective Operation Result Using Ground Station Network", Acta Astronautica, Vol. 61, No. 7-8, pp. 707-711. doi:10.1016/j.actaastro.2006.12.032

Jernes, N, 2011, "NanoCam C1U Datasheet. GS-DS-NANOCAM-6.2, Issue: 6". Gomspace, Aalborg East, Denmark.

Greenland, S. and Clark, C., 2010, "CubeSat Platforms as an On-Orbit Technology Validation and Verification Vehicle", Retrieved in January 27, 2015, from http://www.clyde-space.com/documents/1805

Grossman, E. and Gouzman, I., 2003, "Space Environment Effects on Polymers in Low Earth Orbit", Nuclear Instruments and Methods in Physics Research Section B: Beam Interactions with Materials and Atoms, Vol. 208, pp. 48-57. doi:10.1016/S0168-583X(03)00640-2

Gulzar, K., 2009, "Camera Design for Pico and Nano Satellite Applications", Department of Space Science, Kiruna, Switzerland.

Hardy, A., 2011, "Parylene Coating Can Protect COTS Electronics in Aerospace and Defense Applications", Retrieved in Jan 06, 2015, from http://www.militaryaerospace.com/articles/print/volume-22/

issue-30/opinion/parylene-coating-can-protect-cots-electronics-in-aerospace-and-defenes-applications.html

Hubing, T., 2003, "PCB EMC Design Guidelines: a Brief Annotated List", IEEE International Symposium on Electromagnetic Compatibility, Vol. 1, pp. 34-36. doi: 10.1109/ISEMC.2003.1236559

Johnson, A., 2014, "Prenier EDA Solutions", Retrieved in Jul, 2014, from http://www.eda.co.uk/design_community/design_tips/soldermask.html

Kessel, K., 2005, "NEPP TRO – Lead-Free Soldering for Space Applications Lead-Free Solder Body of Knowledge", NASA, Kennedy Space Center, ITB, Inc., Dayton, USA.

Khurshid, K., Mahmood, R. and ul Islam, Q., 2013, "A Survey of Camera Modules for CubeSats-Design of Imaging Payload of ICUBE-1", Proceedings of the 6th International Conference on Recent Advances in Space Technologies (RAST'13), Istanbul, Turkey.

Kim, D.W., Prins, F.E., Kim, T., Kwong, D.L. and Banerjee, S., 2002, "Charge Retention Characteristics of SiGe Quantum Dot Flash Memories", Proceedings of the Device Research Conference, USA.

King, Y.C., 1999, "Thin Dielectric Technology and Memory Devices", University of California, Berkeley, USA.

Knight, J.C., 2002, "Safety Critical Systems: Challenges and Directions", Proceedings of the 24rd International Conference on Software Engineering, Orlando, USA.

Kuuste, H, 2012, "Estcube-1 Tether End Mass Imaging System Design and Assembly", Institute of Physics Information Technology, Faculty of Science and Technology, University of Tartu, Tartu, Estonia.

LaBel, K.A., Gates, M.M., Moran, A.K., Marshall, P.W., Stassinopoulos, J.B., Seidleck, C.M. and Dale, C.J., 1996, "Commercial Microelectronics Technologies for Applications in the Satellite Radiation Environment", Retrieved in January 27, 2015, from http://nepp.nasa.gov/DocUploads/0754D677-D2AF-4778-B910A6CED2EA7458/apsen.pdf

Litwiller, D., 2009, "CCD vs. CMOS", Photonics Spectra, Vol. 35, No. 1, pp. 154-158.

Marshall Space Flight Center, 2006, "Design Standard for Rigid Printed Circuit Boards and Assemblies", National Aeronautics and Space Administration (NASA), Pennsylvania, USA.

Montenbruck, O., Gill, E. and Kroes, R., 2005, "Rapid Orbit Determination of LEO Satellites Using IGS Clock and Ephemeris Products", GPS Solutions, Vol. 9, No. 3, pp. 226-235.

Pence, L.C., 2010, "Definition of the Flexible Image Transport System (FITS), version 3.0", Astronomy & Astrophysics, Vol. 524, pp. 1-40. doi: 10.1051/0004-6361/201015362

Petkov, M.P., 2003, "The Effects of Space Evironments on Electronic Components", Retrieved in Jan, 06, 2015, from http://trs-new.jpl.nasa.gov/dspace/bitstream/2014/7193/1/03-0863.pdf

Rønning, S.S., 2012, "Optimizing an Infrared Camera for Observing Atmospheric Gravity Waves from a CubeSat Platform", Ph.D. Thesis, Norwegian University of Science and Technology, Trondheim, Norway.

Saundau, R., 2008, "Status and Trends of Small Satellite Missions for Earth Observation", Acta Astronautica, Vol. 66, No. 1-2, pp. 1-12. doi:10.1016/j.actaastro.2009.06.008

Scholz, A., Leyb, W., Dachwaldb, B., Miaua, J. and Juangc, J., 2010, "Flight Results of the COMPASS-1 Picosatellite Mission", Acta Astronautica, Vol. 67, No. 9-10, pp. 1289-1298.

Schor, D., Kinsner, W. and Thoren, A., 2009, "Satellite Ground Station Emulator: an Architecture and Implementation Proposal", Proceedings of the Canadian Conference on Electrical and Computer Engineering, St. John's, Canada.

Selva, D. and Krejci, D., 2012, "A Survey and Assessment of the Capabilities of Cubesats for Earth Observation", Acta Astronautica, Vol. 74, pp. 50-68. doi:10.1016/j.actaastro.2011.12.014

Semiconductor Industry Association, 2014, International Technology Roadmap for Semiconductors (ITRS). Retrieved in January 27, 2015, from http://www.itrs.net/

Seo, H.S., Rhee, J., Han, E.S. and Kim, I.S., 2012, "Thermal Failure of the LM117 Regulator under Harsh Space Thermal Environments", Aerospace Science and Technology, Vol. 27, No. 1, pp. 49-56. doi:10.1016/j.ast.2012.06.006

She, M.M., 2003, "Semiconductor Flash Memory Scaling", University of California, Berkeley, USA.

Stras, L.N., Kekez, D.D., Wells, G.J., Jeans, T., Zee, R.E., Pranajaya, F.M. and Foisy, D.G, 2003, "The Design and Operation of the Canadian Advanced Nanospace eXperiment (CanX-1)", Proceedings of the AMSAT-NA 21st Space Symposium, Toronto, Canada.

Tennyson, R.C., 1995, "Composites in Space Challenges and Opportunities", Proceedings of the 10th International Conference on Composite Materials, Whistler, Canada.

Woellert, K., Ehrenfreund, P., Ricco, A.J. and Hetzfeld, H., 2011, "Cubesats: Cost-Effective Science and Technology Platforms for Emerging and Developing Nations", Advances in Space Research, Vol. 47, No. 4, pp. 663-684. doi:10.1016/j.asr.2010.10.009

16

Kick Solid Rocket Motor Multidisciplinary Design Optimization Using Genetic Algorithm

Fredy Marcell Villanueva[1], He Linshu[1], Xu Dajun[1]

ABSTRACT: In this paper, a multidisciplinary design optimization (MDO) approach of a solid propellant kick rocket motor is considered. A genetic algorithm optimization method has been used. The optimized kick solid rocket motor (KSRM) is capable of delivering a small satellite of 200 kg to a circular low earth orbit (LEO) of 600 km altitude. The KSRM should accelerate from the initial apogee velocity of 5000 m/s up to the orbital insertion velocity of 7560 m/s. The KSRM design variables and the orbital insertion trajectory profile variables were optimized simultaneously, whereas the mass characteristics of the payload deployment module were assigned. A depleted shutdown condition was considered, to avoid the necessity of a thrust termination device, resulting in a reduced total mass of the KSRM. The results show that the proposed optimization approach was able to find the convergence of the optimal solution with highly acceptable value for conceptual design phase.

KEYWORDS: Kick solid rocket motor, Multidisciplinary design optimization, Genetic algorithm.

INTRODUCTION

Using small solid rocket motors (SRM) for space applications became very attractive, because of their advantages compared with liquid propellant driven rocket engines, especially for insertion of small payloads into a circular Low Earth Orbit (LEO). Simplicity, reliability, and easy to fabricate and operate, are some of the main characteristics of the SRMs. In this article, a specific design used for insertion of a payload into a specified orbit commonly known as kick solid rocket motors (KSRM) are analyzed and discussed.

Previous works were performed on analysis of several options of upper stages for different launch vehicles (LV) (McGinnis and Joyner, 2005), and in design and optimization of upper stages for transfer from LEO to geostationary earth orbit (Motlagh and Novinzadeh, 2012).

Casalino *et al.*, (2011) carried out an optimization of a hybrid upper stage configuration with a special emphasis on the grain geometry. He Linshu and Murad (2005) developed an improved method for conceptual design of multistage solid rockets based on depleted shutdown condition, which means that the SRM has to burn all-contained propellant, resulting in a reduced gross mass.

The genetic algorithm (GA) global optimization method is increasingly being used in optimization of aerospace and propulsion systems (Kamran *et al.*, 2009; Tedford and Martins, 2010; Riddle *et al.*, 2007; Rafique *et al.*, 2009). Bayley and Hartfield (2007) used

1.School of Astronautics – Beihang University – Beijing – China

Author for correspondence: Fredy Marcell Villanueva | Beihang University | XueYuan Road No.37, HaiDian District, New Main Building B1011 | Beijing 100191 – China | Email: marcell385@yahoo.com

GA in multidisciplinary design optimization (MDO). Davis (2001), Goldberg (1989), and Fletcher (2000) provided a comprehensive description and step-by-step implementation of the GA in design and optimization of complex systems.

The insertion process of a payload into the orbit of a typical solid propellant LV start after a ballistic flight phase, when the upper stage is in the apogee altitude, which correspond to the required circular orbital altitude. At this point, to reach the required orbital velocity, a kick impulse is necessary, and for this purpose, a KSRM is required. Its propellant should be burnt completely (depleted shutdown condition), avoiding the use of a thrust termination device, resulting in a reduced total mass of the KSRM. The insertion accuracy is maintained by using an attitude control system grouped in a payload deployment module (PDM).

Thus, the objective of this article is to describe the optimization approach of a KSRM for insertion of a small payload into a prescribed circular LEO orbit by using a well-performed GA optimization method as a global optimizer.

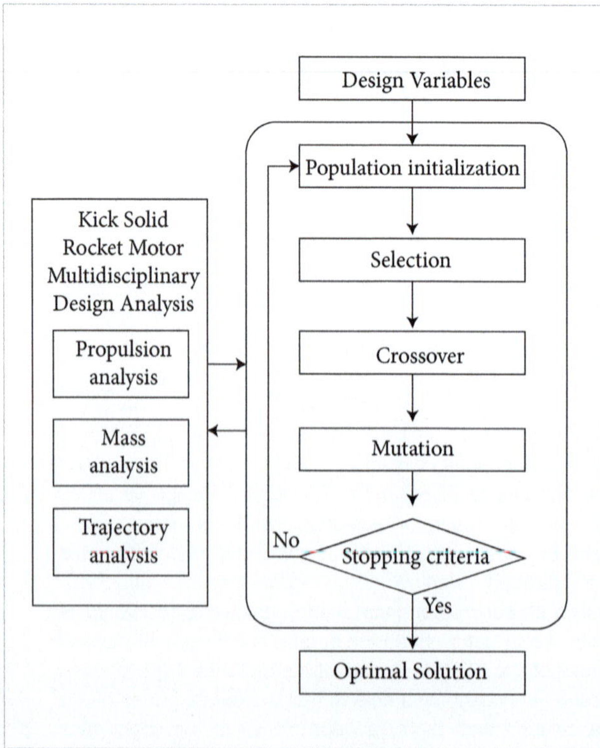

Figure 1. Genetic algorithm optimization approach.

OPTIMIZATION STRATEGY

The MDO approach considered aims to find the KSRM optimal design and the corresponding trajectory to successfully accomplish the specified mission, which is an insertion of a small payload into a prescribed circular LEO orbit. To accomplish this task, a widely used GA optimization method is considered. The main advantage of this method relies on its independency of initial point to perform the optimization.

The GA is considered to be a powerful heuristic global optimization tool in solving complex optimization problems that traditionally has been solved using the approximated analysis. An additional advantage of the GA is the ability to solve discrete and continuous variables.

The operation sequence of the GA optimization method is shown in Fig. 1. Here, the procedure starts with the assignment of the design variables, the algorithm performs several operations as population initialization, selection, crossover and mutation until the optimal solution is reached, whereas all the considered constraints are satisfied.

The drawback of the GA is its computational expense, because a large number of function evaluations are required before finding the optimal solution.

The main characteristics of the GA are presented in Table 1.

Table 1. Genetic algorithm characteristics.

Variables	Characteristics
Generations	200
Population size	100
Stopping criteria	function tolerance $10e^{-6}$
Population type	double vector
Selection	stochastic uniform
Crossover	single point pc=0.8
Mutation	uniform pm=0.25641
Reproduction	elite count=2
Function evaluation	2000

KICK SOLID ROCKET MOTOR MODEL

KICK SOLID ROCKET MOTOR DEFINITION

Mission analysis

The mission is to deliver a small payload to a circular LEO orbit of h_f=600 km altitude. An upper stage composed of a KSRM, a PDM of 100 kg, and, a small payload of 200 kg is considered for this research.

The KSRM should accelerate the upper stage from an initial apogee velocity of V_0=5000 m/s up to the insertion orbital velocity of V_f=7560 m/s, whereas the PDM containing the attitude control system maintains the upper stage witting the allowable insertion accuracy.

KSRM design

There can be many variants of design options; however, for this research effort, a classical configuration was considered. The rocket motor case is considered to be made of high-strength steel, titanium alloy for attachment parts, the ignition system is located in the forward central part of the chamber, and the nozzle has a thrust vector control (TVC) system.

Grain characteristics

For the present analysis, a hydroxyl-terminated polybutadiene (HTPB) grain is selected due to its high performance and suitable for space applications. The composite grain has an internal port and an equivalent length was adopted instead of a complex 3D geometry. This assumption considerably simplifies the model and is acceptable for the conceptual design phase.

PROPULSION ANALYSIS

The propulsion analysis of the KSRM can be calculated using the classical approach. Sutton and Biblarz (2001) and He Linshu (2004a, 2004b) provided a detailed propulsion analysis including the essential parameters, like propellant mass flow rate, burn time, thrust, and nozzle parameters. In this analysis, a constant in time burning surface is considered; a grain geometry shape coefficient k_s is used to represent the constant burning surface of the grain S_b as a function of the equivalent length of grain L_{gn} and diameter D_{gn} as follows:

$$k_s = \frac{S_b}{D_{gn} L_{gn}} \tag{1}$$

The mass of the grain is calculated from the design variables and propulsion analysis, and the burn time t_b, mass of the grain m_{gn}, and mass flow rate $\dot{m}_{gn}$ of the grain are calculated as shown below:

$$m_{gn} = \frac{\pi}{4} \rho_{gn} \eta_v \lambda_{gn} D_{gn}^3 \tag{2}$$

$$t_b = \frac{\pi \eta_v D_{gn}}{4 u k_s} \tag{3}$$

$$\dot{m}_{gn} = \rho_{gn} u \, S_b = \rho_{gn} u \, k_s \lambda_{gn} D_{gn}^{\ 2} \tag{4}$$

$$\lambda_{gn} = \frac{L_{gn}}{D_{gn}} \tag{5}$$

where, u is the burning rate of propellant, ρ_{gn} is the density of grain, $L_{gn}=L_m+0.3151D_m$ is the equivalent length of the grain, $D_{gn}=D_m$ is diameter of the grain, L_m is the rocket motor cylindrical length, λ_{gn} is fineness ratio of the grain (grain length/diameter), and η_v is the grain volumetric loading fraction.

The nozzle throat area A_t, expansion ratio ε, and nozzle exit area A_e are calculated as follows:

$$A_t = \frac{\rho_{gn} u \, S_b}{\Gamma_0 p_{c\max}} \sqrt{R_c T_c} \tag{6}$$

$$\varepsilon = \frac{\Gamma_0}{\left(\frac{p_e}{p_c}\right)^{\frac{1}{\gamma}} \sqrt{\frac{2\gamma}{\gamma-1}\left[1-\left(\frac{p_e}{p_c}\right)^{\frac{\gamma-1}{\gamma}}\right]}} \tag{7}$$

$$A_e = A_t \varepsilon \tag{8}$$

$$\Gamma_0 = \sqrt{\gamma}\left(\frac{2}{\gamma+1}\right)^{\frac{\gamma+1}{2(\gamma-1)}} \tag{9}$$

where, S_b is the burning surface of grain, R_c=326 *J/(kg.K)* is gas constant, T_c=2790*K*° is temperature in the combustion chamber, p_e is exit pressure, p_c is chamber pressure,

$p_{cmax}=1.1p_c$ is the maximum value of chamber pressure, and γ=1.21 the specific heat ratio of the gas.

The vacuum specific impulse I_{sp}^{vac}, and the thrust T can be calculated as shown below:

$$I_{sp}^{vac}=I_{sp}^{a}+\left(\frac{p_e}{p_c}\right)^{\frac{\gamma-1}{\gamma}}\frac{R_c T_c}{g^2 I_{sp}^{a}} \tag{10}$$

$$T=I_{sp}^{vac}\dot{m}_{gn}-p_a A_e \tag{11}$$

where, p_a is the atmospheric pressure, I_{sp}^{a} is average specific impulse, g is acceleration due to gravity, $\dot{m}_{gn}$ is mass flow, and A_e is the nozzle exit area.

MASS ANALYSIS

The mass of the upper stage is represented by the following equations:

$$m_0 = m_{KSRM}+m_{PDM}+m_{PAY} \tag{12}$$

$$m_{KSRM}= m_{st}+\dot{m}_{gn} \tag{13}$$

where, m_0 is the upper stage mass, m_{KSRM} is kick solid rocket mass, m_{PDM} is payload deployment module mass, m_{PAY} is payload mass, and m_{st} is the structural mass of the KSRM.

He Linshu (2004b) provided a detailed calculation of the KSRM structural mass, which is highly acceptable for conceptual design phase, and is shown in the following equation:

$$\begin{aligned} m_{st}=&m_{cy}+m_{c1}+m_{c2}+m_q+m_{j1}+m_{j2}\\ &+m_{in,c1}+m_{in,c2}+m_{in,cy}+m_{noz,ec}+m_{noz,sh}\\ &+m_{noz,in}+m_{ig}+m_{TVC}+m_{cab}+m_{ap} \end{aligned} \tag{14}$$

It is composed of the mass of the motor cylinder, m_{cy}, motor dome ends, m_{c1} and m_{c2}, forward and aft skirts, m_q, forward and aft joints, m_{j1} and m_{j2}, forward and aft insulation liners, $m_{in,c1}$ and $m_{in,c2}$, cylindrical section insulation liner, $m_{in,cy}$, nozzle expansion cone, $m_{noz,ez}$, nozzle spherical head, $m_{noz,sh}$, nozzle insulation, $m_{noz,in}$, igniter, m_{ig}, thrust vector control, m_{TVC}, cables, m_{cab}, and mass of attachment parts, m_{ap}.

The materials selected for the motor case are high-strength steel and titanium alloy, ethylene propylene diene monomer for chamber insulation, and carbon phenolic for the nozzle, whereas the factor of safety was taken as 1.5.

The mass of cylindrical part of the chamber can be calculated by the following relation:

$$m_{cy}=\frac{3K_{cy}\pi f f_p D_{ch}^3 p_c \lambda_{gn}}{8\sigma_b} \tag{15}$$

where, K_{cy}=1.02 is the ratio of the case cylindrical length to rocket motor equivalent length (L_{cy}/L_m), which was obtained from statistics, f is factor of safety, $D_{ch}=D_m$ is diameter of chamber, and σ_b is the ultimate strength of the chamber material.

The mass of the forward motor dome end m_{c1} can be calculated as follows:

$$m_{c1}=\frac{\pi\lambda_e^2 D_{ch}^3 f f_p p_c}{8(\lambda_e^2-1)\sigma\cos\theta_2}\left[1-\frac{1}{1+(\lambda_e^2-1)\sin^2\theta_2}\right] \tag{16}$$

where, λ_e=2 is the chamber ellipsoid ratio, σ is strength ratio $\sigma=\sigma_b/\rho_{cl}$, where, ρ_{cl} is the density of closure material and θ_2 taken as 60°.

The relative pressure in the chamber f_p is determined as shown below:

$$f_p=\frac{p_{c,\max}}{p_c} \tag{17}$$

The aft motor dome end is calculated as follows:

$$\begin{aligned} m_{c2}=&\frac{\pi\lambda_e^2 D_{ch}^3 f f_p p_c}{8(\lambda_e^2-1)\sigma\cos\theta_2}\\ &\cdot\left[\frac{\lambda_e^2 D_{ch}^2-(\lambda_e^2-1)d_n^{\,2}}{\lambda_e^2 D_{ch}^2}-\frac{1}{1+(\lambda_e^2-1)\sin^2\theta_2}\right] \end{aligned} \tag{18}$$

where, d_n=0.4D_{ch} is the diameter of closure rear end opening for nozzle.

The mass of the forward and aft skirts of the SRM are calculated as follows:

$$m_q=\pi D_{ch}^2\rho_q\delta_q\frac{l_{q1}+l_{q2}}{D_{ch}} \tag{19}$$

where, ρ_q is the density of skirt material, δ_q is thickness of skirts, and l_{q1} and l_{q2} are lengths of forward and aft skirts.

The mass of the forward joint to the igniter m_{j1} and the aft joint to nozzle assembly m_{j2} can be calculated as shown below:

$$m_{j1}=\frac{3\pi f f_p p_c d_n^3}{\sigma_j} \tag{20}$$

$$\sigma_j = \frac{\sigma_{bj}}{\rho_j} \tag{21}$$

$$m_{j2} = m_{j1} \tag{22}$$

where, σ_j is the strain of joint, ρ_j is density of the joint material, and σ_{bj} is ultimate strain of joint.

The forward and aft insulation liners $m_{in,c1}$, $m_{in,c2}$ and cylindrical section insulation liner $m_{in,cy}$ can be determined using Eqs. 23 to 25.

$$m_{in,c1} = \frac{1}{4}\rho_{in}\pi D_{ch}^2 R_a t_b \tag{23}$$

$$m_{in,c2} = \frac{\rho_{in}\pi D_{ch}^2 R_a t_b}{4(\lambda_e^2 - 1)}\left[\frac{\lambda_e^2 D_{ch}^2 - (\lambda_e^2 - 1)d_n^{\,2}}{D_{ch}^2}\right] \tag{24}$$

where, ρ_{in} is the density of insulation material and R_a is the rate of ablation of the insulation material.

$$m_{in,cy} = \pi D_{ch} L_{cy} \rho_{in} \frac{(\varepsilon_{in} c_{in} \rho_{in} + \alpha_{gi} c_{cy} \rho_{cy} \delta_{cy})}{\alpha_{gi} c_{in} \rho_{in}} \cdot \left[\sqrt{1 - \frac{2\varepsilon_{in}(\ln\theta_p c_{cy}\rho_{cy}\delta_{cy} + \alpha_{gi} t_b)\alpha_{gi} c_{in}\rho_{in}}{\ln\theta_p(\varepsilon_{in} c_{in}\rho_{in} + \alpha_{gi} c_{cy}\rho_{cy}\delta_{cy})^2}} - 1\right] \tag{25}$$

$$\theta_p = \frac{T_g - T_{cy}}{T_g - T_I} \tag{26}$$

where, L_{cy} is the length of case cylindrical section, ε_{in} is heat transfer coefficient of insulation material, c_{in} is specific heat capacity of insulation, c_{cy} is specific heat capacity of the cylindrical section, α_{gi} is heat transfer coefficient from gas to insulation, ρ_{cy} is density of the case cylindrical section material, T_g is temperature of gas, T_{cy} is allowable temperature of the cylindrical section, and T_I is the initial temperature of the cylindrical section.

The mass of the nozzle expansion cone $m_{noz,ec}$, nozzle spherical head $m_{noz,sh}$, and nozzle insulation $m_{noz,in}$ are expressed using the Eqs. (27), (28), and (29), respectively.

$$m_{noz,ec} = \rho_{ec}\frac{\pi}{4\sin\beta_n}\left(\frac{A_e}{A_t} - 1\right)d_t \cdot \left[f \times 0.67\sqrt[6]{\frac{d_e}{d_t} - 1}\,(1 - S)^{\frac{4}{3}}\sqrt{\frac{P_{c\max}\sigma_{ec}}{E_{ec}}}\right] \tag{27}$$

where, ρ_{ec} is the material density of the expansion cone, β_n is nozzle expansion half angle, S is submerged coefficient of the nozzle, d_e is the nozzle exit diameter, d_t is diameter of throat, σ_{ec} is ultimate strength of the expansion cone material, and E_{ec} is the elastic modulus of expansion cone material.

$$m_{noz,sh} = 3.656\,\rho_{sh} d_t^3 \tag{28}$$

where, ρ_{sh} is the material density of spherical head of nozzle.

$$m_{noz,in} = \rho_{innz} S_{nz} \delta_{innz} \tag{29}$$

where, ρ_{innz} is the material density of the nozzle insulation, S_{nz} is surface area of the nozzle, and δ_{innz} is the thickness of the nozzle insulation.

The igniter mass m_{ig}, can be calculated as follows:

$$m_{ig} = 1.454\left(\frac{7\pi\, d_t^2}{4}\right)^{1.2} \tag{30}$$

The nozzle thrust vector control mass m_{TVC}, is considered as shown below:

$$m_{TVC} = 0.235 m_{noz} \tag{31}$$

The mass of the cables m_{cab} and attachment parts m_{ap} are expressed in Eqs. (32) and (33).

$$m_{cab} = 1.3 L_{cy}\,\rho_c \tag{32}$$

where, ρ_c is the linear density of the cable (kg/m).

And finally, the mass of the attachment parts is calculated as follows:

$$\mathrm{m_{ap}} = (6.13 \cdot 10^{-7})\mathrm{D_m^2 L_{cy}^{1.148}} \tag{33}$$

TRAJECTORY ANALYSIS

For trajectory analysis, a 3 degree of freedom (3DOF) model has been developed and implemented using SIMULINK (Zipfel, 2007; Fleeman, 2001). This model uses the upper stage mass, including the KSRM, PDM and payload mass, and thrust as input parameters. The upper stage is considered as a point-mass in two-dimensional coordinate systems (2D) over a spherical and non-rotating earth, where the Coriolis and centrifugal forces are not considered.

The considered state variables are velocity, flight path angle, altitude, and mass, whereas the control variable is the programmed angle of attack. The trajectory analysis computes the state variables by solving the equation of motion presented in Eq. 34, and evaluating the constraint conditions at every phase of flight.

Figure 2 illustrates the forces acting on the upper stage and below a set of governing equations of motion (He Linshu, 2004a; Xiao, 2001).

$$\begin{aligned}
\frac{dV}{dt} &= \frac{T\cos\alpha}{m} - g\sin\vartheta \\
\frac{d\vartheta}{dt} &= \frac{T\sin\alpha}{mV} - \frac{g\cos\vartheta}{V} + \frac{V}{R_e+h}\cos\vartheta \\
\frac{dh}{dt} &= V\sin\vartheta \\
\frac{dl}{dt} &= \frac{R_e}{R_e+h}V\cos\vartheta \\
\alpha &= \eta + \varphi - \vartheta \\
\eta &= \frac{l}{R_e}
\end{aligned} \tag{34}$$

where, V is the velocity, m is vehicle mass, φ is pitch angle, θ is trajectory angle, ϑ is flight path angle, η is range angle, R_e is radius of earth, h is height above the ground, and l is the range.

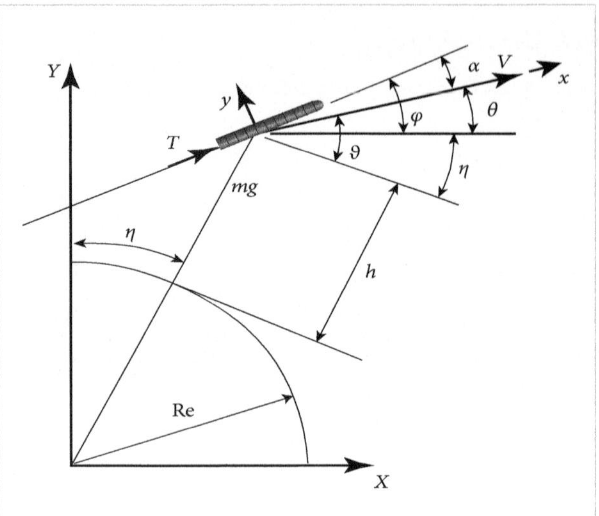

Figure 2. Forces acting on the upper stage.

The axial and normal overload coefficients can be calculated in a body centered velocity coordinate systems (0, x, y) and represented as follows:

$$n_x = \frac{T\cos\alpha}{mg} \leq n_{x\max} \tag{35}$$

$$n_y = \frac{T\sin\alpha}{mg} \leq n_{y\max} \tag{36}$$

The density variation with altitude can be represented as follows:

$$\rho = \rho_0 e^{(-h/\beta)} \tag{37}$$

The variation of gravity with altitude can be represented as follows:

$$g = \frac{\mu}{(R_e+h)^2} \tag{38}$$

where, ρ_0 is the sea level density, β is the density scale height, and μ is the earth gravitational parameter.

The required circular orbital insertion velocity V_f for a given final altitude h_f can be represented as follows:

$$V_f = \sqrt{\frac{\mu}{h_f + R_e}} \tag{39}$$

ORBIT INSERTION PROFILE SEQUENCE

The trajectory optimization was performed considering different constraints that prevent the KSRM, its components, and the payload from failure. The trajectory is modeled considering a depleted shutdown condition at the time of insertion, as follows (He Linshu, 2004b; Qazi and He Linshu, 2005):

Initial Condition

For a typical solid propellant LV orbital payload insertion, the trajectory starts from the time where the upper stage is in the apogee altitude of h_0=600 km, the initial condition is that the flight path angle should be ϑ_0=0 degrees, and the initial velocity V_0=5000 m/s, the angle of attack also should be zero degrees α_0=0.

Pitch over insertion maneuver

After ignition, the upper stage accelerates by the power of the KSRM and flies following the orbital altitude. This phase comprises up to the end of burning time (depleted shutdown condition). A second order curve is fitted from the initial time to the final insertion time; this curve represents the programmed angle for this phase.

Insertion condition

At this time all parameters should comply with the insertion accuracy requirements. The flight path angle at this time should approach zero degrees and within the boundary conditions, $\vartheta_f \approx 0$, the programmed angle of attack is constrained to approach zero $\alpha_f \approx 0$, the orbital velocity should be V_f=7560 m/s corresponding to h_f=600 km altitude, and the normal and axial acceleration should be less than its allowable maximum values.

ORBIT INSERTION PROFILE FORMULATION

The variation of the flight path angle during insertion flight has substantial influence on the injection accuracy in orbit, acceleration loads, and final orbital velocity. It is influenced by a programmed angle of attack. Figure 3 explains the insertion maneuver, and the angle of attack is programmed using the following relations (He Linshu, 2004a; He Linshu, 2004b; Xiao, 2001):

$$\alpha_{prog}(t) = \alpha_{\max} \sin^2 f(t) \tag{40}$$

$$f(t) = \frac{\pi(t - t_1)}{k_m(t_2 - t) + (t - t_1)} \tag{41}$$

$$k_m = \frac{t_m - t_1}{t_2 - t_m} \tag{42}$$

where, α_{max} is the maximum angle of attack, α_{prog} (t) is the programmed angle of attack, k_m is the insertion maneuver variable, t is time of flight, t_m is time corresponding to maximum angle of attack, t_1 is time of start of insertion maneuver, and t_2 is insertion time, coincident in value with the KSRM burning time t_b.

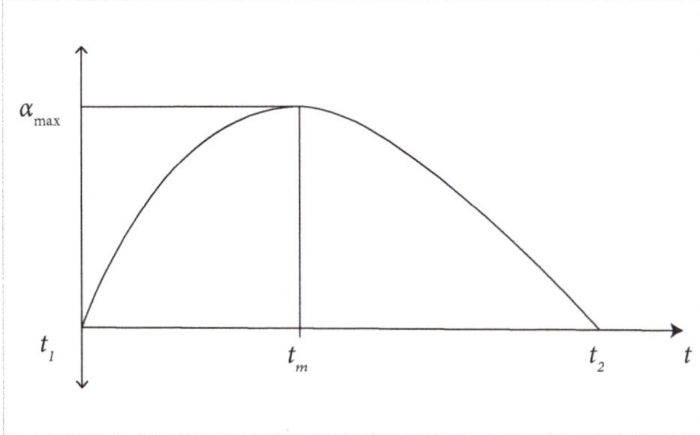

Figure 3. Programmed angle of attack variation.

DESIGN OPTIMIZATION PROBLEM

OBJECTIVE FUNCTION

For the present research effort, the objective is to minimize the KSRM mass m_{KSRM}. The mathematical description of the objective function is as follows:

$$\min m_{KSRM} = f(X) \tag{43}$$

$$g_j(X) \le 0 \tag{44}$$

$$h_k(X) = 0 \tag{45}$$

$$X_{lb} \le X_i \le X_{ub} \tag{46}$$

where, X is the set of variables, X_{lb} is lower bound of variables and X_{ub} is the upper bound of variables.

DESIGN VARIABLES

The variables considered in the KSRM design and the insertion maneuver trajectory can be represented in Eq. (47) and listed in Table 2.

$$X = [L_m, D_m, p_c, p_e, k_s, u, \rho_{gn}, \eta_v, \alpha_{\max}, k_m] \tag{47}$$

Table 2. Design variables.

	Variables	Symbol	Units
X1	Rocket motor cylindrical length	L_m	m
X2	Rocket motor diameter	D_m	m
X3	Chamber pressure	p_c	Pa
X4	Nozzle exit pressure	p_e	Pa
X5	Coefficient of grain shape	k_s	
X6	Grain burning rate	u	m/s
X7	Grain density	ρ_{gn}	kg/m3
X8	Grain volumetric loading	η_v	
X9	Maximum angle of attack	α_{max}	deg
X10	Insertion maneuver variable	k_m	

DESIGN CONSTRAINTS

To prevent the failure of the upper stage, KSRM and PDM, several constraints were selected, and are listed in Table 3.

DESIGN SEQUENCE

This section describes a step-by-step sequence for multidisciplinary design of the KSRM as follows:

Step	Procedure	Reference
1	Define initial variables X	Eq. 47
2	Define constraints C	Table 3
3	Calculate grain mass $\dot{m}_{gn}$	Eq. 2
4	Calculate burning time t_b	Eq. 3
5	Calculate mass flow $\dot{m}_{gn}$	Eq. 4
6	Calculate throat area A_t	Eqs. 6 – 9
7	Calculate specific impulse I_{sp}^{vac}	Eq. 10
8	Calculate thrust T	Eq. 11
9	Calculate KSRM mass m_{KSRM}	Eqs. 13 – 33
10	Set constant values m_{PDM}, m_{PAY}	Mission
11	Calculate upper stage mass m_0	Eq. 12
12	Trajectory conditions $V_0, h_0, \vartheta_0, \alpha_0$	Orbit Insertion Profile Sequence
13	Calculate $\alpha_{prog}(t)$	Eqs. 40 – 42
14	Calculate $\rho(h)$	Eq. 37
15	Calculate $g(h)$	Eq. 38
16	Calculate $V(t)$, $\vartheta(t)$, $h(t)$, $l(t)$,	Eq. 34
17	Calculate overloads n_x, n_y	Eqs. 35, 36
18	Calculate final values V_f, h_f, ϑ_f	Eq. 34
19	Check constraints	Table 3
20	Back to step 1	

OPTIMIZATION RESULT

The optimization results show that the optimized KSRM as well as the upper stage insertion trajectory profile successfully reached the objective function. The optimized values of design variables were obtained, and these variables did not violate the considered design constraints. Table 4 shows the lower bound, upper bound, and optimized values of the design variables.

There are several parameters that characterize the KSRM; however, only the most important were calculated from the optimized design variables and are presented in Table 5.

Table 3. Design constraints.

Constraints		Value	Units
C1	Orbit insertion velocity	V_f=7560±1	m/s
C2	Final altitude	h_f=600±0.1	km
C3	Axial overload	$n_x \leq 12$	
C4	Normal overload	$n_y \leq 2$	
C5	Maximum angle of attack	$\alpha_{max} \leq 10$	deg
C6	Orbit insertion angle	ϑ_f=0±0.02	deg
C7	Fineness grain ratio	$\lambda_{gn} \leq 2$	
C8	Nozzle exit diameter	$d_e \leq 0.95\ D_m$	m
C9	KSRM total length (including nozzle)	$L_{KSRM} \leq 1.8$	m
C10	Burning time	$t_b \leq 50$	s
C11	Nozzle expansion ratio	$\varepsilon \leq 80$	

Table 4. Optimum values of variables.

Variables		Lower Bound	Upper Bound	Optimized Value
L_m	m	0.65	0.90	0.8141
D_m	m	0.65	0.90	0.8140
p_c	Pa	$60e^5$	$70e^5$	$65.901e^5$
p_e	Pa	$0.05e^5$	$0.20e^5$	$0.0581e^5$
k_s		1.50	1.90	1.5050
u	m/s	$6.5e^{-3}$	$8.5e^{-3}$	$6.651e^{-3}$
ρ_{gn}	kg/m^3	1725	1735	1727.01
η_v		0.80	0.86	0.8176
α_{max}	deg	7.5	9.0	8.3601
k_m		0.28	0.45	0.3222

Table 5. KSRM optimum values.

Parameter	Symbol	Unit	Optimized Value
Upper stage mass	m_0	kg	1061.81
Rocket motor mass	m_{KSRM}	kg	761.816
Grain mass	m_{gn}	kg	679.358
Structural mass	m_{st}	kg	82.488
Burning time	t_b	s	45.677
Specific impulse^{vac}	I_{sp}^{vac}	N.s/kg	2544.12
KSRM total length	L_{KSRM}	m	1.733
Nozzle throat diameter	d_t	m	0.062
Nozzle exit diameter	d_e	m	0.5409
Nozzle expansion ratio	ε		76.13

Figures 4 and 5 illustrate the optimized insertion trajectory profile of the upper stage. From Fig. 4 it can be seen that the burning time of the KSRM is 45.67 seconds, the end of burning time shows coincidence with the required insertion parameters. Additionally, the figures evidenced that the payload is inserted at the required altitude of 600 km and the obtained circular orbital velocity is 7560 m/s. The insertion flight path angle is within the required accuracy, whereas the axial and normal overloads were maintained within the constraints limits.

SENSITIVITY ANALYSIS

Monte Carlo analysis is widely used in system and early stage of design. It provides a relatively accurate statistical evaluation of the response distribution under input uncertainties. Monte Carlo sensitivity analyses of the main KSRM parameters were conducted to investigate the effect of uncertainties of design variables over the expected result. For this analysis, a ±1% error was added to every optimized value of design variables. The results are presented in Table 6, and the scatter plot is shown in Fig. 6.

The GA optimization method considered successfully reached the optimal solution, a population of 100 with 200 generation was sufficient to perform the present study; however, several trials had been carried out to obtain the desired accuracy of the optimal solution.

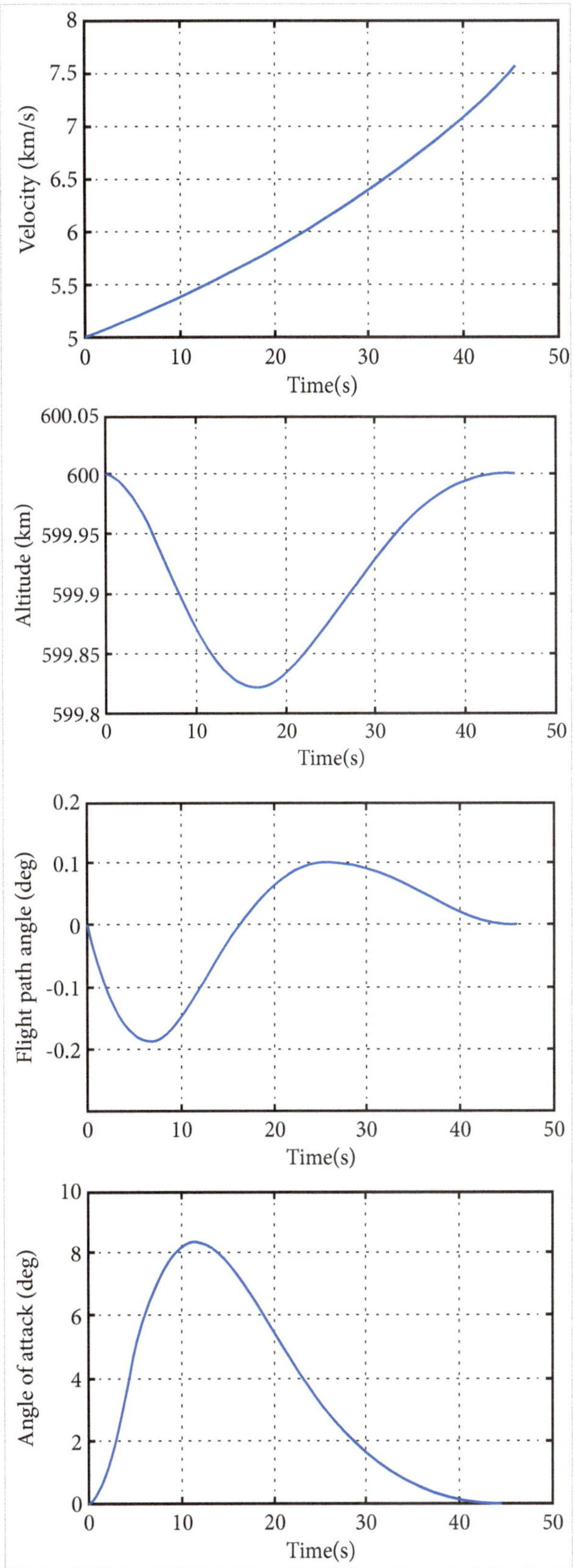

Figure 4. Insertion trajectory profile of the upper stage.

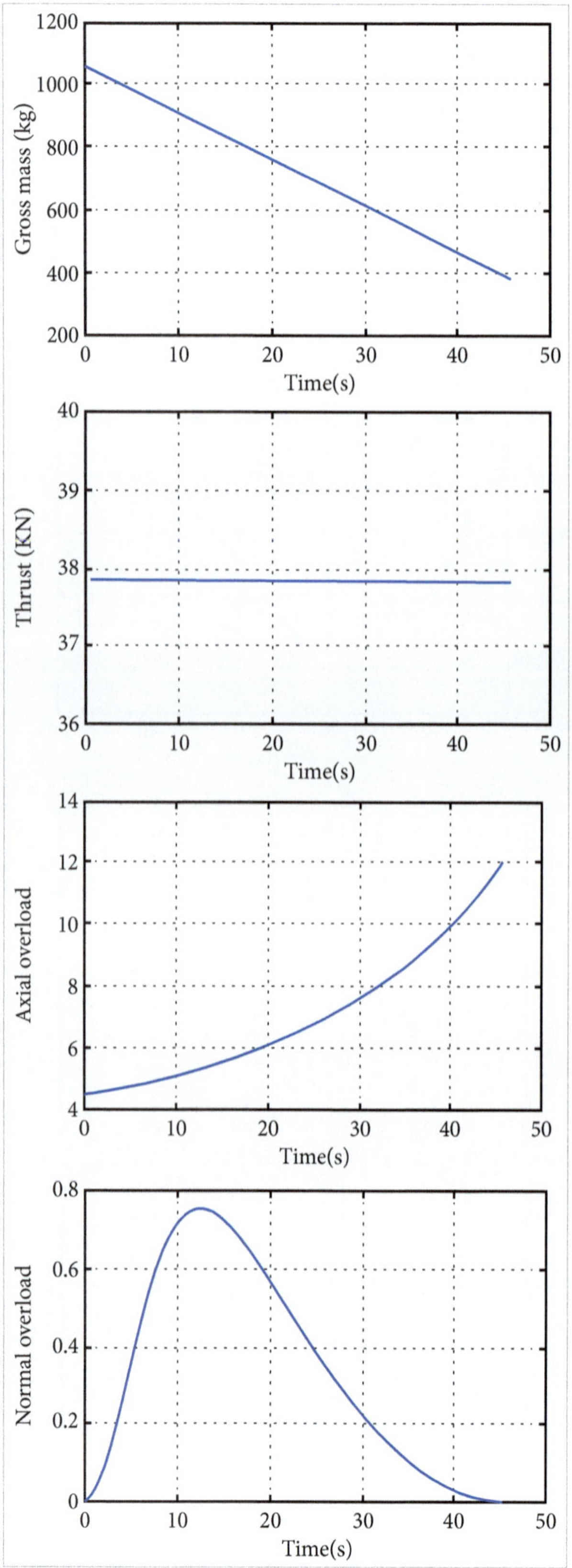

Figure 5. Flight parameters of the upper stage.

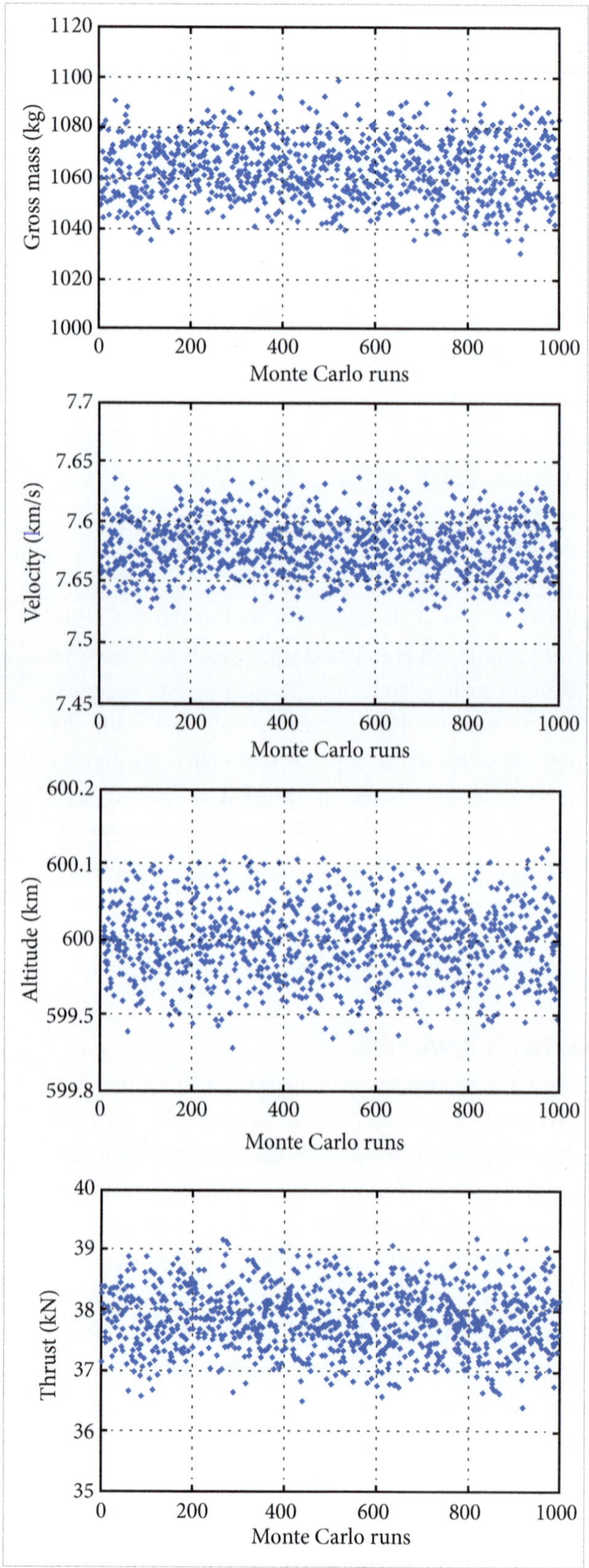

Figure 6. Monte Carlo sensitivity analysis.

Table 6. Monte Carlo results.

Parameter	Unit	Minimum value	Maximum value	Mean value	Standard deviation
Gross mass	kg	1030.6	1098.2	1063.9	11.771
Velocity	m/s	7515.1	7648.2	7579.2	0.0219
Altitude	km	599.86	600.12	600.00	0.0473
Thrust	kN	36.383	39.184	37.846	0.497

CONCLUSION

A GA based optimization approach has been applied to conceptual design and optimization of a KSRM. The advantage of the GA relied on its independency of initial point to calculate the optimum. A 2D dynamic model was developed to simulate the orbital insertion trajectory of the upper stage composed of the PDM of 100 kg, the optimized a payload of 200 kg. A sensitivity analysis was conducted using Monte Carlo method to investigate the variation of the main parameters of the KSRM. The emphasis of this research was to find the optimal KSRM design and the upper stage insertion trajectory profile characteristics required for insertion a small payload into a circular LEO orbit of 600 km.

The results of the KSRM parameters are shown in Tables 4 and 5 and its insertion trajectory profile is shown in Figs. 4 and 5, and Monte Carlo sensitivity analysis shown in Fig. 6 evidenced the validity of the used approach for the early stage of the design process.

ACKNOWLEDGMENTS

Fredy Villanueva wishes to thank Beihang University and China Scholarship Council (CSC), for financial support.

REFERENCES

Anderson M.B., Burkhalter, J. and Jenkins, R., 2001, "Multi-Disciplinary Intelligent Systems Approach to Solid Rocket Motor Design. Part I: Single and Dual Goal Optimization", 37th AIAA/ASME/SAE/ASEE Joint Propulsion Conference and Exhibit, AIAA, paper 2001-3599.

Bayley, D.J. and Hartfield, R.J., 2007, "Design Optimization of a Space Launch Vehicles for Minimum Cost using a Genetic Algorithm", AIAA 43rd AIAA/ASME/SAE/ASEE Joint Propulsion Conference and Exhibit, AIAA, paper 2007-5852.

Casalino, L., Pastrone, D. and Simeoni, F., 2011, "Hybrid Rocket Upper Stage Optimization: Effects of Grain Geometry", AIAA 47th AIAA/ASME/SAE/ASEE Joint Propulsion Conference and Exhibit, AIAA, paper 2011-6024.

Davis, L., 2001, "The Handbook of Genetic Algorithm", Van Nostrand Reingold, New York.

Fleeman, L.E., 2001, "Tactical, Missile Design", AIAA, Reston.

Fletcher, R., 2000, "Practical Methods of Optimization", John Wiley & Sons.

Goldberg, D., 1989, "Genetic Algorithm in Search, Optimization and Machine Learning", first edition, Addison-Wesley-Longman Publishing Co., Reading, MA.

He Linshu, 2004a, "Launch Vehicle Design", Beijing University of Aeronautics and Astronautics Press, Beijing.

He Linshu, 2004b, "Solid Ballistic Missile Design", Beijing University of Aeronautics and Astronautics Press, Beijing.

He Linshu and Murad Y., 2005, "An Improved Method for Conceptual Design of Multistage Solid Rockets based on Depleted Shutdown", Journal of Solid Rocket Technology, Vol. 28, No 2, paper 120-125.

Kamran, A., Lian, G., Godil, J., Siddique, Z., Zeeshan, Q. and Rafique, A., 2009, "Design and Performance Optimization of Finocyl Grain", AIAA Modeling and Simulation Technologies Conference 10-13 August 2009, Chicago, Illinois, AIAA, paper 2009-6234.

McGinnis, P.M. and Joyner, C.R., 2005, "Upper Stage Propulsion Options for Various Launch Vehicle Architectures", AIAA 41st AIAA/ASME/SAE/ASEE Joint Propulsion Conference and Exhibit, AIAA, paper 2005-4180.

Motlagh, J.A. and Novinzadeh, A.B., 2012, "Solid Upper Stage Design Process using Finite Burn Maneuvers for Low Earth Orbit-Geosynchronous Earth Orbit Transfer Phase", Institution of Mechanical Engineers, Part G: Journal of Aerospace Engineering.

Qazi, M. and He Linshu, 2005, "Rapid Trajectory Optimization using Computational Intelligence for Guidance and Conceptual Design of Multistage Space Launch Vehicles", AIAA Guidance, Navigation, and Control Conference and Exhibit, AIAA, paper 2005-6062.

Rafique, A.F., He Linshu, Zeeshan, Q., Kamran, A., Nisar, K. and Wang, X.W., 2009, "Integrated System Design of Air Launched Small Space Launch Vehicle using Genetic Algorithm", 45th AIAA/ASME/SAE/ASEE Joint Propulsion Conference and Exhibit, AIAA, paper 2009-5506.

Riddle, D.B., Hartfield, R.J., Burkhalter, J.E. and Jenkins, R., 2007, "Genetic Algorithm Optimization of Liquid Propellant Missile Systems", AIAA Aerospace Sciences Meeting and Exhibit, AIAA, paper 2007-0362.

Sutton, G.P. and Biblarz, O., 2001, "Rocket Propulsion Elements", Wiley-Interscience, New York.

Tedford, N.P. and Martins, J.R., 2010, "Benchmarking Multidisciplinary Design Optimization Algorithms", Optimization and Engineering, Vol. 11, No 1, paper 159-183.

Xiao, Y.I., 2001, "Rocket Ballistics and Dynamics", Beijing University of Aeronautics and Astronautics.

Zipfel, P.H., 2007, "Modeling and Simulation of Aerospace Vehicle Dynamics", second edition, AIAA Education Series, AIAA, Reston.

An Airplane Calculator Featuring a High-Fidelity Methodology for Tailplane Sizing

Bento Silva de Mattos[1], Ney Rafael Secco[1]

ABSTRACT: The present work is concerned with the accurate modeling of transport airplanes. This is of primary importance to reduce aircraft development risks and because multi-disciplinary design and optimization (MDO) frameworks require an accurate airplane modeling to carry out realistic optimization tasks. However, most of them still make use of tail volume coefficients approach for sizing horizontal and vertical tail areas. The tail-volume coefficient method is based on historical aircraft data and it does not consider configuration particularities like wing sweepback angle and tail topology. A methodology based on static stability and controllability criteria was elaborated and integrated into a MATLAB application for airplane design. Immediate advantages with the present methodology are the design of realistic tail surfaces and properly sized airplanes. Its validation was performed against data of five airliners ranging from the regional jet CRJ-100 to the Boeing 747-100 intercontinental airplane. An existing airplane calculator application incorporated the present tail-sizing methodology. In order to validate the updated application, the Fokker 100 airliner was fully conceptually designed using it.

KEYWORDS: Aircraft design, Tailplane design, Aircraft stability and control.

INTRODUCTION

It is of fundamental importance for any optimization framework tailored to airplane design the use of realistic airplane models. If the disciplines that are embedded in the airplane modeling are not accurate enough, it makes no sense performing any optimization tasks, because the resulting planes may be unviable to develop or deliver an acceptable performance. Thus, the present work is concerned with the development of a computational tool able to properly model transport airplane configurations. An existing computational tool that was christened Aeronautical Airplane (AA) was enhanced with an improved approach for tailplane sizing (Mattos and Magalhães, 2012).

Tail surfaces are used to both stabilize the aircraft and provide control authority that is needed for maneuver and trim. For a conventional aircraft configuration, the tail often has two components, the horizontal and the vertical tails. The primary functions of these components are: take care of airplane trim and stability, and provide control by the elevator and rudder surfaces which are associated with the horizontal and vertical tails, respectively.

With regard to aircraft stability, the purpose of horizontal tail (HT) is to maintain the longitudinal stability; while the vertical ones is responsible for keeping the directional stability. Here, we refer to two stability concepts: static and dynamic. Aircraft static stability is defined as the tendency of an aircraft to return to the original trimmed conditions if diverted by a disturbance. Major disturbance sources are gusts and pilot inputs on the controls. On the other hand, the dynamic longitudinal one is related to the motion of a statically stable airplane, the way that it will return

1.Instituto Tecnológico de Aeronáutica – São José dos Campos/SP – Brazil

Author for correspondence: Bento Silva de Mattos | Instituto Tecnológico de Aeronáutica | Praça Marechal Eduardo Gomes, 50 – Vila das Acácias | CEP: 12.228-900 São José dos Campos/SP – Brazil | Email: bmattos@ita.br

to equilibrium after suffering some kind of disturbance. Basically, there are two primary forms of longitudinal movements with regard to an airplane attempting to return to equilibrium after being disturbed. The first one is the phugoid mode of oscillation, which is a long-period and slow oscillation of the airplane's flight path; the second is a short-period variation of the angle of attack. Usually, this oscillation decreases very quickly with no pilot input (Centennial of Flight, 2011). However, the Centennial of Flight website introduces some misconceptions about Dutch roll, when it states that: "Dutch roll is a motion exhibiting characteristics of both directional divergence and spiral divergence." Therefore, Dutch-roll basic cause is the unbalance between lateral and longitudinal stabilities, when the latter is considerably lower than the first one.

The tail surfaces of transport airplanes are usually designed to obey static stability criteria. If some undesirable dynamic behaviors become evident during a flight test campaign, fixes are provided to overcome the problems. Typical fixes are dorsal and ventral fins, vortex generators, strakes, and in some extreme cases, stablets. The results obtained with an in-house application for tailplane sizing corroborate this statement, as will be shown in the next sections of this work.

As to the controllability topic, it can be stated that an airplane is fully controlled when a flight condition may be changed in a finite time by appropriate control inputs at any new flight condition. In general, a system is considered controllable if it can be transferred from selected initial conditions into chosen final states. The controllability thus describes the influence of external inputs (in general, the controlled variables) to the inner system state. About this, an important distinction must be made between output and state controllability. Output controllability is the notion associated with the system output; the output controllability describes the ability of an external input to move the output from any initial condition to any final one in a finite time interval. No relationship must exist between state and output controllability. The state of a system, which is a collection of its variables values, completely shows the system at any given time. Particularly, no information on the past of a system will help to predict the future, if the states at the present time are known. Complete state controllability (or simply controllability if no other context is given) describes the ability of an external input to move the internal state of a system from any initial one to any other final in a finite time interval.

Concerning transonic airplanes, tail surfaces should be composed of low-thickness and/or higher sweep than that adopted for the wing, in order to prevent strong shocks on the tail in normal cruise. As required for certification regulations and safety purposes, transport airplane must dive in an emergency event occurring at high altitudes. The main reason behind this is to reach another one, where no onboard oxygen is needed for the passengers to breath. In this situation, for high-speed airplanes, the airflow over the wing becomes fully detached and the airplane counts just on the horizontal tailplane to depart from the dive. This provides a good reason for airflow remaining attached to HT at this high-speed condition. Former airplane design teams used to increase the HT quarter-chord sweepback angle relative to that figure of the wing (typically a 5 degrees plus). Although providing higher sweepback angles to HT will mean a longer arm relative to center of gravity (CG) and a higher lift curve slope, modern multidisciplinary design and optimization (MDO) frameworks are able to find out the optimal planform and airfoil shapes for tailplanes. Thus, modern design airplanes not necessarily follow the +5 degrees rule to set such angle for the horizontal tail.

In order to keep weight and drag as low as possible, typical values for the maximum relative thickness for both HT and vertical tail (VT) lie in the range of 8 to 11%. Besides working with lower hinging moments, airfoil shapes for the HT shall present a very low maximum camber in order to avoid the generation of undesired induced drag and to maintain interference drag with the tailcone or the VT as low as possible. It is worth mentioning that the tailcone is a region of high decelerating airflow, and the thickness of boundary layer increases fast and dramatically. Thus, interference drag is a big issue when integrating conventional HTs into the airframe.

Typical aspect ratios for HT vary from four to five. T-tails present sometimes higher aspect ratios (5–5.5) to avoid aft-engine/pylon wake effects (Sadraey, 2009). Usual aspect ratios for VT range from 1.2 to 1.8 with lower values for T-tails. Taper ratios of about 0.4 to 0.6 are typical for tail surfaces, since lower ones would lead to unacceptably small Reynolds numbers (Sadraey, 2009). T-tail vertical surface taper ratios are in the range of 0.85 to 1.0 to provide adequate chord for attachment of the HT and associated control linkages (Sadraey, 2009).

A large variety of tail shapes has been employed on aircraft since the beginning of powered aviation in the early 20th century. These include configurations often denoted by the letters whose shapes they resemble in front view: *T*, *V*, *H*, +, *Y*, inverted *V*. The selection of the particular configuration includes complex system-level considerations.

Most of existing MDO frameworks for aircraft design makes use of tail volume coefficient or scaling factors approach for tailplane calculation. A relatively new example of this can be seen in the paper by Grundlach (1999). The tail volume coefficient approach is based on statistical data and may lead to tailplanes that are unable to comply with stability and controllability requirements. In the present work, a more sophisticated and higher fidelity methodology for the design of tail surfaces was compiled from the specialized literature, then coded using MATLAB platform, validated by comparing calculated tail areas to those from some current airliners, and incorporated into an airplane calculator software. Static stability and controllability criteria were used for the design of HTs and VTs of airplanes under consideration.

METHODOLOGY

AIRPLANE CALCULATOR TOOL

AA is a MATLAB® application that calculates transport airplane characteristics, performance, and layout. User must provide to it information regarding geometry, configuration, topology, wing and tailplanes airfoils, range at given payload, passenger capacity, and engine data of the configuration that is due to be analyzed.

AA makes use of a wing-body full potential code with boundary-layer correction for wings employing the strip sense approach (Karas and Kovalev, 2011). This code is known as BLWF and it is able to automatically generate a multi-block mesh from user-provided configuration parameters. The approximate-factorization algorithm AF2 is used for marching in pseudo time (Holst and Thomas, 1982). Aerodynamics coefficients are calculated using Roskam Class II methodology (Roskam, 2000a) and Torenbeek's formulation to estimate divergence Mach number of wings and HTs (Torenbeek, 1982).

Loads calculations are performed with BLWF for some points in the airplane operational envelope (Fig. 1). Afterwards, the calculated loads are used to estimate the wing structural weight applying Megson's method in some sections of the wingbox (Fig. 2). Elastic deformation is well iteratively, since the loads will vary when the wing deforms (Fig. 3). The procedure that was adopted for the wing-box sizing is described in Videiro (2012).

Figure 1. A full-potential code that accounts for viscous effects is used by Aeronautical Airplane. It is able to handle both low- and high-wing configurations.

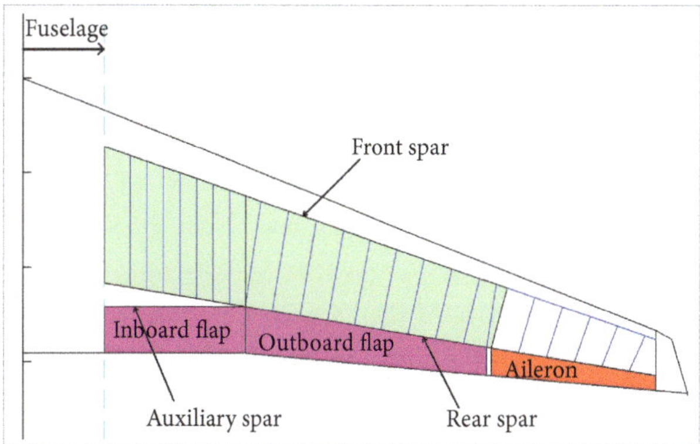

Figure 2. Spars and ribs belong to the wing layout defined by Aeronautical Airplane for a configuration with engines positioned at aft fuselage. The green area shown above indicates availability for fuel storage.

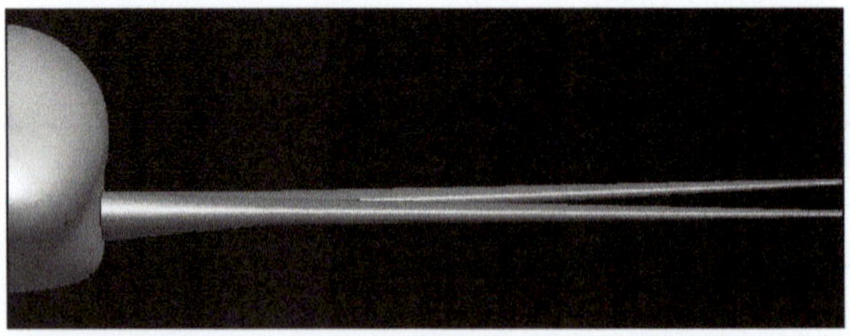

Figure 3. Example of wing elastic bending deformation calculated by Aeronautical Airplane.

BLWF is also used in conjunction with XFOIL for the estimation of the C_{Lmax} for the configuration under study. It is used to calculate the maximum lift coefficient of some sections along the wingspan and BLWF estimates the C_{Lmax} of the configuration using those values in the technique of the critical section.

The turbofan engine model employed by AA is described in Loureiro (2008), who developed a generic deck for turbofan engines based on Benson's work (Benson, 1995). Fan diameter, bypass ratio, fan pressure ratio, engine parts efficiencies, design point, turbine inlet temperature, and overall pressure ratio are some input variables to this model. Typical outputs from the engine module are fuel flow and thrust.

AA also defines the fuselage cross section (Fig. 4) and the design diagram, the latter clustering several Federal Aviation Rules (FAR – from the United States), 25 performance conditions together, enabling a better comprehension of thrust and wing loading figures to fulfill requirements (Fig. 5). Scholz (2011) provided several guidelines that were used to define the fuselage cross section as seen in Fig. 4. The methodology that was employed to generate graphs like the one displayed in Fig. 5 is described in Roskam (2000).

AA calculates fuselage geometry, CG envelope, engine thrust and consumption chart, operational envelope, direct operating cost (DOC), payload-range diagram, airplane flight mechanics, takeoff path, maximum takeoff weight (MTOW) with all airplane component weights, engine emissions, and noise signature.

It must be emphasized that AA does not enforce the airplane under study to obey the requirements in Fig. 4, considering that it is indeed a transport airplane analysis tool, not a design one. If the intent is to design an airplane, AA can be without great effort easily incorporated into a design and optimization framework, considering that modern airplane conceptual design must make use of optimization tools.

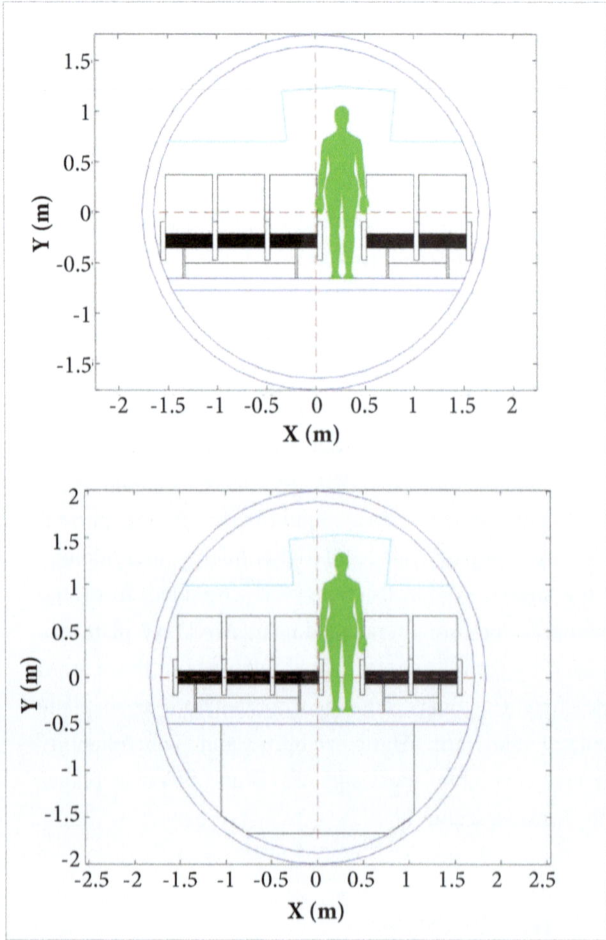

Figure 4. Aeronautical Airplane output of some fuselage cross sections.

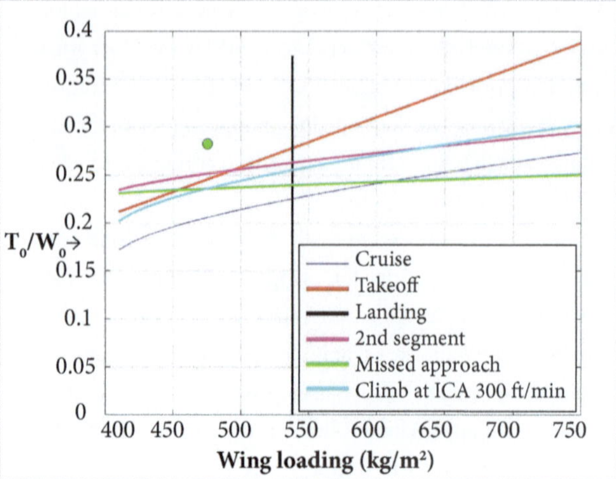

Figure 5. Design diagram of the airplane configuration being investigated (green circle). Criteria like the second segment climb and rate of climb at initial cruise altitude (ICA) are employed to verify whether the configuration comply or not with FAR design requirements (Roskam, 2000).

ENHANCED TAIL SIZING METHOD

Takeoff is a critical condition for tailplanes design, and therefore some relevant aspects of this flight phase are described in the following paragraphs. Figure 6 illustrates the takeoff segments for airplanes. V_1 is the takeoff decision speed, V_R the rotation speed, V_{Lof} velocity of liftoff, and V_2 is the flight speed over a hypothetical 35-feet obstacle. During the initial phase of the takeoff path, the airplane must be accelerated on the ground to V_{EF}, at which point the critical engine must be made inoperative and remain like this for the rest of the takeoff. V_1 means the maximum speed in the takeoff at which the pilot must take the first action, such as: apply brakes, reduce thrust, deploy speed brakes, and stop the airplane within the accelerate-stop distance. It also means the minimum speed in the takeoff, following a failure of the critical engine at V_{EF}, at which the pilot can continue it and achieve the required height above the takeoff surface within its distance (Boeing, 2009).

According to FAR 25.149, V_{mca} is the calibrated airspeed at which:

- critical engine suddenly becomes inoperative;
- control of the airplane is possible to maintain;
- maintain straight flight;
- bank angle shall be lower than five degrees;
- V_{mca} may not exceed 1.2 x V_s.

The conditions for determination of V_{mca} are:

- maximum takeoff thrust on remaining engines;
- most unfavorable CG (usually aft CG), where the tail moment arm is the shortest);
- aircraft trimmed for takeoff (takeoff flap setting);
- maximum TOGW;
- most critical takeoff configurations along the flight path, except for landing gear retracted.

V_{mca} is used for determining the minimum V_R and V_2. The V_2 must also be chosen in order to satisfy the second segment climb requirement. V_{mca} must be lower than the V_2 and considering its importance for aircraft performance and safety operation, it was calculated after the tail surfaces had been obtained. For this purpose, the methodology elaborated by Cavanaugh (2004) was incorporated in the present work.

The main objective of the present paper was the calculation of the HT and VT areas through a higher fidelity method than that offered by the tail volume approach. In order to accomplish this, static stability and controllability criteria were employed for the design of the horizontal and vertical tails. The VT area is obtained by taking the larger one from that calculated employing both criteria, while the HT area is obtained with the procedure described previously. Stability derivatives that are needed for the calculation of HT and VT areas were obtained according to Roskam's methodology (Roskam, 1971). Initial estimated values are required in order to start iterative processes for the HT and VT calculation. MATLAB's *fsolve* minimization algorithm was employed for the calculation of the tail-planes areas. Indeed, no optimization of the tail-plane areas was carried out with *fsolve*. This MATLAB® tool is used to find out the tailplane areas, because they depend on several parameters and variables such as wing location along the fuselage.

The incorporation of a more sophisticated methodology for the tail-plane sizing into AA poses a more complex task in the interactive process for MTOW calculation. The determination of the CG location is needed not only for the tail-plane area measurement, but also to position the main and nose landing gear, which is also dependent on the overall CG location, which will guide the wing placement in the configuration. The modified AA was employed to calculate the characteristics of an airplane similar to Fokker 100, a 107-seater airliner. By varying the quarter-chord wing sweepback angle, the impact on HT and VT areas was obtained.

BASELINE AIRPLANE

Fokker 100 is a medium size twin-turbofan airliner design and market by the extinct Dutch company, Fokker. It is a complete redesign of the Fokker F-38, which rivals the BAE 146. It had longer wings, and fuselage that would seat over 100 passengers and a "glass" cockpit featuring six large displays (World's Aircraft, 2013). In 1988, Fokker flew an uprated Rolls&Royce Tay 650 powered version, which American Airlines ordered 75 aircraft with 75 options. The baseline airplane that was chosen in the present work was Fokker 100 fitted with Rolls&Royce Tay 620 engines. Table 1 presents some relevant data for such airplane.

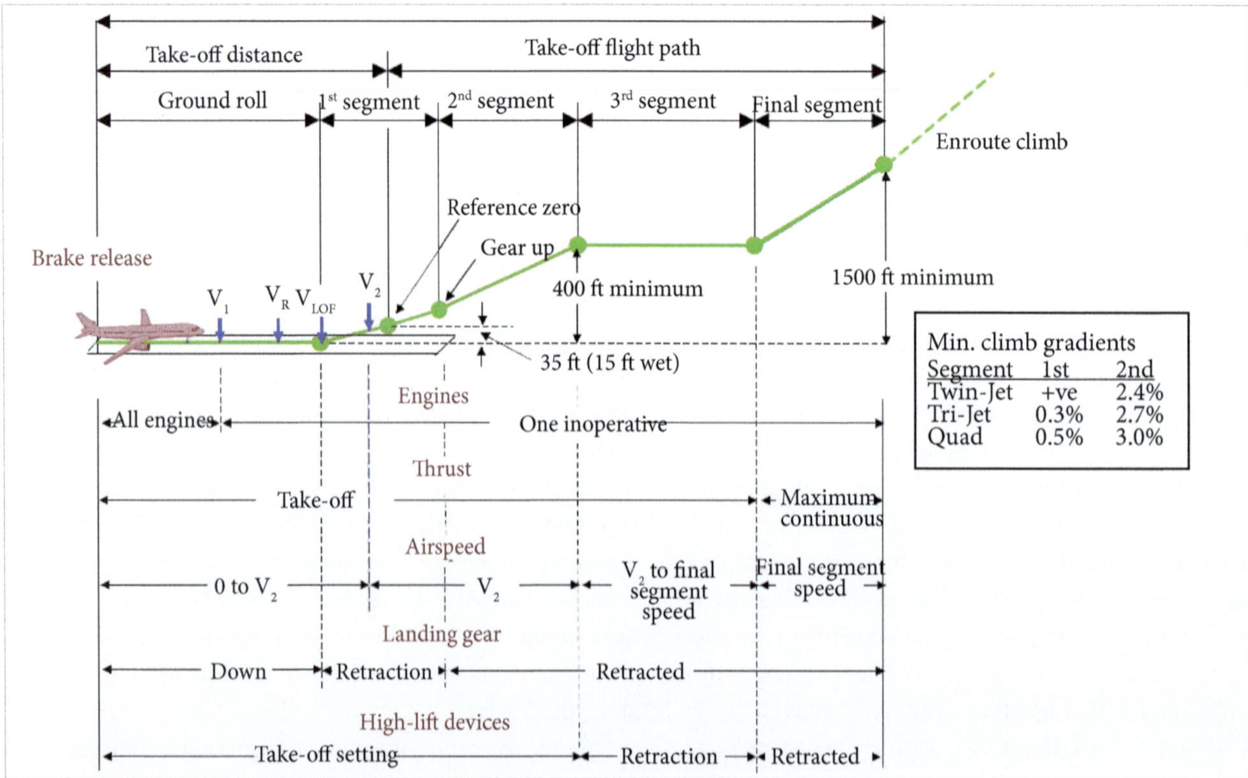

Figure 6. Illustration of take-off segments and required FAR climb requirements.

Table 1. Data for the Fokker 100 fitted with Tay 620 engines.

Performance	
Maximum operating Mach number	0.77
Service ceiling	35,000 ft
Maximum range with maximum payload	1,720 km
Maximum takeoff weight	43,090 kg
Wing	
Aspect ratio	8.43
Reference area	93.5 m^2
Taper ratio	0.235
Mean aerodynamic chord (MAC)	3.80 m
Quarter-chord sweepback angle	17.45°
Horizontal tail	
Area	21.72 m^2
Aspect ratio	4.64
Taper ratio	0.390
Quarter-chord sweepback angle	26.00°
Vertical tail	
Area	12.30 m^2
Aspect ratio	0.89
Taper ratio	0.740
Quarter-chord sweepback angle	41.00°

HORIZONTAL STABILIZER

Design for controllability

The aircraft trim must be guarantee three axes, *x*, *y*, and *z*, namely the lateral, longitudinal, and directional ones, respectively. When the summation of all forces in *x* direction (such as drag and thrust) is zero and of all moments including aerodynamic pitching moment about *y* axis is zero, the airplane is longitudinally trimmed.

HT can be installed at airplane tail-cone or on top of vertical tails. Some airplanes like the Piaggio Avanti business turboprop are fitted with foreplanes that are located at forward fuselage. Such foreplanes are known as canards.

The HT is responsible for maintaining longitudinal trim and making the forces summations to be zero, by generating a necessary lift and contributing in the summation of moments in *y* axis. When the summation of all forces in *y* direction (such as side force) is zero; and of all moments including aerodynamic yawing moment about *z* axis is zero, the aircraft is said to have the directional trim.

In order to trim longitudinally, we obtained the airplane seen in Fig. 7.

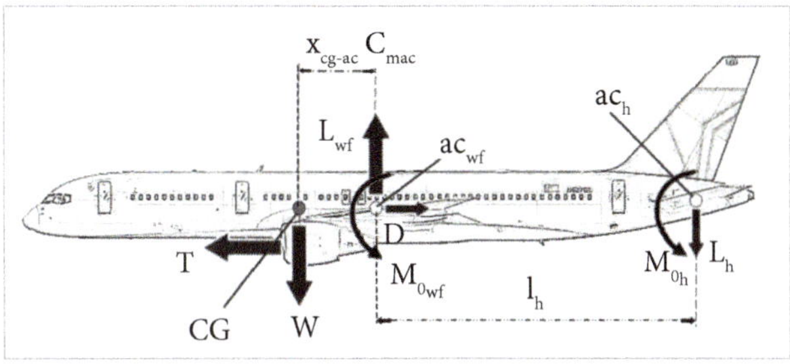

Figure 7. Forces and moments acting for longitudinal trim.

$$M_{0_{wf}} + M_{0_h} + M_{0_e} + L_{wf} x_{cg-ac} c_{mac} - L_h \left(l_h + x_{cg-ac} c_{mac} \right) = 0 \quad (1)$$

where $x_{cg\text{-}ac}$ is the ratio between the distance of aerodynamic center of the wing to CG and the mean aerodynamic chord of the wing (static margin).

Equation 1 can be further developed and we can obtain Eq. 2:

$$\frac{\rho V^2}{2} S_w c_{mac} \left[Cm_{0_{wf}} + Cm_{0_e} + C_{L_{wf}} x_{cg-ac} \right] = \eta_h \frac{\rho V^2}{2} S_h \left[C_{L_h} l_h - c_{mac} Cm_{0_h} + C_{L_h} c_{mac} x_{cg-ac} \right] \quad (2)$$

Finally, an expression for the area ratio is derived (Eq. 3):

$$\frac{S_h}{S_w} = \frac{\left[Cm_{0_{wf}} + Cm_{0_e} + C_{Lwf} x_{cg-ac} \right] c_{mac}}{\eta_h \left[C_{Lh} \left(l_h + x_{cg-ac} c_{mac} \right) - c_{mac} Cm_{0_h} \right]} \quad (3)$$

Another form of Eq. 4 is as follows:

$$\frac{\eta_h S_h l'_h C_{L_h}}{S_w c_{mac}} = Cm_{0_{wf}} + Cm_{0_e} + C_{L_{wf}} x_{cg-ac} \quad (4)$$

with Eq. 5:

$$l'_h = l_h + x_{cg-ac} c_{mac} \quad (5)$$

and $Cm_{0h} \sim 0$, considering that horizontal stabilizers are composed of airfoil sections, which present maximum camber approximately equal to zero.

The combination $S_h l'_h / S_w c_{mac}$ in Eq. 4 is an important nondimensional parameter for the HT design, and is referred to as the "HT volume coefficient" (Eq. 6). The name is originated from the fact that both numerator and denominator have the unit of volume (e.g. m^3). The numerator is a function of HT parameters, while the denominator is of wing parameters (Eq. 6):

$$V_h = \frac{S_h l'_h}{S_w c_{mac}} \quad (6)$$

Table 2 shows typical V_h figures for some aircraft types. The tail volume coefficient is an indication of handling quality in longitudinal stability and control. As the V_h increases, the aircraft tends to be more longitudinally stable, and less longitudinally controllable. The fighter aircraft that is highly maneuverable tend to have a very low tail volume coefficient. Transport airplanes require a higher one because they are tailored to perform in a stable flight, for passenger and crew's comfort.

The V_h parameter is a crucial variable in HT design and is selected at the early stages of aircraft conceptual phase. Although one of the primary functions of the HT is to provide longitudinal stability, the tail volume coefficient

Table 2. Tail volume coefficients for some airplanes (Sadraey, 2009).

Airplane	Type	MTOW (kg)	Wing area (m²)	Overall length (m)	HT area (m²)	V_h
Cessna 172	Light GA (piston powered)	1,100	16.2	7.9	1.94	0.76
Piper PA-46-350P	Light transport (piston powered)	1,950	16.26	8.72	–	0.66
Alenia G222	Military transport (turboprop)	28,000	82	22.7	–	0.85
Fokker 100	Jet airliner (R&R Tay 620)	43,090	93.5	32.5	21.72	1.07
Boeing F/A-18C	Fighter	23,400	46	16.8	–	0.49
Pilatus PC-12	Multipurpose single-engine turboprop	4,100	25.81	14.14	–	1.08
Airbus A340-200	Jet airliner	257,000	363.1	59.39	72.90	1.11
Boeing 747-400	Jet airliner	396,830	525	68.63	136.60	0.81

serves as a significant parameter both in the longitudinal stability and longitudinal trim issues (Sadraey, 2009). On the other hand, this parameter is usually chosen by looking at similar aircraft. In fact, in the aircraft design, each parameter depends on everything else of the configuration in a non-linear and very complex fashion. In modern conceptual design, MDO frameworks have a commonplace in the aircraft industry (Mattos and Magalhães, 2012). Therefore, the horizontal tail, as well as the vertical stabilizer, shall be designed concurrently with wing and fuselage for a clean and efficient design, and picking up a tail volume coefficient makes no sense anymore.

Design for stability

Another design requirement must be examined: aircraft static and dynamic longitudinal stability. The static longitudinal stability is examined through the sign of the longitudinal stability derivative $C_{m\alpha}$ or the location of the aircraft neutral point; the dynamic behavior is associated with autonomous motions such as Dutch roll, spiral divergence and other issues linked to flight quality, which will not be considered in the present work.

For an aircraft with a fixed aft tail, the aircraft longitudinal stability derivative is determined as Eq. 7:

$$Cm_{\alpha} = \frac{dC_m}{d\alpha} = d\left\{\frac{Cm_{0_{wf}} + L_{wf} \times (x_{ac} - x_{cg}) + Cm_{0_h} - L_h \times [l_h + (x_{ac} - x_{cg})]}{\frac{1}{2}\rho V^2 S_w c_{mac}}\right\} \Big/ d\alpha \quad (7)$$

Taking into account that, we have Eq. 8:

$$L_h = C_{L0_h} + C_{L\beta_h}\eta_h S_h\left(1 - \frac{d\varepsilon}{d\alpha}\right),$$

$$C_{m\alpha} = \left(C_{L\alpha_{wf}} + C_{L\alpha_h}\eta_h \frac{S_h}{S_w}\left(1 - \frac{d\varepsilon}{d\alpha}\right)\right)x_{ac-cg} - C_{L\alpha_h}\eta_h \frac{S_h}{S_w}\frac{l_h}{c_{mac}}\left(1 - \frac{d\varepsilon}{d\alpha}\right) \quad (8)$$

Because the tail operates in the downwash field of the wing (for conventional, aft-tail configurations), the effective tail angle of attack is reduced. According to Roskam (1971), the parameter $d\varepsilon/d\alpha$ can be estimated by Eq. 9:

$$\frac{d\varepsilon}{d\alpha} = 4.44\left[k_A k_\lambda k_h \sqrt{\cos\Psi_{w25}}\right]^{1.19} \frac{\left(C_{L\alpha_w}\right)_M}{\left(C_{L\alpha_w}\right)_{M=0}} \quad (9)$$

where:

$$k_A = \frac{1}{AR_w} - \frac{1}{1 + AR_w^{1.7}} \quad (10a)$$

$$k_\lambda = \frac{10 - 3\lambda_w}{7} \quad (10b)$$

$$k_h = \frac{1 - \left|\frac{z_h}{b}\right|}{\sqrt[3]{\frac{2l_h}{b}}} \quad (10c)$$

The relationship $\left(C_{L\alpha_w}\right)_M / \left(C_{L\alpha_w}\right)_{M=0}$ can be calculated by using the following expression (Eq. 11):

$$C_{L\alpha_w} = \frac{2.\pi.AR_w}{2 + \sqrt{AR_w^2.\left(1 + \tan^2\Psi_{50} - M_\infty^2\right) + 4}} \quad (11)$$

When the derivative C_{ma} is negative or the neutral point is behind the aircraft CG, the aircraft is said to be statically longitudinally stable. The limit of the design is found when Eq. 9 is set to zero and therefore Eq. 12 is obtained.

$$\frac{S_h}{S_w} = \frac{C_{L\alpha_{wf}} x_{cg-ac}}{\left\{C_{L\alpha_h}\eta_h\left(1 - \frac{d\varepsilon}{d\alpha}\right)\left[x_{cg-ac} + \frac{l_h}{c_{mac}}\right]\right\}} \quad (12)$$

Obtaining the horizontal tail area

Eqs. 4 and 12 provide two relationships between the area ratio (S_h/S_w) and $x_{cg\text{-}ac}$. These equations can now be combined into a single graph (Scholz, 2011). It should be observed that the aft center of gravity must be positioned at a safe distance to the natural stability limit. For a jet transport aircraft, Roskam established this value as 5% of MAC. However, according to Raymer (1999), this can be further reduced to 3% of MAC. The permitted areas of focus are now between the limit lines of controllability and those of the stability requirement. Between these lines now, the required CG range can thus be fitted to find out the smallest HT surface area. The region of interest lies above the horizontal green line in Fig. 8.

Table 3 shows the CG variation for some airliners. The three-engine DC-10 intercontinental airliner has the lowest CG variation among all airplanes listed in Table 2. Figures for the 737-800 and Boeing 777-200 were obtained from Boeing's training material (Boeing, 2009) and the remaining ones from Chai and Mason (1996).

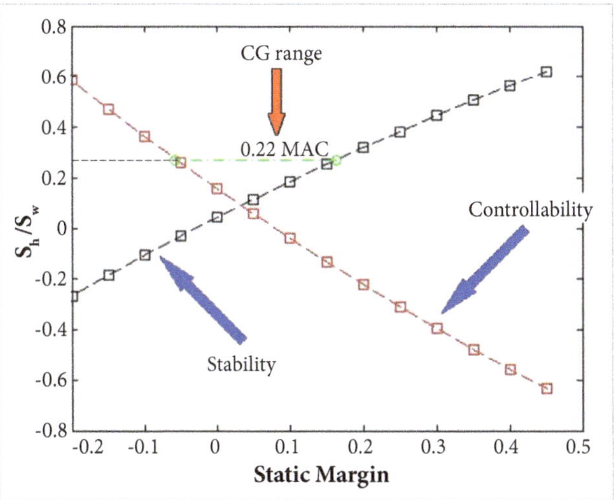

Figure 8. Illustration of the calculation procedure for horizontal tail area ratio. Static margin is the variable $x_{cg\text{-}ac}$.

Table 3. CG variation for some airliners (Chai and Mason, 1996, Boeing, 2011).

Airplane	Fore/aft (% MAC)	Variation (% MAC)
Boeing 737-800	5/36	31
Boeing 767	11/32	21
Boeing 747-400	8.5/33	24.5
Boeing 777-200	14/44	30
DC-10	8/18	10

VERTICAL STABILIZER

Two criteria were adopted for the vertical stabilizer sizing: static stability and controllability. The first condition is related to stability. The tailplane may be sized to fulfill a desired coefficient value that incorporates the variation of yawing moment coefficient with yaw angle, $C_{n\beta}$. The fuselage and VT are the airplane components that significantly impact its directional stability. When an airplane experiences a sideslip angle, in general the fuselage alone will generate a moment that tends to increase the sideslip angle, which is a destabilizing and undesirable condition. The VT plays a major role to the static directional stability. When the airplane experiences a sideslip angle, the VT has the same aerodynamic effects as wings at angle of attack, but at different planes. Thus, it generates a moment relative to airplane's CG, which produces a stabilizing one that tends to neutralize the sideslip angle. The VT usually has a low aspect ratio, which provides a higher stall angle than high-aspect ratio planforms. If stall occurs, a catastrophic situation may result due to the steady increase of the sideslip angle. Ventral fins and stablets are good options to improve Dutch-roll characteristics without an unacceptable weight penalty. These surfaces provide a stable yawing moment at larger sideslip angles. The second consideration for vertical tailplane sizing is associated with controllability. This criterion may be determinant for a multi-engined underwing configuration. Loss of power on the number 1 engine, which are numbered from port to starboard, requires that the pilot simultaneously apply right rudder to correct the resulting yaw moment due to asymmetric thrust condition. Also, it should be applied a rolling moment to the right both to hold the starboard wing low and balance the rolling moment due to the rudder. The low starboard wing produces a side force on the airplane that balances that generated by the rudder deflection.

Engine failure at takeoff poses a critical condition for the VT design. In this case, the remaining engine (or engines) must provide enough thrust to maintain the required rate of climb with the additional drag caused by rudder and aileron deflections. Trim drag due to rudder deflection and to a lesser extent aileron deflection may be critical in meeting FAR.121 climb requirements for the second segment, especially for twin-engine airplanes.

An additional vertical tail-sizing requirement, which is harder to calculate, is to keep directional control while on the ground. The tail must be sized such that the minimum control speed on the ground (V_{MCG}) is less than the takeoff decision speed (V_1). If this is not the case, and V_{MCG} is greater than V_1, then the situation may arise in which critical engine failure occurs above V_1. In such situation, the pilot has to continue the takeoff (because V_1 has been exceeded), but the airplane lacks adequate lateral control while still on the ground. The requirements for V_{MCG} are given in FAR 25.149(e). They are to lose the most critical engine, apply the rudder (without the use of nose wheel steering), and maintain control down the runway with a maximum of 30 feet (~ 10 m) lateral deviation from the runway centerline. In flight test, this is performed at successively lower speeds until the pilot can no longer maintain 30 feet lateral deviation. If additional control power is required for a derivative design (such as having increased engine thrust), improved rudder effectiveness may be achieved by adding vortex generators to the vertical stabilizer (as was done on the L1011 for one customer with particularly short runway, and reduced V_{mca}, requirements). If that does not work, a double-hinged rudder might fix the problem. This was done on the

Boeing 747SP (along with increased tail height) to make up for the shortened fuselage (and thus reduced rudder moment arm), and also on the DC-10.

Design for controllability

The required yawing moment coefficient to maintain steady flight with one failed outboard engine at 1.2 times the stall speed is as specified by FAR 25.149. The remaining outboard engine must be at the maximum available thrust, and the bank angle cannot be larger than five degrees. Figure 9 shows the engine-out situation for a twin-engine configuration. The engine-out constraint is established by constraining the maximum available yawing moment coefficient ($C_{n_{available}}$) to be greater than the required one ($C_{n_{req}}$) for the engine-out flight condition (Eq. 13):

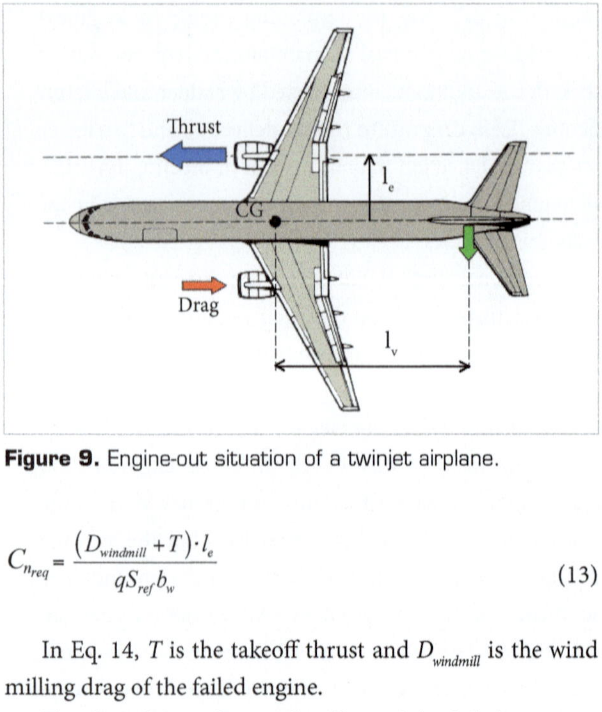

Figure 9. Engine-out situation of a twinjet airplane.

$$C_{n_{req}} = \frac{(D_{windmill} + T)\cdot l_e}{qS_{ref}b_w} \tag{13}$$

In Eq. 14, T is the takeoff thrust and $D_{windmill}$ is the wind milling drag of the failed engine.

The drag due to the wind milling of the failed engine is calculated using the method described in Appendix G-8 of Torenbeek (1982). Thus, we have

$$D_{windmill} = C_{D_{windmill}} qS_{ref} \tag{14}$$

where

$$C_{D_{windmill}} = \frac{0.075d_i^2 + \dfrac{2\pi d_i^2}{1+0.16M_\infty^2}\dfrac{V_n}{V}\left(1-\dfrac{V_n}{V_n}\right)}{S_{Ref}} \tag{15}$$

In Eq. 15, V_n is the nozzle exit velocity and $V_n/V = 0.92$, 0.42 for high- and low-bypass ratio turbofans, respectively. Torenbeek's wind milling drag equation (Torenbeek, 1982) was validated against the flight test data of 747 (Grasmeyer, 1998). Torenbeek's equation for the estimation of wind milling drag coefficient shows relatively good agreement with the flight test data over a range of Mach numbers.

The maximum available yawing moment coefficient is obtained at an equilibrium flight condition with a given bank angle and a maximum rudder deflection (δr). The bank angle is limited to a maximum of five degrees by FAR 25.149, and the aircraft is allowed to have some sideslip (β) (Grasmeyer, 1998). The sideslip angle is found by summing the forces along the y-axis:

$$C_{y\delta a}\delta_a + C_{y\delta r}\delta_r + C_{y\beta}\beta + C_L \sin\phi = 0 \tag{16}$$

We considered in the present work that $\beta = 1$ and ϕ is obtained from Eq. 17:

$$\phi = \arcsin\left[-\frac{\left(C_{y\beta}\beta + C_{y\delta a}\delta_a + C_{y\delta r}\delta_r\right)}{C_L}\right] \tag{17}$$

Considering that $C_{y\delta a} \sim 0$, Eq. 17 can be simplified to Eq. 18:

$$\phi = \arcsin\left[-\frac{\left(C_{y\beta}\beta + C_{y\delta r}\delta_r\right)}{C_L}\right] \tag{18}$$

If ϕ calculated by Eq. 18 is greater than 5^0, we set this value for ϕ in Eq. 16 and then β is obtained instead of fixing a value for ϕ. Thus, if the pilot slightly increases the sideslip angle further than one degree, he/she will have a safe margin for the increase of the $C_{n_{available}}$. In addition, if one engine is inoperative, drag could become a critical issue if the sideslip angle were excessively increased.

After the values for β and ϕ are known, the aileron deflection required to maintain equilibrium flight is obtained by summing the rolling moments about the x-axis:

$$C_{l\delta a}\delta_a + C_{l\delta r}\delta_r + C_{l\beta}\beta = 0 \tag{19a}$$

$$\delta_a = \frac{-C_{l\delta r}\delta_r - C_{l\beta}\beta}{C_{l\delta a}} \tag{19b}$$

The rudder deflection is initially set to the given maximum allowable steady state value, and the bank angle and aileron deflections for equilibrium flight are determined by Eqs. 19a and b. The maximum allowable steady-state deflection is typically 20 to 25 degrees. This allows for an additional five-degree of deflection for maneuvering.

The maximum available yawing moment is found by summing the contributions due to the ailerons, rudder, and sideslip (Eq. 20):

$$C_{n_{avail}} = C_{n\delta a}\delta_a + C_{n\delta r}\delta_r + C_{n\beta}\beta \quad (20)$$

Design for stability

The remained condition for the design of the vertical stabilizer considers that the airplane is experiencing a sideslip angle (Fig. 10). In this situation, the yaw moment balancing when a sideslip angle is present is enforced

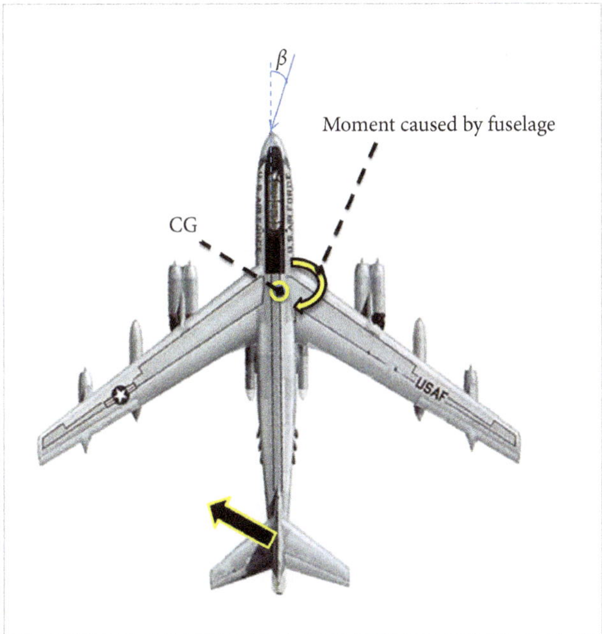

Figure 10. The vertical tail has a stabilizing effect when the airplane is experiencing a sideslip angle.

$$M_{Resulting} = M_{VT} + M_{wing} + M_{fuselage} \quad (21)$$

Eq. 21 can be further developed resulting in Eqs. 22 and 23.

$$\frac{1}{2}\rho V^2 b_w C_{n\beta_{airplane}}\beta = \frac{1}{2}\rho V^2 b_w C_{n\beta_{VT}}\beta + \frac{1}{2}\rho V^2 b_w C_{n\beta_{wing}}\beta + \frac{1}{2}\rho V^2 b_w C_{n\beta_{fuselage}}\beta \quad (22)$$

$$C_{n\beta_{airplane}} = C_{n\beta_{VT}} + C_{n\beta_{wing}} + C_{n\beta_{fuselage}} \quad (23)$$

Terms in Eq. 23 are calculated according to Roskam's methodology (Roskam, 1971):

$$C_{n\beta_{wing}} \cong 0 \quad (25)$$

$$C_{n\beta_{VT}} = -C_{y\beta_{VT}}\frac{(l_v\cos\alpha + z_v\sin\alpha)}{b_w} \quad (26)$$

The stability derivative $C_{y\beta VT}$ in Eq. 26 can be calculated by

$$C_{y\beta VT} = -k_{Cy\beta VT}C_{L\alpha VT}\left(1+\frac{d\sigma}{d\beta}\right)\eta_v\frac{S_{VT}}{S_{Ref}} \quad (27)$$

where

$$k_{Cy\beta VT} = 0.75 \quad \frac{b_v}{d_{f,v}} < 2$$

$$k_{Cy\beta VT} = \frac{1}{6}\frac{b_v}{d_{f,v}} + \frac{5}{12} \quad 2 \le \frac{b_v}{d_{f,v}} < 3.5$$

$$k_{Cy\beta VT} = 1 \quad \frac{b_v}{d_{f,v}} \ge 3.5 \quad (28)$$

and

$$\left(1+\frac{d\sigma}{d\alpha}\right)\eta_v = 0.724 + 3.06\frac{S_v/S_w}{1+\cos\Psi_{1/4}} + 0.4\frac{Z_w}{d} + 0.09AR_w \quad (29)$$

In Eq. 30, Z_w is a vertical distance between the wing quarter-chord at the location of mean aerodynamic chord and the fuselage centerline, positive downwards.

Eq. 24 is analyzed at the beginning and end of the cruise. The most critical condition is then considered.

RESULTS

Before integrating the present methodology into an airplane design application, a numerical tool written in MATLAB® language was developed for tailplane analysis only. The main reason for creating such tool was the validation of the methodology described in the preceding sections. This tool, which was named ITAIL, presents a graphical interface as shown in Fig. 11. Stability derivatives were calculated according to the methods proposed by Roskam (1971). The incorporation of ITAIL into AA will be further described.

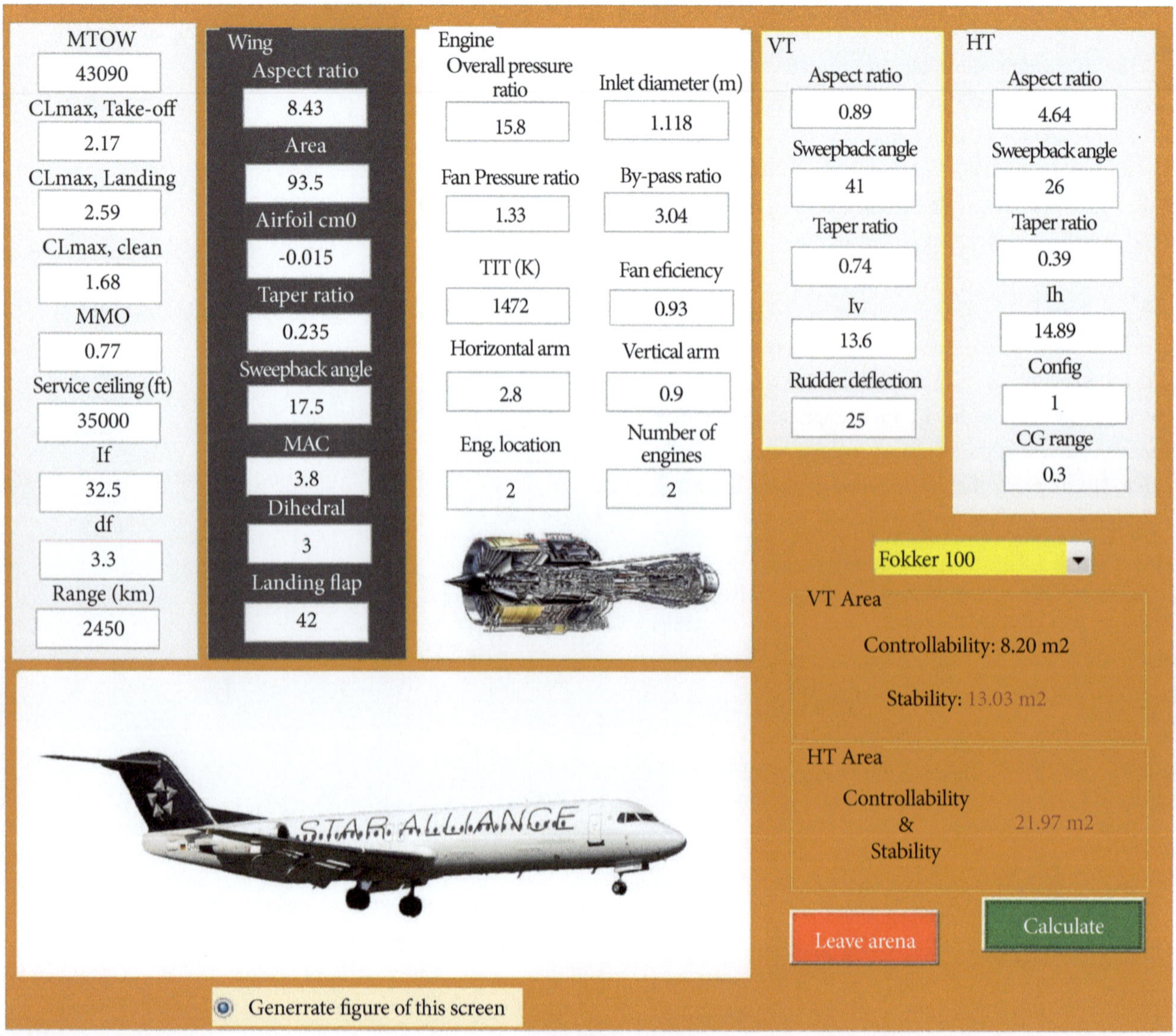

Figure 11. Main panel of ITAIL, a tailplane sizing application written in MATLAB®.

ITAIL

ITAIL offers users the possibility of choosing some airliners from a drop-down menu. A broad range of airliners is represented from the regional jet CRJ-100 to the Boeing 747-100. Some characteristics of the airplanes used in the validation process are listed in Table 4. Both T-tail as well as the conventional tail configuration were considered. Aircraft manufacturers usually bring into market stretched or shortened variants featuring some shared components with a baseline configuration. Usually, stabilizers are designed for the baseline configuration in some cases, taking into account characteristics of the remaining ones. Stretched versions will not be the cause of major concerns because the HT and VT arms will turn these surfaces more effective. However, shortened ones will require larger tailplanes. For this reason, the validation airplanes considered here are baseline configurations from which others were derived. Most airplane data used for ITAIL validation were taken from online sources (Jenkinson, Simpkin and Rhodes, 2011).

Table 5 compares some calculated stability derivatives to values obtained from flight tests (Heffley and Jewell, 1972). Calculated values for the airliners that were taken for validation are shown in Table 6. The highest deviation found was that for the Boeing 737-100 HT area, 6.5%. The

Table 4. Some data belonging to the validation airplanes that were inputted into ITAIL.

Airplane	S_w (m²)	MTOW (kg)	l_h	l_v	Thrust arm (m)	cg range (% MAC)
Boeing 757-200	185.25	115680	19.35	18.97	6.9	22
Fokker 100	93.5	43090	14.89	13.6	2.8	30
Boeing 747-100	511	333400	30.54	28.35	21.3	24
CRJ-100ER	48.35	24091	12.21	9.76	2.2	30
Boeing 737-100	102	44225	11.97	11.04	5.0	30

agreement between calculated and actual figures for the areas can be considered excellent. The factor that determined the Fokker 100 VT area was stability. Considering that this airplane has aft mounted engines, the controllability criterion is not the critical one. A post-design check in order to guarantee that the dynamic stability behavior will be satisfactory is obviously required.

The results presented here corroborates that tailplanes areas of airliners can be usually determined by considering static stability requirements, with undesired dynamic behavior being fixed during flight test campaign if needed. Furthermore, design teams should be aware that aerodynamic phenomena like high interference drag, wake vortex and localized flow separation can turn tailplanes less effective and this issue must be carefully analyzed. Figure 12 displays the V_{mca} and stall speed variation with takeoff weight for the B-757-200 airliner. The kink in the graph can be credited to the fact that solution for the V_{mca} calculation is driven by two kinds of constraints: limitation of rudder or aileron deflection. In fact, the lower boundary represents 60.0% of the MTOW and may not be an operating condition. V_{mca} may become an important driver for the design of the VT if takeoff field length or climb requirements in second segment are very stringent.

Table 5. Comparison of calculated stability derivatives to experimental data (Boeing 747-100).

Derivative	Present work H=12,200 m M=0.90 / α=1.2°	Heffley (1972) H=12,200 m M=0.90 / α=2.4°
$C_{n\beta}$	0.196	0.207
$C_{l\beta}$	-0.129	-0.095
$C_{n\delta a}$	0.007	0.0027
$C_{n\delta r}$	0.137	0.0914
$C_{l\delta r}$	-0.028	-0.0052

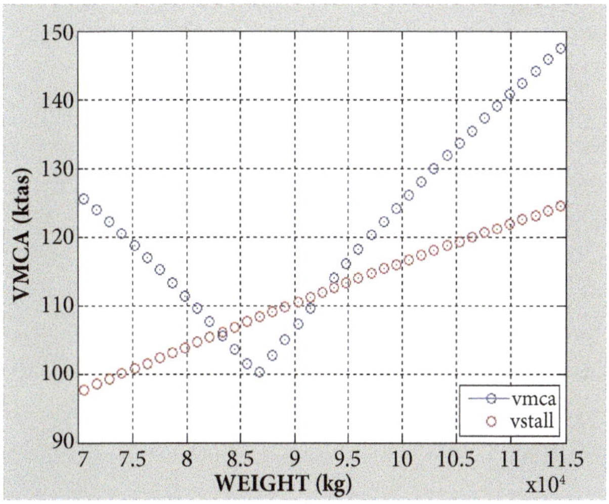

Figure 12. V_{mca} calculation by ITAIL (Boeing 757-200).

Table 6. Calculated tailplane areas for some airliners.

Airplane	HT area (m²)			VT area (m²)		
	Calculated	Actual	Deviation (%)	Calculated	Actual	Deviation (%)
Boeing 757-200 (RB211-535E4)	52.06	50.35	+3.40	35.61	34.27	+3.91
Fokker 100 (R&R Tay 620)	21.97	21.72	+1.15	13.03	12.30	+5.90
Boeing 747-100 (JT9D-7A)	129.28	136.60	-5.35	77.99	77.10	+1.15
Canadair CRJ-100ER (GE CF34-A)	9.67	9.44	+2.44	9.73	9.18	+5.99
Boeing 737-100 (JT8D-7)	19.46	20.81	-6.50	29.77	28.99	+2.70

ENHANCED AERONAUTICAL AIRPLANE

After ITAIL was validated against HT and VT areas found in a large variety of transport airplane areas, its methodology was incorporated into AA. From this point on, the ability of the enhanced AA to correctly calculate VT and HT areas was put under scrutiny. For this purpose, data of the Fokker 100 airliner were input in that application. Establishing a constraint to the static margin to be exactly 12.5% of MAC, the convergence process to determination of the HT area can be seen in Fig. 13. The parameter XLE in the abscissa axis is the distance from the airplane nose of the wing leading-edge point in the fuselage centerline. Figure 14 reveals the geometry variation and the wing repositioning until the convergence was achieved. AA also provides an artistic view of the final configuration, which was displayed in Fig. 14. The HT area that was calculated by AA was 22.1 m², a deviation of 2.3% when compared to the Fokker 100 actual HT area, which is displayed in Table 5.

The VT area obtained with AA was then 13.6 m², which obeys three criteria:

- the criterion that $V_{mca} < 1.2\ V_s$ at MTOW, sea level takeoff resulted in an area of 6.2 m²;
- the controllability requirement demanded an area of 9.7 m²; and
- the static stability imposed the ultimate area of 13.6 m² and drove the sizing.

A study about the impact of the variation of the quarter-chord sweepback angle on the tailplane areas was carried out with AA using the Fokker 100 as baseline airplane. The results displayed in Fig. 15 indicated a decrease of the required HT area of 22% if the sweepback angle varies from 17.5 to 33 degrees. No significant impact on VT area was recorded as expect. Swept-wing airplanes records improved directional stability because a restoring moment arises due to the differential drag forces acting on the port and starboard wings (Fig. 15 at bottom). For this reason, no significant change in VT area is expected as sweepback angle is increased taking into account that for aft-mounted engine configurations the stability criterion is critical. Figure 16 compares the calculated configurations for two configurations with distinct wing quarter-chord sweepback angles, 17.5 and 28 degrees. Static margin was fixed as 12.5% of MAC for both simulations and the vertical tail is the same for both configurations. The Fokker 100 is displayed in green

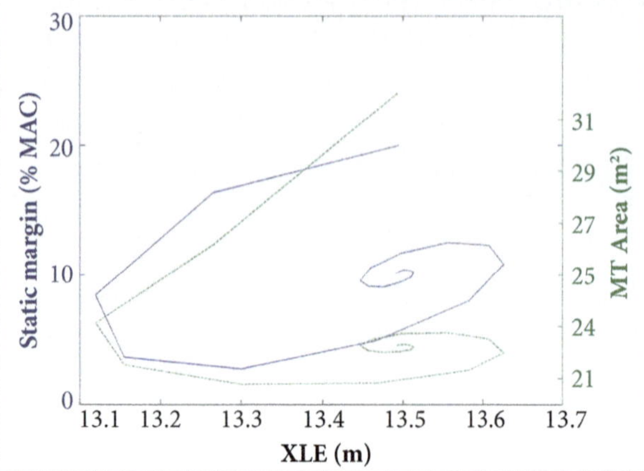

Figure 13. Static margin and horizontal tail area convergence process for the Fokker 100 airliner.

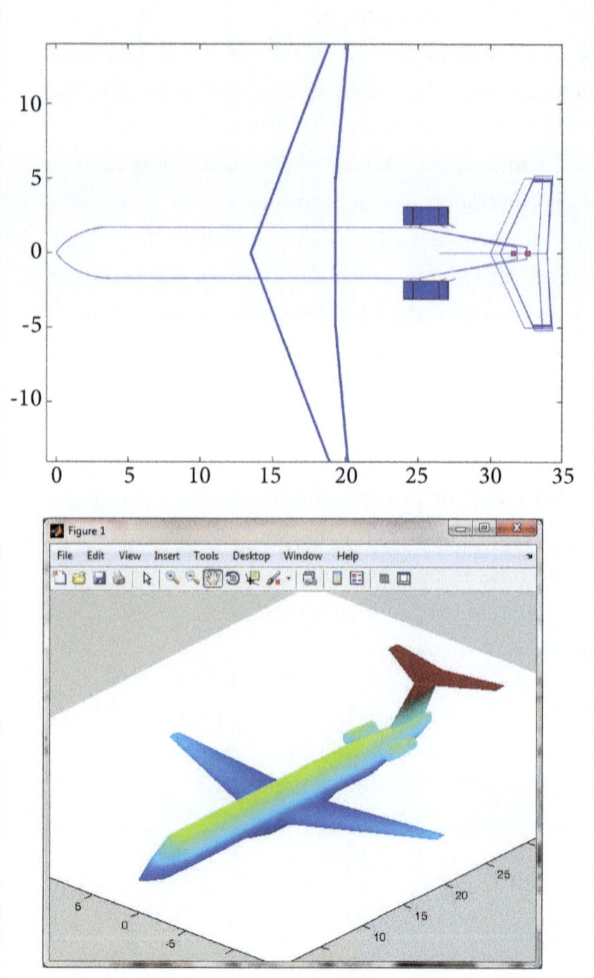

Figure 14. Above: variation of the configuration geometry during the convergence process for the horizontal and vertical tails sizing of the baseline configuration. Wing area and engine data are treated as input variables. Bottom: artistic view of the designed configuration.

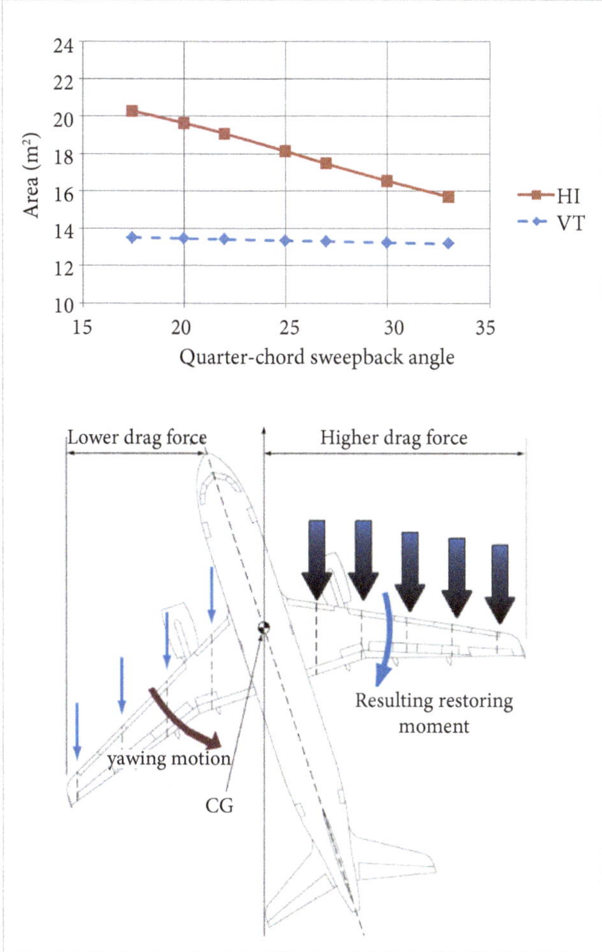

Figure 15. Above: tailplane area variation with wing quarter-chord sweepback angle of the Fokker 100 airplane. Values were obtained with AA calculator. Bottom: swept-wing airplanes produce a restoring moment improving the directional stability.

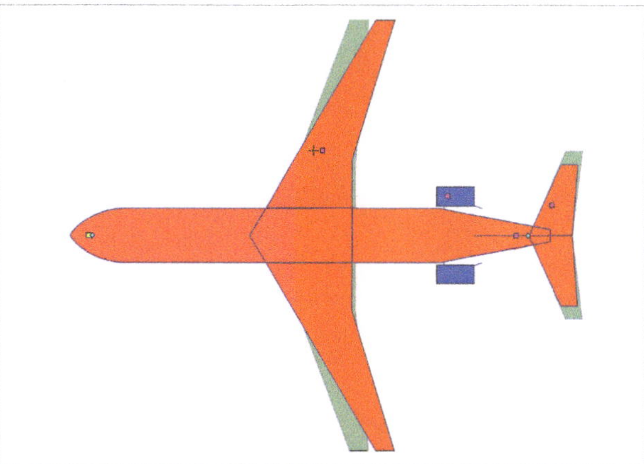

Figure 16. Comparison of the HT for the Fokker 100 (green color) and a derived confi guration with 28 degrees of wing sweepback angle.

colors and features a larger horizontal tail than that of the configuration with higher sweepback angle. The CG location for the empty airplane and the wing aerodynamic center do not change significantly for both configurations. AA also provides an output graph of the short period frequency requirements for longitudinal flight quality in order to verify if the design complies with (Fig. 17).

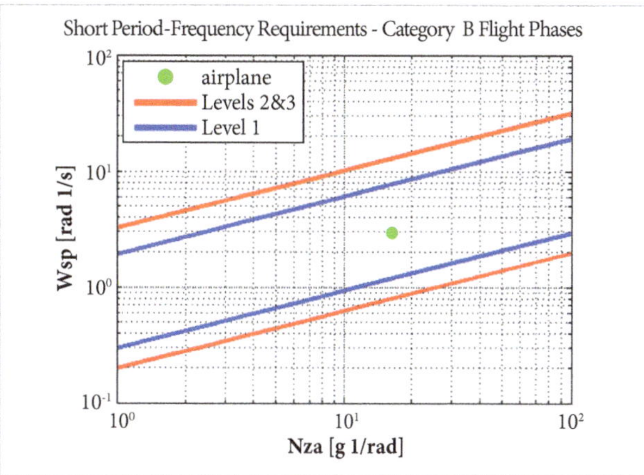

Figure 17. AA analysis for the Fokker 100, which is represented by the circle between the Level 1 boundaries.

CONCLUDING REMARKS

A MATLAB® application called ITAIL was developed to validate a methodology for vertical and HT sizing of transport airplanes. ITAIL employs static stability and controllability criteria for tailplane design. The highest deviation found by the ITAIL application was that for the Boeing 737-100 HT area, i.e. 6.5%. The agreement between calculated and actual figures for the areas can be considered excellent. A post-design check in order to guarantee that the dynamic stability behavior will be satisfactory is required. The results presented here corroborates that tailplanes areas of airliners can be usually determined by considering static stability and controllability requirements,, with undesired dynamic behavior being fixed during flight test campaign if necessary.

The methodology embedded in ITAIL avoids a very arbitrary criterion to the sizing of transport airplanes tail surfaces, i.e. the tail volume coefficient. Thanks to the increase in computing power in the last ten years, conceptual airplane design has steadily become more sophisticated and has incorporated higher-fidelity techniques to model airplane geometry and aeronautical disciplines. In this context, dropping very inaccurate methods like the tail volume coefficient fits well in this trend.

An existing airplane calculator designated AA was enhanced with a higher fidelity approach for sizing horizontal and vertical tailplanes. AA is a MATLAB® application tailored to be incorporated into MDO frameworks. Results obtained with AA using the methodology described in this work and incorporated by ITAIL revealed an excellent agreement with the VT and HT areas of the Fokker 100 airliner fitted with Rolls&Royce Tay 620 engines.

V_{mca} may become an important driver for the design of the VT if takeoff field length or climb requirements in second segment are very stringent. Both ITAIL and AA applications already consider V_{mca} for the VT sizing.

REFERENCES

Benson, T.J., 1995, "An Interactive Educational Tool for Turbojet Engines", Cleveland, Ohio, NASA Lewis Research Center.

Boeing Co., 2009, "Takeoff Performance", Boeing Electronic Training Material, Seattle.

Cavanaugh, M.A., 2004, "A MATLAB m-File to Calculate the Single Engine Minimum Control Speed in Air of a Jet Powered Aircraft", VMCmav1.m User's Manual, Virginia Tech, VA.

Centennial of Flight, "Dynamic Longitudinal, Directional, and Lateral Stability," Retrieved on November 12 2011, from http://www.centennialofflight.gov/essay/Theories_of_Flight/Stability_II/TH27.htm

Chai, S.T. and Mason, W., 1996, "Aircraft Landing Gear Integration in Aircraft Conceptual Design", MAD Report, MAD 96-09-01, Virginia, September.

Grasmeyer, J., 1998, "Stability and Control Derivative Estimation and Engine-Out Analysis," Virginia Tech Department of Aerospace and Ocean Engineering Report, VPI-AOE-254.

Grundlach IV, J.F., 1999, "Multidisciplinary Design Optimization and Industry Review of a 2010 Strut-Braced Wing Transonic Transport," Master of Aerospace Engineering Thesis, Virginia Polytechnic Institute and State University, Blacksburg, Virginia.

Heffley, R.K. and Jewell, W.F, 1972, "Aircraft Handling Qualities Data", NASA CR 2144.

Holst, T.L. and Thomas, S.D., 1982, "Numerical Solution of Transonic Wing Flow Fields", AIAA paper # 82-105.

Jenkinson, L., Simpkin, P., Rhodes, D., 2001, "Civil Jet Aircraft Design", Retrieved on November 2 2011, from http://www.elsevierdirect.com/companions/9780340741528/case-studies/default.htm

Karas, O.V. and Kovalev, V.E., 2011, "BLWF 28 User's Guide", Moscow.

Loureiro, V., 2008, "Optimal design of Transport Airplane with a Realistic Engine Model in the Loop," Undergraduation Thesis, Instituto Tecnológico de Aeronáutica, São José dos Campos, Brazil.

Mattos, B.S. and Magalhães, P.E.C.S., 2012, "Conceptual Optimal Design of Airliners with Noise Constraints", 50th Aerospace Sciences Meeting and Exhibit, Nashville, TN.

Raymer, D.P., 1999, "Aircraft Design: A Conceptual Approach", 3rd Ed, Washington D.C., AIAA.

Roskam, J., 1971, "Methods for Estimating Stability and Control Derivatives of Conventional Subsonic Airplanes," Roskam Aviation and Engineering Corporation, Lawrence, Kansas.

Roskam, J., 2000a, "Airplane Design, Part I: Preliminary Sizing of Airplanes," The University of Kansas, Lawrence, DARcorporation.

Roskam, J., 2000b, "Airplane Design, Part VI: Preliminary Calculation of Aerodynamics, Thrust and Power Characteristics," The University of Kansas, Lawrence, DARcorporation.

Sadraey, M., 2009, "Aircraft Performance Analysis," Verlag Dr. Müller.

Scholz, D., 2011, "PreSTo - Aircraft Preliminary Sizing Tool – From Requirements to the Three-view Drawing", EWADE 2011 – 10th Workshop on Aircraft Design Education, Naples, Italy, May.

Toreenbeek, E., 1982, "Synthesis of Subsonic Airplane Design," Delft University Press, Delft, the Netherlands.

Videiro, P.L., 2012, "Wing Structural Optimization in Aircraft Conceptual Design," undergraduation thesis, Instituto Tecnológico de Aeronáutica, São José dos Campos, Brazil.

"World's Aircraft, "Fokker 100," 2013, Retrieved on June 3 2013, from http://www.aer.ita.br/~bmattos/mundo/airliner/fo100.htm

LIST OF SYMBOLS

AR = Aspect ratio
b = Span
c_{mac} = Mean aerodynamic chord
C_L = Lift coefficient
$C_{L\alpha}$ = Variantion of the lift coefficient with angle of attack
C_{Lmax} = Maximum lift coefficient of wing or airplane configuration
Cm_e = Moment coefficient due to engine
Cm_0 = *Moment coefficient*
$Cm_{0_{wf}}$ = *Moment coefficient of the wing - body combination*
d = *diameter*
D = *Drag force*
M_∞ = *Freestream Mach number*
MAC – Mean aerodynamic chord
l = *length*
L = *Lift force*
q = *Dynamic pressure*
S = *Area*
T = *Engine thrust*
V = *Velocity*
W = *weight*
α = *Angle of attack*
β = *Sideslip Angle*
ε = *Downwash angle at horizontal tail*
η = *Ratio between local and freestream dynamic pressure*
ϕ = *Bank angle*
ρ = *Air density*
Ψ = *Sweepback angle*

SUBSCRIPTS

50 Half chord
ac Aerodynamic
CG Center of gravity
e Engine
h Horizontal tail
mca Minimum control speed in the air
mcg Minimum control speed in the ground
ref Reference (area, length, etc...)
s Stall
w Wing
∞ Freestream

18

Rainy Season Features for the Alcântara Launch Center

Urias Andrade Pinheiro[1], Marcos Daisuke Oyama[2]

ABSTRACT: The rainy season features for the Alcântara Launch Center region (2°S-3°S; 44°W-45°W), located at the northern coast of Brazil, were obtained by using the Climate Prediction Center/National Oceanic and Atmospheric Administration daily precipitation data from 1979 to 2012 accumulated to pentads. The rainy season onset (demise) was defined as the first pentad when precipitation is greater (lower) than the climatological annual average, and this behavior lasts for three out of the four following pentads. The average rainy season features were: 28 January as onset day; 16 June as demise day; 140 days as length; 1527 mm as total precipitation (about 80% of the annual value); and 10.9 mm day^{-1} as intensity (rain rate). The uncertainty on these climatological values due to the use of different precipitation datasets was estimated as few days for the onset/demise days and length, 100 mm for the total precipitation and about 1 mm day^{-1} for the intensity. Except for intensity, the rainy season features showed large interannual variability: standard variation of about one month for onset/demise days, and coefficient of variation of 33 and 40% for length and total precipitation, respectively. The three-week period between 24 March and 13 April belonged to the rainy season of all years. In general, longer (shorter) duration was related to early (late) onset, late (early) demise, and higher (lower) total precipitation. Within the rainy season, on an average, precipitation was lower than 0.1 mm day^{-1} in only four to five days; therefore, the occurrence of "no-rain" days was rather uncommon.

KEYWORDS: Aerospace meteorology, Precipitation, Climatology, Brazilian space program.

INTRODUCTION

The rainy season is usually defined as the period of the year when precipitation is higher (than a given threshold) due to atmospheric conditions that favor the occurrence of rain events. It is described by five features: onset and demise days, length, total precipitation, and intensity (Table 1). Interannual variability of these features can severely affect socio-economic activities; for instance, in the semi-arid Northeast Brazil, people "anxiously await the annual arrival of the rainy season and its promise of an adequate harvest for that year; in the event of drought, [...] agricultural production is compromised and immense human suffering prevails" (Lemos *et al.*, 2002, p. 479).

Here, we focus on the rainy season features for the Alcântara Launch Center [*Centro de Lançamento de Alcântara* (CLA)] region (2°S-3°S; 44°W-45°W; Fig. 1). It is a specific region located at the northern coast of Brazil, and its importance is based on the fact that CLA is the main rocket launching center of the Brazilian space program. For the CLA region, climate is classified as tropical humid (IBGE, 2002), annual precipitation is ~2000 mm, and the seasonal cycle shows maximum (minimum) precipitation in austral autumn (spring) (Pereira *et al.*, 2002). This cycle is closely related to the seasonal latitudinal migration of the Atlantic Intertropical Convergence Zone (AITCZ): during austral autumn, AITCZ attains its southernmost position and directly affects the CLA region (Molion and Bernardo, 2002). Other atmospheric systems, such as coastal squall lines (CSL) (Cohen *et al.*, 1995, 2009; Oliveira, 2012), easterly waves (Alves *et al.*, 2008; Machado *et al.*, 2009), and upper tropospheric cyclonic vortices (UTCV) (Kousky and Gan, 1981; Silva, 2005; Ferreira *et al.*, 2009), also affect the precipitation amount (Barros

1.Instituto Nacional de Pesquisas Espaciais – São José dos Campos/SP – Brazil 2.Instituto de Aeronáutica e Espaço – São José dos Campos/SP – Brazil
Author for correspondence: Marcos Daisuke Oyama | Praça Marechal Eduardo Gomes, 50 – Vila das Acácias | CEP 12.228-901 São José dos Campos/SP – Brazil | Email: marcos.oyama@ymail.com

Table 1. Definition of the rainy season features; p_k is the total precipitation (mm) in day k.

Rainy season feature	Unit	Symbol	Definition
Onset day	Julian day	T_i	–
Demise day	Julian day	T_f	–
Length	Day	L	$L = T_f - T_i + 1$
Total precipitation	mm	P	$P = \sum_{k=T_i}^{T_f} p_k$
Intensity (or average rain rate)	mm day^{-1}	I	$I = P/L$

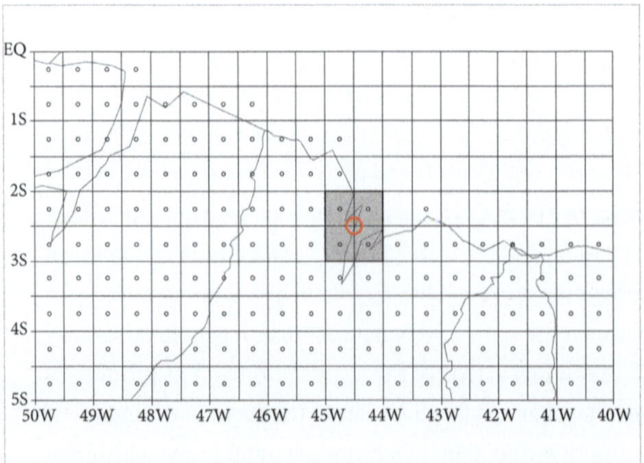

Figure 1. CPC-G precipitation data grid points. Central small circles indicate grid points for which precipitation data exist. The precipitation for the Alcântara Launch Center region is calculated as the average over the gray 1° x 1° box. Alcântara Launch Center position is indicated by the red circle.

and Oyama, 2010). For the CLA region, interannual variability of precipitation is related to sea surface temperature anomalies (SSTA) in tropical Pacific and Atlantic oceans (Kayano and Andreoli, 2006; Marques and Fortes, 2012).

From the aerospace meteorology perspective (Vaughan and Johnson, 2013), knowledge about the rainy season features for the CLA region is particularly important and useful for the planning of rocket launching missions, since absence of rainfall ("no-rain" condition) is usually necessary for rocket-related activities in CLA (Marques and Fisch, 2005). Preliminary information about the rainy season climatology for the CLA region can be derived from the literature: the onset would take place from December to January; the demise, from May to June; and the length would be within the range of five to seven months (Marengo *et al.*, 2001; Liebmann *et al.*, 2007; Silva *et al.*, 2007). More precise information is not available, since the cited studies focused on larger scales (Amazonia or South America) and used different methods for rainy season identification.

In this study, a detailed account on the rainy season features for the CLA region is given. The report is organized as follows. The precipitation dataset and the method to identify the rainy season onset and demise are described in the Data and Methodology section. The climatology and interannual variability of the rainy season features for the CLA region, their sensitivity to the use of different precipitation datasets, a discussion on the meteorological factors related to the rainy season onset, and a brief analysis on the dry days/spells within the rainy season are presented in the Results section. A summary of main findings of the study, as well as possibilities of future work, are given in the Concluding Remarks section.

DATA AND METHODOLOGY

PRECIPITATION DATA

Daily precipitation from the Climate Prediction Center (CPC)/National Oceanic and Atmospheric Administration (NOAA) dataset called "CPC Unified Gauge-Based Analysis of Global Daily Precipitation" (CPC-G), for the period from January 1979 to December 2012 (34 years), are used. The data are based on rain gauge measured precipitation and gridded on a 0.5° x 0.5° mesh using an optimal interpolation (OI) technique (Xie *et al.*, 2007). Daily precipitation for the CLA region is calculated as the area-average precipitation over the four grid points that surround the CLA (Fig. 1). The daily data are accumulated to pentads for rainy season identification.

The choice of OI technique as the interpolation procedure for the CPC-G data is based on a comprehensive assessment over global land areas carried out by Chen *et al.* (2008). They found that, compared to other two widely used interpolation methods, the OI technique consistently performed the best for all situations (regions, seasons, and network densities). Moreover, the OI technique led to a relatively stable performance statistics over regions covered by fewer gauges. This result is particularly suited for this study, as it minimizes the uncertainty related to the sudden drop in the number of gauges used for interpolation over the CLA region in the 2005–2012 period: the number of gauges within the CLA region varies from six to ten in the 1979–1988 period; from five to seven in the 1989–2004 period; and decreases to about two or three in the 2005–2012 period (see also Silva *et al.*, 2007, p. 851).

RAINY SEASON IDENTIFICATION

In the literature, various methods to identify the rainy season features are found (the most common methods are briefly reviewed in Alves *et al.*, 2005, p. 386). Here, we adopt a simple condition adapted from Gan *et al.* (2004), but based solely on pentad precipitation. The rainy season onset (demise) pentad is defined as the first pentad when precipitation is greater (lower) than the climatological annual average, and this behavior lasts for at least three out of the four following pentads. For a given year, the search for the onset pentad starts from pentad 63 (07–11 November) of the previous year and for the demise pentad, from pentad 20 (06–10 April) of the given year (this pentad is in the middle of the climatological rainy season, as will be shown in the next section). The onset (demise) day is defined as the middle day of the onset (demise) pentad, and the other features — length, total precipitation, and intensity — are calculated from expressions given in Table 1.

The identification of the onset pentad is illustrated in Fig. 2. From pentad −10 (which is the pentad 63 of the previous year) to pentad 5, precipitation is lower than the annual average. Pentads 6 and 8, albeit having above-average precipitation, are not the onset pentad, because in only two out of the following four pentads precipitation is greater than the annual average. The onset condition is firstly met by pentad 10.

RESULTS

RAINY SEASON FEATURES

The climatological rainy season features for the CLA region are shown in Table 2. On average, the rainy season extends from the end of January to mid-June, thus lasting for ~4½ months. Compared to the literature, the onset and demise days are within the range of December–January and May–June, respectively, but the length is slightly shorter than the range of five to seven months (Marengo *et al.*, 2001; Liebmann *et al.*, 2007; Silva *et al.*, 2007). The precipitation amount in the rainy season accounts for ~80% of the total annual average. The average intensity, of ~10 mm day^{-1}, is quite high; its value is comparable to the annual mean rain rate over the rainiest parts of Amazonia (Marengo and Nobre, 2009) and may be classified as transition between light and heavy precipitation (Sun *et al.*, 2006).

Except for the intensity, interannual variability of the rainy season features is large: standard deviation for the onset and

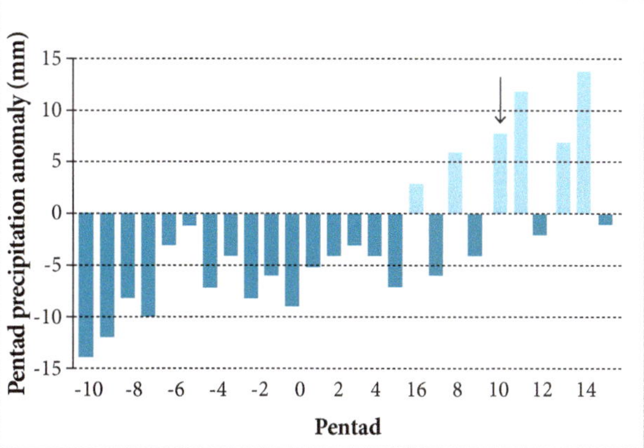

Figure 2. Illustration of the method to identify the rainy season onset. Light (dark) blue bars refer to positive (negative) pentad precipitation anomaly, that is, pentad precipitation greater (lower) than the climatological annual average. Negative pentads refer to pentads of the previous year: pentad p, $p<0$, refers to pentad $p + 73$ of the previous year. The black arrow points to the rainy season onset pentad.

Table 2. Climatological rainy season features for the Alcântara Launch Center region (1979–2012).

Feature	Average	Standard deviation	Coefficient of variation (%)
Onset day	28 January	27 days	–
Demise day	16 June	28 days	–
Length	140 days	46 days	33
Total precipitation	1567 mm	622mm	40
Intensity	10.9 mm day^{-1}	1.8 mm day^{-1}	16

demise days is ~1 month; for the length, ~1½ month; and for the total precipitation, 40% of the average value. The smaller variability of the intensity — 16% of the climatological value — is related to the existence of a clear direct proportionality between total precipitation and length. The linear regression equation between these two features without the intercept term is (Fig. 3a):

$$P = 11.3 \times L \quad (1)$$

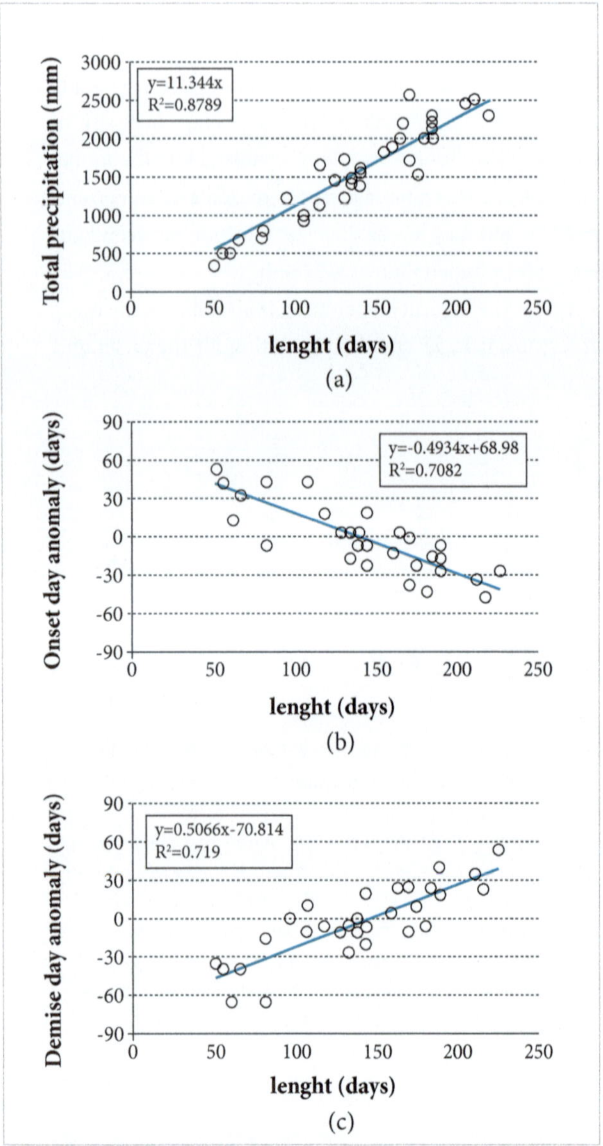

Figure 3. Relation between rainy season length and total precipitation (a), onset day anomaly (b), and demise day anomaly (c). The linear regression line is drawn in blue color. The equation and the coefficient of determination (R^2) are shown. In (a), regression analysis is carried out without the intercept term.

and the coefficient of determination value is very high (~90%). The slope term of Eq. 1, as expected, is close to climatological intensity value.

The temporal evolution of the rainy season from 1979 to 2012 is shown in Fig. 4. Marked changes from one year to the next (e.g., 1986–1988; 1993 and 1994) lead to the high value of standard deviation obtained previously for the onset/demise days and length, and illustrate the lack of lag-1 autocorrelation (almost zero) for these features. There are periods, however, when the rainy season length shows lower variability: for instance, from 1999 to 2004 (from 1979 to 1983, but 1982), the length is longer (shorter) than the average. The three-week period between 24 March and 13 April belongs to the rainy season of all years — this information could be useful for planning purposes. In general (~60% of the years), late (early) onset corresponds to early (late) demise, thus leading to shorter (longer) rainy season length. This behavior is ratified by the linear regression analysis involving these features. Considering length as an independent variable, the equations for onset and demise anomalies are given respectively by (Fig. 3b–c):

$$\Delta T_i = -(0.49 \times L - 69.0); \quad (2)$$

$$\Delta T_f = 0.51 \times L - 70.8 \quad (3)$$

For both equations, the coefficient of determination value is ~70%, which indicates good fitting skill.

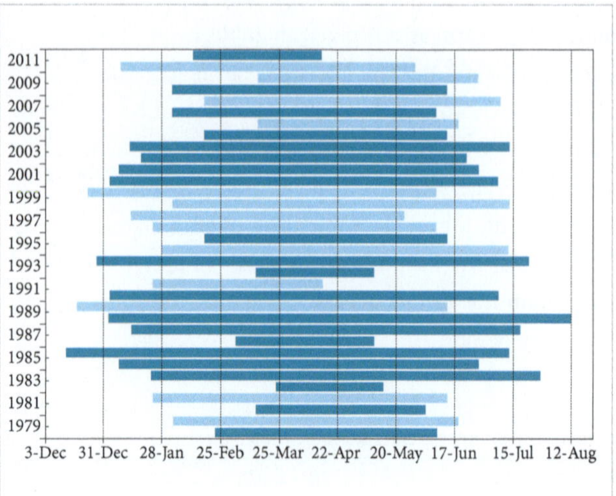

Figure 4. Temporal evolution of the rainy season from 1979 to 2012. The horizontal bars represent the rainy season within each year. Dark (light) blue bars refer to the years when the anomalies of onset and demise days have opposite (the same) sign.

SENSITIVITY TO THE USE OF A DIFFERENT PRECIPITATION DATASET

The climatological rainy season features may be sensitive to the precipitation dataset used. To evaluate the degree of uncertainty related to this sensitivity, the rainy season features for the CLA region from 1997 to 2012 derived using daily precipitation data from a different dataset — the National Aeronautics and Space Administration (NASA) "Global Precipitation Climatology Project (GPCP) 1-Degree Daily Combination" (GPCP-1DD) (Huffman *et al.*, 2001) — are compared to those derived using the CPC-G data. The GPCP-1DD dataset is a satellite-based spatial and temporal downscaling of the GPCP monthly precipitation estimate (outcome of merging microwave and infrared precipitation estimates and rain gauge data) on a regular 2.5° x 2.5° global mesh (Adler *et al.*, 2003), and so does differ from the CPC-G dataset.

The average rainy season features for the 1997–2012 period derived from the two datasets show good overall agreement, as the bias values are low (Table 3). This implies the robustness of the climatological features obtained previously (*cf.* Table 2). The uncertainty may be estimated as few days (1–10 days) for the onset/demise days and length, 100 mm for the total precipitation, and ~1 mm day $^{-1}$ for the average intensity.

The temporal evolution of the onset/demise days and length derived from both datasets are shown in Fig. 5 [the total precipitation and average intensity time series are not shown and are not analyzed because intensity is almost constant (coefficient of variation < 15%) and, consequently, total precipitation is closely proportional to length]. The overall agreement between the time series derived from the two datasets is excellent for the onset day (Pearson linear correlation coefficient, $r \sim 0.9$), and moderate for the demise day ($r \sim 0.5$) and length ($r \sim 0.7$). For the onset day, large differences (~1 month) are found in three years (2006, 2008, and 2009), and the bias value results mostly from these specific differences. For the demise day, systematic differences are found until 2004, the largest absolute differences (1–2 months) are found in three years (1999, 2010, and 2012), and the bias value results mostly from the systematic differences (because the largest differences do not have the same sign and partially cancel each other out). For the length, the time series pattern is similar to the demise day's, and the differences have smaller absolute value, although large differences (>1 month) are still found in two years (1999 and 2010). Therefore, the use of different precipitation datasets may lead to high uncertainty in the rainy season features for specific years, but the average features would be only marginally affected by these temporally localized large differences and, therefore, would be almost the same (i.e., low bias).

The rainy season features derived using the rain gauge data collected at the meteorological station in CLA (this dataset is hereafter referred to as STN) (Marques and Fisch, 2005) are also compared to those derived using CPC-G and GPCP-1DD data. STN dataset should be regarded as the outcome of an independent station, because both CPC-G and GPCP-1DD datasets do not use STN data. The comparison is carried out for the short period from 2003 to 2012, when STN data are more reliable. For this period, the average rainy season features derived using STN data are close to those derived using CPC-G data and GPCP-1DD data (not shown). The temporal evolution of the rainy season features shows good overall agreement among the different datasets (Fig. 5).

- For the period 2005–2012, there is very good agreement. It adds reliability to the results derived using CPC-G data for this period, when uncertainty of CPC-G data are higher due to the small number of stations used for interpolation within the CLA region.
- The largest differences (one to two months) are found in 2003 for the demise day and length. STN data show below-average precipitation during May 2003, and the

Table 3. Average rainy season features for 1997–2012 derived from CPC-G and GPCP-1DD precipitation datasets. Bias refers to GPCP-1DD minus CPC-G-derived values.

Feature	Average value for 1997–2012 derived from		Bias
	CPC-G dataset	GPCP-1DD dataset	
Onset day	28 January	24 January	-4 days
Demise day	14 June	06 June	-8 days
Length (days)	138	134	-4
Total precipitation (mm)	1531	1631	+100
Intensity (mm day $^{-1}$)	11.0	12.4	+1.2

rainy season ends up prematurely, although above-average precipitation is found in the first half of June. For CPC-G and GPCP-1DD data, the below-average precipitation period is much shorter and rainy season demise is postponed to mid-June (not shown). This kind of difference is expected when single station and interpolated data are compared (Silva *et al.*, 2007).

DISCUSSION ON THE METEOROLOGICAL FACTORS

What are the meteorological factors that shape the high interannual variability of the rainy season features for the CLA region? For the 1979–2010 period, Pinheiro (2013) partially addressed this question by focusing on the rainy season onset. The conclusion was that early (late) onset would be mainly related to factors that favor (inhibit) precipitation occurrence over the CLA region: AITCZ located to the south (north) of the mean position and/or with high (low) intensity, negative (positive) SSTA in Nino 3.4 region [according to Trenberth, 1997, El-Niño (La-Niña) events refer to at least six consecutive months when the five-month running means of SSTA in Nino 3.4 region are positive (negative) with absolute value >0.4°C], and negative (positive) cross-equatorial SSTA gradient in tropical Atlantic. Other factors related to atmospheric systems that directly affect the CLA region, such as higher/lower frequency of occurrence of UTCV and/or CSL, would be important to explain the onset day anomalies for specific years (Marques and Fortes, 2012). For instance:

- In 2001, the longest length in the 1997–2012 period for both CPC-G and GPCP-1DD datasets took place (Fig. 5). It resulted mainly from a pronounced early onset related to favorable conditions for precipitation occurrence over the CLA region in January 2001: more intense AITCZ located to the south of the mean position, La-Niña event in tropical Pacific, and negative cross-equatorial SSTA gradient in tropical Atlantic.
- In 1989, the onset day was the same as in 2001 (Fig. 4). Similarly to January 2001, there were a La-Niña event in tropical Pacific and negative cross-equatorial SST gradient in tropical Atlantic in January 1989; however, AITCZ position and intensity were close to the mean. The earliness was also related to higher frequency of occurrence of CSL over the northern coast of Brazil in January 1989.
- In 2010, the latest onset day in the 1997–2012 period for both CPC-G and GPCP-1DD datasets took place (Fig. 5).

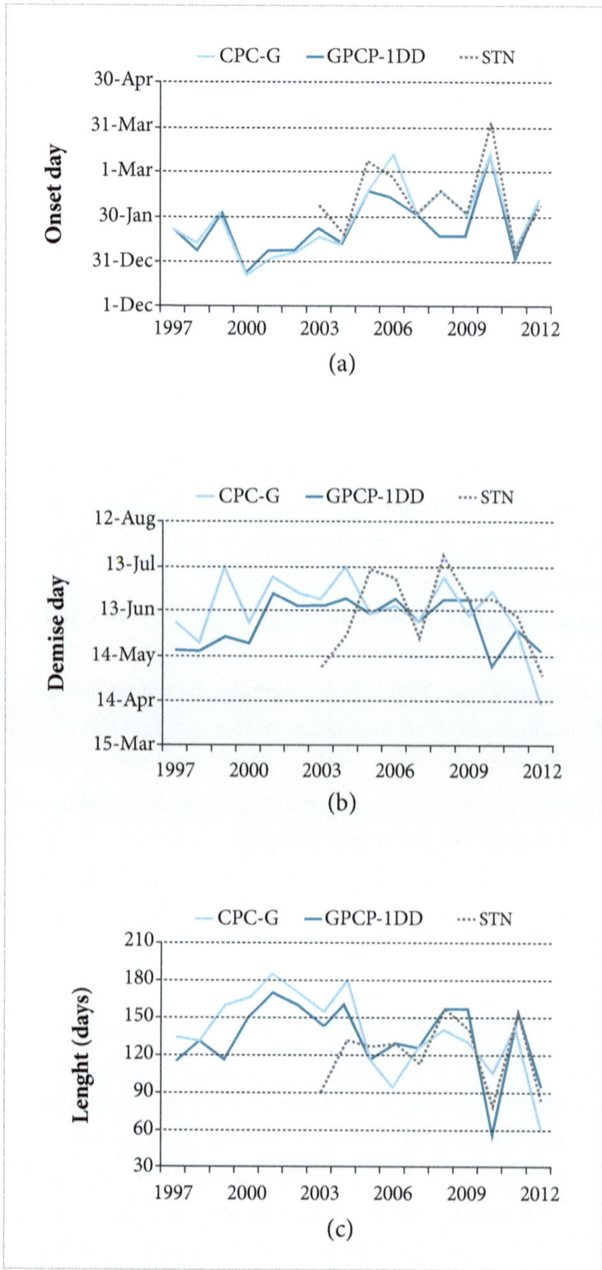

Figure 5. Temporal evolution of the rainy season onset day (a), demise day (b), and length (c) derived from CPC-G (light blue), GPCP-1DD (dark blue), and STN (dotted) precipitation datasets.

The lateness was the result of unfavorable conditions for precipitation occurrence over CLA from mid-February to mid-March of 2010: weak (less intense) AITCZ located to the north of the mean position, El-Niño event in tropical Pacific, positive cross-equatorial SST gradient in tropical Atlantic, and high frequency of occurrence of UTCV over Northeast

Brazil. An example of UTCV that inhibited cloud formation over CLA in 11 March 2010 is shown in Fig. 6.

The analysis for specific years, carried out by Pinheiro (2013), was useful to identify the factors related to rainy season onset, but further studies are needed to assess the relative importance of the factors statistically. Preliminary analysis indicates that AITCZ features (position and intensity), taken together, are able to explain ~20% of the onset day variance. This percentage seems to be consistent, because it is close to the average fraction of monthly precipitation variance explained by AITCZ features between January and February (Oyama and Carvalho, 2012). However, the percentage is rather low due to the difficulty in properly representing the magnitude of the onset day anomalies, even though the anomalies sign could be better predicted from the anomalies of the AITCZ features (proportion correct of about 75%). For the oceanic indices, SSTA in the Nino 3.4 region are able to explain ~5% of the onset day variance; cross-equatorial SSTA gradient in tropical Atlantic, ~25%. These fraction values are consistent with the results of Kayano and Andreoli (2006), who showed that interannual variations of the Northeast Brazil climate are more closely related to the tropical Atlantic variability modes than to the tropical Pacific's. The role of SSTA in the Nino 3.4 region seems to increase for higher onset day variations: in more than half of the years in which the onset day occurs earlier (later) than the first (third) quartile, the rainy season begins under negative (positive) SSTA in Nino 3.4 region. Future work is also needed to unravel the factors that influence the rainy season demise.

DRY DAYS AND DRY SPELLS WITHIN THE RAINY SEASON

In the rainy season, as the atmospheric conditions are favorable to precipitation, the occurrence of dry days (days

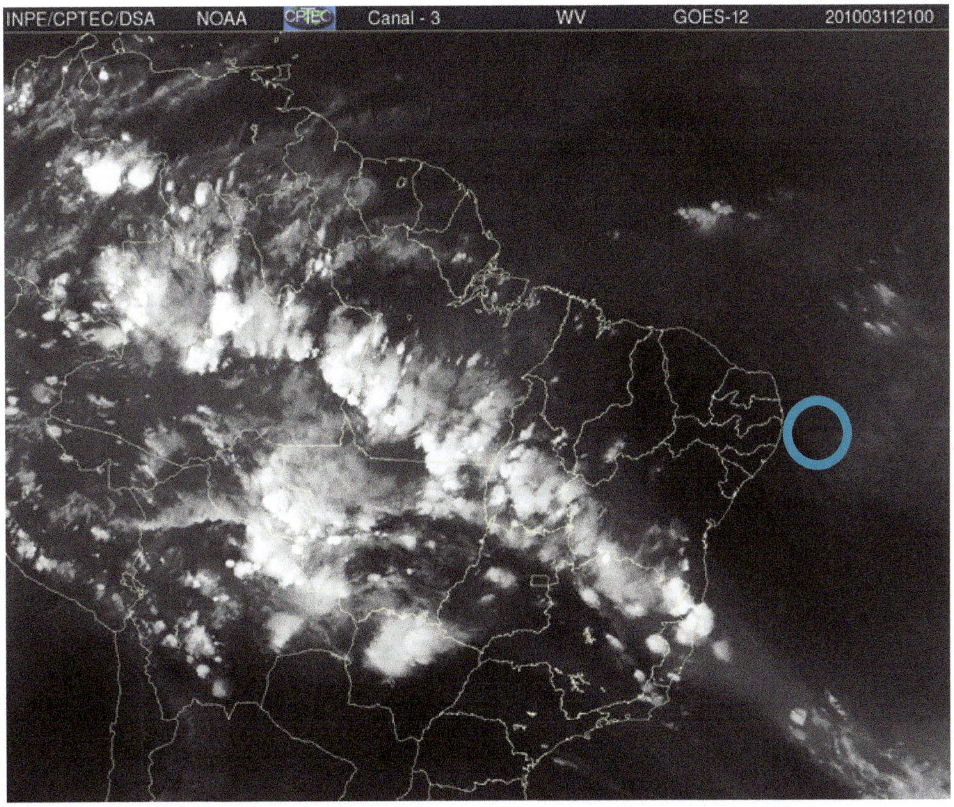

Figure 6. Water vapor satellite image for 21 UTC, 11 March 2010 (source: *Centro de Previsão de Tempo e Estudos Climáticos, Instituto Nacional de Pesquisas Espaciais*, Brazil). The position of the UTCV center, according to the Climanalise bulletin for March, 2010 (http://climanalise.cptec.inpe.br/~rclimanl/boletim/pdf/pdf10/mar10.pdf), is indicated by the blue circle. The subsidence region associated to the UTCV directly affects Northeast Brazil.

without precipitation or with small amounts of precipitation) or dry spells (consecutive dry days) would not be expected. But they do exist and their frequency could be regarded as an additional rainy season feature. Dry days are defined as those with precipitation lower than a given threshold. Here, two thresholds commonly found in the literature (e.g., Vicente-Serrano and Beguería-Portugués, 2003, p. 1107) — 1 and 0.1 mm day $^{-1}$ — are used to identify the dry days within the rainy season for the CLA region.

Considering the higher threshold (1 mm day $^{-1}$), on average, dry days account for about 11% (~15 days) of the rainy season. About half of the dry days are grouped in dry spells, and the amount of dry spells decreases sharply as its length (i.e., the number of consecutive dry days) increases (Fig. 7). The return period is longer than 1 year for dry spells lasting for three days or more. The results presented here are close to the values shown in INMET (2009) for a city near CLA (São Luís).

For aerospace meteorology purposes, the near-zero threshold (0.1 mm day $^{-1}$) is important because it is more related to the "no-rain" condition needed for rocket-related activities in CLA. Using this threshold, the amount of dry days/spells is much lower: on average, there are only four to five dry days within the rainy season, and dry spells of any length have return periods longer than one year (Fig. 7). Therefore, the "no-rain" condition is rather difficult to be met during the rainy season for the CLA region.

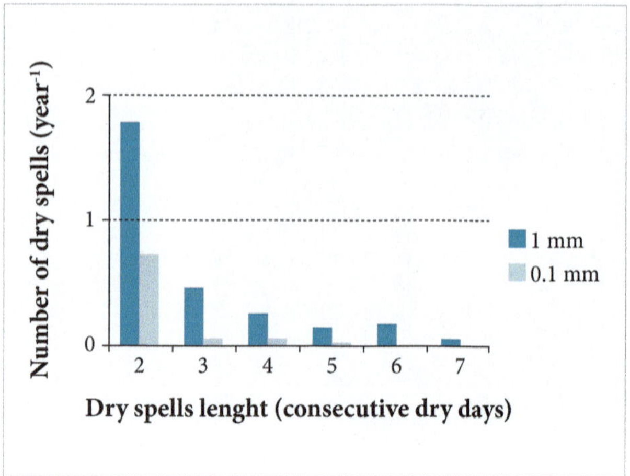

Figure 7. Number of dry spells within the rainy season for the Alcântara Launch Center region per year as function of the dry spell length (i.e., the number of consecutive dry days) for two thresholds: 1 and 0.1 mm day^{-1}.

CONCLUDING REMARKS

The rainy season features — onset and demise days, length, total precipitation, and intensity — for the CLA region were obtained by using the CPC-G daily precipitation from 1979 to 2012 (34 years). For each year, the rainy season was identified objectively from the pentad precipitation data derived from the daily's. The main conclusions of this study were:

- The climatological rainy season features were: 28 January as onset day; 16 June as demise day; 140 days as length; 1527mm as total precipitation; and 10.9 mm d $^{-1}$ as intensity.
- The uncertainty on these climatological values due to the use of different precipitation datasets (e.g., the GPCP-1DD and STN datasets) was estimated as few days for the onset/demise days and length; 100mm for the total precipitation and ~1 mm day $^{-1}$ for the intensity.
- Except for the intensity, the rainy season features showed large interannual variability: standard variation of about one month for onset and demise days, and coefficient of variation of 33 and 40% for length and total precipitation, respectively.
- The three-week period between 24 March and 13 April belongs to the rainy season of all years.
- In general, longer (shorter) duration was related to early (late) onset, late (early) demise, and higher (lower) total precipitation.
- The occurrence of dry spells within the rainy season is rather uncommon; on average, in only four or five days of the rainy season, precipitation is lower than 0.1 mm day $^{-1}$.

The results presented here may be regarded as a first step towards a comprehensive understanding of the rainy season for the CLA region. Follow-up steps could include studies on the meteorological factors/systems and oceanic indices that drive the anomalies of the rainy seasons features (e.g., by expanding the work of Pinheiro, 2013), as well as on the average atmospheric conditions related to the rainy season onset and demise (like Marengo *et al.*, 2001; Barbieri, 2005).

ACKNOWLEDGMENTS

This work is part of the first author's MSc dissertation under the guidance of the second author. The authors are grateful to the three anonymous reviewers for their useful comments and suggestions.

REFERENCES

Adler, R.F., Huffman, G.J., Chang, A., Ferraro, R., Xie, P., Janowiak, J., Rudolf, B., Schneider, U., Curtis, S., Bolvin, D., Gruber, A., Susskind, J., Arkin, P. and Nelkin, E., 2003, "The Version 2 Global Precipitation Climatology Project (GPCP) monthly precipitation analysis (1979–present)", Journal of Hydrometeorology, Vol. 4, No. 6, pp. 1147-1167.

Alves, L.M., Marengo, J.A., Camargo Júnior, H. and Castro, C., 2005, "Início da Estação Chuvosa na Região Sudeste do Brasil: Parte 1, Estudos Observacionais", Revista Brasileira de Meteorologia, Vol. 20, No. 3, pp. 385-394.

Alves, M.A.S., Marques, R.F.C. and Oyama, M.D., 2008, "Detecção de Distúrbios Ondulatórios de Leste a partir de Filtros Temporais", Proceedings of the XV Brazilian Congress of Meteorology, São Paulo, SP, Brazil.

Barbieri, P.R.B., 2005, "Characterization of the rain station in the south and southeastern regions of Brazil associated with the atmospheric circulation" (In Portuguese), MSc Dissertation, Instituto Nacional de Pesquisas Espaciais, São José dos Campos, SP, Brazil, p. 116.

Barros, S.S. and Oyama, M.D., 2010, "Sistemas Meteorológicos Associados à Ocorrência de Precipitação no Centro de Lançamento de Alcântara", Revista Brasileira de Meteorologia, Vol. 25, No. 3, pp. 333-344.

Chen, M., Shi, W., Xie, P., Silva, V.B.S., Kousky, V.E.and Higgins, R.W., Janowiak, J.E., 2008, "Assessing objective techniques for gauge-based analyses of global daily precipitation", Journal of Geophysical Research, Vol. 113, No. D04110.

Cohen, J.C.P., Silva Dias, M.A. and Nobre, C., 1995, "Environmental conditions associated with amazonian squall lines: a case study", Monthly Weather Review, Vol. 123, No. 11, pp. 3163-3174.

Cohen, J., Cavalcanti, I.F.A., Braga, R.H.M. and Santos Neto, L., 2009, "Linhas de Instabilidade na Costa N-NE da América do Sul", in: Cavalcanti, I.F.A.; Ferreira, N.J.; Silva, M.G.A.J. and Silva Dias, M.A.F. (Eds), "Tempo e Clima no Brasil", Ed. Oficina de Textos, São Paulo, Brazil, pp. 75-93.

Ferreira, N.J., Ramirez, M.V. and Gan, M.A., 2009, "Vórtices Ciclônicos de Altos Níveis que Atuam na Vizinhança do Nordeste do Brasil", in: Cavalcanti, I.F.A.; Ferreira, N.J.; Silva, M.G.A.J. and Silva Dias, M.A.F. (Eds), "Tempo e Clima no Brasil", Ed. Oficina de Textos, São Paulo, Brazil, pp. 43-60.

Gan, M.A., Kousky, V.E. and Ropelewski, C.F., 2004, "The South America monsoon circulation and its relationship to rainfall over West-Central Brazil", Journal of Climate, Vol. 17, No. 1, pp. 47-66.

Huffman, G.J., Adler, R.F., Morrissey, M., Bolvin, D.T., Curtis, S., Joyce, R., McGavock, B. and Susskind, J., 2001, "Global precipitation at one-degree daily resolution from multi-satellite observations", Journal of Hydrometeorology, Vol. 2, No. 1, pp. 36-50.

IBGE, 2002, "Mapa Brasil Climas – Escala 1:5.000.000", Retrieved in January 10, 2012, from ftp://geoftp.ibge.gov.br/mapas_tematicos/mapas_murais/clima.pdf

INMET, 2009, "Normais Climatológicas do Brasil: 1961–1990", Retrieved in May 11, 2012, from http://www.inmet.gov.br/portal/index.php?r=clima/normaisClimatologicas

Kayano, M.T. and Andreoli, R.V., 2006, "Relationships between rainfall anomalies over northeastern Brazil and the El Niño-Southern Oscillation", Journal of Geophysical Research, Vol. 111, No. D13101.

Kousky, V.E. and Gan, M.A., 1981, "Upper tropospheric cyclonic vortices in the tropical South Atlantic", Tellus, Vol. 33, No. 6, pp. 538-551.

Lemos, M.C., Finan, T.J., Fox, R.W., Nelson, D.R. and Tucker, J., 2002, "The use of seasonal climate forecasting in policymaking: lessons from Northeast Brazil", Climatic Change, Vol. 55, No. 4, pp. 479-507.

Liebmann, B., Camargo, S.J., Seth, A., Marengo, J.A., Carvalho, L.M.V., Allured, D., Fu, R.; Vera, C.S., 2007, "Onset and end of the rainy season in South America in observations and the ECHAM 4.5 atmospheric general circulation model", Journal of Climate, Vol. 20, No. 10, pp. 2037-2050.

Machado, L.A.T., Ferreira, N.J., Laurent, H. and Diedhiou, A., 2009, "Distúrbios Ondulatórios de Leste", in: Cavalcanti, I.F.A.; Ferreira, N.J.; Silva, M.G.A.J. and Silva Dias, M.A.F. (Eds), "Tempo e Clima no Brasil", Ed. Oficina de Textos, São Paulo, Brazil, pp. 61-74.

Marengo, J.A., Liebmann, B., Kousky, V.E., Filizola, N.P. and Wainer, I.C., 2001, "Onset and end of the rainy season in the Brazilian Amazon Basin", Journal of Climate, Vol. 14, No. 5, pp. 833-852.

Marengo, J.A. and Nobre, C.A., 2009, "Clima da Região Amazônica", in: Cavalcanti, I.F.A.; Ferreira, N.J.; Silva, M.G.A.J. and Silva Dias, M.A.F. (Eds), "Tempo e Clima no Brasil", Ed. Oficina de Textos, São Paulo, Brazil, pp. 197-212.

Marques, R.F.C. and Fisch, G.F., 2005, "As atividades de Meteorologia Aeroespacial no Centro Técnico Aeroespacial (CTA)", Boletim da Sociedade Brasileira de Meteorologia, Vol. 29, No. 3, pp. 21-25.

Marques, R.F.C. and Fortes, M.A.B., 2012, "Estudo da Variabilidade Interanual da Precipitação no Centro de Lançamento de Alcântara (CLA)", Proceedings of the XVII Brazilian Congress of Meteorology, Gramado, RS, Brazil.

Molion, L.C.B. and Bernardo, S.O., 2002, "Uma Revisão da Dinâmica das Chuvas no Nordeste Brasileiro", Revista Brasileira de Meteorologia, Vol. 17, No. 1, pp. 1-10.

Oliveira, F.P., 2012, "Fatores Associados à Iniciação de Linhas de Instabilidade na Região do Centro de Lançamento de Alcântara no Mês de Julho",(in Portuguese), MSc Dissertation, Instituto Nacional de Pesquisas Espaciais, São José dos Campos, SP, Brazil, p. 80.

Oyama, M.D. and Carvalho, M.A.V., 2012, "Influência da Zona de Convergência Intertropical Atlântica na Precipitação da Região do Centro de Lançamento de Alcântara: Estudo Preliminar", Proceedings of the XVII Brazilian Congress of Meteorology, Gramado, RS, Brazil.

Pereira, E.I., Miranda, I., Fisch, G.F., Machado, L.A.T. and Alves, M.A.S., 2002, "Atlas Climatológico do Centro de Lançamento de Alcântara". IAE, São José dos Campos, SP, Brazil (ACA/RT-01/01, GDO-000000/B0047).

Pinheiro, U.A., 2013, "Rainy season over the region of the Alcântara Launch Center" (in Portuguese), MSc Dissertation, Instituto Nacional de Pesquisas Espaciais, São José dos Campos, SP, Brazil, p. 65.

Silva, L.A., 2005, "The influence of upper tropospheric cyclonic vortex (UTCV) on the rainfall over the northeast brazil (NEB) and the associated characteristics" (in Portuguese), MSc Dissertation, Instituto Nacional de Pesquisas Espaciais, São José dos Campos, SP, Brazil, p. 134.

Silva, V.B.S., Kousky, V.E., Shi, W. and Higgins, R.W., 2007, "An improved gridded historical daily precipitation analysis for Brazil", Journal of Hydrometeorology, Vol. 8, No. 4, pp. 847-861.

Sun, Y., Solomon, S., Dai, A. and Portmann, R.W., 2006, "How often does it rain?", Journal of Climate, Vol. 19, No. 6, pp. 916-934.

Trenberth, K.E., 1997, "The definition of El Niño". Bulletin of the American Meteorological Society, Vol. 78, No. 12, pp. 2771-2777.

Vaughan, W.W. and Johnson, D.L., 2013, "Aerospace meteorology: an overview of some key environmental elements", Journal of Aerospace Technology and Management, Vol. 5, No. 1, pp. 7-14.

Vicente-Serrano, S.M. and Beguería-Portugués, S., 2003, "Estimating extreme dry-spell risk in the middle ebro valley (Northeastern Spain): a comparative analysis of partial duration series with a general pareto distribution and annual maxima series with a gumbel distribution", International Journal of Climatology, Vol. 23, No. 9, pp.1103-1118.

Xie, P., Chen, M., Yang, S., Yatagai, A., Hayasaka, T., Fukushima, Y. and Liu, C., 2007, "A gauge-based analysis of daily precipitation over East Asia", Journal of Hydrometeorology, Vol. 8, No. 3, pp. 607-626.

Material Coding for Aircraft Manufacturing Industry

Hong Xia Cai[1], Ming Yu Dai[1], Tao Yu[1]

ABSTRACT: Material coding is the basis for enterprises to perform information management. There are various kinds of materials in the aircraft manufacturing industry. In order to improve the efficiency of aircraft material management, this paper studies the aircraft material coding. The aircraft materials are divided into six categories based on some characteristics. The flexible material coding model is proposed consisting of code fields which indicate the material and code field relations in order to define the constraint relations among code fields. There are three kinds of code fields: class code, property code and flow code, while there are three kinds of relations, including the parallel relation, the subordinate relation and the dependency one. For convenient recognition and operation, the alpha-numeric combination code method is used. The material coding system in which the material code could be automatically generated was finally developed. The system has been applied in the aircraft manufacturing enterprise and it has achieved good results.

KEYWORDS: Aircraft, Material coding, Material classification.

INTRODUCTION

In the information age, it is necessary to establish the information system for enterprises to improve their own competitiveness in the increasingly fierce market. Establishing a unified material code of rules is the bottom for enterprises to perform information management. The unreasonable material code would lead to confusion in management. Depending on the rational material coding, we can optimize material management in order to improve the efficiency of management and to reduce the material inventory cost.

It is not, however, an easy work. The study of information classification and coding started in early 1945, in the United States, and the national material coding system was put forward in 1958. Since the 1960s, countries like Romania and Japan put a large amount of manpower and resources into studying material coding. Recently, many scholars have carried many new researches on how to encode material. Yi *et al.* (2006) and Wang and Wang (2008) introduced coding technology based on the ontology for information integration. Lei *et al.* (2008) discussed code principle structures and characteristics of material classification with group technology. Jiang (2007), Wang and Wang (2008) and Zhao *et al.* (2010) studied the flexible structure and the multi-segment code for the Product Data Management (PDM) system and proposed the information coding model. Material coding management is applied in different industries (Li and Xu, 2012; Xiong *et al.*, 2010), and Xiong *et al.* (2010) suggested that new technologies such as Radio Frequency Identification can be applied in material coding management. Meng and Kong (2013) studied the enterprise material coding in detail, and proposed the purchasing material coding scheme and method,

1.Shanghai University – Shanghai – China

Author for correspondence: Hongxia Cai | Shanghai Key Laboratory of Mechanical Automation and Robotics | School of Mechatronic Engineering and Automation | Shangai University | Yanchang Road 149, Zhabei District | 200072 Shanghai/China | Email: hxcai@staff.shu.edu.cn

based on the K3 auxiliary production software. Wang *et al.* (2011) put forward the flexible classification and coding system for the rapid design of aircraft tooling components. But there have been few researches on the aircraft material coding, especially for the aircraft manufacturing industry. Due to the characteristics of aircraft material and material management requirements, it is of great importance to study aircraft material coding.

ANALYSIS OF AIRCRAFT MATERIAL CODING

Material coding, in a few words, is a material management method which uses symbols or numbers in order to represent specifications or categories of materials, so that it is easy to recognize, track and monitor them. These materials cover not only raw materials and parts, but also semi-finished and finished products, purchased parts, packaging materials, product brochures, even including all the tooling resources, labor insurance supplies and energy for manufacturing or services.

As the special discrete manufacturing industry, the materials in the aircraft industry have their own characteristics: various in kinds and quantities, many self-made parts and design changes, strict quality control and tracking measure. Lots of materials have special customization demand; they must be producted in a single piece or in small batches in order to meet the requirements. Due to its particularity and complexity, material coding is different in the aircraft manufacturing industry.

The general material coding process is shown in Fig. 1. First, sort out all materials. Describe them using a standard pattern and determine the categories for their classification. Then, according to the material classification and its coding standards, it is established a coding structure and a set of encoding rules are generated. Finally, the coding system is developed, so that the material can be automatically coded to be used in the information system.

Because of the particularity of the aircraft manufacturing industry, its material coding presents some difficulties, as shown in Fig. 1, with the red boxes, which are:

AIRCRAFT MATERIAL CLASSIFICATION

Aircraft material classification should consider the variety of the materials. The material coding process covers almost all departments in the enterprise. The materials have complicated properties, and there is a cross impact between the various materials. How to determine the classification boundary and to make a reasonable classification without any ambiguity to distinguish aircraft material is a difficult task. There are two main classification methods, one is a surface classified method and the other one is line classified. The basic principles of

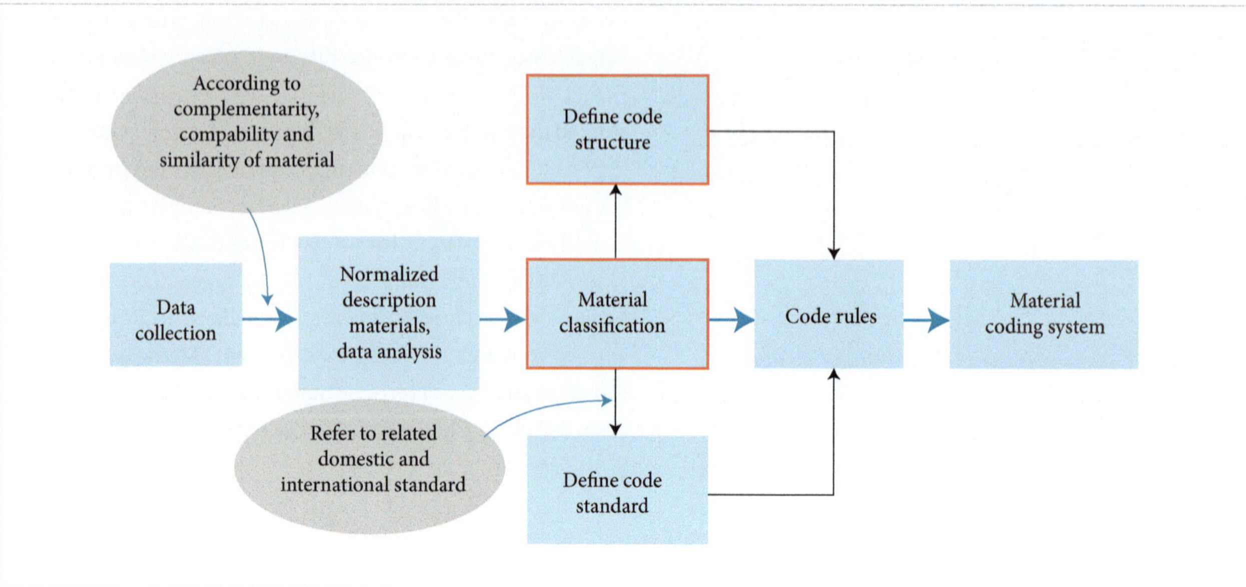

Figure 1. Flow chart of material coding.

classification are meant to ensure that the classification system embraces the features of comprehension, systematization and expansibility. The material code should be separated into several segments. Taking the overall information of material into consideration and arranging them within the different code segments. Taking the natural material attributes as basis. Reserving enough space to code new attributes, in order to ensure the extensibility of material system.

CODING STRUCTURE AND CODING RULES

The coding rules should define how long the code is and what information should be included. Because there are so many material properties, it is difficult that all attributes are represented in the material coding. Arguments will be arisen in different departments. The important information of the material is taken into account, while the changeable or configuration information should be avoided. When the length is determined, we should consider reserving the space for the demand of the product expansion, enterprise extension, long-term planning, etc. Of course, the setting of such long material code is discouraged.

MATERIAL CODING SCHEME

AIRCRAFT MATERIAL CLASSIFICATION

Material classification is the process of seeking common ground while putting aside differences (Wang and Tong, 2006). It distinguishes and classifies materials through certain principles and techniques, according to properties, attributes or characteristics. In order to make sure high efficiency and convenient maintenance is kept, the material classification should follow the principle of "good compatibility, moderate information capacity, simple and standard".

Taking one aircraft manufacturing company as an example, the statistics show that the aircraft material library is very large. Nowadays, there are more than 100,000 kinds of materials, among which are raw materials, non-metallic materials and composite materials, and so on. The aircraft materials can be classified in different ways based on the requirements. According to the production requirement, they are classified into three categories: large parts (including the fuselage, wing, engine hanger, etc.), aircraft related material, non-production aircraft material (including general consumables, tools, and work clothes, etc.). While in inventory management, the classification is: airborne system finished-products, airborne structural components, metal raw materials, parts, standard parts, auxiliary materials and flying materials (referring to the material used on the plane and being taken away on the plane).

The classification methods above are suitable for materials for aircraft manufacture needs, but they do not include all the materials used in the aircraft manufacturing process. The material code is applied to all production and inventory material in the material coding system. Therefore tools, equipments, fittings and others should be contained. Currently, a simple classification is the AB classified control method in the aircraft industry. They are divided into two major groups called A and B: group A is the kind of material for aircraft manufacture needs, such as raw material, parts, standard parts, airborne system finished-products, airborne structural components, and so on, and group B consists of auxiliary materials, including tools, equipment, office suppliers, labor protection, etc.

The main research focuses on a detailed classification of group A. Small classes sharing the same characteristics were grouped into one large class based on the analysis from the early chapter, with normal material classification standard. Thus, metal raw materials, chemical materials, parts and standard parts can fall into one class. Group A is divided into five classes: raw material, airborne system finished-products, airborne structural components, self-made parts and equipment delivered together with the aircraft from suppliers. The aircraft classification is shown in Fig. 2 as follows:

AIRCRAFT MATERIAL CODING STRUCTURE

The material coding structure is generally designed as subordinate relation or material attribute. Also, it can use drawing number as part of the code. The simplest way to do it is sequential coding, which uses Arabic numbers and/or Latin alphabet letters in sequential order as to identify material, such as the department code. The subornation coding has advantages such as clear structure relations, defined information and being easy to remember, but it does not work well to various types of material. Classification coding is easy to search and control, but hard to determine boundary classification. Sequence code is rarely used alone, because it fails in reflecting the relation between the materials. In recent years, with the increasing study on coding technology, it is

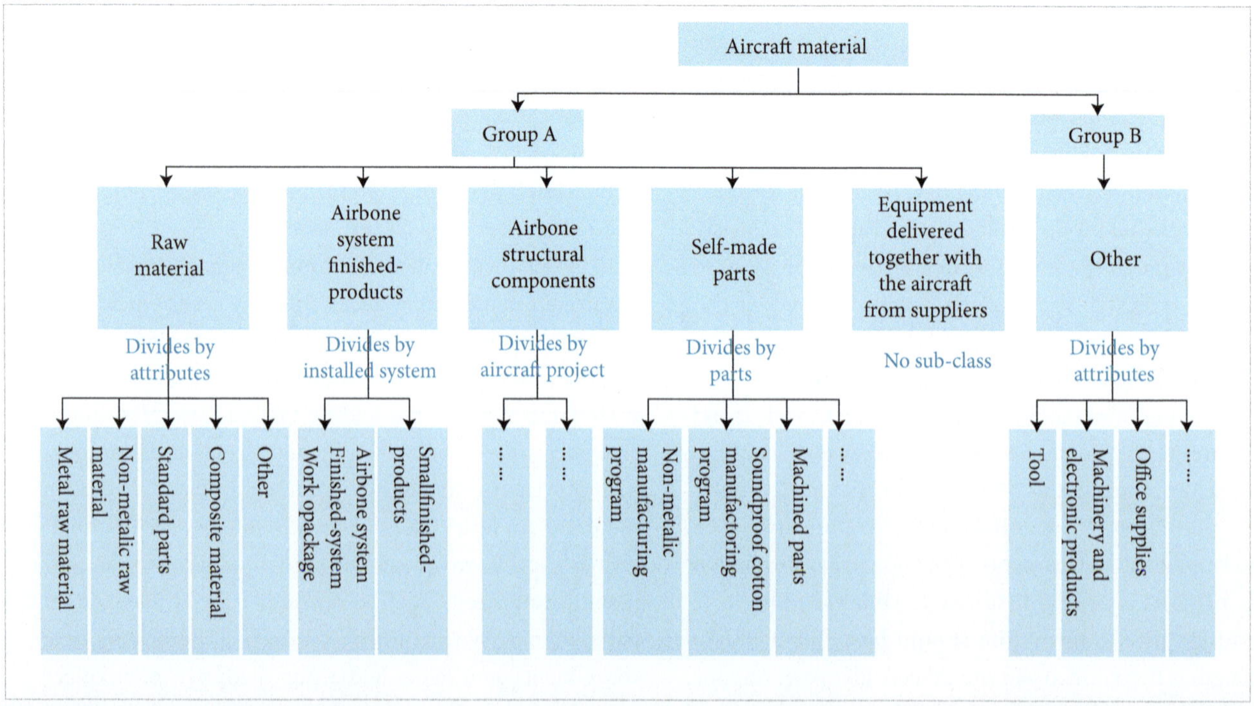

Figure 2. Aircraft material classification.

a trend that code structure is flexible in both code length and code layer relation.

There are some influence factors on flexible code structure. Jiang *et al.* (2008) evaluated the influence factors of material code structure, including code length, features description, structure stability, standard compatibility, organization rationality and structure uniformity, by using an analytic hierarchy process (AHP). The paper concluded that the material code structure should be designed based on the integration of these six factors. Considering one or a few of them and ignoring the other factors will lead to code structure instability. Therefore, the material code structure should be designed in accordance with the enterprise's material properties and its own requirements. It is not feasible to just copy other enterprises' coding system.

The most common material code structure is defined as: routine code, occupy code or flow code (Yi *et al.*, 2006; Zhao *et al.*, 2010). Based on the aircraft material classification and on essential factors for structure, the composite aircraft code structure is proposed, which consists of three kinds of code fields: class code, property code and flow code, as shown in Fig. 3.

The meta-model is composed of code fields, and the code field relation, in which the code field is the basic unit of code, is classified by the meaning of every code segment, and the code field relation defines the constraint relations among code fields. There are three kinds of code fields: class code, property code and flow code. Based on the classification coding principle, each class code, with the same code structure, has a strong correlation, and it includes class value and class name, which cannot be repeated and contains all material object. Property code is to encode material attributes. Flow code is to distinguish the same kind of material with a sequence of numbers. The initial value and step length of flow code is freely defined. Code fields are not independent and they have constraint among them. There are three relationships in our aircraft material coding structure. The parallel relation is simple, like the same level property code. The subordinate relation refers to generating material code in a certain order. The dependency relationship regulates the generation of code values. Direct dependence is where the front class code decides the later code value. Union dependence is that the flow code relies on the class code and on the property code. The property code depends on the class code, called extensive dependency relation.

The coding structure model is the basis of the code rules. This code structure model can meet the flexible coding requirements with high applicability.

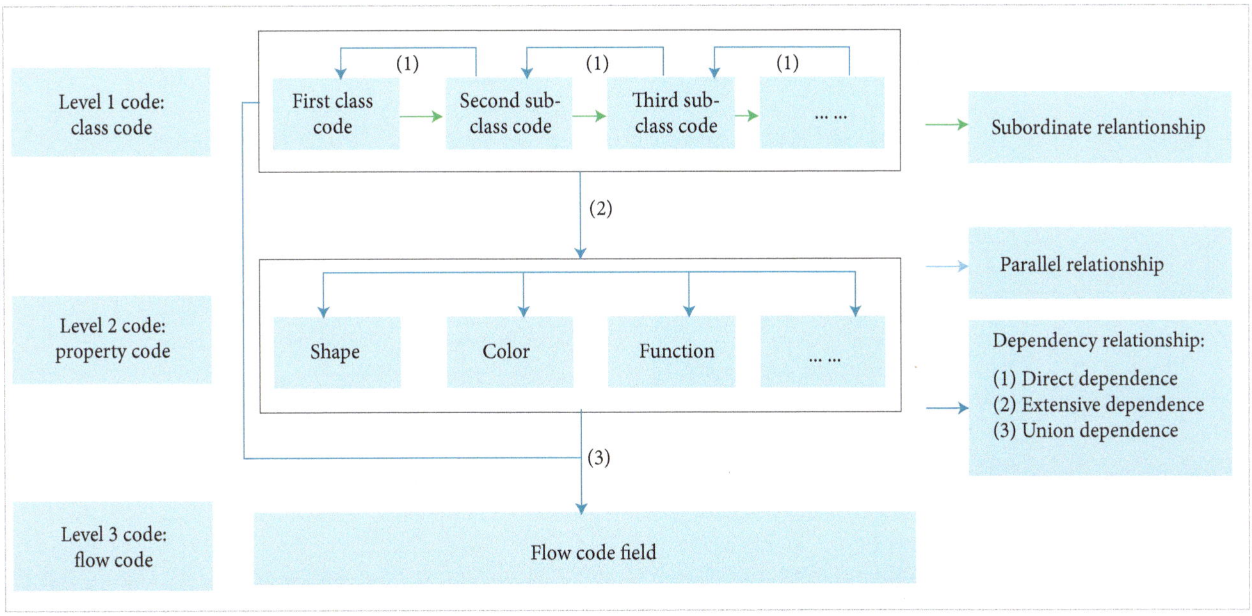

Figure 3. Material code meta-model.

AIRCRAFT MATERIAL CODING RULES

Material coding should follow some basic principles, such as uniqueness, integrity, stability, simplicity, etc. The aircraft material coding rules is established based on the material coding principles, in reference to the national relevant material coding's standard regulations, while considering both the characteristics of the aircraft industry and the aircraft material features.

Generally, there are three coded forms, the number code, the letter code, and the alpha-numeric code. The number and the letter codes have theirs virtues and shortcomings. The letter code has such advantages as explicit material meaning, convenience of recognition, being easy to remember, but complicated and heavy workload when the material library is large. The number code is easy to operate with, it has quick input capacity, and it obviously improves efficiency, however, it is difficult to identify. The hybrid code method, the alpha-numeric encoding form, takes the best of both approaches and is widely used. The aircraft materials are large and include very complex information. For convenient recognition and operation, the alpha-numeric combination code method is used, with numbers from 0 to 9 and letters from A to Z, in random permutation.

Based on the code structure and on the alpha-numeric coding form of aircraft material above, the material coding rule is proposed in Fig. 4.

In the aircraft material code rules, the first class code, the second sub-class code, and the last sub-class code length are two bits. One bit cannot satisfy coding requirements, and it represents poor extensibility. While three bits may bring ill consequences, like over long length and a waste of code space. Property code length is different, and we set it according to the different properties of the aircraft material. Flow code length is determined not only to meet the coding requirements, but also to consider the total code length. Code length with a big gap, will bring management trouble. In the first class code, a small number of equipments from the suppliers use the code "U8" to represent them. And other materials in group B are represented with the number "99".

The characteristics above about code rules are that all materials can be well indicated according to category, function, features and attributes. It is scientific and reasonable, avoiding material coding missing and ambiguity.

MATERIAL CODING SYSTEM

MATERIAL CODING SYSTEM MODEL

The material coding system mainly consists of three parts: definition of code rules, generation of code value, and code management.

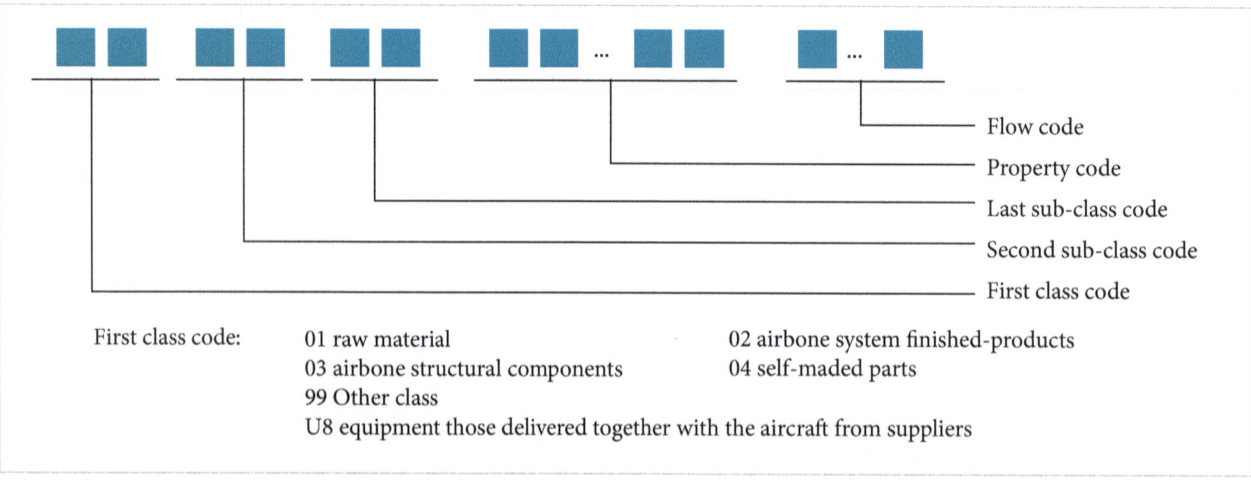

Figure 4. Aircraft material coding rules.

Based on the code structure meta-model, each code field and relation among them should be well-defined in the rule editor. The constraint relationships among code fields should be fixed. Those code rules are put in the code rule library.

Each code should be associated with a corresponding library code. When applying to generate material code, it selects the code rule accordingly, sends the code request to the code generator, and it generates an unique material code. As soon as the code is generated, it is stored in material code library, along with related material information.

Last, the system can enable operators to query or modify code and other management functions in order to maintain conveniece. The aircraft material coding system model is shown in Fig. 5.

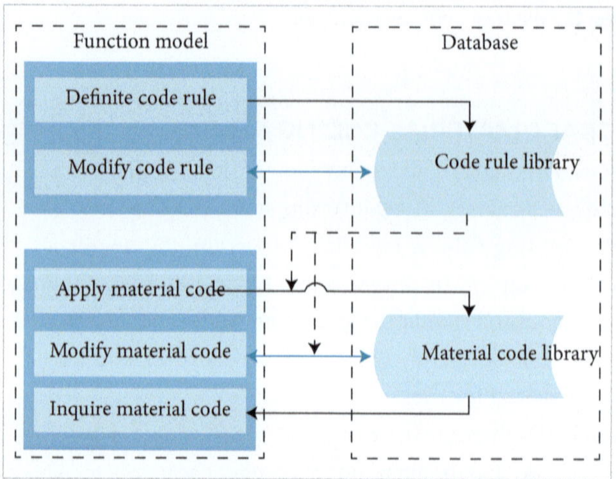

Figure 5. Aircraft material coding system model.

KEY CODE TECHNOLOGY

The key technology of designing coding system is the code metadata model. Based on the code structure meta-model, metadata information model is represented as follows:

The model consists of "Material Object", "Code Rule", "Function" and "Operation". "Material Object" refers to the material and its associated attributes. "Code Rule" includes all defined code rules. "Function" refers to the function of the relation between material and code rule. "Operation" is used to define the code rule's operations. In the model, code rule and code field include their own elements. Code Fields Constraint is the collection of code field relations. Based on the information model, we can build a good database to manage data. Then put it forward in the system processing flow chart in Fig. 6.

CODING SYSTEM IMPLEMENTATION

To have a better management of material coding in Enterprise Information systems for the aircraft manufacturing enterprise, the material code management system is developed, with the Java application platform and the Oracle database technology, based on the Browser/Server structure mode. The system could generate the material code automatically, inquire the material code, search material, and manage material classification. Some main functions and interfaces of the system are discussed below.

Material code application

Input material information and select material classification from the drop-down list shown in the page (Fig. 7). Use Ajax techniques to refresh only material classification items instead of

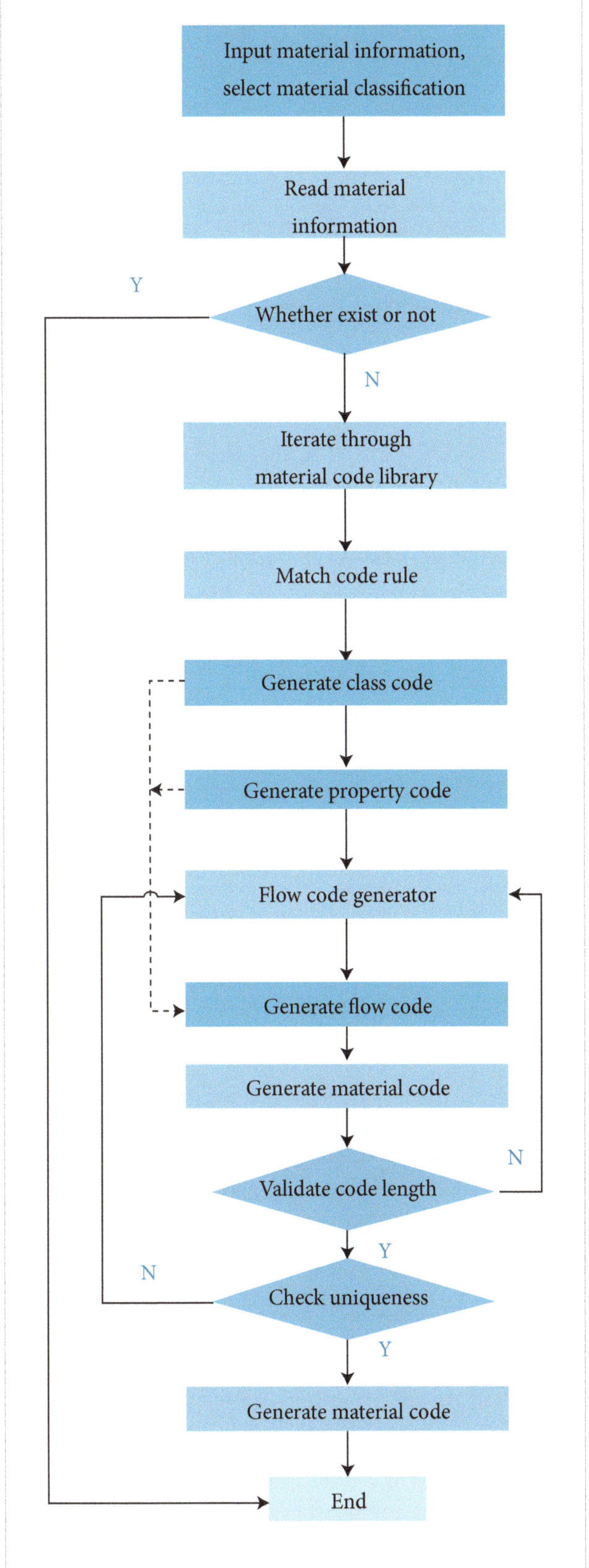

Figure 6. System processing flow chart.

the whole page. When the first class is selected, decide whether the sub-class exits or not, and then update dynamically and display sub-class items.

Material code query

Select the search criteria, such as "classification", "material code", "applicant", show the result in list (Fig. 8). The status is important, which shows the material code is either approved or waiting for approval. And the approver can edit it by clicking the button "approve" or "reject". Through the column "application path", the material classification is easy to be identified.

Material query

This search interface (Fig. 9) shows material information, including material code, name and so on. By selecting the search criteria, those materials which need to be managed and edited are found.

CONCLUSIONS

Material coding is of great significance for the enterprise information management. Taking the characteristic of aircraft materials into consideration, the paper studies material coding technology and proposes an aircraft material coding scheme. Firstly, it classifies aircraft material, secondly establishes code structure and code rule, and finally develops the material coding system which has been applied in the aircraft manufacturing enterprise and achieved good results. This study has greatly improved the coding efficiency and provided a classification method for aircraft material coding and powerful support in this aspect.

ACKNOWLEDGMENT

This work is supported by the Shanghai Science and Technology Committee under Grant No. 12dz1124300& 13521103604. The authors are grateful for the financial support, and they also would like to thank the anonymous reviewers and the editor for theirs comments and suggestions.

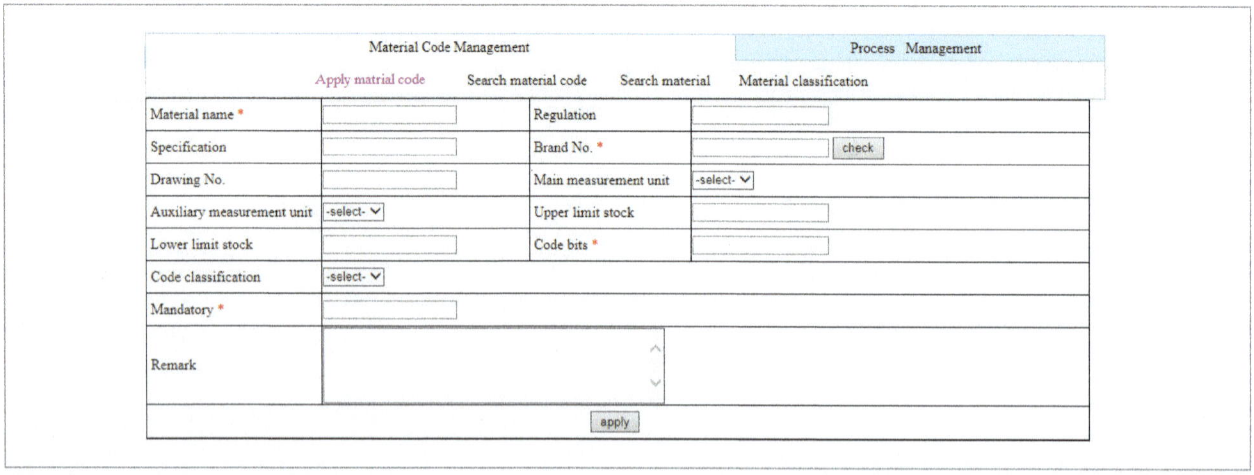

Figure 7. The material code application interface.

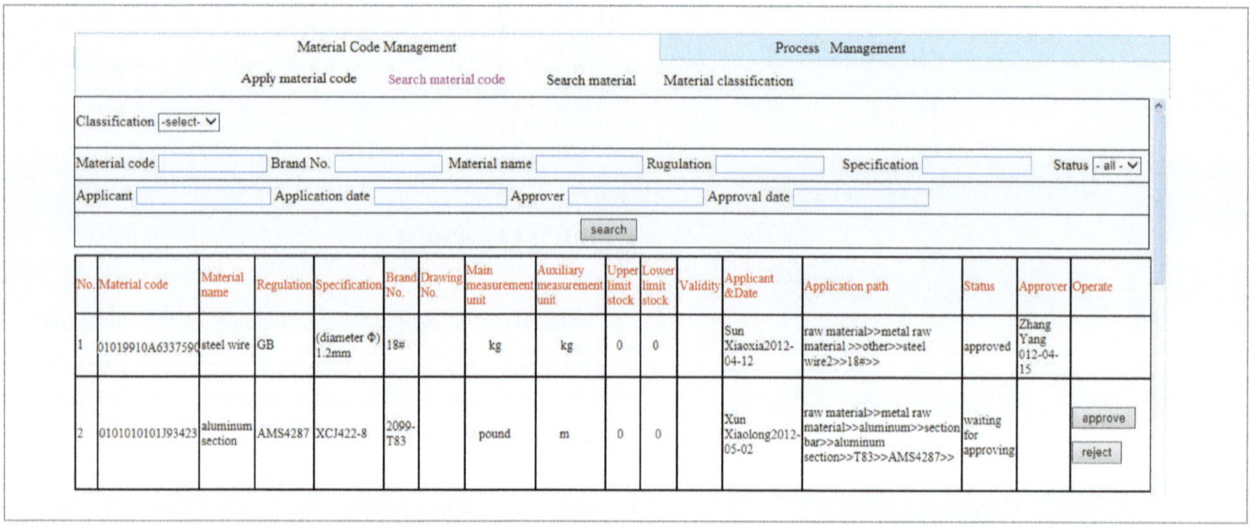

Figure 8. The material code query interface.

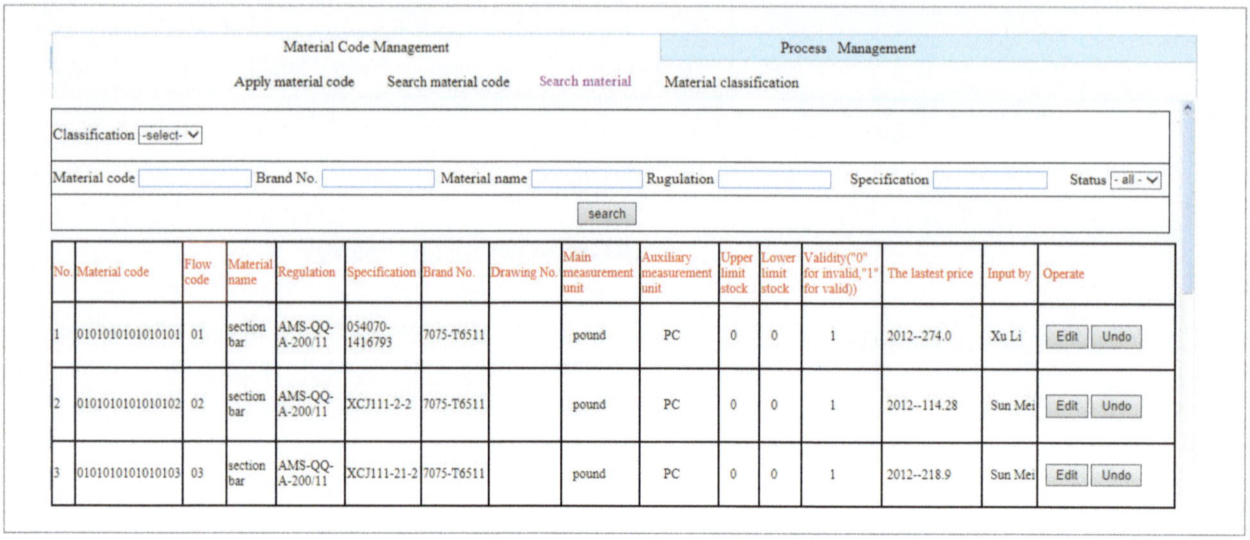

Figure 9. The material query interface.

REFERENCES

Jiang, J., 2007, "An Effective Method Based on Granularity-Structure for Organizing and Utilizing Manufacturing Information Resource", Journal of Northwestern Polytechnical University, Vol.25, No.2, pp. 245-250.

Jiang, J., Wang, J., Guo, L. and Liu, Y., 2008, "Research on Unified Coding Structural Model of Manufacture", Mechanical Science and Technology for Aerospace Engineering, Vol.27, No.1, pp. 13-18 (in Chinese).

Lei, W., Yang, C. and Li, Y., 2008, "The Material Classification Codes System", Group Technology & Production Modernization, Vol.25, No.1, pp. 60-63 (in Chinese).

Li, Y. and Xu, D., 2012, "Development of cutting tools information coding rules for the tools management system", Proceedings - ICIDT 2012, 8th International Conference on Information Science and Digital Content Technology, Vol.2, pp. 397-400.

Meng, X. and Kong, Y., 2013, "Material Classification and Coding based on K3 Manufacture Assistant Software", Shanghai Chemical Industry, Vol.38, No.2, pp.19-21 (in Chinese).

Wang, J. and Wang, B., 2008, "Research on Application of Ontological Information Coding in Information Integration", Global Design to Gain a Competitive Edge, pp.147-155. doi: 10.1007/978-1-84800-239-5_15.

Wang, K. and Tong, S., 2006, "Manufacturing Quality Information Classification based on Group Technology and Quality BOM", IMACS, Multi-conference on "Computational Engineering in Systems Application", pp. 2160-2167.

Wang, R., Zhou, L. and An, L., 2011, "Research on Flexible Classification and Coding Technology for Airplane Fixture Components", Mechanical Engineering & Automation, Vol.1, pp. 21-23 (in Chinese).

Xiong, G., Hu, L., Qin, T., Nyberg, T.R., Wang, F. and Shi Q., 2010, "Design and Improvement of the Material Coding Standardization for Power Group Enterprise", IEEE International Conference on Automation and Logistics, ICAL, pp. 597-602.doi: 10.1109/ICAL.2010.5585353.

Yi, J., Pan, P. and Dong, J., 2006, "Ontology- based PDM coding management middleware", Computer Integrated Manufacturing Systems, Vol.12, No.6, pp.1821-1826 (in Chinese).

Zhao, H., Liu, J., Dong, Y. and Xu, Z., 2010, "Research on Flexible Coding System Model for PDM System", Journal of Engineering Graphics, Vol. 31, No. 5, pp. 34-38 (in Chinese).

Simulation of Ablation in a Sounding Rocket Thermal Protection System Via an Interface Tracking Method with Two Moving Fronts

Humberto Araujo Machado[1]

ABSTRACT: In this work, an interface tracking method is employed to simulate the ablative process in the region near the stagnation point of the VSB-30 sounding rocket Thermal Protection System (TPS). The ablation model considers the presence of two-fronts, the char-melting and the pyrolysis fronts. The results for the proposed model are compared to the ones obtained from the traditional one-front model, which supposes a constant ablation temperature. Results show that the one-front model overestimates the ablation period, mass loss and the internal temperature after the flight. The increase in the accuracy with this model shall provide a better dimensioning of the TPS, reducing its weight and cost.

KEYWORDS: TPS, ablation, moving boundary, computational simulation, sounding rocket.

INTRODUCTION

Sounding rockets are extensively used to provide microgravity environments for scientific experiments. The Institute of Aeronautics and Space (*Instituto de Aeronáutica e Espaço* – IAE) of Brazil has designed, built and launched hundreds of rockets along the past 40 years. The VSB-30 sounding rocket is a two-stage unguided solid propellant rocket used by the Brazilian Space Agency (AEB) and European Space Agency (ESA).

Figure 1 shows a schematic representation of VSB-30. It has a total length of 13 m and a diameter of 0.6 m. It is equipped with two solid propellant motors, namely S31 and S30. S31 acts as a booster during its 15 s burning time, whereas S30 burns for about 30 s, reaching an apogee of 250 km, for a payload mass of 400 kg. VSB-30 was developed by IAE in cooperation with the German Space Agency (DLR). So far, several VSB-30 successful flights occurred, from Alcântara Launch Center (CLA), in Brazil, and from Esrange, Sweden (Garcia *et al.*, 2011), Fig. 2. During its flight, VSB-30 provides 6 minutes of microgravity environment. It is worth mentioning that an S30 motor developed and fabricated by IAE equipped the VS-30-Orion rocket used by Sharp Edge Flight Experiment (SHEFEX) (Turner *et al.*, 2005).

Figure 3 presents the nominal altitude-velocity map for a VSB-30 atmospheric flight. According to Fig. 3, VSB-30

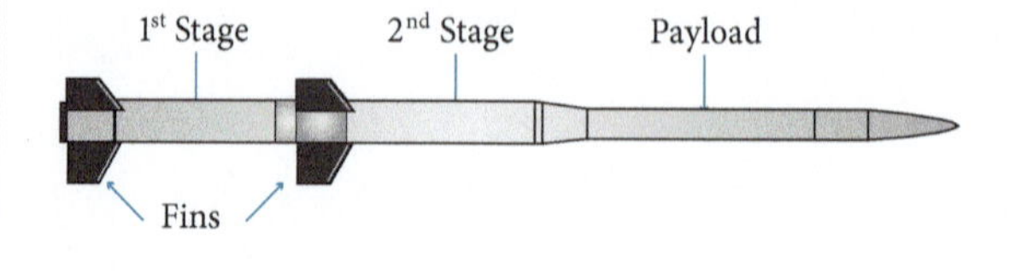

Figure 1. VSB-30 sounding rocket.

1. Instituto de Aeronáutica e Espaço – São José dos Campos/SP – Brazil

Author for correspondence: Humberto Araujo Machado | Praça Marechal Eduardo Gomes, 50 – Vila das Acácias | CEP 12.228-904 São José dos Campos/SP – Brazil | Email: humbaman@uol.com.br

Figure 2. VSB-30 takes off from Esrange, Sweden.

reaches the speed of 7,500 km/h (2,100 m/s) while still flying within the atmosphere. For practical purposes the dense part of the atmosphere is assumed to be 90 km thick. Beyond 90 km, vacuum conditions are assumed.

As a result of such very high speeds, aerodynamic heating arises as a major problem in the VSB-30 design. This problem is larger near the stagnation regions, such as those existing at the nose cap and the fins leading edges, where temperature surpasses 2,100°C (Machado and Pessoa-Filho, 2007). Indeed, it is mandatory to obtain an accurate evaluation of the reached heat flux and the temperatures, in order to correct size the thermal protection. Along the years ablative materials have been effectively used as TPS of space vehicles. This is the case of the nose cap of VSB-30, where a composite material, like Si-phenolic resin, is used (Da Costa *et al.*, 1996; Tick *et al.*, 1965), Fig. 4.

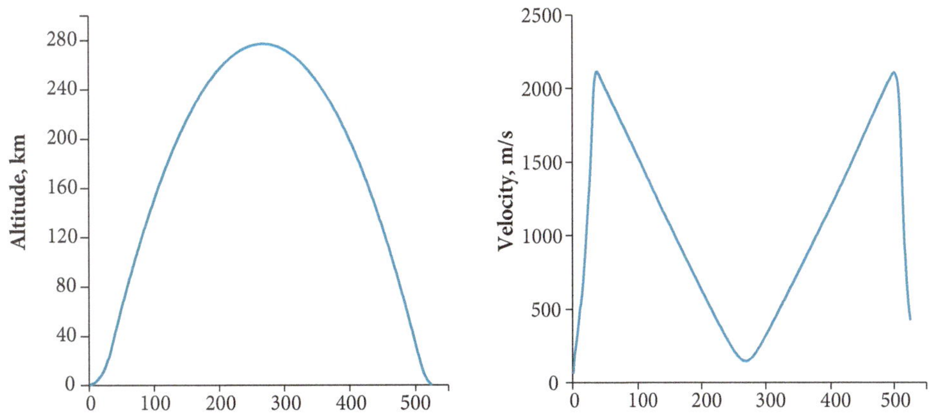

Figure 3. VSB-30 altitude-velocity maps.

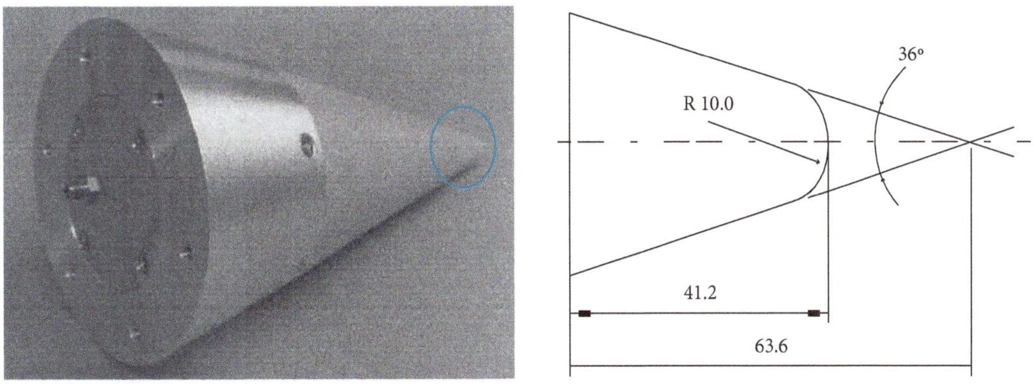

Figure 4. VSB-30 nose cap.

The conventional ablation model considers one front and treats ablation as single phase change process at constant temperature. In this work a model for the ablative process with two moving fronts (char-melting and pyrolysis) is applied in the computational simulation of the ablation in the thermal shield (TPS) of the VSB-30 sounding rocket, in the vicinity of the stagnation point. The objective is to compare the results of the two models, in order to evaluate the difference in design parameters, like external temperature and surface variation.

This work is an extension of the previous work of Machado (2012a), in which the validation of the two-fronts model and the numerical method for ablation can be found. In that work, the two-fronts model has been already compared with some experimental results, presenting a much better agreement than the one-front model. Such procedure will allow a more accurate dimensioning of the rocket thermal protection system, contributing for project optimization.

The interface tracking method proposed by Unverdi and Tryggvason (1992) was employed to solve the moving front problem. This method is based in a hybrid Eulerian-Lagragean approach that presents the advantage of using a regular homogenous mesh and represents the interface as accurate as more points are added to the grid. The equations of mass and energy conservation in their differential form are used to represent the thermal and ablative behavior of the heat shield with the time related to local temperature and thickness variations and solved through the finite volume method.

In the following sections, the physical problem and the method of solution are described, and the final results for the ablation are presented.

PHYSICAL PROBLEM AND MATHEMATICAL MODEL

AERODYNAMIC HEATING

To predict the heat transfer on VSB-30, it is necessary to know the pressure, temperature and velocity fields around the rocket. That can be accomplished by numerically solving the boundary layer equations. However, such a procedure is expensive and time consuming. In the present work a simpler, but reliable, engineering approach is used, which allows to obtain the convective heat transfer coefficient and the representative environment temperature for heat exchange without needing of a numerical solution of the boundary layer equations. The following simplifying assumptions are made:

- zero angle of attack;
- the rotation around its longitudinal axis is neglected;
- atmospheric air is considered to behave as a calorically and thermally perfect gas (no chemical reactions).

Although those hypotheses are not exact, they provide a good approximation to the real situation for the case of a ballistic reentry and for VSB-30 flight in particular, as shown by Machado (2012b). Such hypotheses should be revised in the case of planetary reentry.

The free stream conditions ahead of the nose cap are those given by v_∞, T_∞, p_∞, corresponding, respectively, to velocity, temperature and pressure. By knowing v_∞ and altitude, as function of time, together with an atmospheric model (U.S. Standard Atmosphere, 1976), it is possible to evaluate the free stream properties. For supersonic flow ($M_\infty>1$), a bow shock wave appears ahead of the nose. By using the normal shock relationships (Anderson Jr., 1989), it is possible to calculate v_1, T_1 and p_1 after the shock, in the centerline.

The heat flux over the external surface was calculated through the Zoby's method (Zoby *et al.*, 1981; Miranda and Mayall, 2001). This method employs the heat flux calculation at the stagnation point, obtained through the method of Cohen, as a starting point to integrate the temperature distribution inside the thermal boundary layer over the external surface. That distribution is computed by relating heat transfer to a skin-friction relation based on the momentum thickness through a modified Reynolds analogy form. Corrections for the compressibility (using the Eckert's reference enthalpy relation) and turbulence are added to the calculation. The velocity, temperature and pressure outside the boundary layer are assumed to be that estimated after the shock. The method can be employed in constant and variable entropy-edge conditions and both, reacting and non-reacting gas mixtures. Details of the solution can be found in the work of Machado (2008). Using the Zoby's method, the convective heat transfer coefficient, H, is calculated along the y-coordinate that is measured along the body's surface: $y = 0$ corresponds to the stagnation point, and R is a geometric parameter shown in Fig. 5, in which the red line represents the nose cap surface.

It should be pointed out that such a procedure is performed along the nose cap's surface (following the y-coordinate), for different trajectory times. Therefore, $H = H(y,t)$. The variations of the convective heat transfer coefficient and the adiabatic wall temperature (T_{aw}), also called recovery temperature (T_r), at stagnation point are shown in Fig. 6. An energy balance at the surface, accounting the radiative heat transfer, provides the heat absorbed by the wall.

HEAT CONDUCTION AND ABLATION

Once the convection heat transfer and the adiabatic wall temperature are known, wall temperature distributions can be determined solving the heat transfer equations. VSB-30 nose cap is covered with a composite material (Si-Phenolic resin), which works as an ablative TPS. Until the ablation temperature is reached, a transient heat conduction process occurs. Once the TPS surface reaches the ablation temperature, its thickness is reduced; therefore, a transient, coupled conduction, moving boundary problem appears. In this work, the ablation model proposed for a composite material will consider the presence of two layers: the virgin material and the char layer that appears after the pyrolysis front. Two moving fronts will be accounted for: the pyrolysis front and the ablation front of the char layer, both starting at constant temperatures.

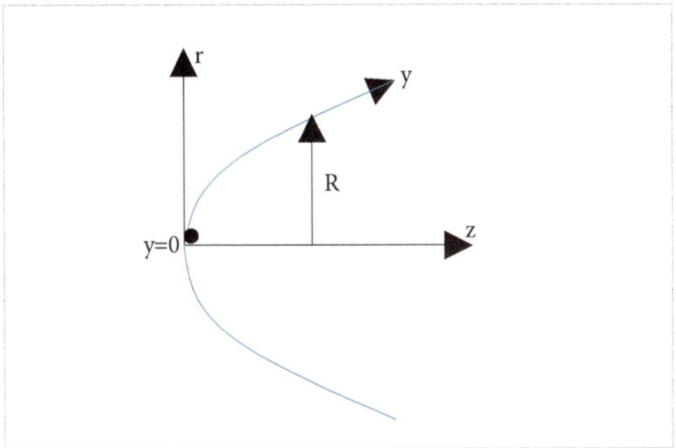

Figure 5. Coordinate system.

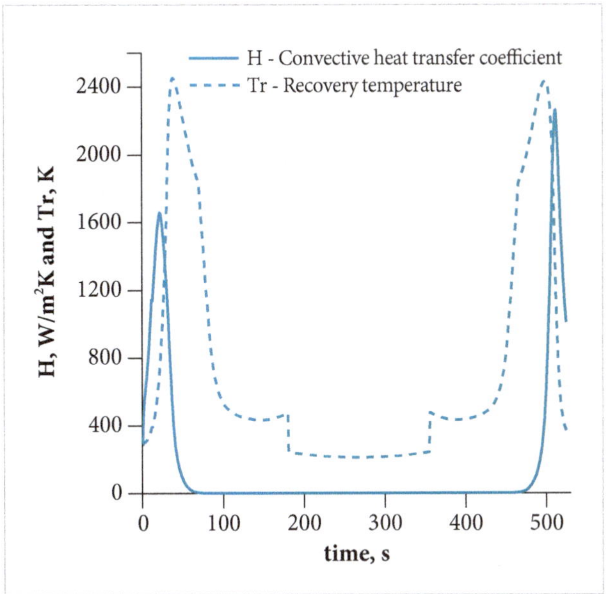

Figure 6. Recovery temperature and convective heat transfer coefficient at stagnation point, obtained through the Zoby's Method, during VSB-30 trajectory.

The set of equations used to represent the physical problem is written according to the interface tracking method (Juric, 1996). The nose cap and the surrounding airflow are represented as parts of a continuous domain of calculation. The application of the energy conservation principle to an infinitesimal volume element in the mathematical domain leads to a partial differential equation for the temperature, namely:

$$\frac{\partial(\rho.C_p.T)}{\partial t} = \nabla(K.\nabla T) + Q \tag{1}$$

where ρ is the density, C_p is the specific heat, T is the temperature, t is the time, K is the thermal conductivity and Q is a source term that accounts for the net heat exchange at the fronts, which will be mathematically represented as interfaces:

$$Q = \int_A q\delta(\boldsymbol{x} - \boldsymbol{x}_F)dA \tag{2}$$

where x is the position in the coordinate system, x_F is the interface position, A is the area, and q is the source term of energy per unit of surface of the interface and must be adapted to the physical model used to represent the behavior of each interface. The following hypothesis will be assumed to build the mathematical model for the ablative and heat conduction processes in the structure:

- solid materials are considered isotropic with constant properties;
- the pyrolisis zone is considered a zero thickness front. Pyrolisis enthalpy and temperature are considered constant;
- the char layer recession occurs through oxidation or sublimation, at constant temperature. The aerodynamic removing of material is neglected;
- absence of melting layer;
- full reaction of the gases and perfect mixing with the air in the boundary layer around the external surface, with negligible influence over the air physical properties;
- air is treated as an ideal gas;

- the flow field around the surface is not affected by the change in the surface geometry and gas injection;
- radiation is absorbed or emitted for surface, but not transmitted.

All these assumptions have being used in previous studies (Machado, 2012a), with successful results in representing the physical process. The second assumption, in particular, was studied by Sias (2009), which has concluded it was accurate enough, when compared to more complex models for ablation. All the chemistry occurring in the ablative process is implicit and is included in the ablation parameters (heat and temperature of pyrolysis). Although the hypothesis of perfect gas behavior for air might yield inaccurate results for the recovery temperature, it occurs in altitudes where the convective heat transfer coefficient approaches to zero, resulting in a negligible heat flux over the surface, and should not affect the ablation. According to these hypothesis, the heat balance in the external surface yields:

$$q = \rho LV + H(t,y)[T_F(t,x_F) - T_{aw}] + \varepsilon\sigma[T_F^4(t,x_F) - T_{aw}^4] \quad (3)$$

where **V** is the interface velocity, L is the heat of ablation of the char layer, H is the convection heat transfer coefficient, T_F is the interface temperature and T_{aw} is the adiabatic wall temperature, also called recovery temperature of the air, ε is the emissivity and σ is the Boltzman constant. One should note that this term might exist in every moving interface. In the pyrolisis front it is simplified, once there is no convection to or from the external flow and the radiative heat transfer is supposed to not occur between the layers (since there is no transmission):

$$q = \rho L_p V \quad (4)$$

In this case, L_p is the heat of pyrolisis. The flux of injection gases is also neglected due its low specific mass, when compared to the solid material. It is remarkable that the specific mass that appears in Eqs. 3 and 4 is the interface specific mass.

Although the airflow is included in the domain, its effects are implicit in the convection coefficient H. As a consequence, this region is considered adiabatic, and the heat capacity and thermal conductivity are assumed to be null. Once ablation temperature (T_A) is reached, the interface condition becomes:

$$T_F - T_A = 0 \quad (5)$$

A similar jump condition appears in the pyrolisis front (the pyrolisis temperature, T_p, replacing T_A).

METHOD OF SOLUTION

The moving boundary problem was solved by the Interface Tracking Method, introduced by Unverdi and Tryggvason (1992), and employed by Juric (1996) in the solution of phase change problems. In this method, a fixed uniform Eulerian grid is generated, where the conservation laws are applied over the complete domain. The interface acts as a Lagragean referential, where a moving grid is applied. The instantaneous placement of the interface occurs through the constant remeshing of the moving grid, and each region of the domain is characterized by the Indicator Function, which identifies the properties of the wall and the air around it.

This method allows for the representation of any geometry used in the TPS, and also the characterization of every layer separately. It is accomplished without a high increase in the computational cost and does not need any pre-processing (construction of unstructured grid or coordinate transformation). In this work, this method is employed to estimate the ablative performance of the TPS, considering a two-dimensional approach in both, the heat conduction and the moving boundary problem. Although here an axisymmetric simulation is performed, the method can be applied to any 2-D geometry. The method is detailed described in the previous work of Machado (2012a), and will be briefly summarized as follows.

The interface is represented as a parametric curve, R(u), where the normal and tangent vectors and curvature are extracted from. The interface points are interpolated by a 4th order Lagrange polynomial that uses two points before and two after the point to be interpolated, which allows one to analytically obtain the geometric parameters and remeshes the curve, keeping the distance d between curve points within the interval $0.9 < d/h < 1.1$, where h is the distance among the fixed grid points, as shown in Fig. 7.

The phase distribution in the domain of calculation is represented with the Indicator Function, $I(x,t)$. For the simple case of only two phases, it varies from 0 (air) to 1 (solid), and it is numerically constructed using the interface curve to determine a source term G(x). The jump of the indicator function across the interface is distributed over the fixed grid points, yielding a gradient field in the mesh:

$$G(x) = \nabla I = \int_A n\delta(x - x_f)dA \quad (6)$$

where n is the normal unitary vector, and the term $\delta(x - x_f)$ should be zero, except over the interface, as represented by the Dirac delta function, δ. However, such a representation is not convenient for a discrete number of points, since the Lagrangian points, x_k, do not necessarily coincide with the Eulerian grid points, $x_{i,j}$. The Distribution Function is used to represent the interface discontinuity as a continuous and smooth function. Such a function is similar to a Gaussian distribution function and its value depends on the distance $|x_{ij} - x_k|$ between the Lagrangean and Eulerian points:

$$D_{i,j}(\boldsymbol{x}_k) = \frac{f[(x_k - x_i)/h].f[(y_k - y_j)/h]}{h^2} \quad (7)$$

where D_{ij} is the Distribution Function for a point k in the Lagrangean mesh with respect to a Eulerian point (x_i, y_j). One should note that increasing h results in a thicker interface. The function f is the probability distribution related to the distance h as:

$$f(x) = \begin{cases} f_1(x) & \text{if } |x| \leq 1 \\ 1/2 - f_1(2 - |x|) & \text{if } 1 < |x| < 2 \\ 0 & \text{if } |x| \geq 2 \end{cases} \quad (8a)$$

$$f_1(x) = \frac{3 - 2|x| + \sqrt{1 + 4|x| - 4x^2}}{8} \quad (8b)$$

The divergence of the gradient field is found by numerical derivation of Poison's equation:

$$\nabla^2 I = \nabla.\boldsymbol{G} \quad (9)$$

Despite being considered constants in each phase, the properties inside the domain must be treated as variable in the formulation. A generic property ϕ (ρ, Cp or K) is expressed as:

$$\phi(x) = \phi_0 + (\phi_1 - \phi_0)\, I(x,t) \quad (10)$$

where f_0 and f_1 are the property values for phases 0 and 1 (according to their values of indicator function), respectively.

The coupling between the moving mesh and the fixed grid is done at each time step, through the Distribution Function, used to represent the source terms in the balance equations and to interpolate the infinitesimal discontinuities into a finite thick region at the interface. In a similar manner, this function is used to interpolate the field variables from the Eulerian grid to the interface. The equations used to distribute the source term in the field and interpolate variables to the interface are:

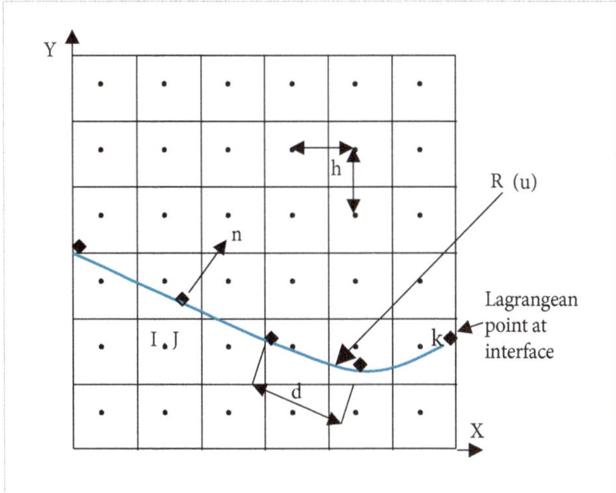

Figure 7. Eulerian and Lagrangean meshes.

$$Q_{i,j} = \sum_k q_k D_{i,j}(x_k) \Delta s_k \quad (11)$$

$$T_k = \sum_{i,j} h^2 T_{i,j} D_{i,j}(x_k) \quad (12)$$

where Ds_k is the average of the straight line distances from the point k to the two points on either side of x_k, and corresponds to the area concerned to the point x_k in the interface surface. Equation 11 is the dicretized form of Eq. 2, where the Dirac delta function was replaced by the distribution function, $D_{i,j}$, which is also done in Eq. 6.

The initial interface shape, $R(u)$, is first specified and then the Indicator Function is constructed. From the initial conditions, the property and temperature fields are determined. Out of the ablative period, the interface temperature keeps bellow the ablation temperature, and the energy equation is solved as a pure heat conduction problem, via the Finite Volume Method, employing the well-known FTCS discretization in an explicit time marching schedule (Patankar, 1980).

As the interface reaches the ablation temperature at a given point, an iterative process starts up, in order to determine the interface velocity at each time step, which must satisfy the temperature condition, Eq. 5, at that interface point. The process goes on as far as the point temperature is equal to ablation temperature. The steps to be followed are:

1. using the current value of interface velocity, V^n (result from the previous time step), the interface points are transported to a new position, calculated explicitly through the equation $V^n = (dx_f/dt).n$, where n is normal unitary vector;
2. density and specific heat are calculated at the new interface position;
3. V^{n+1} is estimated via Newton iterations, using a numerical relaxation schedule, Eq. 13;
4. heat flux q crossing the interface is calculated through Eqs. 3 or 4, depending on the interface, and distributed into the fixed grid through Eq. 11;
5. according to the boundary conditions, energy equation, Eq. 1, is used to obtain the temperature at time step $n+1$;
6. temperature is interpolated to find T_F at the interface, using Eq. 12;
7. the temperature jump condition is tested and if it is lower than the reached tolerance, the fields of viscosity and conductivity are updated for the new position, and one step in time is advanced. If that is not the case, a new estimate for V^{n+1} is calculated and the process returns to step 5.

The convergence criterion used in step 7 is the residual in Eq. 5. Once it has reached the desired tolerance, convergence for interface velocity is assumed. Otherwise, the velocity is corrected via Newton Iterations, given as:

$$V^{n+1} = V^n - \omega.R(T) \tag{13}$$

where ω is a relaxation factor and $R(T)$ is the residual for the temperature jump condition at the interface. Iterations are repeated until $R(T)$ in every point become smaller than the prescribed tolerance. The optimum value for ω is found by numerical experimentation, at the beginning of the calculation. The method was compared with the analytical solution for a simple phase change problem resulting in an excellent agreement (Ruperti Jr., 1991).

In the case of more than one interface, an Indicator Function, I_m, is created for each interface, in order to characterize every region concerned to the interfaces individually. Therefore, in a region m (that corresponds to a specific phase or material), a generic property is estimated as:

$$\phi = \sum_{m=1}^{NFC} \phi_m Ig_m \tag{14}$$

where NFC is the number of interfaces. The Lagrangean grids for all interfaces have the same values for the parameters h and d, shown in Fig. 1, and are constructed from a particular parametric curve $R_m(u)$. Ig_i is the Global Indicator Function for a region m, obtained from the Indicator Function of each interface (calculated as described before). It is given as:

$$Ig_m = I_{m-1} - I_m \tag{15a}$$

$$I_0 = 1 \,;\, I_{NFC} = 0 \tag{15b}$$

If there are more than one moving interface, the source term Q_m for every interface has to be extracted from a modified form of Eq. 2:

$$Q_m = \sum_k \int_A q_m \delta_m (x - x_{Fm,k}) dA_m \tag{16}$$

Actually, according to the numerical method, Eq. 11 will be used to calculate the source term in every interface. The total amount of heat generated will be the summation of the heat sources of all interfaces:

$$Q = \sum_m Q_m \tag{17}$$

The convergence criterion and velocity correction are the same as those for the case of one interface, but they are extended to all interfaces at each time step.

RESULTS

The results were obtained for the region near the stagnation point of VSB-30, Fig. 8, which corresponds to a circular semi-sphere with radius of 280 mm. Note that the Y-coordinate has a different meaning of y-coordinate shown in Fig. 3 (y). Since the flight is considered with zero angle of attack, the problem is considered to be axy-symmetric, and only the half of that region has to be simulated. A 20x20 points grid over a domain of 12x12 mm was employed to simulate the heat transfer and moving boundary problem, with a tolerance of 10^{-6} for the residual in Eq. 5. A resulting 26 points Lagrangean mesh was obtained for the interface used to represent the external surface.

The convergence of the domain representation by the Indicator Function is shown in Fig. 9, for a cutting view of the domain in the diagonal starting from the axes origin.

The more points are added to the mesh the more the thickness reduces. When the number of points reaches 20x20, the interface representation is assumed to be satisfactory, since the rise of the processing time would not compensate the correspondent reduction of interface thickness.

Figure 10 shows the final interface position according with the number of mesh points used, where the results for 15x15 and 20x20 meshes are very close. In Fig. 11, the variation of external temperature with time at the stagnation point shows an even better convergence between these two meshes. Considering these results, the 20x20 mesh was considered refined enough for the purposes of this study.

The results were compared with those from the one-front model (Machado, 2008), using the same numerical parameters. Due to the difference between the models, the composite material (Si-phenoic resin) was characterized in a different way. The properties of resin employed in each model are shown in Table 1.

Figure 12 shows the characterization of every region in the domain through the global indicator function, at $t = 0$. In this case, the two interfaces corresponding to the pyrolisis and carbonization fronts are placed at the same initial position, over the external surface, since the ablative process did not start yet. The colored region corresponds to the interfaces and do not represent exactly the discontinuity, presenting a slope and a finite thickness (about 0.6 mm). This thickness could be reduced through the increase of the number of grid points, which was not considered necessary, according to the convergence analysis.

Figure 13 shows the 3-D view of temperature distribution at various moments. Temperature peaks occur in the external surface, in direct contact with the heated air, during the ascension, at 35 seconds (Fig. 13a) and at 500 seconds during the reentry (Fig. 13c). Between these times, at 100 seconds (Fig. 13b) and after the reentry heating, at 524 seconds (Fig. 13d), a surface cooling occurs. In this case, the external surface becomes a heat sink and yields a "valley" in the temperature profile. One can observe that the temperature of fusion of the char is not reached. Indeed, this surface does not move, since there is no phase change and only the pyrolysis front moves when the pyrolisis temperature of the resin is reached. The region correspondent to the air does not present relevant temperature changes, once it is considered to be adiabatic with zero thermal capacity. Temperature distribution and the external surface position correspondent to these profiles are shown in Fig. 14.

The results were compared to those obtained through the one layer model, where ablation is treated as single-phase change process. Figure 15 shows temperature of internal and external surfaces, both at R=0, along the symmetry axis (at Y=10 mm and Y=0, respectively), what corresponds to the stagnation point line. The two temperature peaks correspond to the ascension and reentry of the vehicle in the atmosphere, including a period between then where a cooling occurs, due the heat losses by radiation, in the absence of convection (H=0, according to Fig. 4). It is noticeable that the ablative process begins when the external surface reaches the pyrolisis temperature, and the ablation front starts to move. The temperature of fusion of the char is not reached at both peaks, even in the points of maximum temperature. For the one layer model, the temperature of ablation is lower, what means the process begins earlier and takes more time. Due the presence of the char layer in the two-fronts model, the temperatures reached in the internal surfaces are lower, what

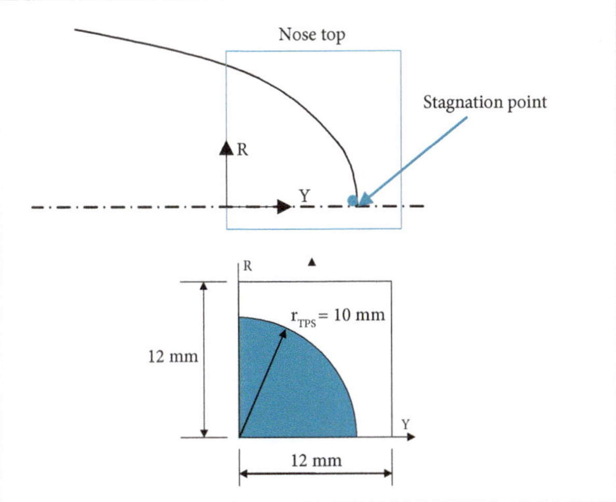

Figure 8. Domain of calculation, regions (TPS in blue) and dimensions.

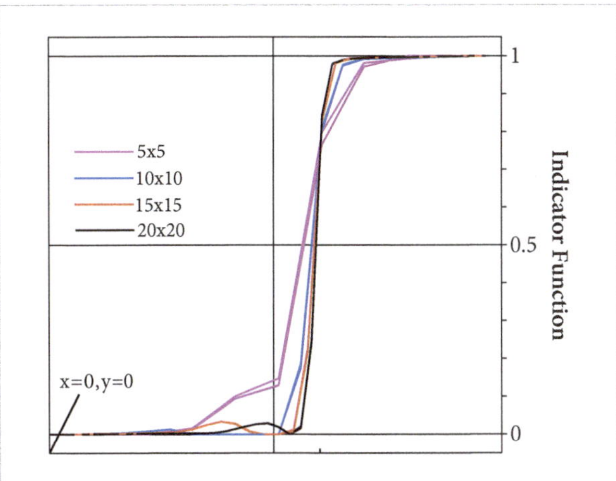

Figure 9. Indicator Function profile observed in the diagonal trough the axes origin, for diverse number of points used in the regular mesh.

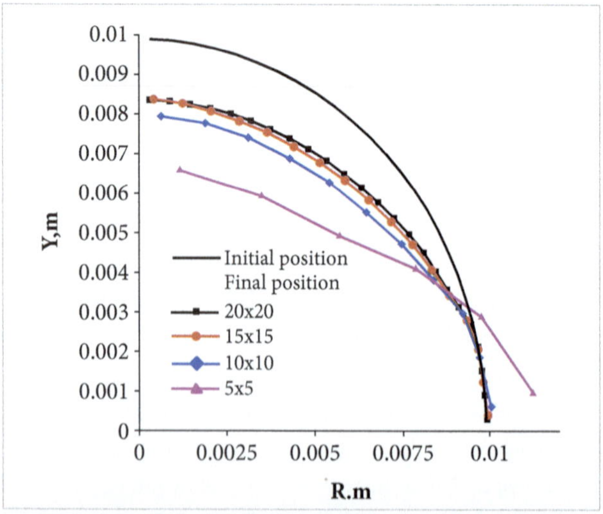

Figure 10. Interface position for the virgin material at the final time, for diverse number of points used in the regular mesh.

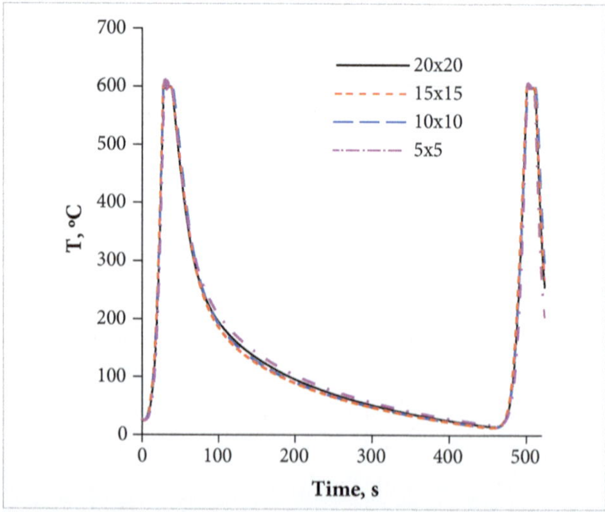

Figure 11. External surface temperature with time, for diverse number of points used in the regular mesh.

Table 1. Properties of Si-phenolic resin.

Property	One-front model	Two-front model	
		Virgin material	Char
Thermal conductivity (W/m°C)	0.485[1]	0.485[1]	0.428[2]
Specific heat, Cp (J/kg °C)	1,256[1]	1,256[1]	879.5[2]
Specific mass, ρ (kg/m³)	1,730[1]	1,730[1]	1,300[2]
Emissivity, ε	0.8[1]	0.8[1]	0.8[3]
Heat of ablation (MJ/kg)	12[4]	–	–
Temperature of ablation (°C)	538[4]	–	–
Heat of pyrolisis (MJ/kg)	–	0.78[5]	–
Temperature of pyrolisis (°C)	–	599[5]	–
Heat of fusion/Sublimation (MJ/kg)	–	–	10.5[3]
Temperature of fusion (°C)	–	–	3,700[3]

[1]Da Costa *et al.* (1996); [4]Gregori *et al.* (2008); [5]Tick *et al.* (1965); [2]Williams and Curry (1992); [3]Sawatimskii (2003).

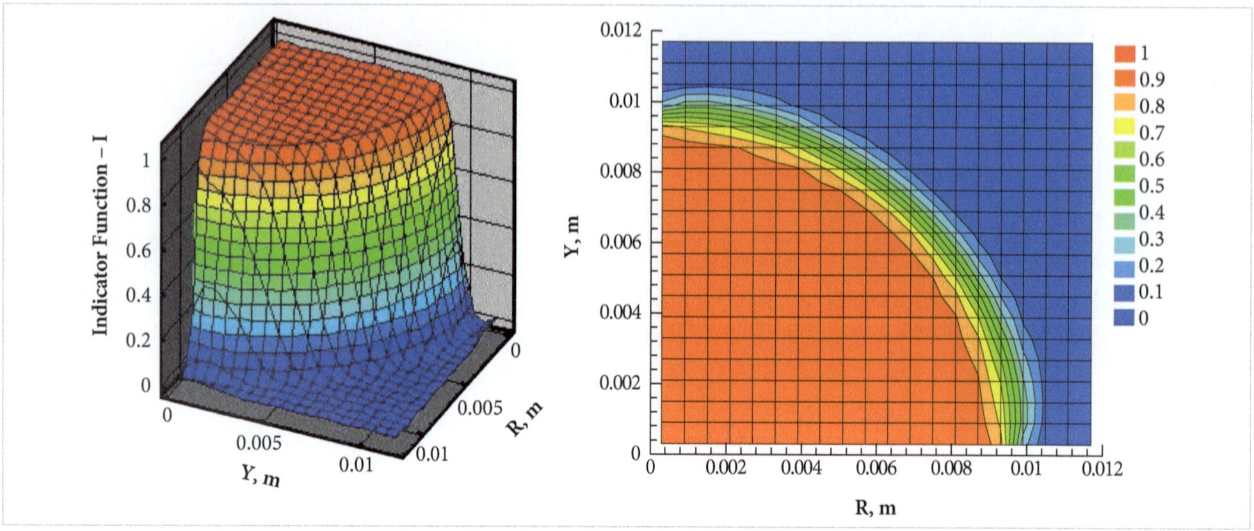

Figure 12. Global Indicator Function for every region: Thermal Protection System in red and air in blue.

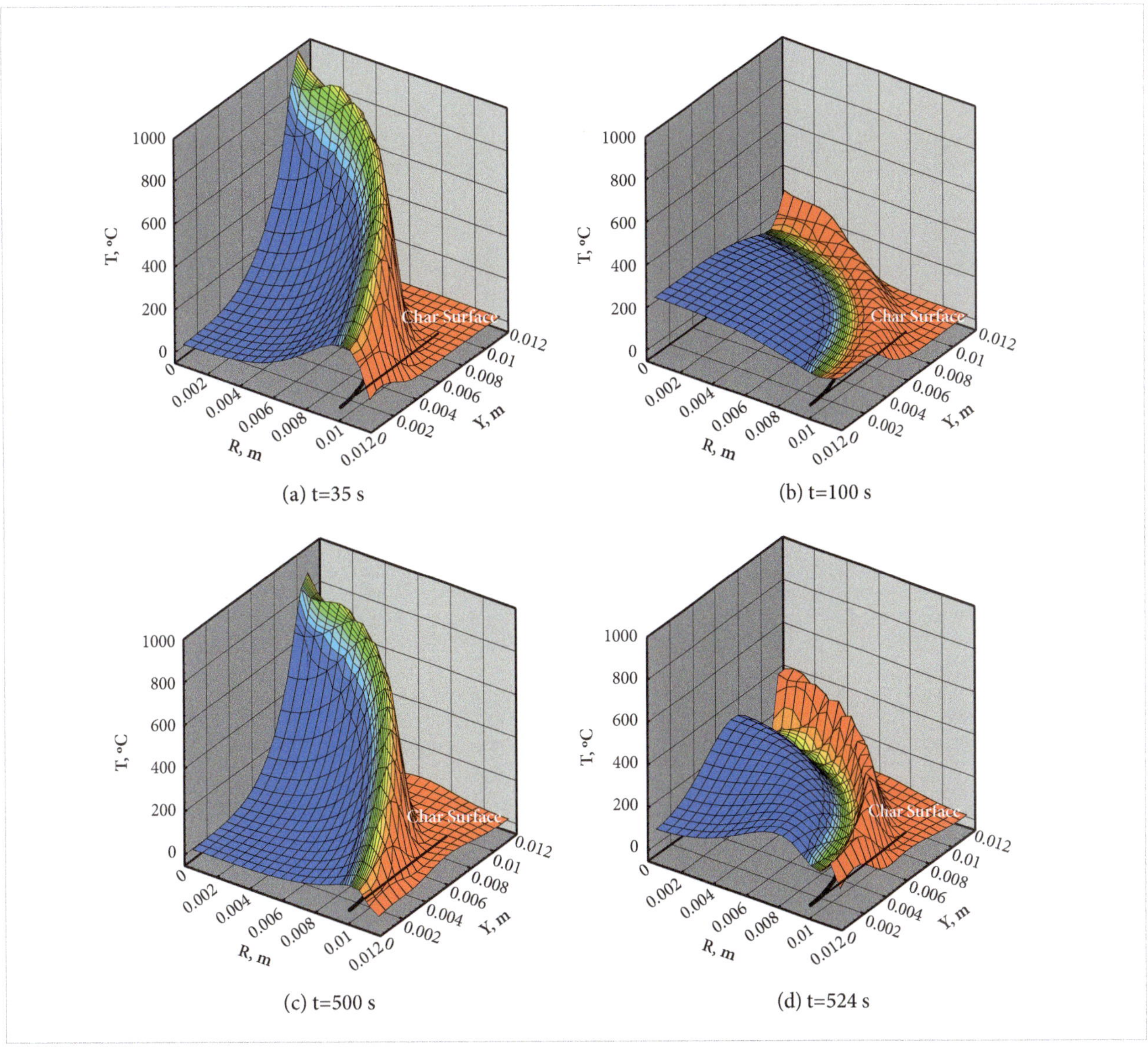

Figure 13. 3-D view of temperature field during the trajectory, with the Global Indicator Function for every region: Thermal Protection System in blue and air in red.

indicates that this layer works as a thermal barrier for the virgin material of the TPS.

Figure 16 shows the interface position with the time, after the two ablation periods (ascension and reentry). According to the results, the ablation is more intense closer the stagnation point for both models. However, the two-fronts model indicates a lower material consumption than the one layer model. The initial position of the external surface keeps constant in the two-fronts model, since the temperature of fusion/sublimation of the char is not reached. The displacement starts after the formation of the pyrolisis front (considered to be initially at 0.1 mm after the external surface, for calculation purposes).

CONCLUSION

In this work, the two-dimensional transient aerodynamic heating and ablation processes in the vicinity of the stagnation point of the VSB-30 sounding rocket TPS were simulated through an interface tracking method considering the presence of two moving fronts, the pyrolisis and the carbonization fronts, and two resulting layers, the virgin material and the char layer. Preliminary results demonstrated that the method is able to capture the temperature peaks and to represent the ablation process as a moving boundary problem, in the presence of more

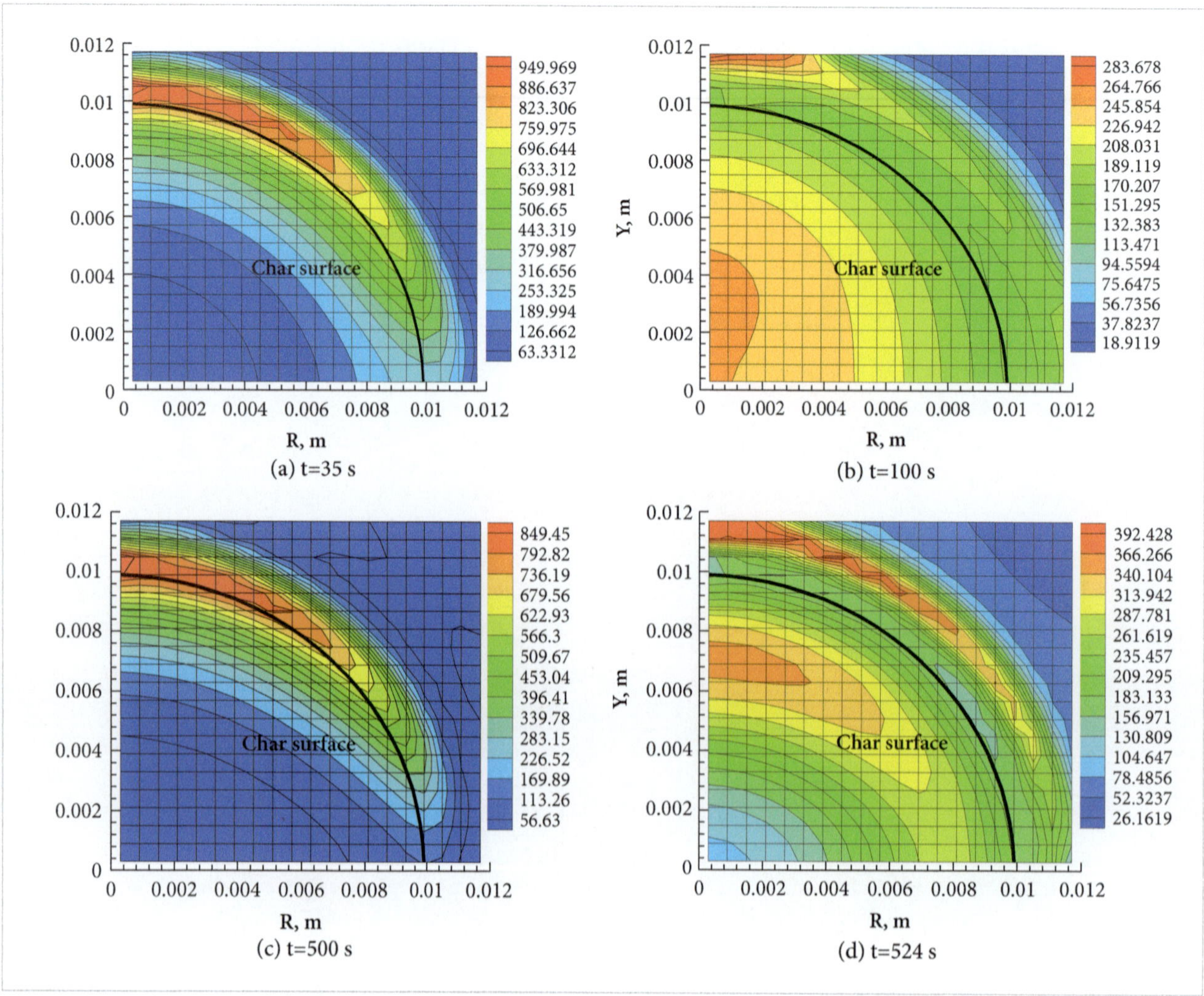

Figure 14. Temperature distribution during the trajectory.

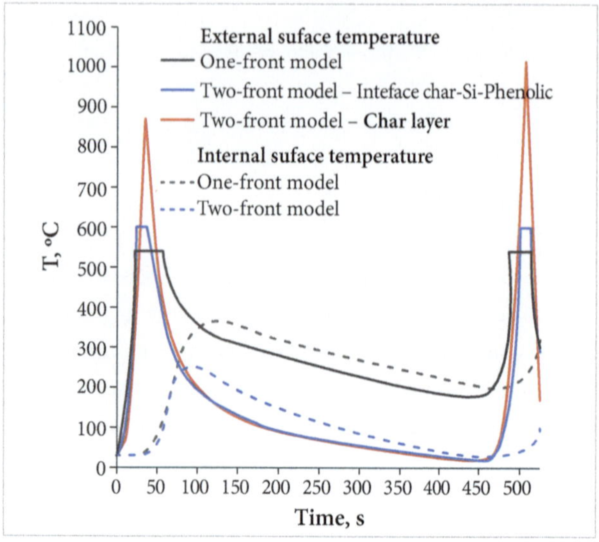

Figure 15. Comparison between the results of the ablation models for temperatures in the stagnation point, *R*=0.

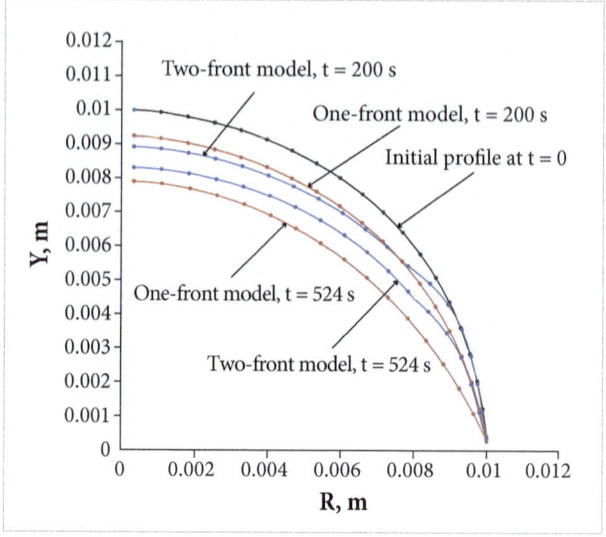

Figure 16. Comparison between the results of the ablation models for interface position with time.

than a single moving front, which allows representing diverse physical simultaneous processes.

When compared with the one-front model, it results in lower temperatures out of the periods of heating, shorter periods of ablation and less consumption of protective material. This analysis can be extended to more regions of the rocket, more layers and other shapes, including more moving fronts, if it is necessary. A more realistic physical model for the ablation in the composite material may now replace the one used in this work. The inclusion of the flow field effects, like injection of mass due to sublimation, shall also be incorporated into the simulation.

Since the two-fronts model has presented better accuracy than the one-front model, as shown in previous studies already mentioned, it seems to be a better option to estimate TPS performance for other vehicles and might provide a more accurate dimensioning and consequently reducing its weight and cost.

ACKNOWLEDGMENTS

The author would like to thank the Conselho Nacional de Desenvolvimento Científico e Tecnológico (CNPq), the Brazilian Federal Agency for Science and Technology, for the financial support during this work.

REFERENCES

Anderson Jr., J.D., 1989, "Hypersonic and High Temperature Gas Dynamics", McGraw-Hill, Blacklick, Ohio, USA.

Da Costa, L.E.V.L., De Mello, F.C. and Pardini, L.C., 1996, "Viability Study of Thermal Protection for SARA Platform", IAE/CTA, Technical note NT-130-ASE-N/96, IAE/CTA, São José dos Campos, São Paulo, Brazil (in Portuguese).

Garcia, A., Yamanaka, S.S.C., Barbosa, A.N., Bizarria, F.C.P., Jung, W. and Sheuerpflug, F, 2011, "VSB-30 Sounding Rocket: History of Flight Performance", Journal of Aerospace Technology and Management, Vol. 3, No.3, pp. 325-330. doi: 10.5028/JATM.2011.03032211

Gregori, M.L., Barros, E. de A., Petraconi Filho, G., Costa, S.F. and Pardini, L.C., 2008, "Properties of Quartz-Phenolic Composites for Thermal Protection Systems", 59th IAC Congress, Glascow, Scotland.

Juric, D., 1996, "Computations of Phase Change," Ph.D. thesis, University of Michigan, Michigan, 166 p.

Machado, H.A. and Pessoa-Filho, J.B., 2007 , "Aerodynamic Heating on VSB-30", ESA Symposium, Visby, Sweden.

Machado, H.A., 2008, "Two-dimensional Simulation of Ablation due to Aerodynamic Heating in a Sounding Rocket", 40th AIAA Thermophysics Conference, Seattle, USA.

Machado, H.A., 2012a, "Simulation of Ablation in a Composite Thermal Protection System via an Interface Tracking Method", Journal of Aerospace Technology and Management, Vol. 4, No. 3, pp. 331-340. doi: 10.5028/jatm.2012.04030312

Machado, H.A., 2012b, "After-Flight Thermal Analysis of the VSB-30 Sounding Rocket Payload", VII National Congress of Mechanical Engineering – CONEM, São Luís, Maranhão, Brazil (in Portuguese).

Miranda, I.F. and Mayall, M.C de M., 2001, "Fluxo de Calor Convectivo em Micro-Satélites em Reentrada atmosférica", Graduate Dissertation, Instituto Tecnológico de Aeronáutica, São José dos Campos, São Paulo, Brazil, 66 p.

Patankar, S.V., 1980, "Numerical Heat Transfer and Fluid Flow", Taylor & Francis, New Delhi.

Ruperti Jr., N.J., 1991, "Solution of a One-Dimensional Ablation Model", Master thesis, Instituto Nacional de Pesquisas Espaciais, São José dos Campos, São Paulo, Brazil (in Portuguese).

Savvatimskii, A.V., 2003, "Melting Point of Graphite and Liquid Carbon", Physics-Uspekhi, Vol. 46, No. 12, pp. 1295-1303. doi:10.1070/PU2003v046n12ABEH001699

Sias, D.F., 2009, "Hibrid Solutions for Heat Transfer on Ablative Thermal Protection Systems", Master thesis, Universidade Federal do Rio de Janeiro, Rio de Janeiro, Brazil (in Portuguese).

Tick, S.J., Huson, G.R. and Griese, R., 1965, "Design of Ablative Thrust Chambers and Their Materials", Journal of Spacecraft and Rocket, Vol. 2, No. 3, pp. 325-331. doi: 10.2514/3.28179

Turner, J., Hörschgen, M., Turner, P., Ettl, J., Jung, W. and Stamminger, A., 2005, "SHEFEX – The Vehicle and Sb-Systems for a Hypersonic Re-entry Flight Experiment", 17th ESA Symposium in European Rocket and Balloon Programmes and Related Research, Sandfjord, Norway.

U.S. Standard Atmosphere, Washington, D.C., U.S. Government Printing Office, 1976.

Unverdi, S.O. and Tryggvason, G., 1992, "A Front-Tracking Method for Viscous, Incompressible, Multi-fluid Flows", Journal of Computational Physics, Vol. 100, No. 1, pp. 25-37.

Williams, S.D. and Curry, D.M., 1992, "Thermal Protection Materials – Thermophysical Property Data", NASA Reference Publication 1289.

Zoby, E.V., Moss, J.N. and Sutton, K., 1981, "Approximate Convective Heat Equations Hypersonic Flows", Journal of Spacecraft and Rockets, Vol. 18, No. 1, pp. 64-70.

Development Status of L75: A Brazilian Liquid Propellant Rocket Engine

Daniel Soares de Almeida[1], Cristiane Maria de Moraes Pagliuco[1]

ABSTRACT: This paper provides an overview of the design and development process of the L75 Liquid Propellant Rocket Engine (LPRE). Being developed at Instituto de Aeronáutica e Espaço (IAE), it is the first Brazilian open-cycle liquid rocket engine pressurized by turbopump, designed to deliver 75 kN of thrust in vacuum as a cryogenic upper stage engine using liquid oxygen and ethanol. The Preliminary Design Review (PDR) was accomplished in December 2011 and December 2012 the project received from *Agência Espacial Brasileira* (AEB) a financial support, through an agreement with *Fundação de Desenvolvimento da Pesquisa* (FUNDEP), to proceeding with the development of models and tests. The main components of the engine, briefly described here, are thrust chamber assembly, gas generator, turbopump assembly, control system and ignition system.

KEYWORDS: Liquid propellant rocket engine, Liquid propulsion, Ethanol.

INTRODUCTION

The knowledge already acquired on technology of solid propellant rocket motors enables Brazil to have propulsion systems for small launch vehicles. In order to obtain commercially feasible vehicles, however, the inclusion of stages with liquid propulsion becomes a necessary condition. To obtain the technology and the know-how to design, manufacture, test and operate a Liquid Propulsion Rocket Engine (LPRE) is a goal to be achieved within the National Program of Space Activity (PNAE, 2012) in order to enable the development of satellite launch vehicles with higher performance and accuracy than solid propulsion.

The development of a LPRE is based on foremost scientific methods and complex technological processes which require a heavy and long-term investment. The experience of other countries shows that the development cycle of a large liquid system is around five to ten years. At the same time, many activities are necessary to involve the broad participation of research centers, universities and industries in the solution of multi-disciplinary problems.

The analysis of successful examples such as occurred in China, India, Japan and South Korea which had access to space after the first three — former Soviet Union, Germany, and the United States of America — shows that foreign partners had always been present in the initial stages of their development. It indicates that an investment in partnership is the key factor to overcome initial difficulties, mistakes and associated risks of the liquid propulsion technology (Niwa and Yoshino, 1997).

The LPRE propels space vehicles better than any other type of chemical propulsion, giving generally a higher specific

1. Instituto de Aeronáutica e Espaço – São José dos Campos/SP – Brazil

Author for correspondence: Daniel Soares de Almeida | Praça Marechal Eduardo Gomes, 50 - Vila das Acácias | CEP: 12228-904 – São José dos Campos/ SP – Brazil | Email: danieldsa@iae.cta.br

impulse compared with solid or hybrid rockets. Usually, the cryogenic propellants have the highest specific impulse in liquid rocket engines.

Some of the most important advantages of LPREs are the very wide range of thrust values to fit specific applications, multiple ignition possibility, fast pulsing, long term operation time and ready to reuse; thrust variation upon command; controlling attitude changes and the capability to be checked and even fully tested before use.

The restart of the LPRE thrust allows an accurate terminal flight velocity, which is important to achieve the required orbit. A remarkably high reliability has been achieved in the production of LPREs besides their lightweight structure, enabling more embedded payload for a mission. Most exhaust gases of LPREs using modern common propellants are non-toxic and environmentally friendly (Sutton, 2006).

The cryogenic propellant liquid oxygen (LOx) and liquid hydrogen (LH_2) produce the higher specific impulse available for all environmentally friendly bipropellant, which permit a considerable increase in cargo delivery for rocket launchers. However, cryogenic components have big issues with evaporation, which limits the duration of the mission flight. In addition to that, LH_2 has the disadvantage of needing big tanks due to its very low density.

The second more energetic and also non-toxic propellant pair is the LOx and kerosene, which has lower specific impulse but higher density and a prolonged service in space rather than LOx/LH_2. LPRE with this propellant combination is suitable for both first and upper launcher stages.

Moreover, although less energetic, the LOx and ethanol propellant pair has the advantages of its ready availability in Brazil, its chemical composition uniformity, being non-toxicity, its cleanliness, its low cost, and it does not suffer pyrolysis or produces significant soot at high temperatures inside the cooling channels of the regenerative cooled combustion chamber (Haeseler *et al.*, 2000).

Engines with higher specific impulse are, in order, those using LH_2 and LOx as propellants and those operating in closed cycle. However, the technological challenges of using LH_2 as propellant and the design of a closed cycle engine would make its development in the country still more complex than that of an open cycle engine using LOx and ethanol.

There are several different designs whereby a turbine can be integrated into a LPRE, and this was identified as different engine cycles. The open cycle has a separate gas generator, where fuel and oxidant are burned at a specific mixture ratio, resulting in the low temperature of the turbine inlet gases to prevent destruction of the turbine blades. This cycle is the simplest, and it often offers the lowest overall cost, provides a low engine structural mass due to lower internal pressure, but gives somewhat not higher performance than the closed cycle, which is more complex in design, manufacturing and testing. Figure 1 shows the L75 scheme, a LPRE with a typical open cycle (Torres *et al.*, 2009).

THE L75 ENGINE CHALLENGE

The current Brazilian Satellite Launcher Vehicle (VLS-1) cannot meet the overall strategic objectives of the space sector. However, it is envisaged the possibility to significantly increase the payload capacity of VLS-1 through partial modification of the vehicle, especially with the replacement of the solid propellant motors by liquid propellant engines in an upper stage.

To achieve this goal, several activities have been conducted in *Instituto de Aeronáutica e Espaço* (IAE) such as the specification, design and construction of liquid rocket engines. One of these projects is the L75 engine.

A timeline of the project with the major milestones is:

- 2008 – Beginning of the L75 project;
- 2009 – System Requirement Review (SRR);
- 2010 – 2012 – FUNCATE (*Fundação de Ciência, Aplicações e Tecnologia Espaciais*) contract: technical service to support development of the project - 20 specialists;
- 2011 – Preliminary Design Review (PDR);
- 2012 – Letter of exchange for Brazilian-German cooperation on the development and tests of a 75 kN engine;
- 2013 – Fuel change from kerosene to ethanol;
- 2013 – 2015 – FUNDEP agreement: to proceed with the development, manufacturing and test of some components.

In 2011, a protocol of intention between Brazilian Space Agency (AEB) and the German Aerospace Center (Deutsches Zentrum für Luft-und Raumfahrt - DLR) was established with the objective of fostering cooperation in space activities. After a visit of a German delegation at IAE in April 2011, a letter of exchange was established in October 2012 for further detailing the envisaged activities within the cooperation. The conductance of a review of the L75 engine was identified as a first major step. Starting in 2012, meetings regarding a specific implementation

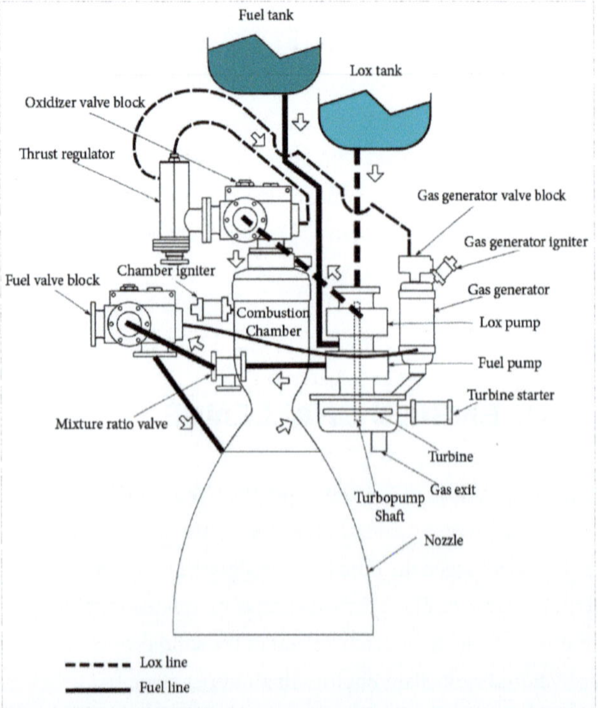

Figure 1. L75 scheme.

agreement between the IAE and the DLR to the development and tests of a 75 kN LOx-ethanol were organized.

The L75 development is being based on the ECSS standards (ECSS, 2004), using development (DM), engineering (EM) and qualification (QM) models engine components, parts and subsystems in order to verify the conformance to requirements. A verification program was established through an engine development plan, according to the L75 engine requirements specifications, considering the verification methods, levels and models.

For the L75 engine project, it is imperative to prioritize the development and mastery of critical space technologies, which are essential to the industrial progress, and the achievement of the necessary national autonomy in such a strategic activity. The allocation of specific system engineering requirements and needs to the L75 engine depends strongly on the type of agreement established between customer — IAE — and supplier — Brazilian enterprises — and the nature and level of complexity of the overall system subject of the agreement. The system engineering organization plans its activities in conformance with the project phases as defined by the management, in accordance to the European Cooperation for Space Standardization (ECSS) standards (ECSS, 2009).

The L75 engine needs, program objectives and requirements identified in the present concept phase are listed below. These requirements are preliminary in nature and not yet complete.

The project needs and objectives identified by IAE are:

- Consolidating the Brazilian space industry, by increasing its competitiveness and innovation capacity;
- Encouraging funding of programs based on public or private partnership;
- Enabling Brazilian suppliers for manufacturing, brazing and welding processes;
- Encouraging the human resources development by training of experts needed in the Brazilian space activities, both domestically and abroad;
- Expanding partnerships with other countries, by prioritizing joint development of technological and industrial projects of mutual interest;
- Developing of the controlling, monitoring, transmission and the data log related to the project;
- Developing of high complexity turbopump technology;
- Developing of an upper stage engine for a launch vehicle capable to carry a payload;
- Developing of staff and enterprises on liquid propulsion activities.

Some tasks are not addressed to the L75 engine project, such as the launcher vehicle, the stage, the propellant tanks systems, the inert gas supply system, and the thrust vector control system.

The critical technologies considered for the development of L75 engine are feasible materials for specific applications; bearings for the high rotation and low temperature operational conditions; seals for cryogenic application; welding and brazing process; development of cryogenic items as valves and regulators; regenerative thrust chamber; system feeding by turbopump; control system; assemblies and integrations; development of heat treatment after brazing for copper alloy and test facilities.

L75 ENGINE MAIN CHARACTERISTICS

The main high-level function of the L75 engine is providing (75.0 ± 5.0) kN of nominal thrust in vacuum with the main operation conditions shown below:

- Specific impulse greater than 2,940 m/s in vacuum;
- Minimum mission burn time of 400 s;
- Global mixture ratio in the range of 1.50 to 1.95;
- Total pressure inside of the combustion chamber in the range of 5 to 7 MPa.

The L75 engine nominal and extreme operational envelopes, a set of physical data in which the propulsion system, subsystem, or component is intended to operate, are shown in Fig. 2.

The results obtained show that the L75 engine in the current configuration has a decent mixture ratio envelope, despite a small margin for thrust variation. The restrictive factors are the thrust chamber maximum mixture ratio, the gas generator maximum and minimum mixture ratios and the turbopump maximum fuel flow.

The L75 engine (Fig. 3) thrust chamber consists of three main components, the combustion chamber, where the propellants undergo a set of chemical reactions at high pressure creating hot gases during the combustion process; the nozzle, where the hot gases are accelerated to supersonic velocities; and the injector head, where the propellants are introduced, broken up into small liquid droplets and distributed over the cross section of the burning region inside the combustion chamber. The thrust chamber cooling is a major challenge because the gas temperature exceeds the melting point of the metallic wall. To overcome this problem, the L75 engine uses a lightweight brazed and welded cooling jacket with a thin inner-wall construction.

The turbopump is one of the most critical components during the development of a LPRE, since it is a high-precision and high-speed rotating machinery operating in critical conditions. The main goal of the turbopump is to feed the thrust chamber and gas generator with the necessary flow and pressure of fuel and oxidizer provided by the pumps, which are driven by the turbine. Its manufacturing is expensive, and its design is complex and demands a great effort of engineering.

There are many different types of valves within the L75 engine project, such as fuel and oxidizer valves, vent valves, gas valves, pressure-regulating valves, check valves, thrust regulator, drain valves and safety valves. Most of the commercially available valves do not fulfill the engine requirements, are too heavy, or are not available for the Brazilian space program. For those reasons, most valves and regulators are being developed.

The L75 engine will operate with a thrust regulator, which is capable to regulate the LOx pressure for the gas generator, modulating the turbine power, which reflects on the turbopump shaft speed and pumps power. It is also equipped with two main valve assemblies, one for LOx and another for ethanol. Both of them are very similar in configuration and their functions cover chill-down/filling bypass, gas purging, pressure safety relief, opening at the start-up sequence and closing at the shutdown sequence.

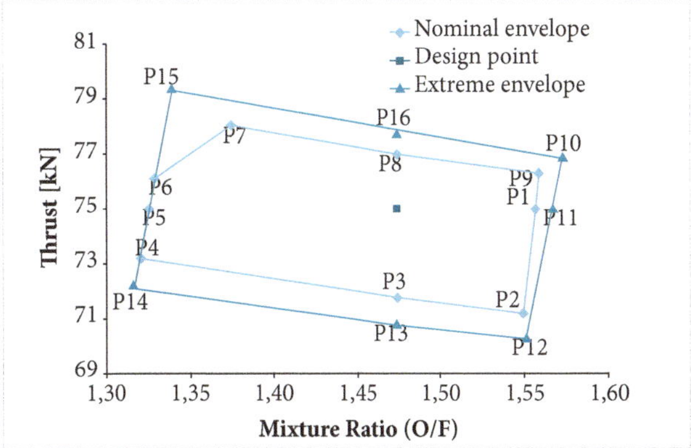

Figure 2. L75 engine envelopes.

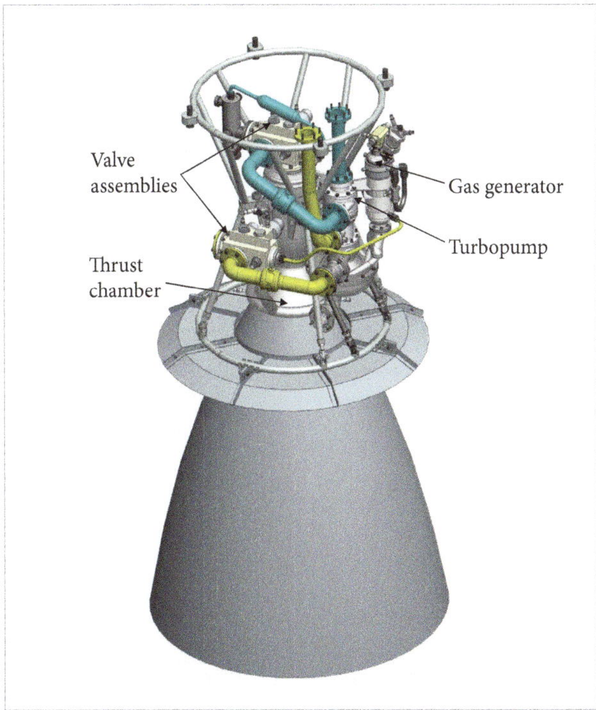

Figure 3. L75 engine.

The aim of the L75 ignition system is to achieve a safe, predictable, and reliable transition to nominal thrust in vacuum. The ignition system devices include a turbine starter, gas generator igniter, thrust chamber igniter and pyrotechnic actuators.

The L75 engine igniter and the turbine starter are based on the operating principles of solid rocket motors, consisting of a metallic structure, an electro-pyrotechnic starter, a reinforcement cartridge, the main cartridge propellant, and the thermal protection and nozzle.

The control system of the L75 engine is responsible for the starting procedure and ensures a safe transition to nominal thrust in vacuum; to achieve the desired thrust by controlling the fuel and oxidizer flow rate and pressure drops; safe shutdown of the firing operation. Furthermore, additionally controlled features are the precision of the automatic thrust control, mixture-ratio control and condition monitoring (safety controls or health monitoring of the engine).

A flow diagram of the L75 engine showing its components is given in Fig. 4.

DEVELOPMENT LOGIC

The development and qualification philosophy dedicated to the L75 engine and its subsystems are presented in Fig. 5, where the hardware model flow is shown.

The development life cycle phasing was elaborated according to ECSS system, but this system can be adapted to specific domains of application by use of tailoring activities necessary to meet specific project needs. The engine development is divided into five main phases: mission identification, feasibility (concept), preliminary design (development), detailed design (engineering) and qualification.

Besides, the development models on thrust chamber, turbopump and gas generator (power pack), valves and regulators and control system level, two DM, one EM and two QM on engine level are considered the minimum necessary for engine development.

The two qualification models will be subject to all essential modifications needed to meet the requirements of the engine specification.

All engine components shall pass by acceptance tests before their integration to the engine.

The selected basic approach is to minimize the overall technical risk and to reduce non-recurring cost during development by use of demonstrated hardware designs as much as possible.

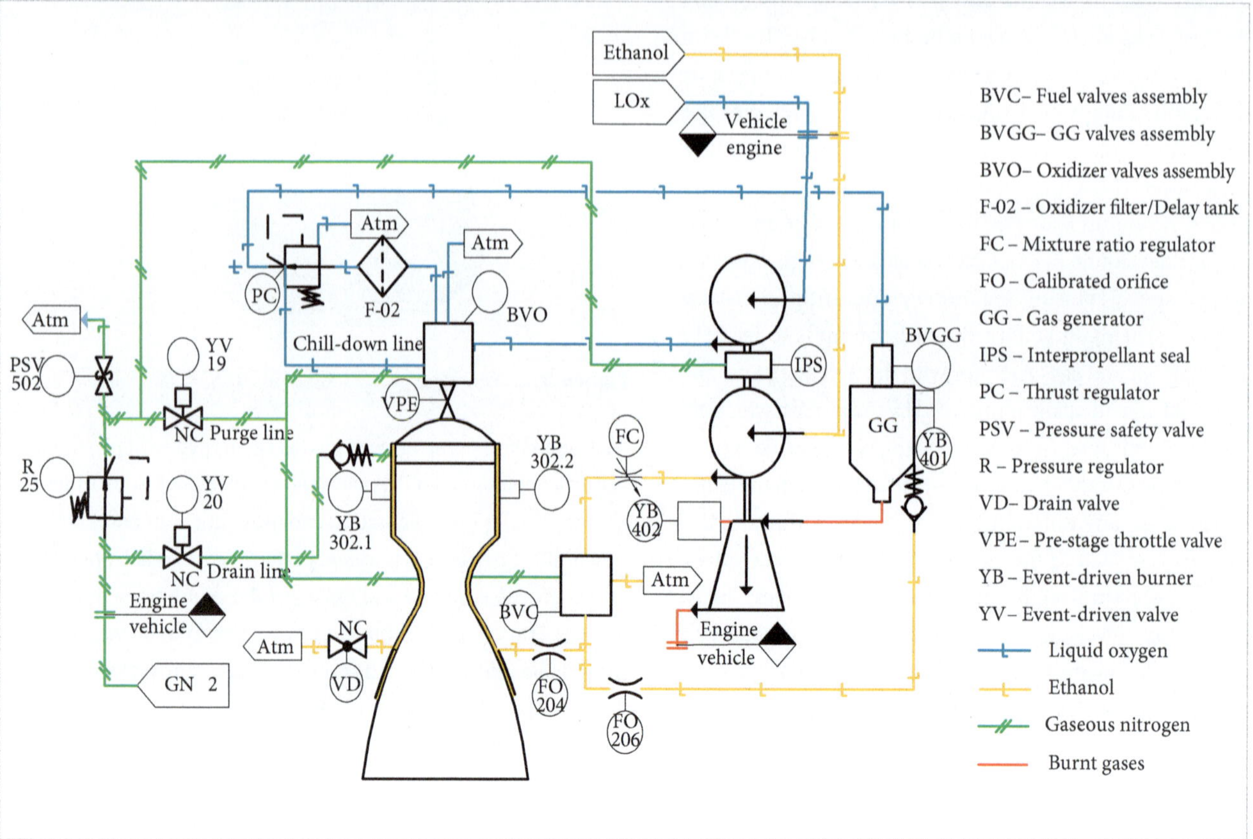

Figure 4. L75 engine flow diagram.

OVERALL DEVELOPMENT STATUS

Until the end of 2012, the L75 project was developed using, in addition to the IAE's human resources, services based in a contracts with FUNCATE. This contract had a technical and service expertise as an objective to support technology development of the L75. In December 2012, the project received from AEB a financial support, through an agreement with FUNDEP, to proceed with the development of the first components, allowing proceeding with the needed tests to confirm the design parameters. This agreement seeks to give more flexibility to the development of the project and involved Brazilian industries in the space program.

GAS GENERATOR

The gas generator is a combustion device used in LPREs for creating gases in suitable conditions, which are used as a working fluid for driving the turbine of the turbopump.

The basic elements of the gas generator (Fig. 5) are: injector element, cylindrical section, convergent section, in addition to fixing structural elements and interfaces. The igniter of the gas generator is part of the ignition system. The L75 gas generator has a mass flow rate of 1.3 kg/s, a mixture ratio of 0.3 and gas temperature of 900 K.

Figure 5. Gas Generator.

The first L75 gas generator was tested in April 2014 on IAE's 20 kN test bench (Fig. 6). These tests are used in order to verify the overall component performance, the combustion process and to validate thermodynamic models used during the design. Figure 7 shows the gas generator relative chamber pressure versus the time during a hot test, demonstrating its stability.

TURBOPUMP

The L75 turbopump (Fig. 8) shaft is supported by two bearings, which are centrally mounted close to the fuel pump, so they can be lubricated and cooled by the fuel. Seals are also provided to limit leakage.

Considering cavitation characteristics and the fuel rich gas on turbine side, the LOx pump is located at one end of the shaft and the fuel pump in the middle, connected by a spline between pumps and a shaft. The gas generator exhaust ducting leads to an impulse type single stage axial flow turbine with partial admission. Between the two pumps, an inter-propellant seal is adopted to avoid any interaction between propellants.

The fluid enters the pumps at low pressure from the tanks. In this way, a tank pressurization system is necessary to maintain the required Net Positive Suction Head (NPSH) to prevent pump cavitation. The pump inlet pressure is usually minimized to reduce the tank size and weight. Once the fluid enters the pump at the inlet, the inducer adds energy to the fluid before passing to the impeller to add sufficient energy to the fluid to suppress cavitation. The rotor significantly adds more kinetic energy prior to directing the flow into the diffuser and the volute.

(a) (b)

Figure 6. (a) Gas generator hardware; (b) Gas generator hot test.

The turbine's (Fig. 9) main characteristics are power of 380 kW; inlet gas pressure of 5.0 MPa; outlet gas pressure of 0.4 MPa; gas flow rate of 1.3 kg/s; temperature in the blades of 800 K, rotation speed of 24,000 rpm. The oxidizer pump works with inlet pressure of 0.4 MPa; outlet pressure of 7.6 MPa; and mass flow rate of 14.5 kg/s. The fuel pump works with an inlet pressure of 0.3 MPa; an outlet pressure of 10.6 MPa; and a mass flow rate of 9.4 kg/s (Fig.10).

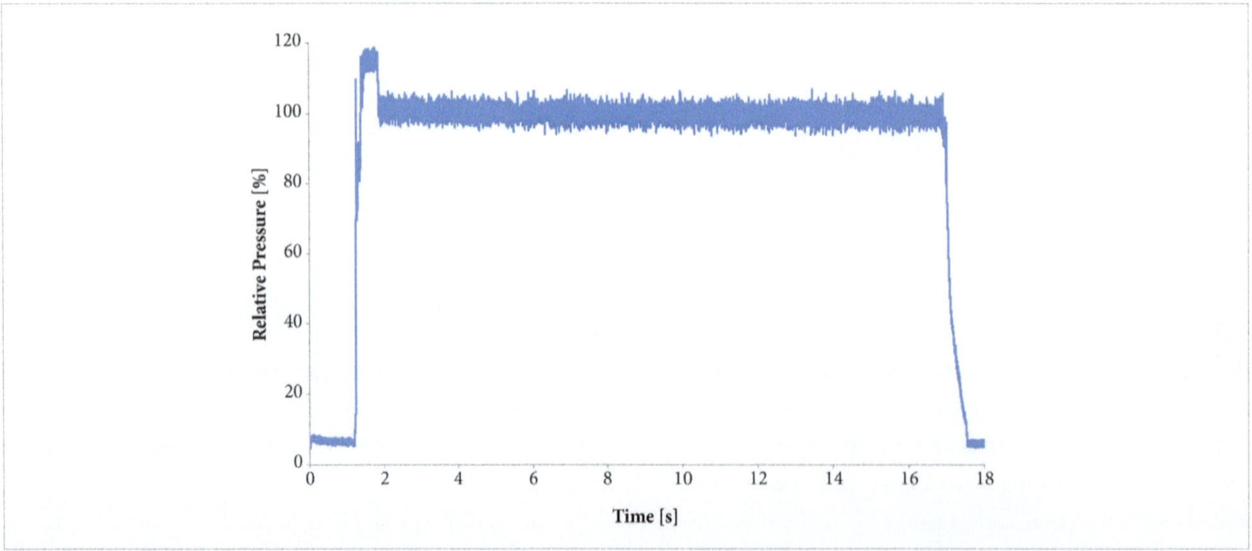

Figure 7. Gas generator relative chamber pressure versus time during a hot test.

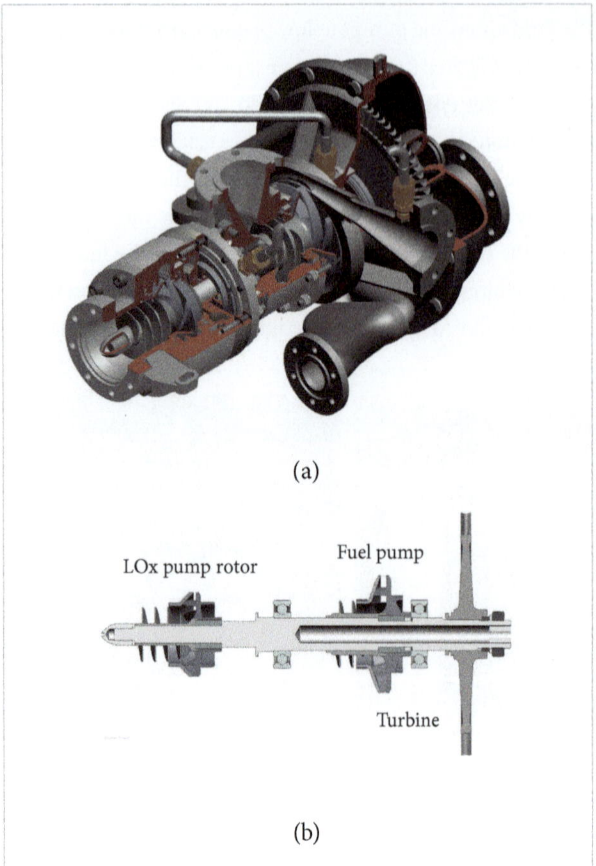

Figure 8. (a) Turbopump; (b) Rotor assembly.

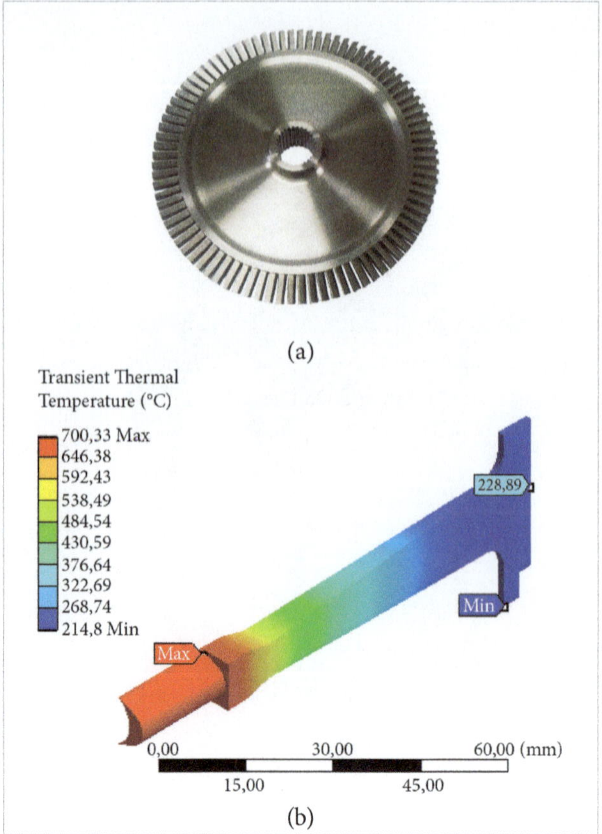

Figure 9. (a) SAE 422 stainless steel turbine disk; (b) Finite element analysis of distribution of temperature on turbine disk.

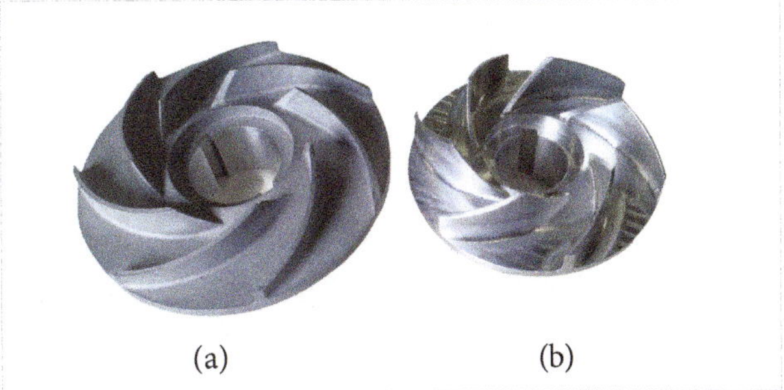

Figure 10. AA 7075 aluminum alloy high-speed rotor pumps (a) ethanol and (b) LOx.

THRUST CHAMBER

The main characteristics of the thrust chamber are: mass flow of 22.8 kg/s; mixture ratio of 1.7; temperature in the combustion chamber of 3400 K; and area expansion ratio of 150. Figure 11 shows the L75 thrust chamber in the short configuration for sea level hot tests.

COMPONENTS

The main characteristics of the L75 thrust regulator (Fig. 12), responsible for the control of the thrust level, are the nominal mass flow rate of 0.3 kg/s (LOx); the inlet pressure of 7.0 MPa and the outlet pressure of 6.0 MPa.

L75 will operate with two main valve assemblies (Fig. 13) with the following parameters: LOx inlet pressure 7.5 MPa; nominal LOx mass flow rate 14.5 kg/s; ethanol inlet pressure 10.6 MPa; ethanol mass flow rate 9.4 kg/s.

The ignition of the L75 engine is a challenging issue. The dominating criterion is reliability, repeatability and robustness of the ignition system (Jean and Dalbies, 2000).

Several tests have already been executed with the thrust chamber and gas generator igniters and the turbine starter, as can be seen in Fig. 14.

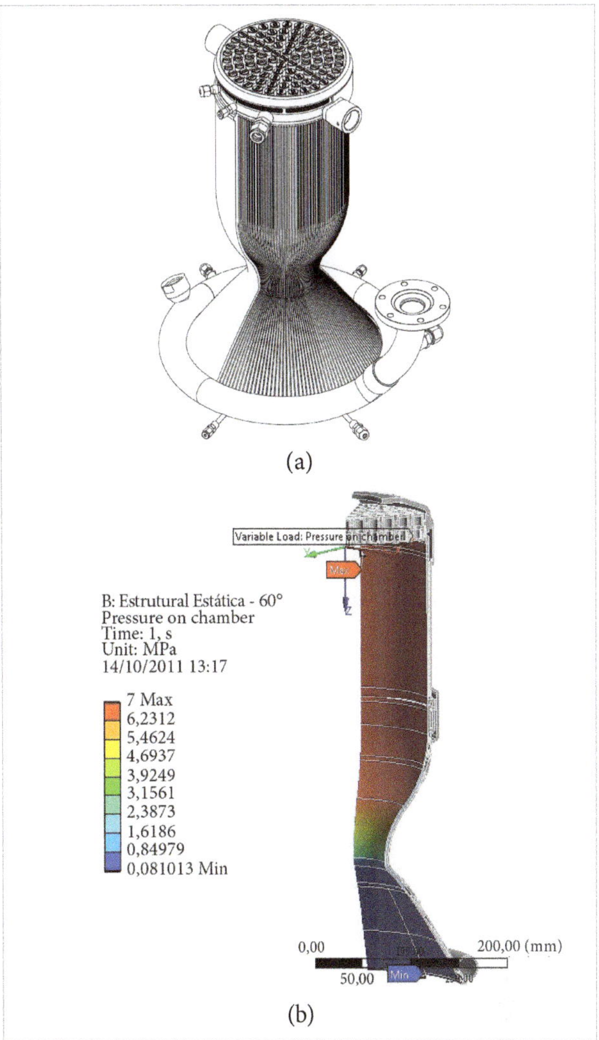

Figure 11. (a) thrust chamber; (b) equivalent stress in MPa (von Mises stress).

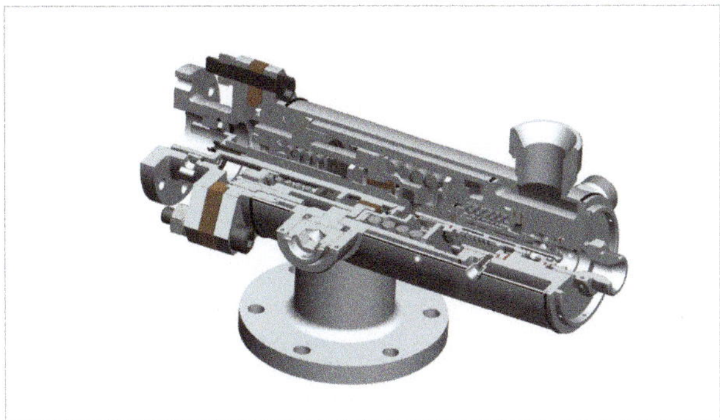

Figure 12. Thrust regulator valve.

TEST FACILITIES

Liquid rocket engines for launch vehicles as well as their subsystems need to be verified and qualified during cold and hot-runs. A high test cadence combined with a flexible test team helps to reduce the cost for test verification during development and qualification as well as during acceptance testing for production. The test facility allows to test subsystems in the same manner as during complete engine system tests and will therefore reduce development time and cost.

The basic strategies for development and verification of the L75 engine are based on appropriate verification methods of the requirements, levels that the component occupies in the product tree, stage of development and definition for the development, engineering and qualification models. One or more models may be required to demonstrate compliance with specified requirements of the design, depending on the level of verification. Acceptance tests will be carried out on component, equipment and engine level. Prior to acceptance test, these items should be checked-out according to the defined assembly or integration procedures.

To confirm adequacy of the design methodology of L75 engine, the construction of test facilities is in progress. The test benches in operation or under implementation are described hereafter.

20 KN FIRING TEST STAND

Designed for firing MFPLs up to 20 kN of thrust using pressurized tanks of LOx and ethanol as a propellant and with a combustion chamber pressure up to 100 bar. This test bench, which currently is used for testing the engines L5 and L15, was adapted for the L75 gas generator hot tests (Fig. 15).

HYDRAULIC TEST STAND

The hydraulic test stand is a facility whose main objective is the characterization of LPRE components, using distilled water as working fluid. It consists basically of the specimens test area, the drive system consisting of pumps and electric motors, the distilled water storage tank, a filtration system, water cooling system and the data acquisition and control systems, which is able to achieve a water flow up to 30 kg/s, working pressure of 35 bar. The system power is 180 HP (132kW). Figure 16 shows the test area of the hydraulic test stand.

PUMPS AND TURBINE COLD TEST STAND

Pumps and turbine cold test stand will allow the testing of the L75 hydraulic pumps and turbine through the operation of individual pumps and turbine, control of components and data acquisition allowing the verification of functional requirements of these pumps and adequacy of the design of the component being tested. Figure 17 shows the scheme of the pumps and turbine cold test stand.

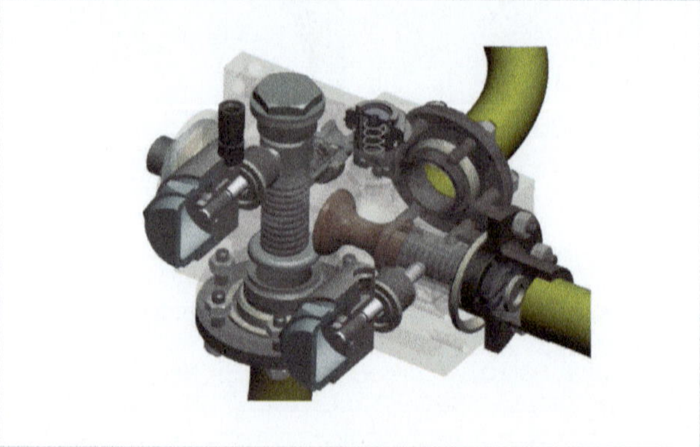

Figure 13. L75 main valve assembly.

Figure 14. Turbine starter hot test.

Figure 15. 20 kN firing test stand.

Figure 16. Hydraulic test stand.

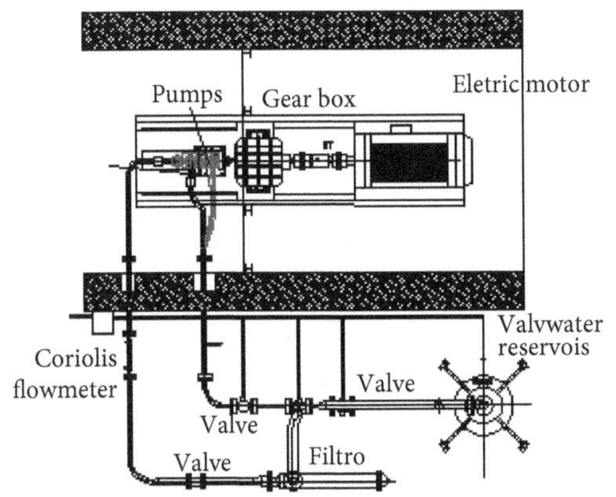

Figure 17. Scheme of the pumps and turbine cold test stand.

FUTURE AND CONCLUSION.

The greatest merit of the L75 project is not only the development of the engine which is, in itself, already a great contribution to formation of a qualified team, implementation of infrastructure for design, manufacturing using qualified Brazilian enterprises and testing, but mainly driving the country on a new technological level to actually be able to open access to space.

The next project critical activities are: manufacture of thrust chamber, turbopump, valves and regulators and cold and hot tests of these models.

Usually, in a rocket engine development, from conception through qualification for flight, the cost of design and manufacture, which is the actual stage of L75 development, corresponds approximately to 20% of the total costs. Therefore, if Brazil indeed masters this long-term technology will be necessary more investments in human resources, infrastructure, manufacturing, etc.

ACKNOWLEDGES

This paper was elaborated with support of a large number of members of the IAE and FUNDEP, and the authors wish to thank the team members.

The authors thank AEB for the financial support.

REFERENCES

ECSS – European Cooperation for Space Standardization, 2004, "ECSS-E-30 part 5.2 draft1 rev. 1", Propulsion for Launchers – Solid and Liquid.

ECSS – European Cooperation for Space Standardization, 2009, "ECSS-M-ST-10C", Project planning and implementation.

Haeseler, D., Götz, A. and Fröhlichs, A., 2000, "Non-toxic propellants for future advanced launchers propulsion systems", AIAA – 2000-3687.

Jean, F. and Dalbies, E., 2000, "Development status of the VINCI engine for Ariane 5 upper stage", AIAA/ASME/SAE/ASEEE 36th Joint Propulsion Conference and Exhibit, Hunstsville – USA.

Niwa, M. and Yoshino, T., 1997, "Liquid Propulsion in Brazil", COBEM 97.

PNAE – Programa Nacional de Atividades Espaciais 2012-2021: AEB - Agência Espacial Brasileira, 2012, Ministério da Ciência e Tecnologia, Brasília, Brazil.

Sutton, G.P., 2006, History of Liquid Propellant Rocket Engines, Virginia, USA: American Institute of Aeronautics and Astronautics, Inc., 911 p.

Torres, M.F.C., Almeida, D.S., Krishna, Y.S.R., Silva, L.A. and Shimote, W.K., 2009, "Propulsão Líquida no IAE: Visão das atividades e perspectivas futuras", Journal of Aerospace Technology and Management, Vol. 1, No. 1, pp. 99-106. doi: 10.5028/jatm.2009.010199106.

One-Stage Power Turbine Preliminary Design and Analysis

Cleverson Bringhenti[1], Jesuíno Takachi Tomita[1], Fernando de Araújo Silva[2], Helder Fernando de França Mendes Carneiro[2]

ABSTRACT: Design and development of gas turbine components are a complex multidisciplinary process. At the beginning of the power class definition and engine configuration it is necessary to conduct a market study. The results obtained are used in gas turbine thermodynamic cycle calculations and analysis in order to define the gas turbine design point. Several possible design points are evaluated during this procedure. After this step, the gas turbine components are designed, including: compressor, combustion chamber and turbine. For industrial gas turbine purposes, it is common to use a free turbine after the gas generator, also commonly named power turbine. In this work, a power turbine was initially designed by meanline techniques, considering internal loss mechanisms, to obtain the main dimensions. The geometries of the components were generated in a 3-D environment to make possible the mesh generation, process to discretize the physical domain into a computational domain and use a 3-D Computational Fluid Dynamics tool. The results from the meanline approach and from the 3-D turbulent flow numerical simulations were compared to verify the turbine operational conditions and its predictions at design and off-design conditions. The gas turbine under study is a project, derived from a low thrust turbojet previously developed by Instituto de Aeronáutica e Espaço. The power turbine project uses the same turbojet gas generator, already designed and currently under tests.

KEYWORDS: Turbine, Design, Analysis, Gas Turbine.

INTRODUCTION

The Instituto de Aeronáutica e Espaço (IAE) is one of the Brazilian government institutes responsible for conducting research and development related to aerospace engineering. IAE has been developing in conjunction with TGM industry and Instituto Tecnológico de Aeronáutica (ITA) a gas turbine project able to be used in two types of applications: a turbojet — in thrust class of 5 kN, and a turboshaft — in a shaft power class of 1.2 MW for power generation application.

The aim is to develop a gas turbine in which the economic feasibility can be justified by its different usages, since its two classes employ the same gas generator. The first version of the designed turbojet is installed in a test facility at IAE and its test protocol is under progress. The turboshaft class was studied after the design of the turbojet engine version. The power turbine preliminary design (Silva, 2012) was finished, but improvements in the aerothermodynamics, heat transfer, stress and strain calculations of the components are being performed.

This work presents the design of the power turbine, its aerothermodynamics, which was performed using a reduced-order numerical design tool and Computational Fluid Dynamics (CFD) simulations to verify the results from the preliminary design made by Silva (2012), providing the data needed for structural and heat transfer analysis, to be done afterwards. An in-house code written in FORTRAN language (Silva, 2012) was used to supply input data for commercial turbomachinery design software AxSTREAM® and to determine a preliminary design of a free power turbine. With the use of AxSTREAM® software the turbine Nozzle Guide Vane (NGV) and rotor

1.Departamento de Ciência e Tecnologia Aeroespacial – Instituto Tecnológico de Aeronáutica – Departamento de Turbomáquinas – São José dos Campos/SP – Brazil.
2.Departamento de Ciência e Tecnologia Aeroespacial – Instituto de Aeronáutica e Espaço – Divisão de Sistemas Aeronáuticos – São José dos Campos/SP – Brazil.

Author for correspondence: Cleverson Bringhenti | Instituto Tecnológico de Aeronáutica – Departamento de Turbomáquinas | Praça Marechal Eduardo Gomes, 50 Vila das Acácias | CEP: 12.228-900 – São José dos Campos/SP – Brazil | Email: cleverson@ita.br

blades were generated by an airfoil stacking procedure and the 3-D geometry was obtained. This 3-D geometry was carefully treated to start the mesh generation process using the software ANSYS TurboGrid®. The CFD solver used in this work was the ANSYS CFX® v.13.

POWER TURBINE DESIGN

The preliminary design process was performed using two different meanline numerical tools. The first is an in-house tool developed by (Silva, 2012) and the second one is commercial software called AxStream®. Meanline means that all fluid properties including the turbomachinery dimensions are calculated at mean blade height. This technique neglects the viscous effects that are accounted using loss modelling. There are several loss models, each one adequate for each case; therefore, there is no general loss model. The average values of all fluid properties from hub-to-tip are very close to the values obtained at mean radius. Thereby, the use of meanline technique is very interesting to obtain fast results and solutions for a first turbomachinery design process.

The main results from the meanline tool developed by (Silva, 2012) are presented in Table 1, in which the loss model developed by Kacker and Okapuu (1982) was used. The power turbine preliminary design was performed using AxSTREAM® software and it was conducted with the Craig and Cox (1970) loss model. Both loss models can be used at design and off-design conditions. For the power turbine studies, both loss models are adequate. For turbomachinery design several parameters must be varied and imposed by the designer during the preliminary sizing. Some design parameters and values were determined using an in-house code developed by Silva (2012) and the results used as an inlet data at the commercial software.

Along the years, several loss models were developed, for example: Ainley and Mathieson (1951), Denton (1978), Dunham and Smith (1968), Dunham and Came (1970), Dunham and Panton (1973), Rody *et al.* (1980) and Smith (1965). There are other models, but only the most commonly used were mentioned. In the present work, the loss models used are not described due to the high number of information, mainly about its calibrations, validation ranges, loss sources, constraints and so on. The reader that would like to understand the models in details it is highly recommended to search for the publication sources above mentioned.

Table 1. Power turbine preliminary data.

Axial channel shape	Nomenclature	Variable mean radius
Divergence angle in the streamwise direction	--	14.1° (stator inlet) 13.5° (stator outlet) 11.4° (rotor inlet) 5.1° (rotor outlet)
η	Isentropic efficiency	90.9%
PW (MW)	Power	1.349
c_f	Flow coefficient	0.64
c_c	Loading coefficient	1.26
RCA	Rotor inlet and outlet axial velocity rate	0.77
Number of stator blades	--	17
Number of rotor blades	--	40
R	Degree of reaction	0.50
α_2	Absolute flow angle rotor inlet	62.4°
β_2	Relative flow angle rotor inlet	9.3°
α_3	Absolute flow angle rotor outlet	12.1°
β_3	Relative flow angle rotor outlet	57.1°

The two-dimensional view of the power turbine stage, in the streamwise direction, calculated by AxSTREAM® making the use of the inlet data, calculated by the in-house code developed by Silva (2012), as presented in Table 1, is shown in Fig. 1a. The blue component is the stator row (or NGV) and the red component is the rotor row.

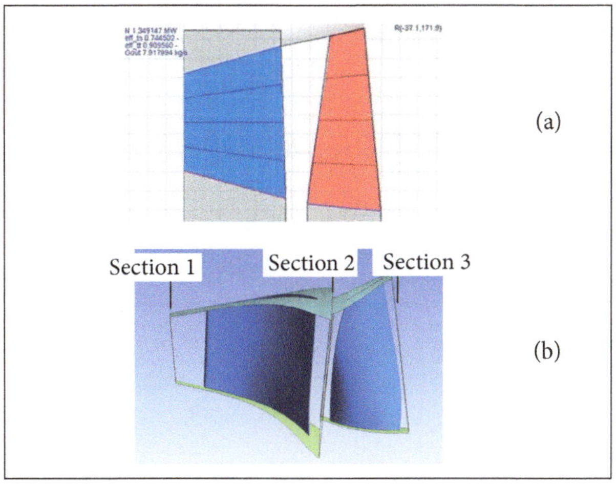

Figure 1. Power turbine meridional view. (a) NGV row in blue and rotor row in red; (b) Different sections.

THE 3-D TURBINE FLOW CALCULATIONS

Since the aim of this study is to discuss the preliminary design results obtained previously Silva (2012), in this paper there are no discussions about numerical schemes, turbulence models, spatial and time marching methods, discretization issues and other topics around the CFD numerical implementations and its programming techniques. Certainly, all software and hardware used in numerical simulations demand high CPU time, currently called high performance computing (or HPC).

In this paper, the methods and models used are available in the open turbomachinery literature and additional details can be found in Craig and Cox (1970), Kacker and Okapuu (1982), Silva *et al.* (2011), Bringhenti *et al.* (2001), Tomita (2009), Silva and Tomita (2011) and Murari *et al.* (2011).

The turbine geometry calculated based on meanline technique was appropriately treated to generate its 3-D geometrical view making use of Computer-Aided Design (CAD) software. It is recognized that the CAD treatments, mesh generation processes and the flow calculations using CFD technique are very time consuming processes.

The use of CFD techniques is very important during the turbomachinery design processes, because some flow features cannot be visualized by the designer from meanline calculations, for example: boundary layer separation, high pressure gradients regions, shock wave position and expansion waves. Thus, in order to calculate the 3-D flow characteristics, the commercial CFD software ANSYS CFX® was used, with the calculations based on the Reynolds-Averaged Navier-Stokes equations (RANS). The turbulence effects were accounted using eddy viscosity turbulence model.

MESH GENERATION PROCESS

The mesh was generated using ANSYS TurboGrid® software. As this software is part of the environment, it also allows a direct link with other ANSYS® package software, such as CFX® for CFD calculation.

Different mesh configurations can be found in the TurboGrid® software. It is important to consider the control volume quality as its orthogonality, skew angles, aspect ratios, smoothing and others as presented in the references Silva *et al.* (2011), Silva and Tomita (2011) and Menter (1983).

Four meshes with different refinements were generated to evaluate the mesh dependency for the particular turbine geometry studied. These meshes compromise hexahedral elements with the H/J/C/L configurations to better accommodate the turbine geometry. At this point, it is important to mention that the y^+ (position from blade surface at perpendicular direction) requirements must be considered, depending on the turbulence model that will be used to calculate the flow eddy viscosity. The Reynolds number was defined relative to the chord, being 4.03 x 10^5 at stator row and 2.68 x 10^5 at rotor row.

The y^+ value consists in determination of the distance between the first centroid (or vertex) of the control volume at wall from the wall surface, ensuring that this distance is consistent with the turbulence model. The y^+ is defined by the equation:

$$y^+ = \frac{\rho \Delta y}{\mu}\left(\frac{\tau_w}{\rho}\right)^{1/2} \quad (1)$$

where:

ρ and μ: density and dynamic viscosity, respectively; Δy: y variation; τ_w: shear stress at wall.

Generally, for $y^+ \leq 2$, a very refined mesh must be generated. For $y^+ > 2$, the CFD software determines the velocity field using wall functions. It is recommended by CFD solver that the best practice to use the wall functions is for y^+ values around $30 \leq y^+ \leq 300$. Different ranges can be found in the literature, but it is highly dependent on the wall functions implemented in each computational program.

Table 2 shows four mesh configurations generated to perform the mesh dependency study. The meshes were generated and the fluid mechanics equations were calculated until the point that there are no fluid properties variation in the solution from CFD calculations. These fluid and flow characteristics variations were considered in specific streamwise locations (section 1, 2 and 3 as shown in Fig. 1b) and from blade hub-to-tip positions. For all cases, the numerical settings as spatial and time integration used in simulations are the same and the two-equation turbulence

Table 2. Mesh configurations data for each case considered.

Mesh	Nomenclature	Number of elements NGV	Number of elements rotor
1	G (coarse)	117,000	76,000
2	F (fine)	274,000	311,000
3	R (refined)	415,000	311,000
4	R+ (very refined)	560,000	399,000

model developed by Menter *et al.* (2004), called Shear-Stress Transport (SST), was used to calculate the flow eddy viscosity. The details are described in the next section.

The Mach number, total pressure and static pressure were compared for all mesh configurations, and the values were calculated using mass averaging at each surface (section). Moreover, the isentropic efficiency, mass flow, expansion ratio and y^+ values were monitored and analyzed for each mesh as shown in Table 3. The reference pressure value is 101,325 Pa.

Table 3. CFD results for the four meshes.

Mesh	G	F	R	R+
η	89.0%	89.5%	90.2%	90.2%
$\dot{m}$	7.182	7.704	7.684	7.677
TPR	1.98	1.97	1.97	1.97
y^+ (ave.)	35 – 68	3 - 34	4 - 70	0.5 - 77
y^+ (max.)	130	68	130	135
Time*	1,000	3,200	4,000	5,400

*CPU time, in seconds per time integration iteration, in a computer with Microsoft Windows 7® 64bits operational system, 3.4 GHz i7 2600 processor with 8GB RAM memory available. $\dot{m}$ mass flow rate; TPR: inlet and outlet total pressure rate.

Figure 1b shows the stator and rotor section numbers used in this study, as well as the blade geometries.

Figures 2 to 4 show the flow characteristics variations (Mach number, static and total pressures) for each mesh in different sections along the turbine blade span. The idea is to evaluate the CFD solutions for different mesh configurations.

In Fig. 2, at section 1, it is possible to observe the impact of the velocity variations in different mesh configurations and its influence on Mach number and consequently on total pressure. The meshes R and R+ presented closely results. Note that, for 0 – 20% of blade span and for 80 – 100% of blade span, the results from different meshes present some interesting differences. This is due to the inadequate mesh resolution close to the wall, where boundary-layer is very important and pronounced.

From Fig. 3, at section 2, the same behavior and explanation can be performed. The meshes R and R+ presented closely results including the mesh F. At this point, it is possible to mention that the mesh G is not recommended for the simulations like the ones used in this work.

In Fig. 4, at section 3, the mesh F presented small differences in the solution when compared with the solutions from meshes R and R+. Again, the meshes R and R+ presented closely results including regions close to the turbine endwall and casing.

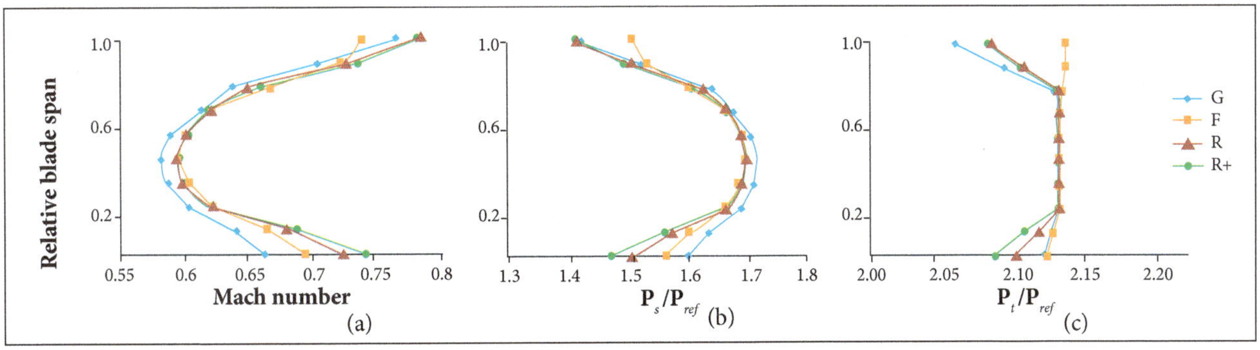

Figure 2. Distribution at section 1. (a) Mach number; (b) Static pressure (Ps); (c) Total pressure.

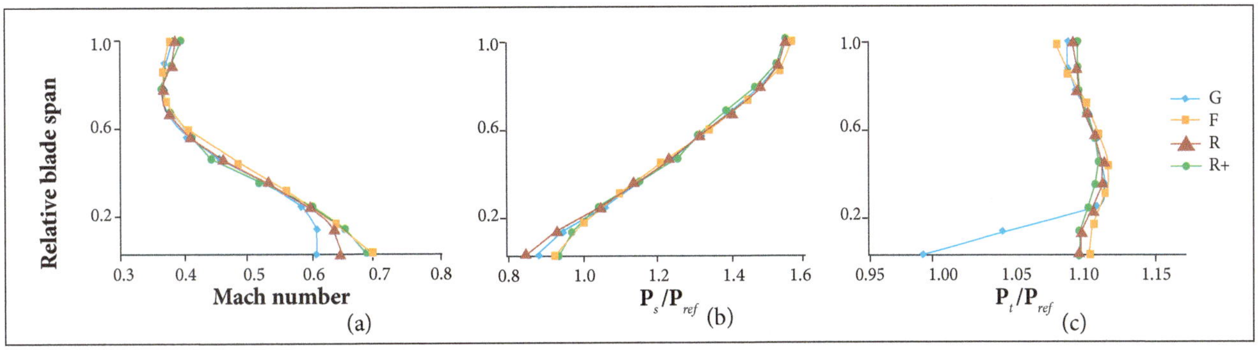

Figure 3. Distribution at section 2. (a) Mach number; (b) Static pressure (Ps); (c) Total pressure.

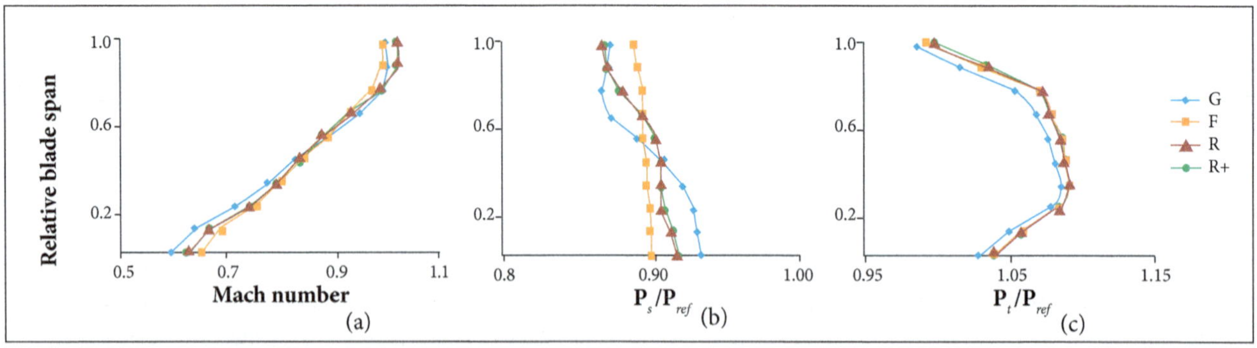

Figure 4. Distribution at section 3.(a) Mach number; (b) Static pressure (Ps); (c) Total pressure.

From these results it is possible to define that the mesh R is more adequate for the numerical simulation of the turbine designed in this study. The mesh R+ is also an alternative, but it is more expensive (CPU time) than mesh R due to the high number of control volumes, becoming the mesh R the best compromise. Figure 5 shows the surface mesh characteristics at stator blade row. Regions close to the wall surfaces need more refinement, the smoothing at wall regions in the computational domain should be carefully built to ensure a good resolution of the boundary-layer and acceptable flow characteristics resolution.

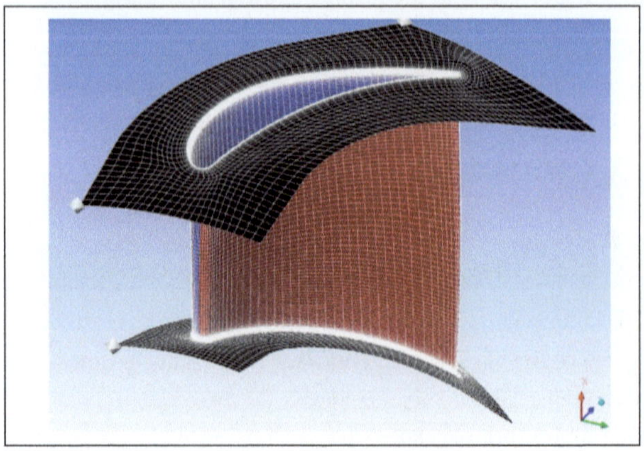

Figure 5. Mesh at stator — pressure side surface.

NUMERICAL ISSUES AND BOUNDARY CONDITIONS

The commercial software CFX® v.13 was used as a CFD solver to perform the flow calculations based on the continuity, Navier-Stokes and energy equations. The compressible version of the solver was set-up. CFX® uses the coupling between the equations and the modified Rhie and Chow (1982) discretization method for the mass terms, similar to the methodology proposed by Majumdar (1998).

For the discretization of convective terms from the Navier-Stokes equations, first and second order upwind numerical schemes are available. The limiters proposed by Barth and Jesperson (1989) are used to control the numerical discretization order, mainly for discontinuously regions, and to avoid numerical instabilities in shock regions when upwind based methods are used. The diffusion terms are discretized making the use of the shape functions. This methodology is usual in finite elements method (FEM) and has shown to be very robust for this purpose. The discretization model applied for the convective terms from Navier-Stokes equations was the second-order upwind scheme.

To accelerate the numerical procedure, the anisotropic algebraic multigrid technique developed by Raw (1996) was used to avoid error propagation due to the irregular elements. Conservation requirements are imposed by the method during the coarsening process. CFX® has only implicit time-marching scheme for time integration. An automatic time-step calculation method was used to ensure good numerical stability during the simulation. Fast convergence with few numbers of iterations can be achieved using this implicit scheme. All simulations were done for steady-state regime.

To account the turbulence effects, the flow eddy viscosity was calculated based on the two-equation Shear Stress Transport (SST) model Menter (1983). This model is adequate for the present analysis according to the recommendations and analysis from Menter *et al.* (2003, 2004), Denton (2010), and CFD Online (2011). The model uses the k-ω formulation at the internal region of the boundary layer and the k-ε model away from the wall. This control is determined using blending functions developed by Menter (1983).

The boundary-conditions were set at several surfaces as described:

- At turbine NGV inlet: total pressure (P_t) = 214,145 Pa, total temperature (T_t) = 990.7 K, turbulent intensity = 5% and α = 19.18° (absolute flow angle);
- At turbine rotor outlet: static pressure (P_s) = 89,891 Pa;
- Rotational speed = 27,920 rpm;
- Periodicity: blade-to-blade;
- At walls: no-slip condition;
- Inter-rows: mixing-plane approach.

The solution from CFD solver was monitored by convergence criteria based on the residue decayment, number of iterations and mass-flow variations (if no changes in the mass-flow was detected, the solution can be considered achieved; this is common in the turbomachinery CFD applications).

Figure 6a shows the residual decayment of the continuity and momentum equations monitored during numerical iterations.

The mass-flow at turbine inlet was monitored during numerical iterations. Figure 6b shows the mass-flow behavior during the numerical simulation. With 150 iterations the solution was considered achieved. An important numerical issue to mention is the very effective initial condition, also called initialization process used in the CFX® solver, in which, before the first iteration, the solver estimates a good distribution of pressure, temperature and velocity fields along the turbine streamwise and blade span directions, to guarantee a good initial condition.

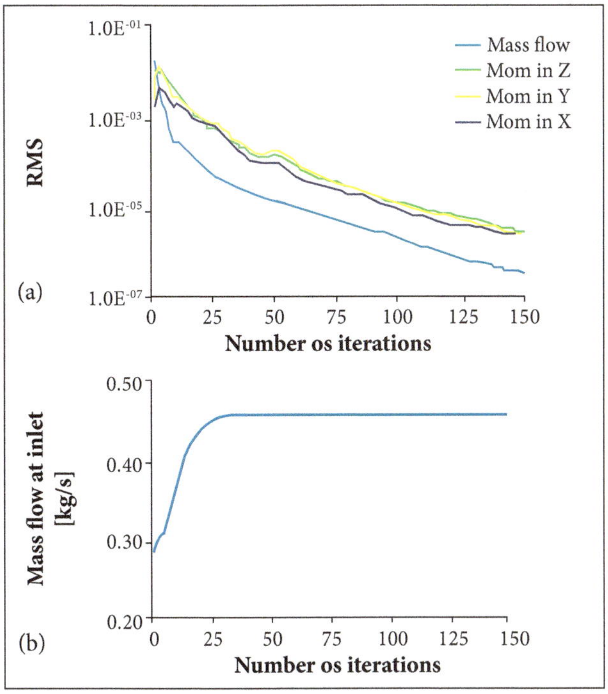

Figure 6. Numerical procedure. (a) History: continuity and momentum equations; (b) Turbine mass flow variation.

DESIGN-POINT ANALYSIS

At design-point condition, the turbine flowfield was calculated using CFD technique and each blade row (NGV and rotor) was analysed separately. The flow characteristics in important and/or critical regions were studied, as near to hub and tip zones, as well as the average values for some parameters at stage inlet and outlet sections. The static pressure distribution along each blade profile is plotted against its chord to evaluate the behavior of the flow along the blade pressure and suction sides. The blade profile is a multi-circular arc (MCA).

The NGV Mach number and entropy contours for 20, 50 and 80% spanwise are shown in Figs. 7 to 9. It is important to mention that the scales considered in each figure are not held constant; this was done for better visualization.

From Fig 7, it is possible to observe high flow acceleration from NGV leading-to-trailing edge, but at 68% of the blade chord a sudden increase in static pressure is observed, causing flow deceleration. The result is an increase in the entropy.

Figure 8 show the same analysis, but for 50% of the NGV blade span. Note that, at suction side, there are quite differences in static pressure values when compared with the 20% of the NGV spanwise. This difference is because the flow velocities are higher in regions close to the NGV hub. The degree of flow deceleration at 50% of the NGV spanwise is lower when compared with 20% of the NGV spanwise. Hence, the internal losses are lower and, consequently, the entropy generation is decreased too.

Figure 9 show the results for 80% of NGV spanwise. Comparing the results of static pressure for different NGV blade span, it is possible to observe that the flow acceleration region, close to the blade leading edge, at blade suction side, changes from around 12 to 18% of the blade chord. The blade angles and thickness chord ratio distributions are responsible for these variations.

Figure 9b shows small Mach number values when compared with Fig. 8b. This is a common flow behavior along the NGV spanwise. The entropy generation are indications that some improvements in the blade design can be necessary to improve the flow characteristics through the NGV channel.

For the blade pressure side, the results show a satisfactory pressure variation. At suction side, close to the hub, there is a supersonic bubble, covering the full flow passage. Except for this bubble influence, the results are in agreement with the prediction (Silva *et al.*, 2012). Figure 10a provides a visualization of the velocity vectors related with the flow acceleration at suction side at 20% of the NGV spanwise.

The total pressure and Mach number distributions, in a midplane between blades of the same row, are shown in Figs. 10b and 10c. In these figures, it is possible to observe the radial variation of the flow velocity and the high velocity region at NGV hub close to the trailing edge. The gas expansion is observed from the total pressure decreasing along the NGV blade chord.

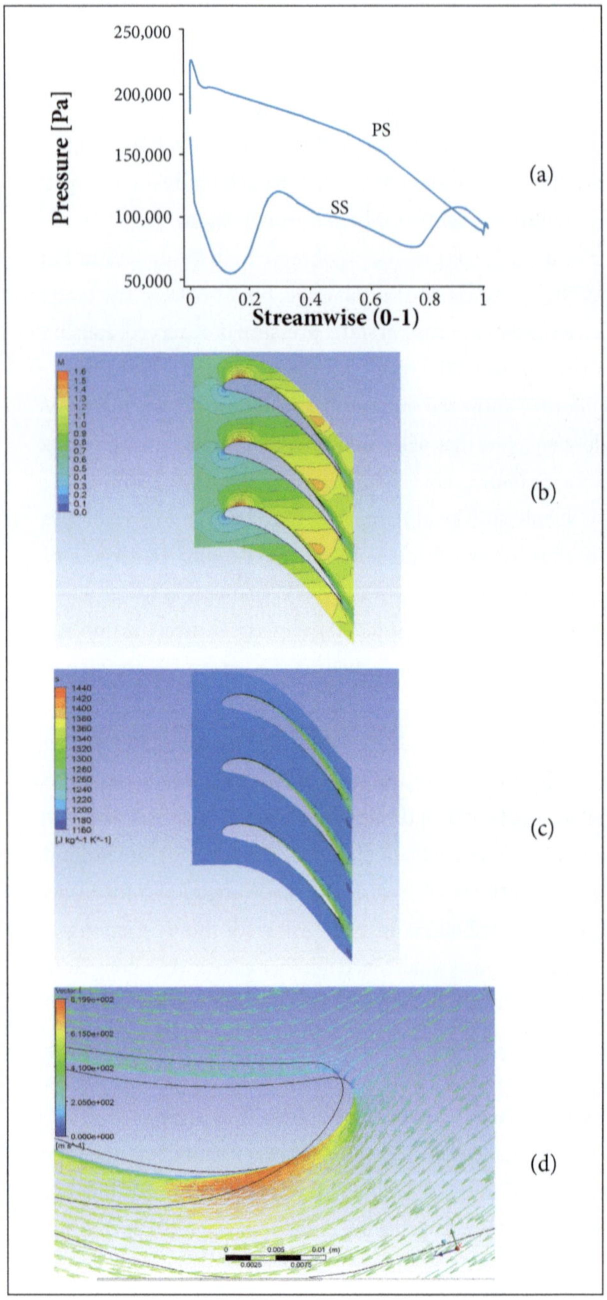

Figure 7. NGV spanwise - 20%. (a) Static pressure variation along the blade chord; (b) Mach number contours; (c) Entropy contours; (d) Velocity vector.

The result also indicates that higher velocities at trailing edge, near to the hub, increase the internal losses due to the association with skin friction.

The total pressure distribution shows the influence of boundary-layer in the turbine hub and casing walls. The total pressure, Mach number and entropy distributions are shown at stator inlet and outlet sections (Figs. 11 to 13).

The influence of boundary-layer should be carefully analyzed during the turbomachinery design process. Depending on the boundary-layer thickness, the influence on the velocity field is expressive, because an increase in velocity changes all flow properties distributions and generates more internal losses, degrading the machine efficiency. The variation of the velocity along the streamwise direction is important due to the influence on the gas expansion process related with the density variation.

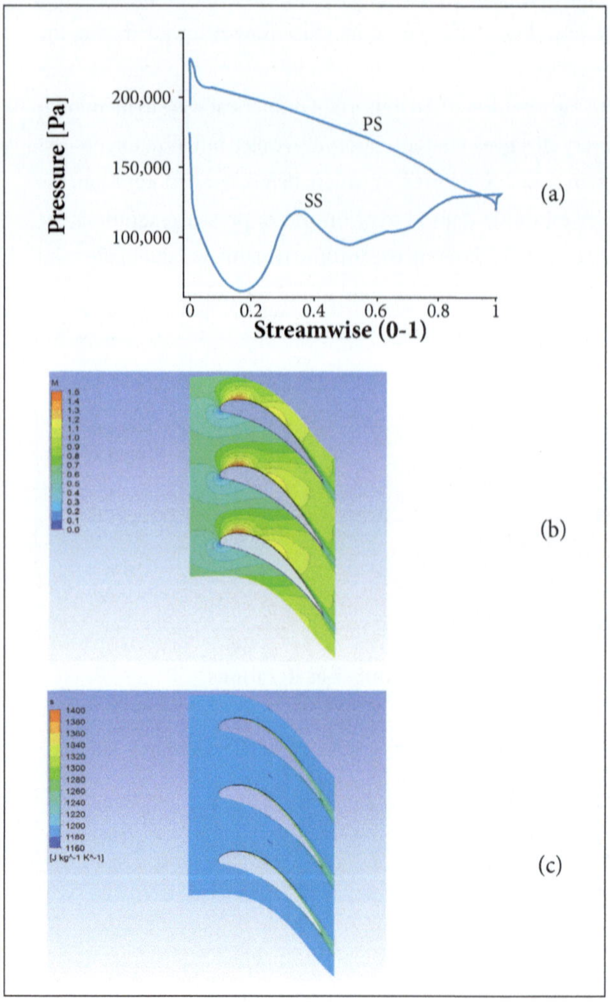

Figure 8. NGV spanwise - 50%. (a) Static pressure variation along the blade chord; (b) Mach number contours; (c) Entropy contours.

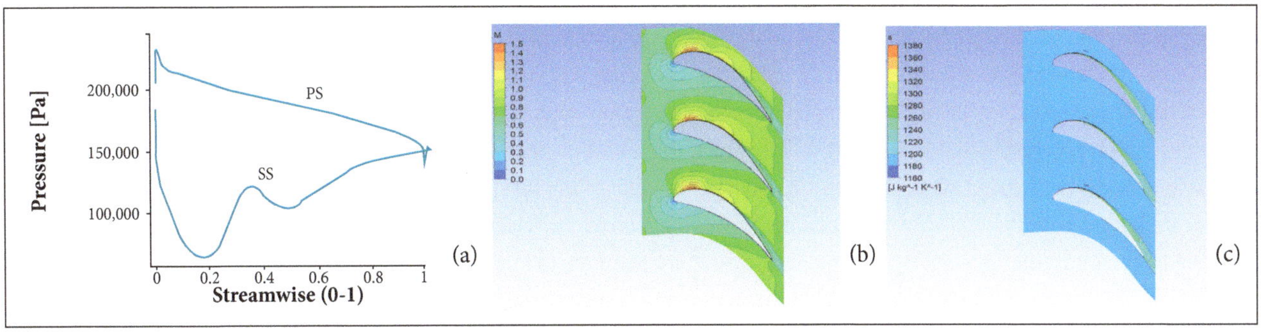

Figure 9. NGV spanwise - 80%. (a) Static pressure variation along the blade chord; (b) Mach number contours; (c) Entropy contours.

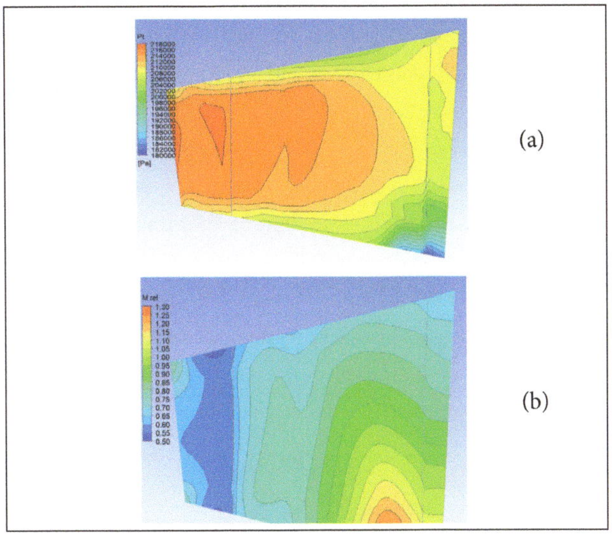

Figure 10. NGV — blade-to-blade midplane. (a) Total pressure distribution contours; (b) Mach number contours.

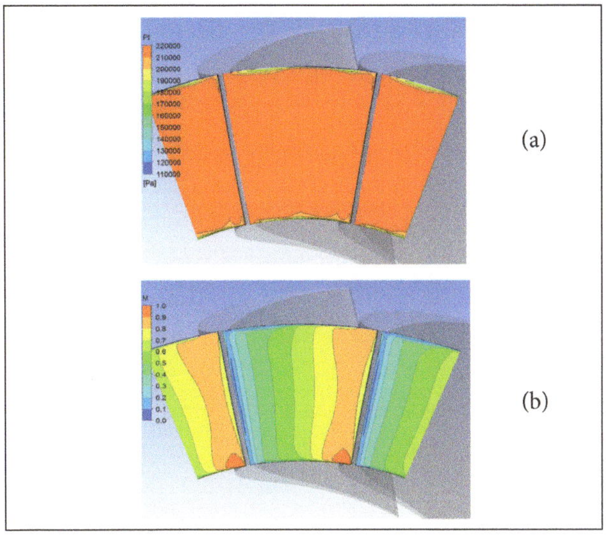

Figure 11. NGV inlet. (a) Total pressure contours; (b) Mach number contours.

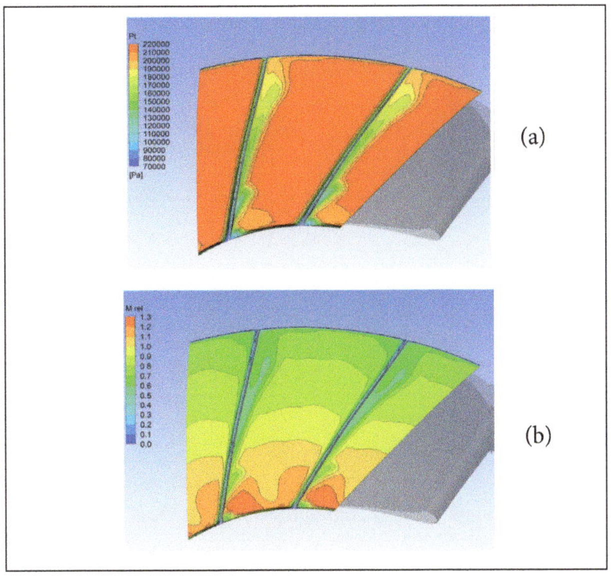

Figure 12. NGV outlet. (a) Total pressure contours; (b) Mach number contours.

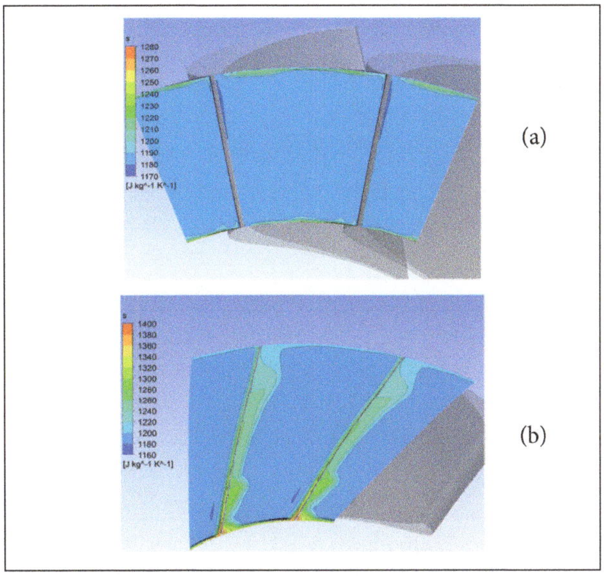

Figure 13. Entropy contours. (a) Stator inlet; (b) Stator inlet.

Entropy generation sources are generated at walls, due to the flow viscous effects and high velocities that are found in the blade-to-blade passage area; the velocity increases because the blade passage area is a convergent nozzle and high flow deflections are needed to improve the energy transfer. Moreover, different outlet blade angles and blade thickness distributions have strong influence on the energy transfer process. The losses are associated with the flow velocities and their growth is proportional to the velocity magnitude squared. As aforementioned, at the NGV domain, the flow has higher velocities close to the hub regions and its losses can be observed in Fig. 13b, in which high entropy generation can be observed.

The same analysis was performed for rotor blade row as presented in the Figs. 14 to 16. The static pressure distribution along the blade chord at suction and pressure sides, Mach number and entropy contours are presented. The same spanwise percentages were used: 20, 50 and 80%. Figure 14a shows a good variation of static pressure at blade surfaces for 20% of the rotor spanwise.

It is possible to observe that the Mach number is lower than unity and the turbine is unchoked at the design-point. This is a preliminary design strategy to maintain a certain design flexibility during the preliminary analysis. Sometimes, to obtain more power or pressure ratio, it is necessary to handle some possibilities. Hence, it is recommended that the design parameters not start with the limit values. Unlike NGV, at rotor hub, the velocity magnitude is lower than rotor tip values due to the tip speed. Figure 14b shows a good Mach number behavior of the flow along the rotor passage. Figure 14c shows good characteristics around the entropy evaluation. There is no separation or serious problems close to the rotor trailing edge at 20% of rotor spanwise.

Figure 15 present the results for 50% of rotor spanwise. The flow velocity across the rotor passage for this position is greater than 20% of rotor spanwise. Mach number reaches the unity. At this position the flow is choked and a high energy transfer from fluid to rotor is achieved.

Though the value of Mach number for 50% of rotor spanwise reaches the unity, the behavior of the flow for this blade position is acceptable. The entropy increases only at rotor trailing edge. This is normal condition due to the flow mixture in this region.

The characteristic of the static pressure at 80% of the rotor spanwise is also acceptable, as shown in Fig. 16a. High Mach number can be observed at 80% of the rotor spanwise; this is due to the high tip speed values at this position, and supersonic

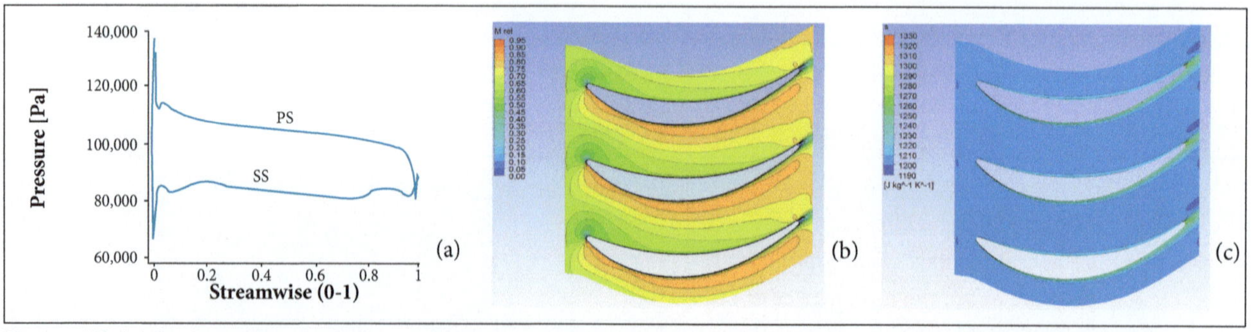

Figure 14. (a) Static pressure variation at 20% of the rotor spanwise; (b) Mach number contours at 20% of the rotor spanwise; (c) Entropy contours at 20% of rotor blade span.

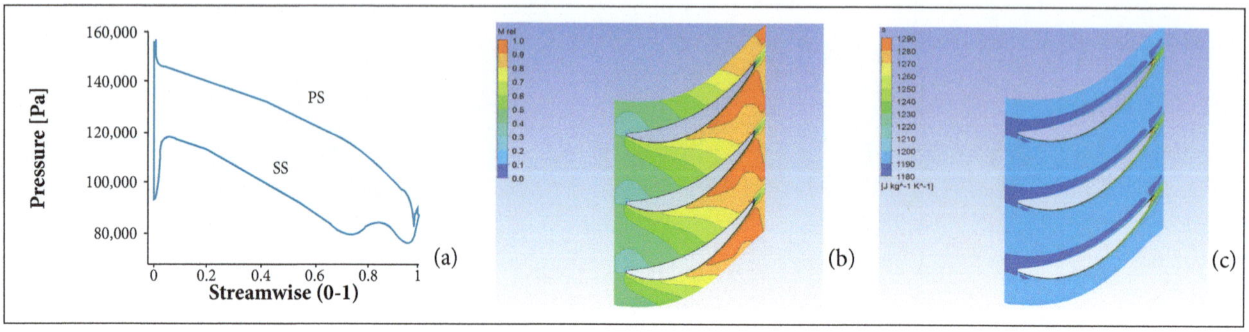

Figure 15. (a) Static pressure variation at 50% of the rotor spanwise; (b) Mach number contours at 50% of the rotor spanwise; (c) Entropy contours at 50% of rotor blade span.

flow can be reached as presented in Fig. 16b. In the present work, the tip Mach number reached the value of 1.2.

Some geometrical modifications at rotor blade in the suction side, around at 80% of rotor spanwise, can be necessary to improve the flow characteristics in this region. The wake formation can be minimized changing the blade outlet angle and its thickness.

The entropy contours verified in Fig. 16c is due to the mixing region. In general, the results from CFD simulations at 80% of the rotor spanwise are acceptable.

For the suction side, as at stator, the results show satisfactory behavior for the flow property distributions. At rotor, there is a larger difference in flow behavior between the three radial positions, as shown in Figs. 14 to 16. Near the hub, the flow is almost subsonic and the parameter variations are softly and small. This is a consequence of the use of low degree of reaction at this region.

Considering the analysis at hub, the flow properties increase along the surface, becoming very smooth until approximately 60% of the chord, where it starts to suffer influence from the suction side, as also occurs at NGV row. However, for the rotor, this variation increases until the trailing edge, what is a consequence of the smaller blade-to-blade passage due to the higher number of blades. The highest Mach number value, approximately 1.0, is obtained close to the tip. All these results are in agreement with the prediction described by Silva (2012).

At blade suction side, mainly close to the tip, flow properties present great variation until the throat region (approximately 60% of the chord). At this point, the flow properties start to oscillate, with a sequence of deceleration, acceleration and deceleration again, as can be seen in Figs. 16a and 16b. The maximum Mach number found in both accelerations does not exceed 1.2. These results are also in agreement with the predictions made by Silva (2012).

In Fig. 17a it is possible to see the influence of the tip clearance on the rotor flowfield. The radial variation of the pressure and the velocity in this location are high. The clearance must be evaluated with a very careful process due to the high machine degradation related with the leakage flow. This leakage flow creates a secondary flow that interacts with the main flow. This mixing process generates high entropy and energy dissipation. Figure 17a shows the streamlines in this region. It is very common the structures formation, as scrapping vortex, that generally causes the drop in turbine efficiency.

Figure 17b shows the entropy contours at rotor row outlet with interesting vertical formations at rotor tip clearance region. In the same figure it is possible to observe the secondary flow

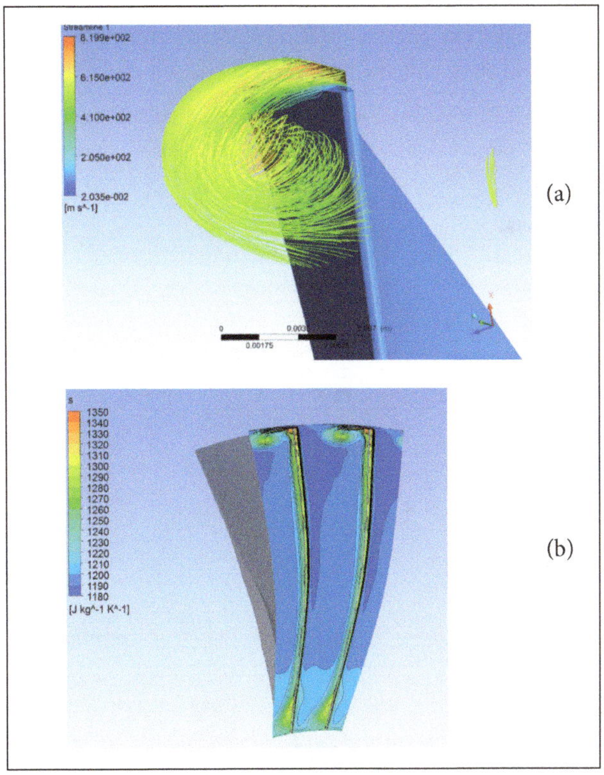

Figure 17. (a) Streamlines due to the rotor tip clearance; (b) Entropy contours at rotor outlet.

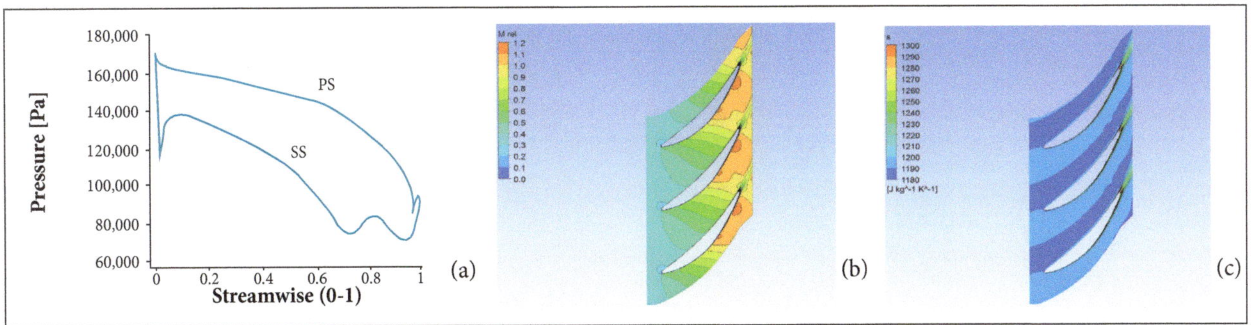

Figure 16. (a) Static pressure variation at 80% of the rotor spanwise; (b) Mach number contours at 80% of the rotor spanwise; (c) Entropy contours at 80% of rotor blade span.

growing and acting in the main flow, close to the hub, and creating at rotor leading edge horseshoe vortex. The effects of tip clearance in axial flow turbines that operate with high pressure ratios are severe for efficiency and gas expansion process. Several studies have been conducted to improve turbine performance decreasing the effects of tip clearance (Silva and Tomita, 2013; Ameri *et al.*, 1998; Tallman, 2004; Lee and Choi, 2010; Tallman and Lakshminarayana, 2001; Azad *et al.*, 2000; Bunker, 2014; Thulin *et al.*, 1982; Sang and Byoung, 2008; Krishnababu *et al.*, 2009).

The results shown in the previous figures present a satisfactory flowfield within the axial flow turbine domain. The power turbine stage performance can be evaluated through the parameters calculated from the CFD techniques, as shown in Table 4. For efficiency, power and mass flow rate in parenthesis, the AxSTREAM® predicted different values and are also shown the error.

Table 4. Turbine stage performance parameters.

Parameter	Value	
η (%)	90.1 (90.9 – 0.8%)	
PW (MW)	1.238 (1.349 – 8.2%)	
m (kg/s)	7.717 (7.918 – 2.5%)	
	Inlet	Outlet
T_t (K)	990	852
T_s (K)	923	815
P_t (Pa)	214,784	108,760
P_s (Pa)	162,619	90,724
M_{rel}	--	0.85
V (m/s)	--	482
β (°)	--	51.9
M	0.66	0.51
C (m/s)	390	273
α (°)	19.1	1.1

V: relative velocity; C: absolute velocity; M: Mach number.

Generally, 1-D methods use loss modeling developed for reduced-order design tools, as meanline. These models should be calibrated for each design purpose. Some three-dimensional effects are not possible to account using 1-D tools, but, in general, the results agree for both techniques.

OFF-DESIGN POINT ANALYSIS

The off-design performance analysis was conducted only at the nominal speed of 27,920 rpm for comparison with the predicted results, varying the expansion ratio during the simulations. The results are presented in Fig 18 together with the preliminary study and its respective values.

Figure 18a shows the variation of mass-flow with expansion ratio, in which the results from meanline and 3-D CFD are quite different. The CFD results show that the turbine choke condition is close to 6.85 kg/s of mass flow. This value is lower than that the choke condition determined in the meanline tool. This is due to the boundary-layer growth along the turbine streamwise direction that is, generally, underestimated by reduced-order tools. With high boundary-layer thickness, the velocity in the main flow increases, causing choking conditions. Meanline tool generally uses very simple models to account the blockage along the streamwise directions and in some tools this value is estimated by the designer. The expansion ratio for both tools at design-point presented slightly values with a difference around 0.11. Based on the CFD results, the turbine is choked at design-point conditions. This is normal and is in agreement with turbine design methodology and means that high energy transfer occurs in the gas expansion process.

It is clear that a calibration of some coefficients used in the loss modeling should be adjusted to improve the turbine prediction for performance calculations. This is a very common action into the turbomachines design, in which, for several reasons, these

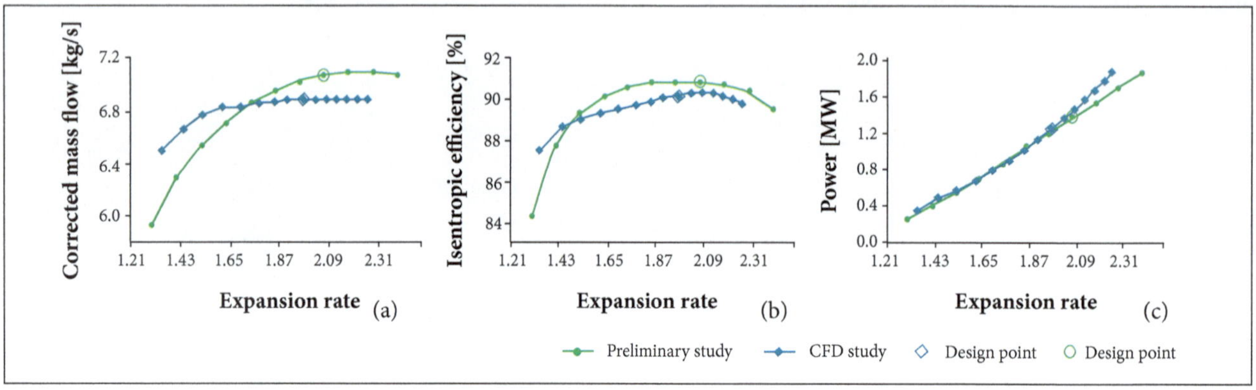

Figure 18. (a) Corrected mass flow; (b) Isentropic efficiency; (c) Power versus expansion rate.

calibration issues are considered as proprietary information for each industry.

Figure 18b shows that the peak efficiency at design-point operation condition is slightly different for both tools with a difference of 0.6, in which the efficiency from CFD solution is lower. This is because the loss mechanisms determined by meanline are underpredicted. Generally, 3-D CFD considering viscous flows presents more accurate results due to the better representation of boundary-layers and fluid friction. The efficiency determined by CFD tool is calculated based on mass-averaging process.

Figure 18c shows a good agreement between both numerical tools until the expansion rate of 1.98. The power determined by CFD tool is lower at design-point (around 160 kW lower than meanline tool). This can be explained due to the internal losses underpredicted by meanline tool. In general, all obtained results were well represented using the meanline tools, that are very fast to perform the calculations, comparing CPU time to calculate the 3-D turbulent flow by CFD and simple correlations.

CONCLUSION

In this study an in-house code written in FORTRAN language Silva (2012) was used to supply input data for commercial turbomachinery design software AxSTREAM® and to determine a preliminary design of a free power turbine. With the use of AxSTREAM® software the turbine NGV and rotor blades were generated by an airfoil stacking procedure and the 3-D geometry was obtained. This 3-D geometry was carefully treated to start the mesh generation process using the software ANSYS TurboGrid®. Four meshes with different number of control volumes were generated to study the mesh dependence from CFD solver. The CFD solver used in this work was the ANSYS CFX® v.13. Based on the mesh dependence study the mesh R was chosen to perform all other simulations of the turbine design-point and off-design point operational conditions.

The 3-D turbulent flow within the turbine computational domain was performed using RANS method applied on the fluid mechanics equations (continuity, momentum and energy). The numerical issues and boundary conditions were described in details. The simulation results from meanline and CFD tools were compared at design and off-design conditions. For both, the results are physically correct and are in agreement with the behavior of turbine flowfield.

Improvements in blade profile can be performed to increase the turbine efficiency decreasing the internal losses as shown from CFD results. Future work can be developed to calibrate the loss model to be applied in this free power turbine design. The CFD tool can be used aiming at improvements in the loss modelling calibration process.

The numerical tools and the methodology used in the present work can be used for axial flow turbine design purposes.

ACKNOWLEDGEMENTS

Márcio Teixeira de Mendonça, TGM Turbinas, Fundação de Amparo à Pesquisa do Estado de São Paulo (FAPESP), Conselho Nacional de Desenvolvimento Científico e Tecnológico (CNPq), and Coordenação de Aperfeiçoamento de Pessoal de Nível Superior (CAPES) are acknowledged for their support to the research carried out at IAE and the Center for Reference on Gas Turbines of ITA.

REFERENCES

Ainley, D.G. and Mathieson, G.C.R., 1951, "A Method of Performance Estimation for Axial Flow Turbines", British ARC, R&M 2974, Retrieved in May 7, 2015, from http://naca.central.cranfield.ac.uk/reports/arc/rm/2974.pdf

Ameri, A.A., Steinthorsson, E., Rigby, D.L., 1998, "Effect of Squealer Tip on Rotor Heat Transfer and Efficiency", ASME Journal of Turbomachinery, Vol. 120, No. 4, pp. 753-759. doi: 10.1115/1.2841786

Azad, G.S., Hart, J.-C., Teng, S., Boyle, R.J., 2000, "Heat Transfer and Pressure Distributions on a Gas Turbine Blade Tip", Proceedings of the ASME Turbo Expo, Munich, Germany, GT2000-0194.

Barth, T.J. and Jesperson, D.C., 1989, "The Design and Application of Upwind Schemes on Unstructured Meshes", AIAA Paper 89-0366.

Bringhenti, C., Barbosa, J.R. and Carneiro, H.F.F.M., 2001, "Variable Geometry Turbine Performance Maps for the Variable Geometry Gas Turbine", Proceedings of the 16th Brazilian Congress of Mechanical Engineering, COBEM, Uberlândia, Brazil.

Bunker, R.S., 2014, "Turbine Heat Transfer and Cooling: an Overview", Proceedings of the ASME Turbo Expo, San Antonio, USA, GT2013-94174.

CFD Online, 2011, "Best Practice Guidelines for Turbomachinery CFD", Retrieved in Jan 19, 2012, from http://www.cfd-online.com

Craig, H.R.M. and Cox, H.J.A., 1970, "Performance Estimation of Axial Flow Turbines", Proceedings of the Institution of Mechanical Engineers, Vol. 185, No. 1, pp. 407-424. doi: 10.1243/PIME_PROC_1970_185_048_02

Denton, J.D., 1978, "Throughflow Calculations for Transonic Axial Flow Turbines", ASME, Journal of Engineering for Gas Turbines and Power, Vol. 100, No. 2, pp. 212-218. doi: 10.1115/1.3446336

Denton, J.D., 2010, "Some Limitations of Turbomachinery CFD", Proceedings of the ASME Turbo Expo 2010, Glasgow, Scotland, GT2010-22540.

Dunham, J. and Came, P.M., 1970, "Improvement to the Ainley/Mathieson Method of Turbine Performance Prediction", ASME, Journal of Engineering for Gas Turbines and Power, Vol. 92, No. 3, pp. 252-256. doi:10.1115/1.3445349

Dunham, J. and Panton, J., 1973, "Experiments on the Design of a Small Axial Turbine", IMEch Conference Publication, No. 3, pp. 56-65.

Dunham, J. and Smith, D.J.L., 1968, "Some Aerodynamic Aspect of Turbine Design", Proceedings of AGARD Conference, Advanced Components for Turbojet Engines, AGARD CP 34, Part 2, pp. 21.1 - 21.15.

Kacker, S.C. and Okapuu U., 1982, "A Mean Line Prediction Method for Axial Flow Turbine Efficiency", ASME, Journal of Engineering for Gas Turbines and Power, Vol. 104, No. 1, pp. 111-119. doi: 10.1115/1.3227240

Krishnababu, S.K., Newton, P.J., Dawes, W.N., Lock, G.D., Hodson, H.P., Hannis, J. and Whitney, C., 2009, "Aerothermal Investigations of Tip Leakage Flow in Axial Flow Turbines - Part I: Effect of Tip Geometry and Tip Clearance Gap", Journal of Turbomachinery, Vol. 131, No. 1, pp. 011006-1-011006-14. doi: 10.1115/1.2950068

Lee, S.W. and Choi, M.Y., 2010, "Tip Gap Height Effects on the Aerodynamic Performance of a Cavity Squealer Tip in a Turbine Cascade in Comparison with Plane Tip Results: Part 2 - Aerodynamic Losses", Experiments in Fluids, Vol. 49, No. 3, pp. 713-723. doi: 10.1007/s00348-010-0849-5

Majumdar, S.R., 1998, "Under Relaxation on Momentum Interpolation for Calculation of Flow with Nonstagarred Grids", Numerical Heat Transfer, Vol. 13, pp. 125-132.

Menter, F.R., 1983, "Zonal Two-Equation k-ω Turbulence Model for Aerodynamic Flows", AIAA Paper, Washington, USA.

Menter F.R., Kuntz, M. and Langtry, R., 2003, "Ten Years of Industrial Experience with the SST Turbulence Model", Proceedings of the 4th International Symposium on Turbulence, Heat and Mass Transfer, Antalya, Turkey.

Menter, F.R., Langtry, R. and Hansen, T., 2004, "CFD Simulation of Turbomachinery Flows - Verification, Validation and Modeling", Proceedings of the European Congress of Computational Methods in Applied Sciences and Engineering, Jyvaskyla, Finland.

Murari, S., Sathish, S., Shraman, G. and Liu, J.S., 2011, "CFD Aerodynamic Performance Validation of a Two-Stage High Pressure Turbine", Proceedings of the ASME Turbo Expo 2011, Vancouver, Canada, GT2011-45569.

Raw, M.J., 1996, "Robustness of Coupled Algebric Multigrid for the Navier-Stokes Equations", AIAA Paper 96-0297.

Rhie, C.M. and Chow, W.L., 1982, "A Numerical Study of the Turbulent Flow Past an Isolated Airfoil with the Trailing Edge Separation", AIAA Paper 82-0998.

Rody, F., Varetti, M. and Tomat, R., 1980, "Low Pressure Turbine Testing", AGARD CP 293, pp. 30.1-30.13.

Sang, W.L. and Byoung, J.C., 2008, "Effects of Squealer Rim Height on Aerodynamic Losses Downstream of a High-Turning Turbine Rotor Blade", Experimental Thermal and Fluid Science, Vol. 32, No. 8, pp. 1440-1447.

Silva F. A., 2012, "Preliminary Design and Analysis of a Axial Turbine Stage for Conversion of a Turbojet to Turboshaft" (in Portuguese), Master Thesis, Instituto Tecnológico de Aeronáutica, São José dos Campos, Brazil, 183p.

Silva, F.A., Bringhenti, C., Carneiro, H.F.F.M., 2012, "Power Turbine Preliminary Design Used to Convert a Small Turbojet to Turboshaft", Proceedings of the CONEM 2012, VII Congresso Nacional de Engenharia Mecânica, São Luís, Brazil.

Silva, D.T. and Tomita, J.T., 2011, "Axial Turbomachinery Flow Simulations with Different Turbulence Models", Proceedings of the 21st Brazilian Congress of Mechanical Engineering, Natal, Brazil.

Silva, L.M. and Tomita, J.T., 2013, "A Study of the Heat Transfer in a Winglet and Squealer Rotor Tip Configurations for a Non-Cooled HPT Blade Based on CFD Calculations", ASME TurboExpo, GT2013-95164.

Silva, L.M., Tomita, J.T. and Barbosa, J.R., 2011, "A Study of the Influence of the Tip-Clearance of an Axial Turbine on the Tip-Leakage Flow using CFD Techniques", Proceedings of the 21st Brazilian Congress of Mechanical Engineering, Natal, Brazil.

Smith, S.F., 1965, "A Simple Correlation of Turbine Efficiency", Journal of Royal Aeronautical Society, Vol. 69, pp. 467-470.

Tallman, J.A.A., 2004, "Computational Study of Tip Desensitization in Axial Flow Turbines Part 2: Turbine Rotor Simulations with Modified Tip Shapes", Proceedings of the ASME Turbo Expo, Vienna, Austria, GT2004-53919.

Tallman, J.A.A. and Lakshminarayana, B., 2001, "Methods for Desensitizing Tip Clearance Effects in Turbines", Proceedings of the ASME Turbo Expo, New Orleans, USA, GT2001-0486.

Thulin, R.D., Howe, D.C. and Singer, I.D., 1982, "Energy Efficient Engine - High-Pressure Turbine Detailed Design Report" (NASA CR-165608), Cleveland, USA.

Tomita, J.T., 2009, "Three-Dimensional Flow Calculations of Axial Compressors and Turbines Using CFD Techniques", Ph.D. Thesis, Instituto Tecnológico de Aeronáutica, São José dos Campos, Brazil.

Permissions

All chapters in this book were first published in JATM, by Departamento de Ciência e Tecnologia Aeroespacial (DCTA); hereby published with permission under the Creative Commons Attribution License or equivalent. Every chapter published in this book has been scrutinized by our experts. Their significance has been extensively debated. The topics covered herein carry significant findings which will fuel the growth of the discipline. They may even be implemented as practical applications or may be referred to as a beginning point for another development.

The contributors of this book come from diverse backgrounds, making this book a truly international effort. This book will bring forth new frontiers with its revolutionizing research information and detailed analysis of the nascent developments around the world.

We would like to thank all the contributing authors for lending their expertise to make the book truly unique. They have played a crucial role in the development of this book. Without their invaluable contributions this book wouldn't have been possible. They have made vital efforts to compile up to date information on the varied aspects of this subject to make this book a valuable addition to the collection of many professionals and students.

This book was conceptualized with the vision of imparting up-to-date information and advanced data in this field. To ensure the same, a matchless editorial board was set up. Every individual on the board went through rigorous rounds of assessment to prove their worth. After which they invested a large part of their time researching and compiling the most relevant data for our readers.

The editorial board has been involved in producing this book since its inception. They have spent rigorous hours researching and exploring the diverse topics which have resulted in the successful publishing of this book. They have passed on their knowledge of decades through this book. To expedite this challenging task, the publisher supported the team at every step. A small team of assistant editors was also appointed to further simplify the editing procedure and attain best results for the readers.

Apart from the editorial board, the designing team has also invested a significant amount of their time in understanding the subject and creating the most relevant covers. They scrutinized every image to scout for the most suitable representation of the subject and create an appropriate cover for the book.

The publishing team has been an ardent support to the editorial, designing and production team. Their endless efforts to recruit the best for this project, has resulted in the accomplishment of this book. They are a veteran in the field of academics and their pool of knowledge is as vast as their experience in printing. Their expertise and guidance has proved useful at every step. Their uncompromising quality standards have made this book an exceptional effort. Their encouragement from time to time has been an inspiration for everyone.

The publisher and the editorial board hope that this book will prove to be a valuable piece of knowledge for researchers, students, practitioners and scholars across the globe.

List of Contributors

Ronaldo Rodrigues Ferreira
Universidade Federal do Rio Grande do Sul - Porto Alegre/RS - Brazil
Instituto de Informática, Universidade Federal do Rio Grande do Sul | Avenida Bento Gonçalves, 9500, Campus do Vale - Bloco IV - Agronomia | CEP 91.509-900 Porto Alegre/RS - Brazil

Rafael Baldiati Parizi
Universidade Federal do Rio Grande do Sul - Porto Alegre/RS - Brazil

Luigi Carro
Universidade Federal do Rio Grande do Sul - Porto Alegre/RS - Brazil

Álvaro Freitas Moreira
Universidade Federal do Rio Grande do Sul - Porto Alegre/RS - Brazil

Paula Cristiane Pinto Mesquita Pardal
Instituto de Ciência e Tecnologia - São José dos Campos/SP - Brazil
Rua Eng. João Fonseca dos Santos, 158, Apto. 163B - Vila Adyana | CEP 12.243-620 São José dos Campos/SP - Brazil

Hélio Koiti Kuga
Instituto Nacional de Pesquisa Espacial - São José dos Campos/SP - Brazil

Rodolpho Vilhena de Moraes
Instituto de Ciência e Tecnologia - São José dos Campos/SP - Brazil

Raed Kafafy
International Islamic University Malaysia - Selangor - Malásia
IIUM-Mechanical Engineering Department | Jalan Gombak, 50728 | Selayang/Selangor - Malásia

Muhammad Hanafi Azami
International Islamic University Malaysia - Selangor - Malásia

Moumen Idres
International Islamic University Malaysia - Selangor - Malásia

Rene Francisco Boschi Gonçalves
Instituto Tecnológico de Aeronáutica - São José dos Campos/SP - Brazil
Departamento de Química, Instituto Tecnológico de Aeronáutica | Praça Mal. Eduardo Gomes, 50 -

Koshun Iha
Instituto Tecnológico de Aeronáutica - São José dos Campos/SP - Brazil

José Atílio Fritz Fidel Rocco
Instituto Tecnológico de Aeronáutica - São José dos Campos/SP - Brazil

Luca De Filippis
Politecnico di Torino - Torino - Italy

Enrico Gaia
Thales Alenia Space Italia S.p.A - Torino - Italy

Giorgio Guglieri
Politecnico di Torino - Torino - Italy
Department of Mechanical and Aerospace Engineering, Politecnico di Torino | C.so Duca degli Abruzzi 24, Torino - 10129 - Italy

Marco Re
Politecnico di Torino - Torino - Italy

Claudia Ricco
Università degli Studi di Torino - Torino - Italy

Jeremy Straub
University of North Dakota - Grand Forks/ND - USA
3950 Campus Road, Stop 9015, Grand Forks/ ND | 58202-9015 | USA

Erick Braga F. Galante
Instituto Militar de Engenharia - Rio de Janeiro/RJ - Brazil
Departamento de Engenharia, Química - SE/5 - Praça Marechal Tibúrcio 80, CEP 23070-200 - Rio de Janeiro/RJ - Brazil

Assed Haddad
Universidade Federal do Rio de Janeiro - Rio de Janeiro/RJ - Brazil

Dieter Boer
Universitat Rovira i Virgili - Tarragona - Spain

Danielle Bonifácio
Indústria de Material Bélico do Brasil - Magé/RJ - Brazil

Fernanda Francisca Maia
Instituto Nacional de Pesquisas Espaciais - São José dos Campos/SP - Brazil

Leonardo Henrique Gouvea
Instituto Nacional de Pesquisas Espaciais - São José dos Campos/SP - Brazil

Luis Gustavo Ferroni Pereira
Instituto Nacional de Pesquisas Espaciais - São José dos Campos/SP - Brazil

Ricardo Vieira
Instituto Nacional de Pesquisas Espaciais - São José dos Campos/SP - Brazil

Fernando de Souza Costa
Instituto Nacional de Pesquisas Espaciais - São José dos Campos/SP - Brazil
Laboratório Associado de Combustão e Propulsão - INPE | Rodovia Presidente Dutra, Km 40 | CEP 12.630-000 Cachoeira Paulista/SP - Brazil

Nelson Alex Roso
Instituto Tecnológico de Aeronáutica - São José dos Campos/SP - Brazil
Praça Marechal Eduardo Gomes 50, São José dos Campos / SP | CEP: 12.228-901 - Brazil

Romero da Costa Moreira
Instituto Tecnológico de Aeronáutica - São José dos Campos/SP - Brazil

José Edimar Barbosa Oliveira
Instituto Tecnológico de Aeronáutica - São José dos Campos/SP - Brazil

Glenn Baxter
RMIT University - Melbourne/VIC - Australia
School of Aerospace, Mechanical and Manufacturing Engineering, Building 57, Level 3, 115 Queensberry Street - Carlton South/VIC - Australia 3053

Kyriakos Kourousis
RMIT University - Melbourne/VIC - Australia

Graham Wild
RMIT University - Melbourne/VIC - Australia

Lester de Abreu Faria
Instituto Tecnológico de Aeronáutica - São José dos Campos/SP - Brazil
Praça Marechal Eduardo Gomes, 50 - Vila das Acácias | CEP 12.228-900 São José dos Campos/SP - Brazil

Luciano Barbosa Magalhães
Comando-Geral de Operações Aéreas - Rio de Janeiro/RJ - Brazil

Roberto d'Amore
Instituto Tecnológico de Aeronáutica - São José dos Campos/SP - Brazil

Tingting Yu
Shanghai University - Shanghai - China

Hongxia Cai
Shanghai University - Shanghai - China
Shanghai Key Laboratory of Intelligent Manufacturing and Robotics | Shanghai University | Yanchang Road 149, Zhabei District | 200072 - Shanghai - China

Valentino Lau
Instituto Nacional de Pesquisas Espaciais - São José dos Campos/SP - Brazil
Avenida dos Astronautas, 1.758, CEP: 12.227-010 - São José dos Campos/SP - Brazil

Fabiano Luis de Sousa
Instituto Nacional de Pesquisas Espaciais - São José dos Campos/SP - Brazil

Roberto Luiz Galski
Instituto Nacional de Pesquisas Espaciais - São José dos Campos/SP - Brazil

Evandro Marconi Rocco
Instituto Nacional de Pesquisas Espaciais - São José dos Campos/SP - Brazil

José Carlos Becceneri
Instituto Nacional de Pesquisas Espaciais - São José dos Campos/SP - Brazil

Walter Abrahão dos Santos
Instituto Nacional de Pesquisas Espaciais - São José dos Campos/SP - Brazil

Sandra Aparecida Sandri
Instituto Nacional de Pesquisas Espaciais - São José dos Campos/SP - Brazil

Fabiano Armellini
Polytechnique Montréal - Montréal - Canada
Department of Mathematical and Industrial Engineering - Polytechnique Montréal | 2900 Boulevard Edouard-Montpetit-Montréal/QC H3T 1J4 | Canada

Paulo Carlos Kaminski
Universidade de São Paulo - São Paulo/SP - Brazil

Catherine Beaudry
Polytechnique Montréal - Montréal - Canada

Juan Sebastián Triana
Universidad Sergio Arboleda - Bogotá - Colombia
Calle 74 14-14 14-14 | Bogotá - Colombia

Sebastian Bautista
Universidad Sergio Arboleda - Bogotá - Colombia

Freddy Alexander Díaz González
Universidad Sergio Arboleda - Bogotá - Colombia

Fredy Marcell Villanueva
School of Astronautics - Beihang University - Beijing - China
Beihang University | XueYuan Road No.37, HaiDian District, New Main Building B1011 | Beijing 100191 - China

He Linshu
School of Astronautics - Beihang University - Beijing - China

Xu Dajun
School of Astronautics - Beihang University - Beijing - China

Bento Silva de Mattos
Instituto Tecnológico de Aeronáutica - São José dos Campos/SP - Brazil
Praça Marechal Eduardo Gomes, 50 - Vila das Acácias | CEP: 12.228-900, São José dos Campos/SP - Brazil

Ney Rafael Secco
Instituto Tecnológico de Aeronáutica - São José dos Campos/SP - Brazil

Urias Andrade Pinheiro
Instituto Nacional de Pesquisas Espaciais - São José dos Campos/SP - Brazil

Marcos Daisuke Oyama
Instituto de Aeronáutica e Espaço - São José dos Campos/SP - Brazil
Praça Marechal Eduardo Gomes, 50 - Vila das Acácias | CEP 12.228-901 São José dos Campos/SP - Brazil

Hong Xia Cai
Shanghai University - Shanghai - China
Shanghai Key Laboratory of Mechanical Automation and Robotics | School of Mechatronic Engineering and Automation | Shangai University | Yanchang Road 149, Zhabei District | 200072 Shanghai/China

Ming Yu Dai
Shanghai University - Shanghai - China

Tao Yu
Shanghai University - Shanghai - China

Humberto Araujo Machado
Instituto de Aeronáutica e Espaço - São José dos Campos/SP - Brazil
Praça Marechal Eduardo Gomes, 50 - Vila das Acácias | CEP 12.228-904 São José dos Campos/SP - Brazil

Daniel Soares de Almeida
Instituto de Aeronáutica e Espaço - São José dos Campos/SP - Brazil
Praça Marechal Eduardo Gomes, 50 - Vila das Acácias | CEP: 12228-904 - São José dos Campos/ SP - Brazil

Cristiane Maria de Moraes Pagliuco
Instituto de Aeronáutica e Espaço - São José dos Campos/SP - Brazil

Cleverson Bringhenti
Departamento de Ciência e Tecnologia Aeroespacial - Instituto Tecnológico de Aeronáutica - Departamento de Turbomáquinas - São José dos Campos/SP - Brazil
Instituto Tecnológico de Aeronáutica - Departamento de Turbomáquinas | Praça Marechal Eduardo Gomes, 50 Vila das Acácias | CEP: 12.228-900 - São José dos Campos/SP - Brazil

Jesuíno Takachi Tomita
Departamento de Ciência e Tecnologia Aeroespacial - Instituto Tecnológico de Aeronáutica - Departamento de Turbomáquinas - São José dos Campos/SP - Brazil

Fernando de Araújo Silva
Departamento de Ciência e Tecnologia Aeroespacial - Instituto de Aeronáutica e Espaço - Divisão de Sistemas Aeronáuticos - São José dos Campos/SP - Brazil

Helder Fernando de França Mendes Carneiro
Departamento de Ciência e Tecnologia Aeroespacial - Instituto de Aeronáutica e Espaço - Divisão de Sistemas Aeronáuticos - São José dos Campos/SP - Brazil